Mobile and Handheld Computing Solutions for Organizations and End-Users

Wen-Chen Hu
University of North Dakota, USA

S. Hossein Mousavinezhad
Idaho State University, USA

Managing Director:	Lindsay Johnston
Editorial Director:	Joel Gamon
Book Production Manager:	Jennifer Yoder
Publishing Systems Analyst:	Adrienne Freeland
Assistant Acquisitions Editor:	Kayla Wolfe
Typesetter:	Alyson Zerbe
Cover Design:	Nick Newcomer

Published in the United States of America by
Information Science Reference (an imprint of IGI Global)
701 E. Chocolate Avenue
Hershey PA 17033
Tel: 717-533-8845
Fax: 717-533-8661
E-mail: cust@igi-global.com
Web site: http://www.igi-global.com

Library of Congress Cataloging-in-Publication Data

Mobile and handheld computing solutions for organizations and end-users / Wen- Chen Hu and S. Hossein Mousavinezhad, editors.
    p. cm.
 Includes bibliographical references and index.
 Summary: "This book discusses a broad range of topics in order to advance handheld knowledge and apply the proposed methods to real-world issues for organizations and end users"--Provided by publisher.
   ISBN 978-1-4666-2785-7 (hbk.) -- ISBN 978-1-4666-2786-4 (ebook) -- ISBN 978-1-4666-2787-1 (print & perpetual access) 1. Mobile computing. 2. Portable computers. I. Hu, Wen Chen, 1960- II. Mousavinezhad, S. Hossein, 1947-
  QA76.59.M59 2012
  004.165--dc23
                         2012032562

British Cataloguing in Publication Data
A Cataloguing in Publication record for this book is available from the British Library.

The views expressed in this book are those of the authors, but not necessarily of the publisher.

Table of Contents

Section 1
Mobile Security

Section 2
Mobile Evaluations and Analyses

Nan Jing, Bloomberg L. P., USA
Yong Yao, IBM Silicon Valley Lab, USA
Yanbo Ru, Business.com Inc., USA

Kamel Rouibah, Kuwait University, Kuwait
T. Abbas H., Kuwait University, Kuwait

John Garofalakis, University of Patras, Greece
Antonia Stefani, University of Patras, Greece
Vassilios Stefanis, University of Patras, Greece

Section 3
Mobile Applications

Christian R. Prause, Fraunhofer FIT, Germany
Marc Jentsch, Fraunhofer FIT, Germany
Markus Eisenhauer, Fraunhofer FIT, Germany

Ching-Long Yeh, Tatung University, Taiwan
Chun-Fu Chang, Tatung University, Taiwan
Po-Shen Lin, Tatung University, Taiwan

Jie Sun, Ningbo University of Technology, China
Yongping Zhang, Ningbo University of Technology, China
Jianbo Fan, Ningbo University of Technology, China

Section 7
Mobile Green Computing, Location-Based Services (LBS), and Mobile Networks

Detailed Table of Contents

Section 1
Mobile Security

David Kuo, San Jose State University, USA
Daniel Wong, San Jose State University, USA
Jerry Gao, San Jose State University, USA, & Tsinghua University, China
Lee Chang, San Jose State University, USA

The wide deployment of wireless networks and mobile technologies and the significant increase in the number of mobile device users has created a very strong demand for emerging mobile commerce applications and services. Barcode-based identification and validation solutions are considered an important part of electronic commerce systems, particularly in electronic supply chain systems. This paper reports a mobile-based 2D barcode validation system as part of mobile commerce systems. This barcode-based validation solution is developed based on the Data Matrix 2D-Barcode standard to support barcode-based validation in mobile commerce systems on mobile devices. The paper demonstrates its application by building a mobile movie ticketing system.

Roel Peeters, Katholieke Universiteit Leuven, Belgium
Dave Singelée, Katholieke Universiteit Leuven, Belgium
Bart Preneel, Katholieke Universiteit Leuven, Belgium

Designing a secure, resilient and user-friendly access control system is a challenging task. In this article, a threshold-based location-aware access control mechanism is proposed. This design uniquely combines the concepts of secret sharing and distance bounding protocols to tackle various security vulnerabilities. The proposed solution makes use of the fact that the user carries around various personal devices. This solution offers protection against any set of (t-1) or fewer compromised user's devices, with t being an adjustable threshold number. It removes the single point of failure in the system, as access is granted when one carries any set of t user's devices. Additionally it supports user-centered management, since users can alter the set of personal devices and can adjust the security parameters of the access control scheme towards their required level of security and reliability.

Chapter 3

Radio Frequency Identification (RFID) has been applied in various high security and high integrity settings. As an important ubiquitous technique, RFID offers opportunities for real-time item tracking, object identification, and inventory management. However, due to the high distribution and vulnerability of its components, an RFID system is subject to various threats which could affect the system's abilities to provide essential services to users. Although there have been intensive studies on RFID security and privacy, there is still no complete solution to RFID survivability. In this paper, the authors classify the RFID security techniques that could be used to enhance an RFID system's survivability from three aspects, i.e., resilience, robustness and fault tolerance, damage assessment and recovery. A threat model is presented, which can help users identify devastating attacks on an RFID system. An RFID system must be empowered with strong protection to withstand those attacks and provide essential functions to users.

Chapter 4

Security has become the Achilles' heel of many organizations in today's computer-dominated society. In this paper, a configurable intrusion detection and response framework named Mobile Agents based Distributed (MAD) security system was proposed for enterprise network consisting of a large number of mobile and handheld devices. The key idea of MAD is to use autonomous mobile agents as lightweight entities to provide unified interfaces for intrusion detection, intrusion response, information fusion, and dynamic reconfiguration. These lightweight agents can be easily installed and managed on mobile and handheld devices. The MAD framework includes a family of autonomous agents, servers and software modules. An Object-based intrusion modeling language (mLanguage) is proposed to allow easy data sharing and system control. A data fusion engine (mEngine) is used to provide fused results for traffic classification and intrusion identification. To ensure Quality-of-Service (QoS) requirements for end users, adaptive resource allocation scheme is also presented. It is hoped that this project will advance the understanding of complex, interactive, and collaborative distributed systems.

Chapter 5

Today's IT systems are ubiquitous and take the form of small portable devices, to the convenience of the users. However, the reliance on this technology is increasing faster than the ability to deal with the simultaneously increasing threats to information security. This paper proposes metrics and a methodology for the evaluation of operational systems security assurance that take into account the measurement of security correctness of a safeguarding measure and the analysis of the security criticality of the context in which the system is operating (i.e., where is the system used and/or what for?). In that perspective, the paper also proposes a novel classification scheme for elucidating the security criticality level of an IT system. The advantage of this approach lies in the fact that the assurance level fluctuation based on the correctness of deployed security measures and the criticality of the context of use of the IT system or device, could provide guidance to users without security background on what activities they may or may not perform under certain circumstances. This work is illustrated with an application based on the case study of a Domain Name Server (DNS).

Section 2
Mobile Evaluations and Analyses

Chapter 6

Nan Jing, Bloomberg L. P., USA
Yong Yao, IBM Silicon Valley Lab, USA
Yanbo Ru, Business.com Inc., USA

Context-aware advertising is one of the most critical components in the Internet ecosystem today because most WWW publisher revenue highly depends on the relevance of the displayed advertisement to the context of the user interaction. Existing research work focuses on analyzing either the content of the web page or the keywords of the user search. However, there are limitations of these works when being extended into mobile computing domain, where mobile devices can provide versatile contexts, such as locations, weather, device capability, and user activities. These contexts should be well categorized and utilized for online advertising to gain better user experience and reaction. This paper examines the aforementioned limitations of the existing works in context-aware advertising when being applied for mobile platforms. A mobile advertising system is proposed, using location tracking and context awareness to provide targeted and meaningful advertisement to the customers on mobile devices. The three main modules of this comprehensive mobile advertising system are discussed, including advisement selection, advertisement presentation, and user context databases. A software prototype that is developed to conduct the case studies and validate this approach is presented.

Chapter 7

Kamel Rouibah, Kuwait University, Kuwait
T. Abbas H., Kuwait University, Kuwait

This study develops a model to assess the consumer acceptance of Camera Mobile Phone (CMP) technology for social interaction. While there has been considerable research on technology adoption in the workplace, far fewer studies have been done to understand the motives of technology acceptance for social use. To fill in this gap, this study develops a model that is based on the following theories: the technology acceptance model, the theory of reasoned action, the attachment motivation theory, innovation diffusion theory, and the theory of flow. The first research method used was a qualitative field study that identified variables that most drive CMP acceptance and build the research model using a sample of 83 consumers. The second method was a quantitative field study. Data was collected from a sample of 240 consumers in Kuwait and used to test the proposed model. The results reveal two types of use: "social use" and "use before shopping," explaining 32.3% and 30% of the variance respectively. Most importantly, the study reveals that personal innovativeness, attachment motivation and social norms have an important effect on CMP acceptance. The implications of this study are highly important for both researchers and practitioners.

Chapter 8

John Garofalakis, University of Patras, Greece
Antonia Stefani, University of Patras, Greece
Vassilios Stefanis, University of Patras, Greece

Business to consumer m-commerce services are here to stay. Their specifics, as software artifacts, indicate that they are primarily and most importantly user-driven; as such user perceived quality assessment should be an integral part of their design process. Mobile design processes still lack a formal and systematic quality control method. This paper explores m-commerce quality attributes using the external quality characteristics of the ISO9126 software quality standard. The goal is to provide a quality map of a B2C m-commerce system in order to facilitate more accurate and detailed quality evaluation. The result is a new evaluation framework based on decomposition of m-commerce services to three distinct user-software interaction patterns and mapping to ISO9126 quality characteristics.

Section 3
Mobile Applications

Chapter 9

Christian R. Prause, Fraunhofer FIT, Germany
Marc Jentsch, Fraunhofer FIT, Germany
Markus Eisenhauer, Fraunhofer FIT, Germany

Thousands of small and medium-sized companies world-wide have non-automated warehouses. Picking orders are manually processed by blue-collar workers; however, this process is highly error-prone. There are various kinds of picking errors that can occur, which cause immense costs and aggravate customers. Even experienced workers are not immune to this problem. In turn, this puts a high pressure on the warehouse personnel. In this paper, the authors present a mobile assistance system for warehouse workers that realize the new Interaction-by-Doing principle. MICA unobtrusively navigates the worker through the warehouse and effectively prevents picking errors using RFID. In a pilot project at a medium-sized enterprise the authors evaluate the usability, efficiency, and sales potential of MICA. Findings show that MICA effectively reduces picking times and error rates. Consequentially, job training periods are shortened, while at the same time pressure put on the individual worker is reduced. This leads to lower costs for warehouse operators and an increased customer satisfaction.

Chapter 10

Ching-Long Yeh, Tatung University, Taiwan
Chun-Fu Chang, Tatung University, Taiwan
Po-Shen Lin, Tatung University, Taiwan

The trend of services on the web is making use of resources on the web collaborative. Semantic Web technology is used to build an integrated infrastructure for the new services. This paper develops a distributed knowledge based system using the RDF/OWL technology on peer-to-peer networks to provide the basis of building personal social collaboration services for e-Learning. This paper extends the current tools accompanied with lecture content to become annotation sharable using the distributed knowledge base.

Chapter 11

Jie Sun, Ningbo University of Technology, China
Yongping Zhang, Ningbo University of Technology, China
Jianbo Fan, Ningbo University of Technology, China

Driving is a complex process influenced by a wide range of factors, especially complex interactions between the driver, the vehicle, and the environment. This paper represents the complex situations in smart car domain. Unlike existing context-aware systems which isolate one context situation from another, such as road congestion and car deceleration, this paper proposes a context model which considers the driver, vehicle and environment as a whole. The paper tries to discover the inherent relationship between the situations in the smart car environment, and proposes a context model to support the representation of situations and their correlation. The detailed example scenarios are given to illustrate our idea.

Section 4
Mobile Human Computer Interaction (HCI)

Chapter 12

Karin Leichtenstern, Augsburg University, Germany
Elisabeth André, Augsburg University, Germany
Matthias Rehm, University of Aalborg, Denmark

There is evidence that user-centred development increases the user-friendliness of resulting products and thus the distinguishing features compared to products of competitors. However, the user-centred development requires comprehensive software and usability engineering skills to keep the process both cost-effective and time-effective. This paper covers that problem and provides insights in so-called user-centred prototyping (UCP) tools which support the production of prototypes as well as their evaluation with end-users. In particular, UCP tool called MoPeDT (Pervasive Interface Development Toolkit for Mobile Phones) is introduced. It provides assistance to interface developers of applications where mobile phones are used as interaction devices to a user's everyday pervasive environment. Based on found tool features for UCP tools, a feature study is described between related tools and MoPeDT as well as a comparative user study between this tool and a traditional approach. A further focus of the paper is the tool-supported execution of empiric evaluations.

Chapter 13

Panos Markopoulos, Eindhoven University of Technology, The Netherlands
Vassilis-Javed Khan, NHTV Breda University of Applied Sciences, The Netherlands

The Experience Sampling and Reconstruction Method (ESRM) is a research method suitable for user studies conducted in situ that is needed for the design and evaluation of ambient intelligence technologies. ESRM is a diary method supported by a distributed application, Reconexp, which runs on a mobile device and a website, enabling surveying user attitudes, experiences, and requirements in field studies. ESRM combines aspects of the Experience Sampling Method and the Day Reconstruction Method aiming to reduce data loss, improve data quality, and reduce burden put upon participants. The authors present a case study of using this method in the context of a study of communication needs of working parents with young children. Requirements for future developments of the tool and the method are discussed.

a ubiquitous environment rely on the interaction among devices. In order to support the development of applications in this context, the heterogeneity of communication protocols must be abstracted and the functionalities dynamically provided by devices should be easily available to application developers. This paper proposes a Device Service Oriented Architecture (DSOA) as an abstraction layer to help organize devices and its resources in a ubiquitous environment, while hiding details about communication protocols from developers. Based on DSOA, a lightweight middleware (uOS) and a high level protocol (uP) were developed. A use case is presented to illustrate the application of these concepts.

Chapter 17

Been-Chian Chien, National University of Tainan, Taiwan
Shiang-Yi He, National University of Tainan, Taiwan

Developing pervasive context-aware systems to construct smart space applications has attracted much attention from researchers in recent decades. Although many different kinds of context-aware computing paradigms were built of late years, it is still a challenge for researchers to extend an existing system to different application domains and interoperate with other service systems due to heterogeneity among systems This paper proposes a generic context interpreter to overcome the dependency between context and hardware devices. The proposed generic context interpreter contains two modules: the context interpreter generator and the generic interpreter. The context interpreter generator imports sensor data from sensor devices as an XML schema and produces interpretation scripts instead of interpretation widgets. The generic interpreter generates the semantic context for context-aware applications. A context editor is also designed by employing schema matching algorithms for supporting context mapping between devices and context model.

Section 7
Mobile Green Computing, Location-Based Services (LBS), and Mobile Networks

Chapter 18

Ji Gu, University of New South Wales, Australia
Hui Guo, University of New South Wales, Australia

Instruction prefetching is an effective way to improve performance of the pipelined processors. However, existing instruction prefetching schemes increase performance with a significant energy sacrifice, making them unsuitable for embedded and ubiquitous systems where high performance and low energy consumption are all demanded. This paper proposes reducing energy overhead in instruction prefetching by using a simple hardware/software design and an efficient prefetching operation scheme. Two approaches are investigated: Decoded Loop Instruction Cache-based Prefetching (DLICP) that is most effective for loop intensive applications, and the enhanced DLICP with the popular existing Next Line Prefetching (NLP) for applications of a moderate number of loops. The experimental results show that both DLICP and the enhanced DLICP deliver improved performance at a much reduced energy overhead.

Chapter 19

Interactive Rendering of Indoor and Urban Environments on Handheld Devices by Combining
Visibility Algorithms with Spatial Data Structures 341

Wendel B. Silva, University of Utah, USA

Maria Andréia F. Rodrigues, Universidade de Fortaleza – UNIFOR, Brazil

This work presents a comparative study of various combinations of visibility algorithms (view-frustum culling, backface culling, and a simple yet fast algorithm called conservative backface culling) and different settings of standard spatial data structures (non-uniform Grids, BSP-Trees, Octrees, and Portal-Octrees) for enabling efficient graphics rendering of both indoor and urban 3D environments, especially suited for low-end handheld devices. Performance tests and analyses were conducted using two different mobile platforms and environments in the order of thousands of triangles. The authors demonstrate that navigation at interactive frame rates can be obtained using geometry rather than image-based rendering or point-based rendering on the cell phone Nokia n82.

Chapter 20

Design and Implementation of Binary Tree Based Proactive Routing Protocols for Large
MANETS 359

Pavan Kumar Pandey, Aricent Technologies, India

G. P. Biswas, Indian School of Mines, India

The Mobile Ad hoc Network (MANET) is a collection of connected mobile nodes without any centralized administration. Proactive routing approach is one of those categories of proposed routing protocol which is not suitable for larger network due to their high overhead to maintain routing table for each and every node. The novelty of this approach is to form a binary tree structure of several independent sub-networks by decomposing a large network to sub-networks. Each sub-network is monitored by an agent node which is selected by several broadcasted regulations. Agent node maintains two routing information; one for local routing within the sub-network and another for routing through all other agent node. In routing mechanism first source node checks for destination within sub-network then source sends destination address to respective parent agent node if destination is not available in local routing, this process follows up to the destination node using agent mode. This approach allowed any proactive routing protocol with scalability for every routing mechanism. The proposed approach is thoroughly analyzed and its justification for the connectivity through sub-networks, routing between each source to destination pair, scalability, etc., are given, which show expected performance.

Preface

This book, *Mobile and Handheld Computing Solutions for Organizations and End-Users,* collects high-quality research papers and industrial and practice articles in the areas of mobile and handheld computing from academia and industry. It includes research and development results of lasting significance in the theory, design, implementation, analysis, and application of mobile and handheld computing, and other critical issues. Twenty excellent articles from fifty world-renowned scholars and industry professionals are included in this book. The chapters are classified into seven sections: (1) mobile security, (2) mobile evaluations and analyses, (3) mobile applications, (4) mobile human computer interaction, (5) mobile health, (6) pervasive computing, and (7) mobile green computing, location-based services, and mobile networks. This preface introduces this book by giving the current trends of mobile and handheld computing, the organization of this book, and a summary.

INTRODUCTION

Computer technologies evolve fast and constantly. Organizations and end-users have to study the computer trend carefully before they adapt any computer products or technologies. Otherwise, they might have to update or redo their systems in a short time. On the other hand, developers have to use the trendy technologies and components to build their products. Products without following the trend will be abandoned by users in no time. For example, PDAs (personal digital assistants) were very popular in the past. However, a PDA without a phone capability can hardly be found these days because smartphones, which have both functions of PDAs and cellular phones, have replaced PDAs completely. Other than personal computers and servers, organizations and end-users have another option, mobile and handheld computing, for solving their problems nowadays. Mobile and handheld computing provides the features of convenience and mobility that are not possible with personal computers or servers. This book presents various solutions of the contemporary mobile and handheld computing for organizations and end-users.

The evolution of computing started from mainframes in 1940s, when the computers were mainly for governments and large corporations. The creation of personal computers in 1980s completely changed the way of computing. Computers are no longer for corporations only. They are for individuals and families too. Today, it is the turn of mobile and handheld computing based on various market research; e.g., three summaries of the research are given as follows:

- In 2011, there were 488 million smartphones shipped worldwide. By 2016, the IDC, a market research company, forecasts 1.16 billion smartphones will be shipped annually (Blagon, 2012). It is about $(1,160-488)\div 488=138\%$ increase in five years.
- Canalys (2012) estimated that 488 million smartphones were sold in 2011, compared with 415 million personal computers, which included tablet PCs such as iPad. It was the first time that smartphones outsold PCs in a year.
- Since the economic meltdown in 2008, unemployment is a big headache for most countries. Mobile and handheld computing has created many employment opportunities. According to IDC (2012), the world's mobile worker population will reach 1.3 billion, accounting 37.2% of the total workforce by 2015.

From the above forecasts, it is clear that the mobile and handheld computing has come of age. Organizations and end-users need to weigh every solution, including mobile and handheld computing, for their problems these days.

This book provides rich solutions of mobile and handheld computing for organizations and end-users. The preface includes four sections and the rest of this preface is organized as follows:

Section 2: Contemporary methodologies and technologies of mobile and handheld computing. This section discusses contemporary methodologies, technologies, and theories of mobile and handheld computing. It includes the following subjects: (1) location-based services, (2) mobile payment methods, (3) mobile computing for big data, and (4) mobile cloud computing. They are the most trendy research subjects of mobile and handheld computing nowadays. Each subject will be introduced in this section.

Section 3: Organization of this book. This book consists of twenty chapters divided into seven sections: (1) mobile security, (2) mobile evaluations and analyses, (3) mobile applications, (4) mobile human computer interaction, (5) mobile health, (6) pervasive computing, and (7) mobile green computing, location-based services, and mobile networks. Each chapter will be briefly introduced in this section.

Section 4: Summary. The last section summarizes this preface and brings the book to you.

CONTEMPORARY METHODOLOGIES AND TECHNOLOGIES OF MOBILE AND HANDHELD COMPUTING

Mobile and handheld computing are useful and have been applied to a wide variety of areas such as business and healthcare. This section gives the contemporary methodologies, technologies, and theories of mobile and handheld computing for organizations and end-users. It includes the following four subjects: (1) location-based services, (2) mobile payment methods, (3) mobile and handheld computing for big data, and (4) mobile cloud computing.

Location-Based Services (LBS)

The emerging smartphones create many kinds of applications that are not possible or inconvenient for PCs and servers, even notebooks. One of the best-seller applications is location-based services, which provide services based on the geographical position of a mobile handheld device. Plenty of LBS applications can be found from the mobile app stores. Examples of LBS applications include: (1) finding a nearby ethnic restaurant or specific store when visiting a new town, (2) scanning the barcode of a product, checking its prices, and finding a store in the neighborhood selling the product with a low price, (3) sending or receiving location-aware coupons or invitations, and (4) locating interesting persons such as a date in an area.

At the same time, many innovative methods and technologies have been applied to LBS research. One of the LBS research projects is to use spatial trajectory prediction to locate mobile objects. Inertia has a moving object follow a path or trajectory that resists any change in its motion. Human travel patterns normally have the similar inertia feature. For example, the vehicles on a highway usually stay on the highway or people tend to walk towards a popular destination such as a mall or a park. This research tries to find the anticipated locations/paths of moving objects by using spatial trajectory prediction. One example of this research is as follows. Assume you have several friends who supply their mobile locations to you constantly. Now, you want to reach any one of them, but the problem is their locations are dynamic instead of static. Using a method of spatial trajectory prediction, we may be able to predict our friends' forthcoming locations and find the ones who are close to us in the next moment.

This research tries to locate mobile objects by using spatial trajectory prediction. The proposed method consists of the following five steps:

1. **Location Data Collection:** Mobile users' locations (including longitudes, latitudes, and times) need to be sent to the system from time to time. In order to do this, the mobile users need to install the app in their devices.

2. **Location Data Preparation:** Location data supplied by the GPS (Global Positioning System) is usually not stable and may contain noises. Before it is used by the system, the location data must be processed.

3. **Location Data Storage and Indexing:** The amount of location data collected could be huge because many locations of numerous objects need to be saved. Additionally, the vast amount of location data makes the indexing and searching more difficult. Innovative methods have to be used to store, index, and search location data.

4. **Spatial Trajectory Mining and Prediction:** Route projection is used to predict the future trajectory. However, this approach suffers from a problem of unrealistic prediction. It is because the prediction is based on route projection and the projection may not have a road.

5. **Using Spatial Trajectory Prediction to Locate Moving Objects:** Once the predicted locations are found, finding the nearest objects becomes a trivial task. Euclidean metric can be used to find the distance.

It is believed that the number of smartphones sold will surpass the number of feature phones sold in the near future. Compared to feature phones, smartphones are able to perform many more advanced functions such as mobile Web browsing, mobile office, and mobile gaming. One of the mobile applications, location-based services, has attracted a great attention recently. A location-based service is a service based on the geographical position of a mobile handheld device. The proposed research proposes location-based research, which uses spatial trajectory prediction to locate mobile objects, whose locations are constantly changed.

Mobile Payment Methods

Mobile handheld devices such as smartphones are extremely popular and convenient these days. People carry them anytime, anywhere and use them to perform daily activities like making phone calls, checking emails, and browsing the mobile Web. An up-and-coming app is mobile payment. For example, a smartphone with an NFC (near field communication) function allows its owner to perform small-amount transactions such as paying parking fees and buying drinks or snacks from vending machines. According to Gartner (2012), worldwide mobile payment transaction values will surpass $171.5 billion in 2012, a 61.9 percent increase from 2011 values of $105.9 billion. The number of mobile payment users will reach 212.2 million in 2012, up from 160.5 million in 2011. Two common kinds of mobile payment methods are

- **Macropayments:** This kind of payment is commonly used by electronic commerce and involves significant amounts of money, for example more than US $10.00. Payments by credit cards are the most common method for macropayments.
- **Micropayments:** These usually involve small amounts of below about US $10.00, which are too small to be economically processed by credit cards. The amounts are usually charged to users' phone bills or accounts.

A typical macropayment/micropayment scenario proceeds as follows, as illustrated in Figures 1 and 2 for macro and micro -payments, respectively, where the number is the order of steps taken:

1. A mobile user submits his/her credit-card or personal information to the mobile content via a handheld device.
2. A third-party processor, instead of the content provider, verifies and authorizes the transaction.
3. The third-party processor routes verification and authorization requests to the card issuing bank or mobile carrier.
4. The user pays his/her monthly credit-card or phone bill, which includes the fees for the mobile transactions.
5. The bank pays the mobile content provider or the mobile carrier pays the mobile content provider directly or through a bank after deducting a transaction fee.

Figure 1. A typical macropayment scenario

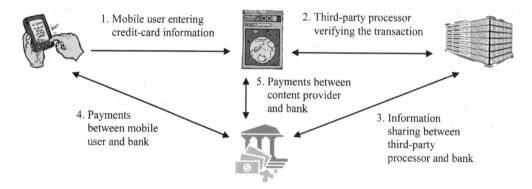

Figure 2. A typical micropayment scenario

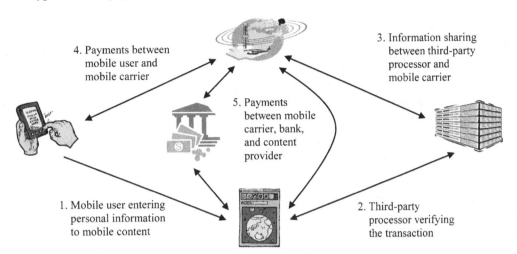

4. Payments between mobile user and mobile carrier

3. Information sharing between third-party processor and mobile carrier

5. Payments between mobile carrier, bank, and content provider

1. Mobile user entering personal information to mobile content

2. Third-party processor verifying the transaction

Mobile and Handheld Computing for Big Data

Big data has existed for a long time, but it did not catch a great attention until recently because it did not prevail. However, owing to the high popularity of computers and great IT advancements, big data is everywhere today. Tremendous amount of data is generated every day, everywhere, from fields such as business, research, and sciences. For example, at least one million Web pages are added to the Internet every day, a massive amount of genetic data is created from various genome projects, or vast astronomical data is recorded after studying numerous galaxies. The growth of information size is not linear, but exponential. Traditional methods such as files and relational databases are not able to handle this kind of data anymore because of its vast size and high complication. Other methods have to be created or used to manage the big data, which is complex, unstructured, or semi-structured.

Big-data methods include data capture, storage, search, sharing, analysis, and visualization. Mobile and handheld computing could also take great part in helping managing big data. There are many ways that mobile and handheld computing can be applied to big data management; for example:

- **Data Capture:** Increasing wealth of data is generated from mobile phones and sensors; for example, images from freeway traffic sensors and GPS (global positioning system) data from moving objects such as vehicles and hikers.
- **Data Sharing:** Big data is made easily and conveniently to share and distribute via mobile handheld devices. For example, mobile users can easily access the weather data from anywhere, anytime.
- **Data Visualization:** Discover hidden behavioural patterns through statistical techniques, focusing on more methodological contributions related to data mining or developing novel ways for data visualization.

Mobile Cloud Computing (MCC)

Mobile cloud computing has come a long way since the start of the World Wide Web. In the beginning, it was the Web services, which are a software system designed to provide data and services to other applications over a network, so the data and services could be used widely and conveniently. Web services allow an application to be converted into a Web application, so the application can be published, found, and used through the Web. Cloud computing comes next to the Web services. It is even more useful than Web services. Cloud computing is the remote use or access of the storage or software on the Internet from a computer to perform various operations. The storage and software could be a shared pool of configurable computing resources such as networks, servers, storage, applications, and services. By using cloud computing, the resources can be improved and updated fast and conveniently and the end-user operations would not be disrupted by the change. One of the cloud-computing examples is to file tax via TurboTax on the Internet. The TurboTax software and your tax data are all stored and accessed remotely via the Internet.

With the prevalence of smartphones, cloud computing has been extended to mobile cloud computing, which is the remote use or access of the storage or software on the Internet from a handheld device to perform various operations. One of the mobile cloud-computing examples is to send or read emails via Gmail on the Internet through a mobile handheld device. The mail software and data are all stored and accessed remotely via the Internet through a mobile handheld device. According to the Juniper Research (2010), the market for cloud-based mobile applications will grow 88% from 2009 to 2014 with $9.5 billion. Mobile cloud computing research includes emerging and future trends in research and application that integrate the cloud computing paradigm into mobile devices, mobile applications, security and privacy, and mobile services, evaluating the impact of mobile applications on cloud computing techniques. Three research areas of mobile cloud computing worthy mentioned are (1) security and privacy, (2) mobile human computer interaction, and (3) mobile apps and services. Mobile cloud computing requires to use a mobile network and the Internet, which are vulnerable. Without rigorous security and privacy protection, users would be reluctant to use it. Mobile users perform mobile cloud computing via handheld devices, which have small screens and limited functions compared to desktop computers. The service providers must supply user-friendly interface for mobile users to perform cloud computing seamlessly and conveniently. Finally, mobile handheld devices have introduced new research and application areas for cloud computing; e.g., location-aware cloud computing. Hence, more opportunities are created by mobile cloud computing.

ORGANIZATION OF THE BOOK

This book provides timely, critical technologies, methodologies, and theories of mobile and handheld computing to mobile IT workers and students. It contains twenty chapters divided into seven sections: (1) mobile security, (2) mobile evaluations and analyses, (3) mobile applications, (4) mobile human computer interaction, (5) mobile health, (6) pervasive computing, and (7) mobile green computing, location-based services, and mobile networks, each of which will be briefly introduced next.

Mobile Security

Mobile security is a branch of computer technology applied to mobile handheld devices for protection of devices and its data from theft, corruption, or natural disaster. Without rigorous mobile security, mobile applications would not be prevalent because users would hesitate to use them. Chapters 1-5 are related to mobile security.

Chapter 1: A 2D Barcode Validation System for Mobile Commerce. Barcode-based identification and validation solutions have been considered as an important part of electronic commerce systems. Mobile commerce is believed to be the next generation of electronic commerce. This chapter reports the research, architecture, design, security, and issues of implementing a 2D barcode validation system for mobile commerce systems. The proposed barcode-based validation solution is developed based on the Data Matrix 2D-Barcode standard. Furthermore, the chapter also demonstrates its application by building a mobile movie ticketing system.

Chapter 2: Threshold-Based Location-Aware Access Control. This chapter proposes a threshold-based location-aware access control mechanism, which uniquely combines the concepts of secret sharing and distance bounding protocols to tackle various security vulnerabilities. Additionally it supports user-centered management, since users can alter the set of personal devices and can adjust the security parameters of the access control scheme towards their required level of security and reliability.

Chapter 3: Survivability Enhancing Techniques for RFID Systems. Radio Frequency Identification (RFID) has been applied to various high security and high integrity settings and RFID security techniques could be used to enhance an RFID system's survivability. This research classifies the RFID security techniques according to three aspects: (1) resilience, (2) robustness and fault tolerance, and (3) damage assessment recovery. This chapter also presents a threat model which can help users identify devastating attacks on an RFID system. The results show an RFID system must be empowered with strong protection to withstand those attacks and provide essential functions to users.

Chapter 4: Mobile Agent Based Network Defense System in Enterprise Network. A configurable intrusion detection and response framework named Mobile Agents based Distributed (MAD) security system is proposed for enterprise network consisting of a large number of mobile and handheld devices. The key idea of MAD is to use autonomous mobile agents as lightweight entities to provide unified interfaces for intrusion detection, intrusion response, information fusion and dynamic reconfiguration. This chapter includes three major contributions: (1) an object-based intrusion modeling language (mLanguage) is proposed to allow easy data sharing and system control, (2) a data fusion engine (mEngine) is used to provide fused results for traffic classification and intrusion identification, and (3) to ensure Quality-of-Service (QoS) requirements for end users, adaptive resource allocation scheme is also presented.

Chapter 5: Security Assurance Evaluation and IT Systems' Context of Use Security Criticality. Today's IT systems and devices are operating in contexts that are fast evolving, with regular emergence of new threats for which existing security measures were not designed to deal with, or their inappropriate deployment makes them less effective in thwarting. This chapter proposes metrics and a methodology for the evaluation of operational system' security assurance that take into account the measurement of security correctness of a safeguarding measure and the analysis of the security criticality of the context in which the system is operating. This research also proposes a novel classification scheme for elucidating the security criticality level of an IT system. The advantage of this approach could provide guidance to users without security background on what activities they may or may not perform under certain circumstances.

Mobile Evaluations and Analyses

Evaluations and analyses are important part of product development. Launching a product without a thorough evaluation and analysis could be risky. For example, many smartphones have been developed and put on the market, but only a handful of them survive. This book includes the following three chapters of mobile evaluations and analyses:

Chapter 6: Modeling and Analyzing User Contexts for Mobile Advertising. Existing research work in context-aware advertising mainly focuses on analyzing either the content of the Web page or the keywords of the user search. Mobile devices can provide versatile contexts, such as locations, weather, device capability and user activities, which are far beyond just page content and search keywords. When context-aware advertising is extended into mobile computing domain, these contexts should be well categorized and utilized to gain better user experience and reaction. This research proposes a mobile advertising system that is using location tracking and context awareness to provide targeted and meaningful advertisement to the customers on mobile devices.

Chapter 7: Effect of Personal Innovativeness, Attachment Motivation, and Social Norms on the Acceptance of Camera Mobile Phones: An Empirical Study in an Arab Country. This research develops a model to assess the consumer acceptance of Camera Mobile Phone (CMP) technology for social interaction. The model is based on the following theories: (1) the technology acceptance model, (2) the theory of reasoned action, (3) the attachment motivation theory, (4) innovation diffusion theory, and (5) the theory of flow. Two research methods were used in this chapter. The first one was a qualitative field study that was used to identify the variables that most drive CMP acceptance and build the research model using a sample of 83 consumers. The second method was a quantitative field study. Data was collected from a sample of 240 consumers in Kuwait and was used to test the proposed model. The results show that personal innovativeness, attachment motivation and social norms have an important effect on CMP acceptance.

Chapter 8: A Framework for the Quality Evaluation of B2C M-Commerce Services. Specifics of business to consumer m-commerce services indicate that they are primarily and most importantly user-driven. Therefore, user perceived quality assessment should be an integral part of their design process. Mobile design processes still lacks a formal and systematic quality control method. This chapter explores m-commerce quality attributes using the external quality characteristics of the ISO9126 software quality standard. Its goal is to provide a quality map of a B2C m-commerce system in order to facilitate more accurate and in detail quality evaluation. The result is a new evaluation framework based on decomposition of m-commerce services to three distinct user-software interaction patterns and mapping to ISO9126 quality characteristics.

Mobile Applications

Mobile and handheld computing has been applied to a wide variety of applications like mobile commerce and mobile learning. This section includes the following three mobile applications: a mobile support system, personal annotation management, and smart cars.

Chapter 9: MICA—A Mobile Support System for Warehouse Workers. This chapter presents a mobile assistance system, MICA, for warehouse workers to apply the new Interaction-by-Doing principle, which is an enhancement of Interaction-by-Movement. MICA effectively reduces picking times and error rates by unobtrusively navigating the worker through the warehouse and effectively preventing picking errors using RFID. Consequentially, job-training periods are shortened, while at the same time, pressure put on the individual worker is reduced. This leads to lower costs for warehouse operators and an increased customer satisfaction.

Chapter 10: Ontology-Based Personal Annotation Management on Semantic Peer Network to Facilitating Collaborations in E-Learning. The trend of services on the Web is making the use of resources on the Web collaborative and the Semantic Web technology is used to build an integrated infrastructure for the new services. In this chapter, the authors attempt to take advantages of both technologies, collaborative services in Web 2.0 and semantic integrated infrastructure in Semantic Web, to develop a platform for managing personal annotations. On the proposed platform, user can annotate lecture contents and share the annotated results using Web 2.0 services, like wiki, blog, instant messaging, etc., or reading tools, for example, MS PowerPoint, Adobe Acrobat.

Chapter 11: A Petri-Net Based Context Representation in Smart Car Environment. Driving is a complex process influenced by a wide range of factors, especially complex interactions between the driver, the vehicle and the environment. This chapter aims at the representation of complex situations in smart car domain. Unlike existing context-aware systems which isolate one context situation from another, such as road congestion and car deceleration, this chapter proposes a context model which considers the driver, vehicle and environment as a whole. It tries to discover the inherent relationship between the situations in the smart car environment, and proposes a context model to support the representation of situations and their correlation.

Mobile Human Computer Interaction (HCI)

Mobile HCI is the study of how users interact with mobile handheld devices. Smartphones were available a long time ago, but they were never prevailing until the introduction of iPhones in 2007. One of the major reasons that makes iPhones extremely popular is their revolutionary user interfaces including voice recognition, finger gestures, and sliding menus. Two chapters related to mobile HCI research are introduced next.

Chapter 12: Tool-Supported User-Centered Prototyping of Mobile Applications. It is known that user-centered development can increase the user-friendliness of resulting products and thus the distinguishing features compared to products of competitors. However, the user-centered development requires comprehensive software and usability engineering skills to keep the process both cost-effective and time-effective. This paper introduces a user-centered prototyping tool called MoPeDT (Pervasive Interface Development Toolkit for Mobile Phones) that provides assistance to interface developers of applications where mobile phones are used as interaction devices to a user's everyday pervasive environment.

Chapter 13: Sampling and Reconstructing User Experience. The Experience Sampling and Reconstruction (ESRM) method is a research method suitable for user studies conducted in situ that are needed for the design and evaluation of ambient intelligence technologies. ESRM is a diary method supported by a distributed application, Reconexp, which runs on a mobile device and a website, enabling us to survey user attitudes, experiences, and requirements in field studies. ESRM combines aspects of the Experience Sampling Method and the Day Reconstruction Method aiming to reduce data loss, improve data quality and reduce burden put upon participants. This article presents a case study of using this method in the context of a study of communication needs of working parents with young children. Requirements for future developments of the tool and the method are discussed.

Mobile Health

Mobile health is the practice of medicine and public health supported by mobile handheld devices, e.g., delivery of healthcare information anytime, anywhere, and real-time monitoring patients' vital signs. This book includes the following two mobile-health chapters.

Chapter 14: Mobile E-Health Information System. A mobile e-health information system (MEHIS) aims to speed up the operations of healthcare in medical centers and hospitals. However, the proper implementation of MEHIS involves integrating many subsystems for MEHIS to be properly executed. A typical MEHIS can consist of many components and subsystems, such as appointments and scheduling; admission, discharge, and transfer (ADT); prescription order entry; dietary planning; and smart card sign-on. This chapter describes the development of a MEHIS with open-source Eclipse, using currently available healthcare standards. They discuss the issues of building a mobile e-Health information system which can help achieve the goal of ubiquitous and mobile applications for the personalization of e-Health.

Chapter 15: Integration of Health Records by Using Relaxed Acid Properties between Hospitals, Physicians, and Mobile Units like Ambulances and Doctors. This research proposes an architecture of integrating stationary health units with mobile health units. In central databases, the consistency of data is normally implemented by using the Atomicity, Consistency, Isolation and Durability (ACID) properties of a Data Base Management System (DBMS). This is not possible if mobile databases are involved and the availability of data also has to be optimized. The so called relaxed ACID properties across different locations are therefore used in this research. The objective of designing relaxed ACID properties across different database locations is to make it possible for all the involved locations to operate in disconnected mode. At the same time, it gives the users a view of the data that may be inconsistent across different locations but anyway better than the data in a centralized database with low availability for the users.

Pervasive Computing

Chapter 16: DSOA: A Service Oriented Architecture for Ubiquitous Applications. The services provided by a ubiquitous environment rely on the interaction among a variety of devices. In order to support the development of applications in this context, the heterogeneity of communication protocols must be abstracted and the functionalities dynamically provided by devices should be easily available to application developers. This chapter proposes an extension of the SOA architecture, denominated DSOA (Device Service Oriented Architecture), to model a smart space taking into account the requirements outlined above. Based on DSOA, the authors also developed a lightweight multi-platform communication interface, called uP, and the middle-ware uOS to support the development of smart applications. At the end, a use case is presented to illustrate the application of these concepts.

Chapter 17: A Generic Context Interpreter for Pervasive Context-Aware Systems. Pervasive context-aware systems for smart space applications have attracted much attention from researchers. In this chapter, a generic context interpreter is proposed to overcome the physical context dependence problem between context and sensor devices in the framework of context-aware computing. The proposed interpreter consists of two modules: (1) the context interpreter generator and (2) the generic interpreter. The context interpreter generator imports sensor data from sensor devices as an XML schema and produces interpretation scripts instead of interpretation widgets. The generic interpreter then generates the semantic context for context-aware applications. An interface tool, the context editor, is also designed by employing automatic XML schema matching schemes for supporting smart context mapping between devices and context model.

Mobile Green Computing, Location-Based Services (LBS), and Mobile Networks

The last section includes the following three chapters, two mobile applications and a mobile network. Mobile green computing is the practice of designing, manufacturing, using, and disposing of mobile handheld devices, and associated subsystems for reducing the use of hazardous materials, maximizing energy efficiency, and promoting the recyclability or biodegradability of products. Two chapters related to mobile green computing are included in this book.

Chapter 18: Reducing Power and Energy Overhead in Instruction Prefetching for Embedded Processor Systems. Instruction prefetching is an effective way to improve the performance of pipelined processors. However, existing instruction prefetching schemes increase performance with a significant energy sacrifice, making them unsuitable for embedded and ubiquitous systems where high performance and low energy consumption are all demanded. This research reduces energy overhead in instruction prefetching by using a simple hardware/software design and an efficient prefetching operation scheme. Two approaches are investigated in this chapter: (1) Decoded Loop Instruction Cache based Prefetching (DLICP) that is most effective for loop intensive applications, and (2) the enhanced DLICP with the popular existing Next Line Prefetching (NLP) for applications of a moderate number of loops.

Chapter 19: Interactive Rendering of Indoor and Urban Environments on Handheld Devices by Combining Visibility Algorithms with Spatial Data Structures. This work presents a comparative study of various combinations of visibility algorithms including: (1) view-frustum culling, (2) backface culling, and (3) a simple yet fast algorithm, called conservative backface culling. Different settings of standard spatial data structures are also employed including (1) non-uniform Grids, (2) BSP-Trees, (3) Octrees, and (4) Portal-Octrees for enabling efficient graphics rendering of both indoor and urban 3D environments, especially suited for low-end handheld devices. The results demonstrate that navigation at interactive frame rates on low-end handheld devices can be obtained using geometry rather than image-based rendering or point-based rendering.

Chapter 20: Design and Implementation of Binary Tree Based Proactive Routing Protocols for Large MANETS. The Mobile Ad hoc Network (MANET) is a collection of connected mobile nodes without any centralized administration. Proactive routing approach is not suitable for larger network due to their high overhead to maintain routing table for each and every node. This approach forms a binary tree structure of several independent sub-networks by decomposing a large network to sub-networks. The source node checks for destination within the sub-network in the proposed routing mechanism. If the destination is not available in the local routing, then the source node sends the destination address to respective parent agent node. This process is repeated up to the destination node. The proposed method allows any proactive routing protocol with scalability for every routing mechanism.

SUMMARY

Personal computers and servers used to be the only solutions for the problems of organizations and end-users. Since the introduction of iPhone in 2007, the solutions have been extended to including mobile and handheld computing. According to Apple Inc. (2012), more than 25 billion apps have been downloaded from its App Store by the users of the more than 315 million iPhones and iPads worldwide. The App Store offers more than 550,000 apps in 2012 and examples of the apps include: (1) *business*: word processors, voice recording, tax filing, and offices, (2) *entertainment*: video games, music, movies, and ringtones, (3) *finance*: banking, budgets, expense trackers, and spreadsheets, (4) *information*: news, sports, weather reports, and wiki, and (5) *social networking*: Facebook, dating, texting, and tweeting. The apps cover almost every area of daily life. When organizations and end-users search for solutions for their problems, they no longer ignore mobile and handheld computing. They either develop their own mobile solutions, or purchase and customize the apps available in the app stores.

Mobile and handheld computing plays an important role in helping organizations and end-users solve their problems, but at the same time, people have a hard time following the newest mobile technologies and apply them to their problems. It is because new mobile technologies and methods are invented every day. This preface introduces four trendy subjects of mobile and handheld computing: (1) location-based services, (2) mobile payment methods, (3) mobile and handheld computing for big data, and (4) mobile cloud computing. They are the latest and popular subjects. Location-based services are unique for mobile and handheld computing. They provide new opportunities unseen before; for example, Foursquare helps users find where their friends are and collect location-aware coupons. Many methods and technologies are created to make mobile payment possible and convenient. One of the technologies is NFC (near field communication) based on wireless RFID technology. NFC is a set of standards for establishing communication among devices by using interacting electromagnetic radio fields, which require to bring the devices together or into close proximity. With the great advancement of computers and sensors, a huge amount of complicated data is collected and stored every day. Traditional methods such as relational databases and file systems are no longer able to handle the big data. Innovative methods and technologies have to be used to manage the big data. Finally, cloud computing has attracted a great attention these days. Applications become more useful and convenient by using cloud computing and mobile cloud computing makes applications even more useful and convenient.

This book provides the latest research and applications of mobile and handheld computing for organizations and end-users. It includes twenty chapters divided into seven sections: (1) mobile security, (2) mobile evaluations and analyses, (3) mobile applications, (4) mobile human computer interaction, (5) mobile health, (6) pervasive computing, and (7) mobile green computing, location-based services, and mobile networks. These chapters keep the readers aware of the contemporary technologies and methods of mobile and handheld computing and help readers apply them to real world problems. The editors want to thank the contributing authors for their excellent works. Without their contributions, this book is not possible. We hope you will enjoy reading this book and be able to make use of it.

Wen-Chen Hu
University of North Dakota, USA

S. Hossein Mousavinezhad
Idaho State University, USA

REFERENCES

Apple Inc. (2012, March 5). *Apple's app store downloads top 25 billion*. Retrieved June 12, 2012, from http://www.apple.com/pr/library/2012/03/05Apples-App-Store-Downloads-Top-25-Billion.html

Blagon, J. (2012, March 29). *IDC forecasts 1.16 billion smartphones shipped annually by 2016*. Retrieved July 21, 2012, from http://www.theverge.com/2012/3/29/2910399/idc-smartphone-computer-tablet-sales-2011

Canalys. (2012, February 3). *Smart phones overtake client PCs in 2011*. Retrieved July 15, 2012, from http://www.canalys.com/newsroom/smart-phones-overtake-client-pcs-2011

Gartner. (2012, May 29). *Gartner says worldwide mobile payment transaction value to surpass $17.5 billion*. Retrieved from July 7, 2012, from http://www.gartner.com/it/page.jsp?id=2028315

IDC. (2012, January 5). *Mobile worker population to reach 1.3 billion by 2015, according to IDC*. Retrieved from April 11, 2012, from http://www.idc.com/getdoc.jsp?containerId=prUS23251912

Juniper Research Ltd. (2010, January 2). *Mobile cloud applications & services*. Retrieved June 21, 2012, from http://www.juniperresearch.com/reports/mobile_cloud_applications_and_services

Section 1
Mobile Security

Chapter 1
A 2D Barcode Validation System for Mobile Commerce

David Kuo
San Jose State University, USA

Daniel Wong
San Jose State University, USA

Jerry Gao
San Jose State University, USA & Tsinghua University, China

Lee Chang
San Jose State University, USA

ABSTRACT

The wide deployment of wireless networks and mobile technologies and the significant increase in the number of mobile device users has created a very strong demand for emerging mobile commerce applications and services. Barcode-based identification and validation solutions are considered an important part of electronic commerce systems, particularly in electronic supply chain systems. This paper reports a mobile-based 2D barcode validation system as part of mobile commerce systems. This barcode-based validation solution is developed based on the Data Matrix 2D-Barcode standard to support barcode-based validation in mobile commerce systems on mobile devices. The paper demonstrates its application by building a mobile movie ticketing system.

INTRODUCTION

The wide deployment of wireless networks and mobile technologies and the significant increase in the number of mobile device users have created a very strong demand for emerging mobile commerce applications and services. According to

Gao, Prakash, and Jagatesan (2007), 2D barcodes can be used to support pre-sale, buy-and-sell, and post-sale activities for mobile commerce transactions. For example, 2D barcodes can be used as advertisements, coupons, or promotional materials that can be captured and decoded by users with mobile devices. Moreover, 2D barcodes enable

DOI: 10.4018/978-1-4666-2785-7.ch001

mobile devices to become a point-of-sale device that reads the barcode and facilitates payment transactions. After a payment transaction, 2D barcodes can be used by customers as a receipt or proof of purchase to gain access to the purchased goods and services with their mobile phones. Recently, people have gradually realized the importance of 2D barcodes and their great application value in M-Commerce because of the following (Gao, Prakash, & Jagatesan, 2007):

- 2D barcodes provide a new effective input channel for mobile customers carrying mobile devices with inbuilt cameras.
- 2D barcode is becoming a popular approach to present semantic mobile data with standard formats.
- 2D barcodes support a new interactive and efficient approach between mobile customers and wireless application systems.
- 2D barcode technology can be and is being used in diverse applications in mobile commerce.

Similar to RFID-based technology and solutions, barcode-based identification and validation solutions have been considered an important part of electronic commerce systems, particularly in electronic supply chain systems. However, although there are many benefits of using 2D barcodes, they are not widely utilized in the United States, especially in mobile commerce. This paper reports the research, architecture, design, security, and issues of implementing a 2D barcode validation system to encourage readers to adopt this technology into mobile commerce systems. This barcode-based validation solution is developed based on the Data Matrix 2D-Barcode standard. Furthermore, the paper also demonstrates its application by building a mobile movie ticketing system.

This paper is structured as follows. The next section covers the basics of 2D barcodes and related supporting technologies. Furthermore, it reviews the related work and applications in mobile commerce and presents a 2D barcode based validation system, including the system architecture, functional components, and used technologies as well as its secure 2D barcode-based framework. In addition, it reports its application in a movie ticketing prototype system in mobile commerce. Finally, the conclusion remarks are given in the last section.

UNDERSTANDING 2D BARCODES AND SUPPORTING TECHNOLOGIES

Although there are a number of widely used 2D barcodes today, different barcodes and standards are used in different countries and industry segments. Figure 1 shows the samples of three popular types of 2D barcodes. Quick Response

Figure 1. 2D barcode samples

a) QR Code   b) PDF 147   c) Data Matrix

(QR) code is mostly used in Japan. It can encode not only binary and text data, but also Japanese characters. Japanese companies use them to encode product information. Data Matrix 2D barcodes are popularly used in United States (IEEE Computer Society, 2005). They are printed on parcel labels on product packages to track the shipment and identify of products in supply chains. Organizations using Data Matrix include NASA, United State Postal Office, and the US Department of Defense (IEEE Computer Society, 2005). Both QR code and Data Matrix have been standardized by International Standard Organization (ISO). Other 2D barcodes include ColorCode, VS Code, Visual Code, Shot Code, etc (Kato & Tan, 2007b).

Data Matrix Standard

As described in (International Organization for Standardization, 2000), "Data Matrix is a two-dimensional matrix symbology, which is made up of square modules arranged within a perimeter finder pattern." According to its standard specification (the ISO/IEC 16022 specification), a Data Matrix symbol is composed of three parts: the encoded data, four borders, and the quite zone. Each part contains either white or black solid squares called modules. A black module represents a binary one, and a white module represents zero. To facilitate scanning devices to locate the symbol, a Data Matrix symbol contains an L shape solid modules to define the orientation, boarders, and size of the shape. The other two sides are represented by broken modules alternating between white and black. The whole symbol is surrounded within white modules marked as the quite zone. A Data Matrix symbol uses Reed-Solomon Error Checking and Correction (ECC) level 200 for error detection and correction. The two types of ECC levels supported by Data Matrix are ECC 000–140 and ECC 200. The maximum data that a Data Matrix code can encode is 2,335 alphanumeric characters, 3,116 numbers, or 1,556 bytes. Because of its advantages

and compact size, the Data Matrix barcode is one of the 2D barcodes that is often used with mobile devices. Using 2D barcodes (like Data Matrix), we need to understand its standard, encoding and decoding processes, and its error detection and correction rules.

Encoding and Decoding Processes

According to the ISO Data Matrix specification (International Organization for Standardization, 2000), an encoding process for creating a Data Matrix symbol consists of the following steps:

1. Evaluate the given data stream to determine the best encoding scheme and convert the data into codewords.
2. Generate the error checking and correction codewords.
3. Place codewords in the symbol.

A decoding process is needed to enable hardware devices and software programs to locate and decode Data Matrix barcodes (or symbols). This process includes the following steps:

1. Find the candidate areas that might contain the L-shape finder.
2. Find the finder in these candidate areas.
3. Find the two lines that are outside but closest to the finder.
4. Find the two opposite sides of the finder with alternating black and white modules.
5. Find the line that passes through alternating black and white modules for each side.
6. For each side, find the number of data modules.
7. Find the centerline that passes through the alternating modules for each side.
8. Use the centerline to find all the data modules from left to right and bottom to top until reaching to the borders.

Error Correction and Codeword Generation

The ECC 200 symbols use Reed-Solomon for error checking. The Reed-Solomon code is given by the equation $R\text{-}S(n,k) = (2^m - 1, 2^m - 1 - 2t)$, where k is the number of codewords of actual encoded data, n is the total number of codewords, m is the number of bits in a codeword, and t is the number of error codewords that can be corrected. 8 is a popular value for m since it's the number of bits in a byte. The number of checking codewords is calculated as $2t = n - k$. The checking codewords are the reminder of the data codewords divided by the generator polynomials. If encoded data is damaged or distorted, the checking codewords can be used to restore the data (Sklar, 2002).

2D Barcode Scanners

As the article in (Honeywell Imaging and Mobility, 2008) points out, choosing the right 2D barcode scanner can be challenging when developing a mobile 2D barcode application. Not all types of barcode scanners are able to read 2D barcodes from the screen of mobile devices. For example, laser barcode scanners cannot read barcodes from LCD screens. This issue is a major obstacle for the success of m-ticketing and m-voucher solutions. Most retailers today do not have the right equipment to read barcodes from mobile devices.

Due to the popularity and advantages of using 2D barcodes with mobile devices, more barcode scanners are built with the imaging technology. These scanners are often called imagers. A 2D barcode imager works similar to a digital camera. It is equipped with an imaging sensor. It reads a barcode by first capturing its image and then decodes its data. This feature allows the imagers to decode barcodes from self-illuminating displays. Moreover, most imagers support both 1D and 2D barcodes, making them suitable for mixed barcode environments. The scanner interface has also been evolved in the past few years. The latest 2D barcode scanners support keyboard simulation and the UBS interface (URL: www.idautomation. com/imagers). This allows the scanner to simulate the keyboard through the USB interface, entering the decoded data as if it was entered by the keyboard. Figure 2 shows a 2D barcode scanner that is able to scan barcodes from LCD displays. A 2D barcode imager normally needs to be attached to a PC or a laptop for it to work with the barcode application. However, in many cases, mobile barcode imagers are needed. For example, to scan 2D barcodes at the entrance of a ball game, it's impossible for a barcode inspector to carry a laptop. Therefore, another type of 2D barcode scanner is built into a PDA. The barcode reader and the client application can be combined into one device. PDA 2D barcode scanners increase the mobility for the barcode inspectors, but their costs are much higher.

Figure 2. Metrologic Elite MS7580 Genesis area-imaging scanner

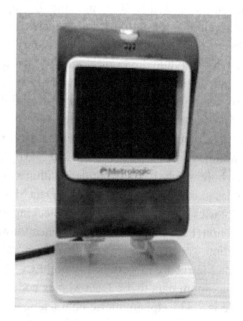

RELATED WORK IN MOBILE COMMERCE

Barcodes have been widely used for identifying products and delivering information (Vandenhouten & Seiz, 2007). The invention of 2D barcodes has significantly increased the security and data capacity for a barcode. As mobile devices are built with more sophisticated functions selling at more affordable prices, there are new mobile applications with 2D barcodes. Most cell phones today are built in with a camera. One type of 2D barcode applications is to use camera-enabled cell phones as barcode readers to decode the barcode content. This allows users to collect more information just by a click of a button (Falas & Kashani, 2007). The other type of usage is to use the mobile device as a carrier of 2D barcodes, which are validated directly from the mobile device at the point of use. For such applications, a comprehensive 2D barcode processing, delivering, and validation system is needed. This is very useful in mobile-commerce where digital tickets, coupons, and invoices can be delivered to mobile devices as 2D barcodes. They can be used anytime and anywhere for mobile commerce applications.

According to Gao, Prakash, and Jagatesan (2007), there are different types of 2D barcode mobile applications.

- As discussed in (Kato & Tan, 2007b), in Japan, 2D barcodes can be seen on websites, street signs, product packages, stores, magazines, and ads. Most camera-enabled cell phones in Japan are built with a barcode reader. A survey shows that in year 2006, 82.6% of the respondents had used their barcode readers on their cell phones with QR codes. This allows users to collect information without tedious data entry. This type of application can also be used in zoos or museums for users to collect detailed information on an animal or a historical art piece (O'Hara et al., 2007a, 2007b).

- There is an increasing trend to use 2D barcodes in mobile-commerce. 2D barcodes can be utilized in pre-sale, buy-and-sale, and post-sale activities, where purchase information, invoices, and promotional materials such as coupons and advertisements can be encoded into 2D barcodes and delivered to customers' cell phones through emails or Multimedia Messaging Service (MMS). Customers can receive them instantaneously and use them anytime anywhere with their cell phones. Purchase information and invoices can also be encoded into 2D barcodes. Many electronic stores offer store pickup services that customers can purchase items online and pick them up at a local store. After the transaction is complete, the store sends a 2D barcode receipt to the customer. When the customer picks up the purchased items from a store, the store staff only needs to scan the barcode to verify the purchase information. This has greatly increased the convenience for users.

- 2D barcodes can also be used for wireless payments (Gao, Kulkarni, Ranavat, Chang, & Hsing, 2009). Payment transaction information (such as credit card data) can be encoded into 2D barcodes and used at retail stores, taxi, payment terminals, and even mobile internet payment. However, using 2D barcodes for payments has security issues that need to be addressed. Since 2D barcodes are just electronic image files, they can be easily transferred and copied. Therefore, 2D barcode payment systems need to be carefully designed and rigorously tested to ensure a secure environment for using these barcodes.

The rest of this section reviews some 2D barcode-based application systems and validation solutions for mobile commerce.

Airline 2D Barcode Boarding Passes. As described in (Bouchard, Hemon, Gagnon, Gravel, & Munger, 2008) and (Mobiqa, 2008), airlines are using mobile 2D barcode validation systems as an alternative to paper-based boarding passes to speed up the boarding time. During the check-in process, the airline staff sends the boarding information as a 2D barcode to the passenger's cell phone through MMS. When the passenger passes security checkpoints or boards on a plane, the security staff scans the barcode to reveal the passenger's boarding information and identity for ID validation against their passport. As a result of this system, it decreases the time spent during the boarding process and helps the airline to save costs on paper and printing devices. However, for such a system, there is an important issue that needs to be addressed. The system has to make sure the barcodes can be scanned and validated from different cell phone models. Each cell phone model has different screen size, color, resolution, brightness, contrast…etc. These factors can affect the barcode reading. To resolve this issue, the system created an adaptive model to generate the barcode based on different characteristics of the mobile screen. The study found that to achieve the best result is to make each barcode module at least 4 pixels. The International Air Transport Association (IATA) has approved to use 2D barcodes as boarding passes. By 2010, every airline will implement a mobile 2D barcode boarding pass solution. Today, both Air Canada and Northwest Airlines have already implemented this technology (URL: http://www.aircanada.com/en/travelinfo/traveller/mobile/mci.html).

Sports Game Ticketing Services. Another type of mobile barcode validation systems can be seen in concert and game ticketing services. When combined with mobile payment services, the whole transaction from purchasing to receiving the ticket can be done using a mobile phone. Mobiqa (2007a), a mobile barcode solution provider, partnered with PayPal to provide a mobile payment and ticket delivering service for rugby games. This solution solved the problem that game goers cannot receive their game tickets due to postal strikes during that time. It allows users to purchase tickets using the PayPal Mobile service. Using the PayPal Mobile service solves the security concern when the users entering their credit card information using their cell phones. Once the transaction is complete, the ticket information is encoded into a barcode and sent to the user's mobile device. When the user arrives at the venue, the staff scans the barcode from the user's mobile device using a PDA scanner and the barcode is validated with the payment system.

2D Barcode Medical Prescription System. As mentioned previously, one advantage of 2D barcodes is that they can encode more data compared to 1D barcodes. This advantage has been utilized in pharmaceutical science. The Taiwan government developed a 2D barcode prescription system (2DBPS) for its National Health Insurance (NHI) (Wang & Lin, 2008). With this system, doctors' medical prescriptions are encoded as 2D barcodes and given to patients. When the patient arrives at a pharmacy, the pharmacist only needs to scan the barcode to validate the prescription with the back-end server. This system has several advantages. Firstly, it reduces human errors occurred in manual data entry. The pharmacist only needs to scan the barcode and the prescription information is automatically entered into the system. Drugs can be dispensed more accurately. Secondly, it saves time for doctors, patients, and pharmacists by reducing manual labors such as writing prescription and entering prescription data. Lastly, it increases patient's privacy because medical prescription data are presented as a 2D barcode. However, from the literature that we found about this system, it only supports paper-based 2D barcode prescriptions. Supporting mobile devices can be a future improvement, and it will not be difficult to implement since the 2D barcode-processing infrastructure has already been developed. The only two issues that need to be addressed and implemented are the barcode delivery method to mobile devices and the capability of scanning barcodes from different mobile devices.

AN OVERVIEW OF A 2D BARCODE-BASED VALIDATION SOLUTION

In mobile-commerce, 2D barcodes can be used to present different types of commerce information. To support coupon redemption, product delivery (or pick-up), and merchandise check-out in mobile commerce applications, we need a 2D barcode-based validation solution to integrate with electronic payment systems (or mobile POS-based terminals). To use 2D barcodes, the users just need to bring their mobile devices to a point-of-use terminal. The merchant staff uses a 2D barcode scanner along with the mobile client to read the barcode off the screen of the mobile device. Each barcode is validated against the server through the Internet, and then the user identity and purchasing record are verified. If the result is fine, the merchant will accept the user's barcode. This section describes the system infrastructure and related components.

System Infrastructure and Architecture

The 2D Barcode Validation System consists of three portions, the server, the client, and the end user. They are connected by three types of networks as shown in Figure 3. First, the Internet supports the internet-based communications between the server system and its clients through HTTP, SOAP and SMTP protocols. Second, the wireless network supports the wireless communications between the server system and mobile devices with a Wi-Fi connection. The end user can use these devices to visit the server's web store, for example, to make 2D barcode ticket purchases. Third, the cellular network supports the communications for mobile phones through the HTTP protocol and MMS. It allows the end user to receive 2D barcodes in MMS messages.

The 2D Barcode Validation System includes a server, a mobile client, a USB 2D barcode imager, and a backend database server. As shown in Figure 4, the 2D Barcode Validation System has a 4-tier architecture to increase the independency between different layers and reduce their change impacts. The user interface layer includes a point-of-use GUI client application and a web client running in the web browser of a PC or a mobile

Figure 3. System infrastructure

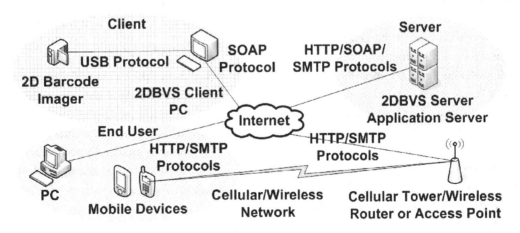

Figure 4. System architecture

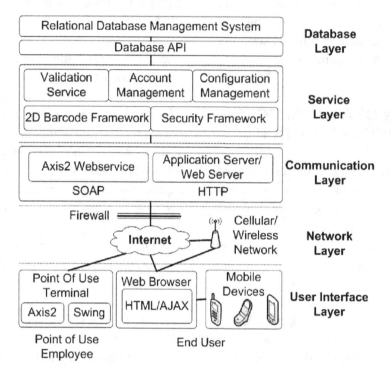

device. The communication layer provides communication protocols including HTTP and SOAP. The service layer provides the application functional services of 2D barcode framework, security framework, barcode validation, and account management. The data base layer consists of entity models, database access libraries, and a relational database management system.

Server Components

The server consists of the 2D barcode web application running on a J2EE application server. The web application contains the mobile/web store, 2D barcode framework, validation web services, and data access services. As shown in Figure 5, the server includes the following service components:

- **2D Barcode Web Store:** Provides an interface for users to purchase 2D barcodes.
- **Validation Service:** Allows the client to validate barcodes.

- **User Service:** Manages user accounts and authentications.
- **Order and Payment Services:** Manage orders and payment processes.
- **Event Service:** Manages event and ticket records.
- **Encryption Service:** Abstracts the security framework and the encryption logic.
- **2D Barcode Service:** Contains the encoding and decoding algorithms. It works in conjunction with Device Configuration Manager to produce the best image for each different type of mobile devices. Based on different mobile device profiles, the barcode service encodes different sizes of images to best display for the mobile device. This ensures interoperability for the 2D barcode imager to accurately read the 2D barcode from different mobile device screens.

Figure 5. Server components

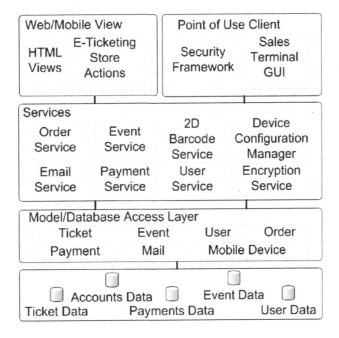

Client Components

The client system consists of a GUI application and a 2D barcode imager that allows barcode inspectors to validate and display the content of the issued 2D barcodes. It is installed in the merchant stores where 2D barcodes need to be validated. It consists of the components shown in Figure 6. Each component of the client system is described in details as follows:

- **Client GUI:** This GUI interface supports client users to interact with the system.
- **2D Barcode Validation Handler:** Responsible for validating 2D barcodes with the server.
- **Secure Session Establisher:** Responsible to establish a secure session between the client and the server.
- **Security Framework Client:** The security framework on the client system. It contains the libraries and interfaces for encrypting

Figure 6. Client components

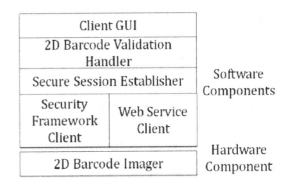

and decrypting data and generating certificates to establish secure sessions.

- **Web Service Client:** The client web service interface that communicates with the web services provided by the server.
- **2D Barcode Imager:** The 2D barcode scanner hardware, which can be used to scan 2D barcodes from mobile devices and feed them to the system.

Security Components

In the security encryption framework, all purchasing information (such as invoices or tickets) is encoded into a 2D barcode using two different security solutions. The details are reported in (Gao, Kulkarni, Ranavat, Chang, & Hsing, 2009). An asymmetric encryption algorithm, Rivest, Shamir, and Adleman (RSA), is used to ensure that all the barcodes are generated by the server application. Another asymmetric encryption approach Elliptic Curve Cryptography (ECC) is used to provide a secure channel between the server and the client. A symmetric encryption scheme Advanced Encryption Standard (AES) is used for authenticat-

ing the actual barcode owner by encrypting the barcode information using a passphrase provided by the owner.

The security framework mainly uses the security components from the previous Secure Mobile Payment System developed by the SJSU graduate students (Gao, Kulkarni, Ranavat, Chang, & Hsing, 2009). As shown in Figure 7, it was modified to support the underlying encryption and decryption. The framework contains various cryptography utilities for ECC, RSA, and symmetric cryptographies. Table 1 summarizes the security methods used in the proposed system.

Used Technologies

Several technologies are used in developing the system. They are summarized below.

- **Client Technologies:** We used Java and Java Swing API and NetBeans GUI framework to create client GUI components. A Metrologic Elite MS7580 Genesis area-imaging scanner (at $366.47) is used with the PS2 keyboard wedge interface. As shown in (URL: www.totalbarcode.com), this scanner supports both 1D and 2D barcodes. In addition, the scanner supports USB and RS232 interfaces and uses the imaging technology to read barcodes from LCD screens. We have successfully tested this scanner with different LCD displays.

Figure 7. Security components

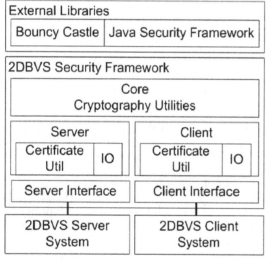

Table 1. Usage of Security Methods

Security→ Target↓	Password Based	Symmetric (AES)	Asymmetric (RSA)	Asymmetric (ECC)	Barcode Based
Client User/Consumer	x				
Client System		x	x	x	
Server System		x	x	x	
Secure Session				x	
Insensitive Data			x		x
Sensitive Data		x			x

- **Server Technologies:** The server is developed based on the Java EE platform. The Bouncy Castle Crypto API is used to implement cryptographic algorithms to encrypt and decrypt the barcode data. Moreover, the Axis2 framework is used to implement the validation web services. JSON (JavaScript Object Notation) is used throughout the system for passing the data.
- **Middle-Tier Technologies:** The system uses a J2ME 2D barcode encoding and decoding algorithm provided by www. drhu.org, and it's ported to the J2SE environment to support barcode encoding for movie tickets in our demo system. The other used technologies are JavaMail API, Servlet, JSP, Struts 2, Ext JS 2.0, Spring Framework, Apache Velocity, Tiles, and Quartz. The Apache Tomcat application server is used to support Internet communications with clients.
- **Data-Tier Technologies:** The MySQL relational database management system (RDBMS) is used as the back-end database. Java Persistence API and Hibernate are used to support the communications with the database.

A 2D BARCODE SOLUTION

Secure 2D Barcode Framework

The secure 2D barcode framework allows users to encode and decode information. It consists of two parts: a) the 2D barcode encoding and decoding library, and b) the encryption library. The barcode library only encodes and decodes Data Matrix barcodes and its algorithm is based on ISO/ IEC 16022 specification. The encryption library provides asymmetric and symmetric encryptions.

The asymmetric encryption method uses the RSA algorithm to encrypt the data using the server's private key to prevent malicious users creating fake barcodes. The 2D barcode validation client system has the server's public key for data decryption. The server's public key is encrypted by the client user's password. Therefore, even though a malicious user compromises the client system, without knowing the client user's password, no one can decrypt the data. This method is more suitable for insensitive data and the point-of-use terminals that need a faster validation process since it does not require the user to enter a passphrase. The symmetric encryption uses AES algorithm. It allows the user to encrypt data using a passphrase before it is encoded into a barcode. Therefore, this method is more suitable for sensitive data.

The secure 2D barcode framework provides the following function features: 2D Data Matrix barcode encoding and decoding, barcode image conversion, symmetric and asymmetric encryption and decryption, and web service-based functionalities.

2D Barcode Validation

The 2D barcode validation process consists of two steps: 1) establish a secure session 2) and validate the barcode through the secure session. The process of establishing a secure session is represented in Figure 8, and it uses the Public-Key Infrastructure (PKI) approach. The server acts as the Certificate Authority (CA). To set up a secure session with the server, the client first generates an ECC key pair and sends a certificate request to the server. The server responses back with CA and client certificates, and then the client exchanges its shared secret with the server. The shared secrets are used to setup the Ecliptic Curve Integrated Encryption Scheme (ECIES) as discussed in (Certicom, 2000). After the ECIES is ready, the client can use it to encrypt the client user's username and password and send them to the server to complete the login process. If the login information is successfully authenticated,

Figure 8. Secure session setup process

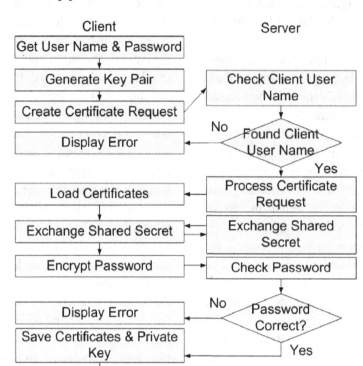

the client encrypts and saves the certificates and the private key and the secure session is deemed ready. Otherwise, the client discards these security data and asks the client user to enter the username and password again.

To validate a barcode, as described in Figure 9, the client system first gets the decoded data from the GUI and decrypts the barcode content. After the decryption, the client system displays the decrypted data on the screen and sends it to the server for validation using the established secure session. Then, the server searches for the barcode record in database to check if the given barcode is a valid one. If the validation result is fine, a successful response will be returned. Otherwise, the server sends a descriptive error message back to the client. The barcode inspector can proceed to verify the barcode information such as the user's name and the product information. If everything is correct, the client system sends

the barcode usage record to the server to prevent future uses.

A 2D BARCODE VALIDATION APPLICATION: MOBILE MOVIE TICKETS

Mobile Movie Tickets System

To demonstrate the usage of our proposed solution, a 2D barcode ticketing prototype system is implemented. Based on the 2D barcode-ticketing scenario, the system allows users to purchase movie tickets through a storefront website, called Movie Express, and delivers the tickets to their mobile devices. It supports two types of purchases: movie tickets and ticket orders. Their scenario diagrams are provided in Figure 10 and Figure 11.

Figure 9. Barcode validation process

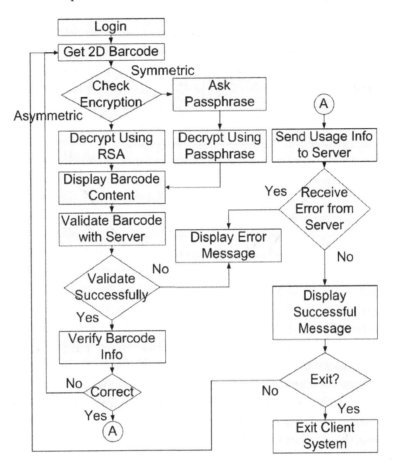

Figure 10. Movie ticket purchase scenario diagram

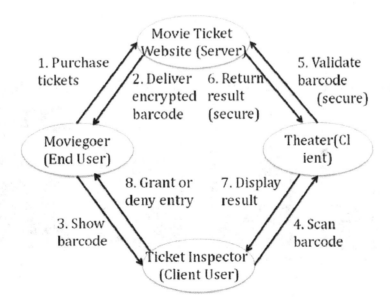

Figure 11. Ticket order purchase scenario diagram

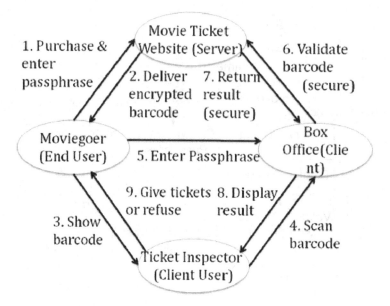

For movie ticket purchases, users can use a web browser or an internet-enabled mobile client to access the 2D barcode ticket storefront website as shown in Figure 12. After registering and logging into the store, the user can purchase various movie tickets (see (a) and (b)) and they will receive the ticket barcode by email (see (c)) or MMS. If multiple tickets are purchased, the website allows the user to send each ticket to a different email address or cell phone. When arriving at the theater, the user shows the ticket barcode to the ticket inspector for validation. After scanning the barcode (as shown in (d)), the ticket information will be displayed on the client system. Meanwhile, the client system validates this barcode with its server to make sure the ticket is valid. If the barcode passes the validation, the ticket inspector proceeds to verify the ticket information and let the user enter the theater. Finally, the ticket inspector uses the client system to send the ticket usage

Figure 12. Movie ticket purchase and barcode delivery

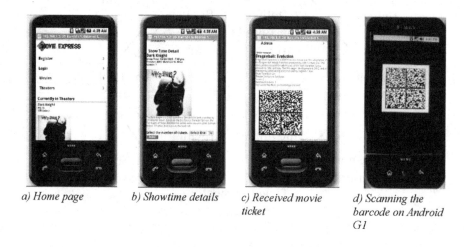

a) Home page   b) Showtime details   c) Received movie ticket   d) Scanning the barcode on Android G1

information to the server to prevent the same barcode being used again.

For ticket order purchases, users can select the "Will Call" style for their movie ticket orders. During the purchase, the user is asked to enter a passphrase to encrypt the order, and its corresponding 2D barcode is sent to them. After receiving the order barcode, the user can pick-up the tickets at the box office. The box office staff scans the barcode and asks the user to enter the passphrase to decrypt the barcode data. After the order information stored in the barcode has been successfully decoded, decrypted, validated, and verified, the staff gives the ordered tickets to the user to complete the process.

Performance Benchmarks

To evaluate the performance of the prototype system, a series of performance testing have been conducted for the main services and different security schemes. The testing environment and data are described in Table 2. Each performance result is the average of 5 runs. The ciphertext is encoded using the Base64 encoding scheme.

The system's main services include the certificate request process, ECC shared secret exchange, client user authentication, barcode valida-

tion, and barcode redemption. The performance testing of these functions are conducted on the server. The performance chart and the result are shown in Figure 13.

Security is an important feature of the system. However, it takes a toll on the performance. In order to understand the performance impact that the security feature causes, the performances of different types cryptographic algorithms and approaches have been tested. The performance chart and result are shown in Figure 14. To reach the same security level, a 128-bit key is used for AES, a 256-bit key is used for ECC, and a 3072-bit key is used for RSA (Krasner, 2004). For the barcode data ciphers, the performances of three different algorithms (AES, RSA with AES, and RSA) have been compared. Initially in the system, RSA was used to encrypt insensitive data in the barcode with a 480-bit key since the plaintext of a barcode ticket is about 480 bits. However, RSA possesses some issues that it is almost never used to encrypt messages in practice (Ferguson, Schneier, & Kohno, 2010). First, the encrypting message is limited to the RSA key size. Second, RSA operations are expensive in terms of computation. Based on our testing result, RSA encryption is almost 11 times slower than AES encryption when it encrypts the given plaintext for 250 times. Fur-

Table 2. Performance testing environment and data

Environment	MacBook Mac OS X 10.6.3 with 2.4 GHz Intel Core 2 Duo processor and 4 GB 667 MHz DDR2 SDRAM		
Plaintext (Length: 60)	52\nDavid Kuo\nDark Knight\nAMC Metreon 16 Imax\n20100520\n2330\n1		
Ciphertext Using AES (Length: 88)	YXxhmZboMl4E2sOWfq0jyPx9u0m9cOCpb8wI+Pc4iWg1hg2ONz0c0ygnoar/FJaGkvNb6UMIUv-7FUYCOJiegQA==		
Ciphertext Using ECC (Length: 112)	Y4TOs7rfMUqqSSHkokmZQOgy8NuDFX9P/u2lHFh6vJZAlrcMyUhq/D9GnPXaDkzwePYwQ4ToB/MuhmZz+KGR2pHB2jqlyu2qC0dk6dZJUiZIP8CZ		
Ciphertext Using RSA-480 (Length: 80)	KnQewbd4+NQ07H86jNEfZfKObA1N8HD+q2t4flo5bG25QHiJHVvKO3vFr5yWMb4ksJFT2Y9OIOLhUBxI		
Displayed Text	52 David Kuo Dark Knight AMC Metreon 16 IMax 20100520 2330 1	Description	Movie ticket ID Owner name Movie name Theater name Show date Show time Screen number

Figure 13. The performance chart of main services

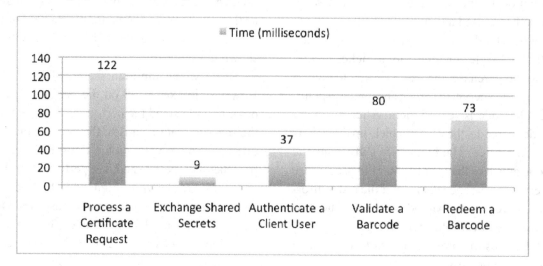

thermore, a 480-bit RSA key is less secure compared to AES with a 128-bit key. Therefore, a combination of RSA and AES is used to replace pure RSA. The plaintext is first encrypted with a 128-bit AES key and then the key is encrypted with a 128-bit RSA private key. Both encrypted

AES key and the ciphertext are encoded into a barcode. This approach has increased the performance by 74%.

For the secure channel between the client and server, we compared the performance of ECC and RSA. The performance chart and the result are

Figure 14. The performance chart of AES, RSA with AES, and RSA

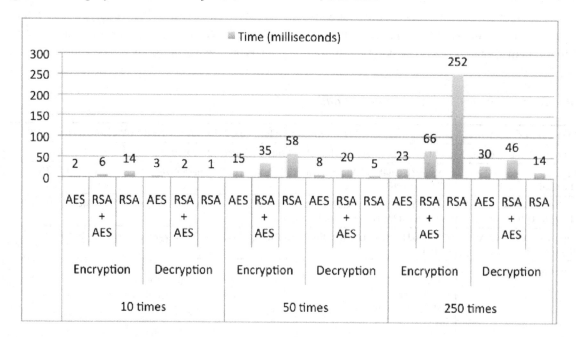

shown in Figure 15. The performance of RSA is significantly worse compared to ECC when it encrypts the given plaintext for 50 times.

CONCLUSION

Using 2D barcodes with mobile devices has many benefits. It provides mobility, security, and convenience. Although there are numerous mobile 2D barcode technologies in the market, there are no academic research reports or papers discussing about complete 2D barcode-based validation solutions and systems. This paper introduces a 2D barcode validation system for mobile commerce applications. It provides a secure way to encode ticket and order information into 2D barcodes, which can be easily transported through email or MMS.

Mobile commerce is becoming a lucrative market for merchants, banks, and service businesses. 2D barcodes have become the de-facto standard for carrying information in mobile commerce. The proposed validation solution provides a secure way to process and validate 2D barcodes for mobile commerce. It works effectively while we apply in the movie ticketing prototype. Similarly, it also can be used for e-validation in other situations, such as retail-oriented check-out, product delivery and pick-up, station-based payment, terminal-based coupon check-up. For the future research and development direction, we are working on an integrated, SOA-based 2D barcode mobile commerce solution that includes advertising, purchasing, payment, and validation subsystems.

ACKNOWLEDGMENT

This research project was funded by Sun Microsystems in 2009, and Dr. Gao was supported by Tsinghua National Laboratory for Information Science and Technology (TNLIST) in 2010.

Figure 15. The performance chart of ECC and RSA

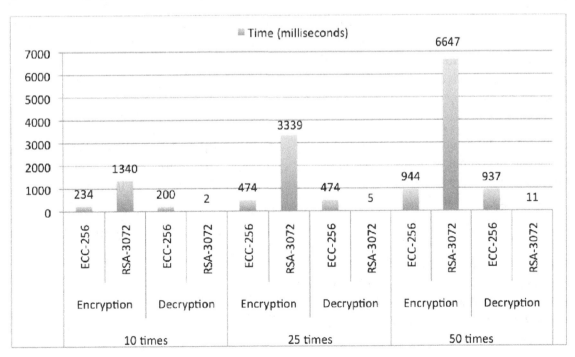

REFERENCES

Anderson, R. (2001). *Security engineering: A guide to building dependable distributed systems*. New York, NY: John Wiley & Sons.

Bouchard, T., Hemon, M., Gagnon, F., Gravel, V., & Munger, O. (2008, March). Mobile telephones used as boarding passes: Enabling technologies and experimental results. In *Proceedings of the Forth International Conference on Autonomic and Autonomous Systems* (pp. 255-259).

Certicom Research. (2000). *Standards for efficient cryptography – SEC1: Elliptic curve cryptography*. Retrieved from http://www.secg.org/collateral/sec1_final.pdf

Falas, T., & Kashani, H. (2007). Two-dimensional bar-code decoding with camera-equipped mobile phones. In *Proceedings of the Fifth Annual IEEE International Conference on Pervasive Computing and Communication Workshops* (pp. 297-600).

Ferguson, N., Schenier, B., & Kohno, T. (2010). *Cryptography engineering: Design principles and practical applications*. Indianapolis, IN: Wiley.

Gao, J. Z., Kulkarni, V., Ranavat, H., Chang, L., & Hsing, M. (2009). A 2D barcode-based mobile payment system. In *Proceedings of the 3rd International Conference on Multimedia and Ubiquitous Engineering* (pp.320-329).

Gao, J. Z., Prakash, L., & Jagatesan, R. (2007). Understanding 2D-barcodes technology and applications in m-commerce – design and implementation of a 2D barcode processing solution. In *Proceedings of 31st Annual International Computer Software and Applications Conference*, Beijing, China (Vol. *2*, pp. 49-56).

Hansen, F. A., & Grønbæk, K. (2008). Social web applications in the city: A lightweight infrastructure for urban computing. In *Proceedings of the Nineteenth ACM Conference of Hypertext and Hypermedia* (pp. 175-180).

Honeywell Imaging and Mobility. (2008). *Mobile ticketing technology*. Retrieved from http://www.airport-int.com/categories/mobile-ticketing/mobile-ticketing-choosing-the-right-technology-platform-is-critical-to-your-programs-success.asp

IATA. (2007). *Standard paves way for global mobile phone check-in*. Retrieved from http://www.iata.org/pressroom/pr/2007-11-10-01.htm

IEEE Computer Society. (2005). The 2D data matrix barcode. *IEEE Computing & Control Engineering Journal*, 16(6), 39.

International Organization for Standardization. (2000). *ISO/IEC 16022: Information technology – international symbology specification – data matrix*. Retrieved from http://www.iso.org/iso/catalogue_detail.htm?csnumber=29833

Kato, H., & Tan, K. T. (2007a). First read rate analysis of 2D-barcodes for camera phone applications as a ubiquitous computing tool. In *Proceedings of the IEEE Region 10 Conference* (pp. 1-4).

Kato, H., & Tan, K. T. (2007b). Pervasive 2D barcodes for camera phone applications. *IEEE Pervasive Computing/IEEE Computer Society and IEEE Communications Society*, 6(4), 76–85. doi:10.1109/MPRV.2007.80

Krasner, J. (2004). *Using elliptic curve cryptography (ECC) for enhanced embedded security financial advantages of ECC over RSA or Diffie-Hellman (DH)*. Retrieved from http://embedded-forecast.com/EMF-ECC-FINAL1204.pdf

Menezes, A., Oorschot, P., & Vanstone, S. (1996). *Handbook of applied cryptography*. Boca Raton, FL: CRC Press.

Mobiqa. (2007a). *Case study: Scotland rugby league world cup qualifier: Mobiqa team up with PayPal mobile to offer the world's first end to end mobile ticket purchase and delivery service.* Retrieved from http://www.mobiqa.com/live/files/RugbyLeaguecasestudy.pdf

Mobiqa. (2007b). *Case study: Village cinemas Czech Republic: Village cinemas became the first cinema chain in Europe to offer the mobile ticketing service to film fans.* Retrieved from http://www.mobiqa.com/cinema/files/Village-Cinemacasestudy.pdf

Mobiqa. (2008). *Case study: Northwest Airlines revolutionise air travel.* Retrieved from http://www.mobiqa.com/airlines/files/NWAcasestudy.pdf

O'Hara, K., Kindberg, T., Glancy, M., Baptista, L., Sukumaran, B., & Kahana, G. (2007a). Collecting and sharing location-based context on mobile phones in a zoo visitor experience. *Computer Supported Cooperative Work, 16*(1-2), 11–44. doi:10.1007/s10606-007-9039-2

O'Hara, K., Kindberg, T., Glancy, M., Baptista, L., Sukumaran, B., Kahana, G., et al. (2007b). Social practice in location-based collecting. In *Proceedings of the SIGCHI Conference on Human Factors in Computing Systems* (pp. 1225-1234).

Sklar, B. (2002). *Reed-Solomon codes.* Retrieved from http://www.informit.com/content/images/art_sklar7_reed-solomon/elementLinks/art_sklar7_reed-solomon.pdf

Stui, M. (2005). *The use of bar code SMS in mobile marketing, advertising, CRM.* Retrieved from http://www.adazonusa.com/theuseofbarcodesmsinmobilemarketingadvertisingcrm-a-3.html

Vandenhouten, R., & Seiz, M. (2007, September). Identification and tracking goods with the mobile phone. In *Proceedings of the International Symposium on Logistics and Industrial Informatics* (pp. 25-29).

Wang, W. L., & Lin, C. H. (2008). A study of two-dimensional barcode prescription system for pharmacists' activities of NHI contracted pharmacy. *Yakugaku Zasshi, 128*(1), 123–127. doi:10.1248/yakushi.128.123

This work was previously published in the International Journal of Handheld Computing Research (IJHCR), Volume 2, Issue 2, edited by Wen-Chen Hu, pp. 1-19, copyright 2011 by IGI Publishing (an imprint of IGI Global).

Chapter 2
Threshold–Based Location–Aware Access Control

Roel Peeters
Katholieke Universiteit Leuven, Belgium

Dave Singelée
Katholieke Universiteit Leuven, Belgium

Bart Preneel
Katholieke Universiteit Leuven, Belgium

ABSTRACT

Designing a secure, resilient and user-friendly access control system is a challenging task. In this article, a threshold-based location-aware access control mechanism is proposed. This design uniquely combines the concepts of secret sharing and distance bounding protocols to tackle various security vulnerabilities. The proposed solution makes use of the fact that the user carries around various personal devices. This solution offers protection against any set of (t-1) or fewer compromised user's devices, with t being an adjustable threshold number. It removes the single point of failure in the system, as access is granted when one carries any set of t user's devices. Additionally it supports user-centered management, since users can alter the set of personal devices and can adjust the security parameters of the access control scheme towards their required level of security and reliability.

INTRODUCTION

Contactless smartcards are often used to enforce access control for secure facilities and buildings. These security tokens contain identifying information and a secret key, used to identify the user carrying the smartcard. When the user approaches the building, he puts his contactless smartcard close to a reader installed in the proximity of the door. Both devices will then carry out a challenge-response protocol, in which the user's smartcard authenticates itself to the reader (in some scenarios, mutual authentication is required). If the protocol finishes successfully, the user is granted access. Besides access to buildings, similar mechanisms are employed to enter a car

DOI: 10.4018/978-1-4666-2785-7.ch002

(Microchip KeeLoq), to use public transport (Octopus Cards, OV-chipkaart, Oyster Online), and even for payments with contactless credit cards (Mastercard PayPass, Visa Paywave).

Although widely used, this conventional access control solution has some important drawbacks, such as several security vulnerabilities. First, the use of a single security token introduces a single point of failure in the system. If this token gets stolen, an unauthorized adversary could get access to a secure building or resource. Security tokens and smartcards could also be compromised or cloned. A recent example of the latter was the MIFARE attack discovered by Gans et al. (2008).

A second security vulnerability is relay attacks, which are also known as mafia fraud attacks. These are man-in-the-middle attacks where a verifier (e.g., the reader next to the door of a building) is tricked in believing that a prover (e.g., the smartcard) is in its close vicinity by an adversary surreptitiously forwarding the signal between the verifier and an out-of-range prover (Kim et al., 2009). Such an attack is important in the setting of access control systems, particularly when challenge-response protocols are employed, and should definitely be avoided.

In addition, both reliability and user-friendliness could be improved in conventional access control systems. For each system the user is enrolled to, and this can be a relatively high number, he has to carry around a separate smartcard or security token. The legitimate user that does not carry around the security token automatically cannot get access. Furthermore, revocation of a particular token is often a cumbersome and relatively slow process. This is illustrated by the following plausible scenario. When initiating the revocation process, the user first informs the facility manager. Second, revocation lists are updated and distributed. Third, the user gets a new token or smartcard. Since such a revocation process is slow, it also poses a security risk: there is a grace period in which the adversary can still use the token before the revocation lists are updated.

Fortunately, both security vulnerabilities can be tackled by introducing several countermeasures. The single point of failure can be removed by sharing the secret over a set of user's personal devices. The vulnerability against relay attacks can be solved by using distance bounding protocols. In addition, secret sharing also provides reliability to the user and allows for, through the mechanism of resharing, user-centered access control. It hence automatically improves the user-friendliness of the system.

Secret Sharing

The concept of secret sharing was first introduced by Shamir (1979). Instead of storing a secret on one device, the secret is divided into k pieces, each stored on different devices. Let t be a threshold number chosen by the user. This parameter directly relates to the security level of the scheme. The main characteristic of secret sharing is that the key can easily be reconstructed from any t pieces, but even complete knowledge of (t-1) pieces reveals absolutely no information about the secret. These shares are evaluations of a unique polynomial of degree (t-1). For (t-1) or fewer shares, one does not obtain any information at all about the secret, as one cannot reconstruct the unique underlying polynomial. For t or less stolen devices, there is hence no need for revocation as the adversary obtains no information about the secret. Secret sharing also allows for user-centered access control, since the user can decide which devices get a piece of the key, how large the threshold value t should be, and when the secret shares should be updated (i.e. when the resharing process should take place).

Outline

In the introduction, we showed that conventional access control mechanisms suffer from several security, reliability and usability issues. We put forward the idea of employing distance bounding protocols in combination with secret sharing on personal devices to solve several of these problems. In the next section, the general principles of distance bounding protocols are discussed more detail. Next we describe our threshold-based access control scheme, which uses distance bounding protocols to tackle relay attacks. We also describe how a user can manage the access control configuration. To demonstrate that our proposed access control scheme solves the identified vulnerabilities, we discuss the security properties of our solution more in detail.

DISTANCE BOUNDING PROTOCOLS

To enhance entity authentication protocols, location information can be incorporated. Entities which are in a specific location or within a certain range of a particular device will be granted some privileges, in contrast to all other entities. Distance bounding protocols, which have been introduced by Brands and Chaum (1994), can be used to cryptographically enforce the notion of "proximity." The concept of proximity-based authentication is depicted in Figure 1. Authentication requests originating from devices that are located within

Figure 1. The concept of proximity-based authentication: an authentication request from device A is accepted, and one from device B is rejected

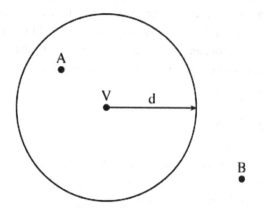

the range d of the verifier V are accepted, all other requests are rejected.

Distance bounding protocols combine physical and cryptographic properties to enable a verifying party (i.e. the *verifier*) to determine an upper bound on the distance between itself and a *prover*, who claims to be within a certain range.

The main building block of a distance bounding protocol is a challenge-response protocol which is executed n times. The number of rounds n is a security parameter. During each of the n rounds, which are also called *rapid bit exchanges*, the verifier measures the time between sending a challenge and receiving the response. The highest round trip time is then selected and multiplied with the propagation speed of the communication medium. This gives an estimate, i.e. an upper bound, on the distance between prover and verifier.

Let us now focus more on this distance estimation. Although the principle of measuring the round trip time during n rounds of the protocol is rather straightforward, one should take into account some practical details such as:

- **Processing Delay:** It is important to notice that the round trip time is not equal to the propagation delay. For example, it takes some time to compute and transmit (or receive) the response. The variation on the processing delay should hence be as small as possible compared to the propagation time, because we are only interested in the latter. The uncertainty on the distance estimation directly depends on this variation. Unfortunately, in some settings it is not possible to determine or control the processing delay exactly. To solve this issue, one often tries to minimize the processing delay (e.g., by using very simple hardware operations in the protocol), such that its variation can be neglected compared to the time of flight. Another strategy is to increase the propagation delay by selecting an appropriate communication medium. One should however take into account that accuracy is typically not the main design requirement in the setting of proximity-based authentication. One wants to compute an upper bound on the distance, not the exact location of the prover.

- **Communication Medium:** Two common communication technologies are ultrasound and RF communication. Both have their specific (dis)advantages. The former is rather slow, and the processing delay can hence be neglected. The accuracy of the timing measurements is also not very critical in this scenario. The technology is however susceptible to *wormhole attacks*, where an adversary, which is physically located between prover and verifier, forwards the signal between two distant points using a faster communication medium. RF communication is resistant to these attacks, but imposes more strict requirements on hardware, which could be a bottleneck. The round trip time should be measured very accurately, and the processing delay should be fixed and/or very small. These requirements can however be realized in practice. A prototype system has been implemented by Rasmussen and Capkun (2010) where the prover is able to receive, process, and transmit signals in less than 1 ns.
- **Transmission Format:** Clulow et al. (2006) have shown that one should also optimize the choice of transmission format when designing distance bounding protocols, to preclude certain physical attacks. They recommend using a communication format in which only a single bit is transmitted and the recipient can instantly react on its reception. In each of the n rounds, a single-bit challenge-response protocol should be carried out.

- **Mobility:** Since distance bounding protocols are carried out in mobile networks, prover and/or verifier could be mobile. This is however not a problem, because of two reasons. First, the execution time of the n fast bit exchanges is rather short. In realistic use case scenarios, the prover and/or verifier will have barely moved. This small execution time also limits the effect of a mobile adversary. Second, small variations in the round trip time, which could be caused by movement or by a cheating prover, are cancelled out by selecting the maximum round trip time. This maximum is not affected when prover and verifier are closer to each other in a few rounds.

In contrast to many distance estimation solutions, distance bounding protocols particularly focus on security. The verifier which carries out this protocol with a remote prover, wants to be able to check both the prover's identity, as the correctness of the latter's claim to be in the proximity of the verifier. To design a protocol that achieves these requirements, one should take into account some important cryptographic principles. It should be impossible for the prover to send the response before receiving the challenge from the verifier. Otherwise, the prover can pretend to be closer than he really is. This implies that the response should depend on the (random) challenge. A second requirement is that the prover cryptographically identifies itself during the execution of the distance bounding protocol (e.g., by using a secret key). Otherwise, the verifier does not know which entity is within a certain range.

The typical layout of a distance bounding protocol is depicted in Figure 2. Note that there are three distinct phases. In the first phase, prover and verifier can exchange some random nonces, create commitments, etc. The second and most important phase of the protocol is the n rounds

Figure 2. Three phases of distance bounding protocols

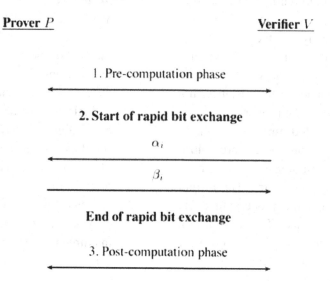

Prover P **Verifier** V

1. Pre-computation phase

2. Start of rapid bit exchange

α_i

β_i

End of rapid bit exchange

3. Post-computation phase

of rapid bit exchanges. The protocol ends with a post-computation phase, in which signatures can be computed, commitments being opened, etc. In some protocols, the first and/or third phase is omitted.

By employing the principles discussed above, one can design a distance bounding protocol which precludes one or more of the following attacks:

- **Distance Fraud Attacks:** One wants to prevent a dishonest prover claiming to be closer than he really is. This attack is called distance fraud attack and is conceptually shown in Figure 3. Distance bounding protocols are particularly designed to prevent this type of attack.

- **Mafia Fraud Attacks:** These were first described by Desmedt (1988). In this attack scenario, both prover and verifier are honest, but a malicious intruder is perform-

ing the fraud. This is a man-in-the-middle attack where the intruder I is modeled as a malicious prover $\overline{P}$ and verifier $\overline{V}$ that cooperate, as shown in Figure 4. The malicious verifier $\overline{V}$ interacts with the honest prover P and the malicious prover $\overline{P}$ interacts with the honest verifier V. The physical distance between the intruder and the verifier is small. This attack enables the intruder to identify himself to V as 'P being close to V,' without either P or V noticing the attack. Drimer and Murdoch (2007) have presented a practical mafia fraud attack on the United Kingdom's EMV payment system Chip & Pin.

- **Terrorist Fraud Attacks:** These are an extension of mafia fraud attacks (Desmedt, 1988). The intruder (being close to the verifier) and the prover will collaborate in this

Figure 3. Distance fraud attack

Prover ←————————————→ Verifier

Figure 4. Mafia fraud attack

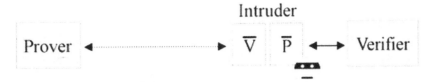

attack. It is important to note that the intruder must not know the private key of the prover, since the latter does not fully trust the former. If the intruder would know this private key, then it is impossible to make a distinction between the intruder and the prover, and as a result, terrorist fraud attacks can no longer be prevented. They would be the same party from a cryptographic point of view. The concept of a terrorist fraud attack is shown in Figure 5.

In this article, we will not focus on terrorist fraud attacks, and only concentrate on secure distance bounding protocols which prevent distance and mafia fraud attacks. Although the idea has been introduced more than fifteen years ago by Brands and Chaum (1994), it is only quite recently that distance bounding protocols attracted the attention of the research community. Hancke and Kuhn (2005) pointed out that distance bounding protocols should be designed to cope well with substantial bit error rates during the rapid single bit exchanges, as these are conducted over noisy wireless ad hoc channels. They incorporated this important requirement in the design of their RFID distance bounding protocol. Singelée and Preneel (2007) have proposed a noise resilient extension of the Brands-Chaum

protocol that provides mutual entity authentication. Recently, various other distance bounding protocols have been proposed in the literature. A short overview can be found in Avoine and Tchamkerten(2009), Rasmussen and Capkun (2009), and Kim et al.(2009).

LOCATION-AWARE ACCESS CONTROL

We focus on a location-aware access control mechanism where access is granted based on the combination of identity and proximity information. The setting is illustrated in Figure 6, where access control is enforced to enter a particular building. The user has a group of personal devices that will carry out a proximity-based authentication protocol with a verifying entity, e.g., a reader placed next to the door of the building. The former will be denoted by the prover, the latter by the verifier. If the protocol finished successfully, the user can enter the building. Only one of the user's devices will communicate directly with the verifying entity. This device is denoted by the gateway device in the rest of the article. The gateway device is also responsible for initiating the access control mechanism with the verifier. The other devices of the user are called end-

Figure 5. Terrorist fraud attack

Figure 6. Distributed access control setting

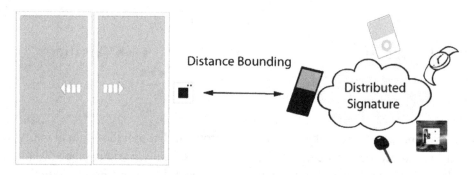

devices. They will also contribute in the access control scheme, but will not communicate directly with the verifier. We envision the user's mobile phone to act as gateway device, but the user can also choose to use one of his other devices. The only requirement for the gateway device is that it should have sufficient computational resources (to carry out cryptographic algorithms such as hash functions and digital signatures) and that it has an adequate user-interface. Note that it is not required to use the same gateway device in different instances of the protocol. However, in each run of the protocol, there is always exactly one gateway device; all other user's devices act as end-device. Since only the gateway device will communicate directly with the verifying entity, the end-devices do not need to be able to carry out a distance bounding protocol. The gateway device can use onboard or separate dedicated hardware (connected to the device) to perform the rapid bit exchanges during the run of the protocol.

As already discussed, by carrying out a distance bounding protocol, a verifying party can determine an upper bound on the distance between itself and a prover, who claims to be within a certain range. Cryptographic distance bounding protocols often require the prover and the verifier to compute a non-probabilistic function on a known input and a cryptographic key, known to both the prover and the verifier. Typically a pseudorandom function such as HMAC (Bellare et al., 1996) or CBC-MAC (International Organization for Stan-

dardization, 1994) is used. Such functions cannot be used in our access control mechanism, as we have chosen for a distributed solution where a gateway device and several end-devices need to collaborate to successfully complete the protocol. Instead, we will distribute the cryptographic key among the user's devices, and replace the pseudo-random function by a cryptographic function which can be partially evaluated by each of the user's devices[1]. Each share of the cryptographic key is stored on a separate device. As a consequence, in contrast to the implementation of conventional access control systems, the adversary needs to compromise at least t devices instead of just one security token. Afterwards, the gateway device can then combine several of these partial evaluations to obtain the evaluation of the cryptographic function. An interesting function that has all these properties is the computation of an RSA signature (Rivest et al., 1978).

Through the mechanism of resharing, we also support user-centered management. Users can alter the set of personal devices. They can add or remove devices from this set, and can tune the security parameters of the access control scheme towards the required level of security and reliability.

Adversarial Model and Assumptions

Each device has its own public-private key pair. Public keys of potential gateway devices are known to all devices. All devices' public keys are known to devices that can contribute in resharing. How to register the public key with other devices is out of the scope of this article. One can use for example pairing protocols to authenticate the public key. These protocols are standardized in the ISO/IEC 9798-6 standard (International Organization for Standardization, 2005).

The goal of the adversary is to get unauthorized access (e.g., enter a building). To achieve this objective, the adversary can perform passive and/or active attacks. The adversary, which is computationally bounded, can largely extend his communication range, and send/receive messages from a large distance. However, we assume that the adversary cannot increase the propagation speed of the communication medium. This implicitly implies that RF communication is used, or that the adversary cannot carry out a wormhole attack when a slower communication medium, such as ultra-sound, is used. The latter however requires additional countermeasures, which will not be further discussed. Also the influence of multipath propagation, interference with other wireless signals, etc. are not discussed in this article.

There are two scenarios that we need to investigate: the prover being located at a large distance from the verifier, and the prover being in the proximity of the verifier. The adversary (i.e. the person in control of the attacks) is assumed to be physically close to the verifier in both scenarios (otherwise, an attack would not make much sense from a practical point of view). Let us first focus on the first scenario. To get unauthorized access, the adversary needs to carry out a mafia fraud attack and forward all messages to a proxy device that is hidden in the neighborhood of the prover. An adversary can also compromise a subset of the user devices and use these devices to perform the mafia fraud attack (i.e. sign particular data). This subset is however assumed to be strictly smaller than the threshold number t. As we will show later in the article, our solution is resistant to such mafia fraud attacks. In the second scenario, where the user's devices are close to the verifier,

we assume that the user can physically verify the presence of the adversary. In this setting, other (even non-technical) attacks are more likely to take place (e.g., such as trying to sneak into the building when the prover enters). We hence mainly need to focus on the first scenario.

Our Threshold-Based Solution

We will now present our threshold-based access control solution that uses distance bounding protocols. When carrying out our scheme, several steps need to be performed. Initially, each new user needs to be registered during the enrollment phase. After this phase, which only needs to be done once for each new entity, the user is ready to actively use the access control mechanism. Access control is carried out by conducting a distance bounding protocol based on digital signatures. One of the main components of this protocol is a distributed RSA signature generation. For security and usability reasons, the private key used to compute the distributed RSA signature needs to be reshared at regular intervals.

Enrollment Phase

Before a new user can actively use the access control mechanism, he first has to enroll. This enrollment phase is rather similar to the initialization phase of a conventional access control scheme, but there are some important differences. Instead of generating a shared secret key and storing it on the user's contactless smartcard, the verifier will generate a shared private RSA key that will be distributed among the personal devices of the user. There are two options. The verifier can send the private key to the gateway device, which will then share the key among the user's devices; each device will receive a share d_i. The device's shares are constructed by evaluating a polynomial of degree (t-1), with the polynomial's constant factor equal to the secret to be shared, in their corresponding identities (Shamir, 1979). Next the gateway device should delete the private key. Another solution is that the verifier itself acts as the trusted dealer, since it knows the private key.

The initial secret sharing phase goes as follows. The trusted dealer first broadcasts the public RSA parameters e,N. Next it computes and distributes a random number v and the initial verification keys $v_1 \ldots v_k$ (with $v_i = v^{d_i}$). Each device i then gets its initial share d_i of the private key over a private channel. Techniques on how to construct a private channel are out of the scope of this article. The verification keys are stored by all the user's devices. These are used during resharing and to verify the gateway device's knowledge of its share, needed for the distributed signature generation.

Distance Bounding Using Digital Signatures

As already discussed, we have opted to use a threshold-based distance bounding protocol to enable a verifying party to check that a threshold number of particular devices are within a certain range. A group of personal devices will collaborate during the proximity-based authentication process. We started from the Hancke-Kuhn protocol (2005), as it is noise resilient and does not require the computation of a signature at the end of the protocol. We slightly modified the protocol in such way that instead of a pseudo-random function, prover and verifier need to compute an RSA signature S on a message M, which depends on the nonces exchanged between prover and verifier. We will use RSA signatures as defined in PKCS #1 version 2.1 (RSA Laboratories, 2002), but for simplicity reasons denote the encoded message as M. The resulting distance bounding protocol is depicted in Figure 7.

The protocol works as follows. Initially the user has to confirm that he wants to start the access control mechanism with a specific verifier, by performing a particular action (e.g., pressing a button) on the gateway device. After this approval by the user, the distance bounding protocol can start. One should note that in most use case

Figure 7. Threshold-based distance bounding protocol

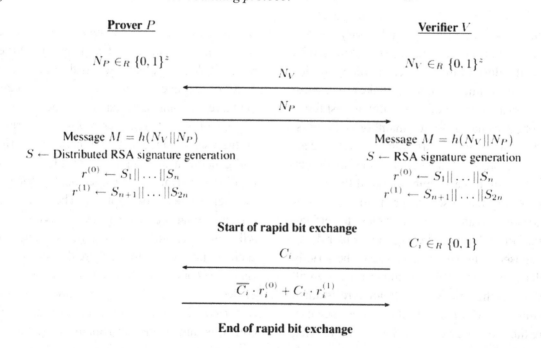

scenarios, only one verifier will be in close proximity. There is hence no risk that a protocol run will be carried out with two verifiers in parallel, which could cause ping pong effects. If two verifiers are close by, the prover selects the verifier with whom he wants to carry out the protocol. After this initialization phase, both prover and verifier are ready to carry out the protocol. Only these two parties will execute the protocol, messages from other entities will be ignored.

When the protocol is initiated, the prover and verifier first exchange a random nonce, N_P and N_V respectively. Both parties then compute an RSA signature S, using the private key d which is known to both parties, on the message M, which is the result of applying a cryptographic hash function on the concatenation of N_V and N_P.

$$S = M^d \bmod N, \quad \text{for} \quad M = h(N_v \parallel N_p)$$

The verifier can directly compute the RSA signature on the message. However, at the prover's side the private key d is shared among the user's personal devices. The RSA signature on the message is generated in a distributed way. More details on the distributed RSA signature generation can be found in the next section.

After applying a deterministic reduction[2] function RED on the signature, the result is split in two n-bit sequences $r^{(0)}$ and $r^{(1)}$.

$$RED(S) = r^{(0)} \parallel r^{(1)}$$

Then a series of n rapid bit exchanges is performed. In each round, the verifier sends a random single bit challenge C_i to the prover. If this challenge equals zero, then the prover responds with the i-th bit of $r^{(0)}$. If the challenge equals one, then the prover sends the i-th bit of $r^{(1)}$. In each round, the verifier measures the time between sending C_i and receiving the corresponding response. The maximum round trip time is selected and this

measurement determines an upper bound on the estimation of the distance between prover and verifier. If at least (n-x) of the responses sent by the prover are correct (the security parameter x denotes the number of allowed bit errors during the rapid bit exchange), the protocol succeeds.

Distributed RSA Signature Generation

Desmedt and Frankel (1991) proposed the first (non-robust) threshold RSA signature scheme. Later robust, but less practical threshold RSA signatures schemes were proposed by Frankel et al. (1997) and Rabin (1998). The first practical, robust threshold RSA signature scheme was presented by Shoup (2000). We will use this technique to generate the signature S on the message M.

The gateway device will initiate the distributed RSA signature generation. This device broadcasts the message M together with a zero knowledge proof (denoted by ZKP) of its share. The ZKP consists of a Schnorr signature (1989) (c,z) on the message and the public RSA key e. The gateway device chooses r at random in the interval:

$$\{0 \ldots 2^{L(N)+2L_1-1}\}$$

with L(N) the bitlength of N and L_1 a secondary security parameter.

$$v' = v^r; \quad c = H\left(v', M, e, N\right); \quad z = d_i c + r$$

The end-devices will first verify the ZKP by computing v' using the public verification key v_i, and checking c:

$$v' = v^z v_i^{-c}; \quad c = H(v', M, e, N)$$

Note that by carrying out this ZKP, one has the guarantee that the distributed signature generation is initiated by a (trusted) gateway device, and not by a (hidden) proxy device controlled by the adversary. Next, the end-devices broadcast their partial signatures S_i encrypted with the public key of the gateway device.

$$S_i = M^{2" d_i}$$

The result is that only the gateway device can combine the partial signatures (using a subset of t devices, with t being the secret sharing threshold) into the signature on the message M. The Lagrange multipliers are used to interpolate through points on the polynomial of degree (t-1). Let λ_i be the Lagrange multipliers multiplied with

$$\Delta = \prod_{i=1}^{k} i$$

The latter is necessary to get integer values, since the order of the subgroup $\Phi(N)$ is unknown.

$$w = \prod_i S_i^{\lambda_i}, \text{ for } \lambda_i = \Delta \prod_{j \neq i} \frac{j-i}{j}$$

We now have $w^e = M^{4\Delta^2}$. Since e and $4\Delta^2$ are coprime, we can find values α and β for which $\alpha 4\Delta^2 + \beta e = 1$ using Euclid's algorithm.

$$S = w^\alpha M^\beta$$

The gateway device can verify the correctness of the signature S using the public RSA parameters e,N.

$$S^e \bmod N = M$$

If the verification of the signature does not hold, the subset contains one or more cheating devices. The gateway device then selects a different subset of t devices and recomputes the signature. This process is repeated until the verification of S succeeds.

Resharing

By carrying out secret resharing, new shares of the secret are generated and the old shares are rendered useless. This means that an adversary is forced to break the scheme within the time frame between two consecutive instances of resharing. Resharing also allows to go from a (t,k)-secret sharing to a (t', k')-secret sharing, with t' the new threshold number and k' the new number of participants. The threshold number t determines the level of security, i.e., the number of devices an adversary needs to compromise within the available timeframe. The number of devices k together with the threshold number t determines the level of reliability, since the legitimate user will still be able to use the scheme when the combined number of devices that are not present and compromised by an adversary is less than (k-t). This is very important from a usability point of view.

Because of the resharing mechanism, our access control solution supports user-centered management. The set of personal devices, of which a subset of at least t devices is needed during the authentication phase, can be changed; hence partial signature rights can be revoked and/or granted. Resharing is typically required when a device gets stolen, or when a user purchases a new device. From the moment a device is identified as compromised, it should be excluded from the set of devices that share the private RSA key. It will hence not receive a new share and its old share will be rendered useless.

Secret resharing, without reconstruction of the secret, was first described by Desmedt and Jajodia (1997) and Frankel et al. (1997). We will use the techniques proposed by Wong et al. (2002), where new participants can verify the validity of their shares.

The resharing mechanism works as follows. Basically, every contributing device constructs a polynomial of degree $(t'-1)$, with t' the new threshold number. The number of contributing devices needs to be at least t, since this many shares are needed to reconstruct the original shared secret. Commitments to the coefficients c_{ij} of the constructed polynomial are broadcast.

$$f_i(x) = d_i + c_{i1}x + \ldots + c_{i(k'-1)}x^{t'-1},$$
$$\text{for} \quad \forall j \in \{1,\ldots,t'\} : C_{ij} = v^{c_{ij}}$$

Subshares (evaluations of the polynomial $f_i(x)$) are handed out to the set of participating devices:

$$\forall j \in \{1,\ldots,k'\} : d_{ij} = f_i(j)$$

The commitments to the coefficients of the polynomial and verifications keys allow validating these subshares.

$$v^{d_{ij}} = v_i \prod_{l=1}^{t'-1} C_{il}^{j^1}$$

Contributing devices, for which all sent subshares validate, are added to the set of qualified dealers. The size of this set should be at least t to continue. The subshares from qualified dealers are combined into the new shares. Next, new verification keys are broadcasted.

$$d'_j = \frac{\sum d_{ij}\lambda_i}{\Delta} ; v_{j'} = v^{d'_j}$$

The validity of these new verification keys, hence the new shares, can be tested by combining them:

$$v^{\Delta} = \left(\prod v_{i'}^{\lambda_i}\right)^e$$

Please note that Δ and λ_i in the last equation need to be calculated for the new set of participants, if the set of participants is changed in this resharing instance.

DISCUSSION

Security Analysis

Our threshold-based location-aware access control scheme improves the resilience to various important security vulnerabilities. We will now briefly discuss its security properties, although without giving formal proofs.

Please note that if an adversary succeeds in compromising the threshold number t of devices at one particular point in time, he can reconstruct the private RSA key, and hence completely break the access control scheme.

Relay Attacks

One of the main components in our threshold-based location-aware access control scheme is the slightly modified distance bounding protocol of Hancke and Kuhn. This protocol is employed to prevent relay attacks. The adversary cannot conduct a protocol run simultaneously with the prover and the verifier, as this would increase the time of flight (and hence cause the distance bounding protocol to fail).

To analyze the resilience to relay attacks, we will now discuss the security properties of the distance bounding protocol more in detail. An adversary, who wants to authenticate himself successfully to the verifier, can follow several attack strategies. The first strategy is the *no-ask* strategy. The adversary just guesses all the responses during the rapid bit exchange phase. His success probability would then be $1/2^n$. However, the best strategy is the *ask-in-advance* strategy. Before discussing the attack more in detail, one should note that our distance bounding protocol starts by the prover and verifier exchanging some random nonces. Since the time is not measured during this stage of the protocol, the adversary does not yet interfere. The attack, which consists of two consecutive phases, will take place during the fast bit exchanges.

In the first phase, the adversary carries out a protocol run with the honest prover. During each of the n rounds, he guesses the challenges C_i in advance. To finish the protocol run, the adversary needs to trick the user in approving that the distance bounding protocol is being carried out. Let us for now assume that the adversary succeeds in this goal (we will come back to this issue later in this article). At the end of the protocol run, the adversary will hence have received n responses r_i.

In the second phase of the attack, the adversary is now located close to the verifier and performs a protocol run with the latter. In each round of the fast bit exchanges, the verifier sends a random challenge, and expects a correct response from the adversary. There are two scenarios. If the verifier sends a challenge C_i equal to the challenge used in the first phase of the attack (i.e. if the adversary has guessed the challenge correctly), then the adversary replies with the response r_i received from the prover. This response will always be correct. If the adversary has guessed the challenge C_i wrongly, he cannot use the response r_i. As a result, he will send a random response to the verifier. In this case, he has a 1/2 probability of replying with a correct response. Since each scenario occurs with a probability of 1/2, it is relatively easy to compute the overall success probability of the relay attack (Hancke & Kuhn, 2005). By following the *ask-in-advance* strategy, the adversary has, in each of the n rounds, a probability of 3/4 to send a correct response. If the attack succeeds, the adversary is able to wrongfully convince the verifier that an entity in possession of the private RSA key is in the vicinity.

The attack probability slightly improves when one incorporates the effect of bit errors due to noise (Singelée & Preneel, 2007). This is however outside the scope of this article.

User Interaction Improves Security

As explained above, an adversary needs to trick the user in carrying out a run of the distance bounding protocol without being in the vicinity of the verifier. This goal can be achieved by compromising at least one device with a user interface (i.e. a gateway device), since these devices can initiate a protocol instance.

A compromised gateway device could carry out mafia fraud attacks, where it requests the other devices in the network to compute a distributed RSA signature. To mitigate this security risk, one could require the user to approve this request on a predefined number of end-devices with a user interface. Analogue to the gateway device, these devices need to proof knowledge of their shares. This means an adversary would have to compromise at least this predefined number of devices with a user interface to carry out a successful mafia fraud attacks.

By varying this predefined number, one can change the adversary's success probability (and hence the security level of the protocol). Peeters et al. (2009a) showed that by having a user verifying his request at only a small number of devices with a user-interface, the adversary's probability of success already reduces drastically. This is hence an interesting solution to further improve the security level of the system at a limited cost (in terms of user friendliness).

User-Centered Configuration

End users will be in control of the security of their system, and be able to adjust the security parameters of the access control scheme towards their preferences. Through the mechanism of resharing, they can alter the set of personal devices and/or change the threshold number t. We recommend that the default value of the threshold number should be equal to

$$t = \frac{k+1}{2}$$

This parameter choice gives the best trade-off between security and reliability. A higher threshold number would make the system more secure, but the number of active devices that the user needs to carry would also increase, resulting in a higher probability to be denied access (i.e. more devices are needed to successfully authenticate).

Let us illustrate this by observing three concrete scenarios where resharing takes place: adding a device, removing a device and refreshing the shares of the private key. A preliminary usability study conducted by the authors (Peeters et al., 2009b) showed that resharing in these three concrete scenarios is perceived by the end users as three different operations. From a security perspective, adding and removing devices always needs authorization by the end user, as an adversary could gain an advantage by performing these operations. For example, the adversary could try to add devices that are under his control, or remove devices that are not under his control. Furthermore, adding a new device provides an additional technical challenge as the device is not yet known to the group of other personal devices. The public key of the new device needs to be securely transferred to these personal devices and the public key needs to be authenticated. More details on how to perform resharing that is authorized by the end user can be found in (Peeters et al., 2009a).

Compromised devices, e.g. stolen or lost devices, can be removed from the group without these needing to be present. This should be done as soon as possible after the detection of the compromise. Refreshing the shares helps to improve security when the end user is in possession of devices of which the adversary has obtained the share, but which are no longer controlled by the adversary. For example, if one leaves his mobile phone unattended during a lunch break, an adversary could extract the value of the share and afterwards could put the phone back in such a way that the legitimate owner will never notice it has been compromised. Refreshing the shares solves this issue. Note that refreshing does not help against actively compromised devices (e.g., on which root kits or other malware is installed). If the attacker is still in control of the device, he can obtain the new value of the share. However note that the access control system itself remains secure as long as fewer than t devices are compromised.

CONCLUSION

There is an evident need for the design of innovative secure and user-friendly access control systems. Conventional systems entail several security vulnerabilities. By using a single security token, one introduces a single point of failure in the system. Users cannot authenticate themselves without their token. Moreover, an adversary that steals or compromises a token will get the access privileges of the corresponding user, until revocation has taken place. Challenge-response protocols conducted in wireless networks are also vulnerable to relay attacks.

In addition to these security vulnerabilities, conventional access control mechanisms often suffer from usability issues. Carrying out changes (such as revoking or updating keys) to centrally managed systems tends to be a rather slow and cumbersome process. Users can also not adjust the security properties of the access control scheme, as this is enforced by the system itself.

In this article, we proposed a threshold-based location-aware access control mechanism which combines the concepts of secret sharing and distance bounding protocols. Its main component is a distributed RSA signature generated by t user's devices: one gateway device and (t-1) end-devices. The gateway device interacts directly with the verifier and combines the partial signatures. We particularly envision the user's mobile phone to act as gateway device in our access control scheme.

We demonstrated that our solution solves the dependency on a single token, and is resistant to relay attacks. It offers protection against any set of (t-1) or fewer compromised user's devices. Compared to conventional access control mechanisms, our solution could offer an increased level of user-friendliness as it supports user-centered management. Users can vary the set of personal devices, of which at least t need to be present during authentication. This threshold number determines the security level of our access control scheme and can be freely adapted by the user.

To further support the claim that our proposed access control scheme improves user-friendliness, there is a need for a large usability study in which quantitative data is gathered. This will be part of future research.

ACKNOWLEDGMENT

This work is funded by the Katholieke Universiteit Leuven, and supported in part by the Concerted Research Action (GOA) Ambiorics 2005/11 of the Flemish Government, by the IAP Programme P6/26 BCRYPT of the Belgian State (Belgian Science Policy), and by the Flemish IBBT projects. Roel Peeters is funded by a research grant of the Institute for the Promotion of Innovation through Science and Technology in Flanders (IWT-Vlaanderen).

REFERENCES

Avoine, G., & Tchamkerten, A. (2009). An efficient distance bounding RFID authentication protocol: Balancing false-acceptance rate and memory requirement. In P. Samarati, M. Yung, F. Martinelli, & C. A. Ardagna (Eds.), *Proceedings of the 12th International Conference on Information Security* (LNCS 5735, pp. 250-261).

Bellare, M., Canetti, R., & Krawczyk, H. (1996). Keying hash functions for message authentication. In N. Koblitz (Ed.), *Proceedings of the 16th Annual International Cryptology Conference on Advances in Cryptology* (LNCS 1109, pp. 1-15).

Brands, S., & Chaum, D. (1994). Distance-bounding protocols. In T. Helleseth (Ed.), *Proceedings of the Workshop on the Theory and Application of Cryptographic Techniques* (LNCS 765, pp. 344-359).

Clulow, J., Hancke, G. P., Kuhn, M. G., & Moore, T. (2006). So near and yet so far: Distance bounding attacks in wireless networks. In L. Buttyán, V. D. Gligor, & D. Westhoff (Eds.), *Proceedings of the 3rd European Workshop on Security and Privacy in Ad Hoc and Sensor Networks* (LNCS 4357, pp. 83-97).

Desmedt, Y. (1988). Major security problems with the ``unforgeable'' (Feige)-Fiat-Shamir proofs of identity and how to overcome them. In *Proceedings of SecuriCom* (pp. 15-17).

Desmedt, Y., & Frankel, Y. (1991). Shared generation of authenticators and signatures. In J. Feigenbaum (Ed.), *Proceedings of Advances in Cryptology* (LNCS 576, pp. 457-469).

Desmedt, Y., & Jajodia, S. (1997). *Redistributing secret shares to new access structures and its applications* (Tech. Rep. No. ISSE-TR-97-01). Washington, DC: George Mason University.

Drimer, S., & Murdoch, S. (2007). Keep your enemies close: Distance bounding against smartcard relay attacks. In *Proceedings of the 16th USENIX Security Symposium* (pp. 87-102).

Frankel, Y., Gemmell, P., MacKenzie, P. D., & Yung, M. (1997). Optimal resilience proactive public-key cryptosystems. In *Proceedings of the IEEE Symposium on Foundations of Computer Science* (pp. 384-393). Washington, DC: IEEE Computer Society.

Gans, G. K., Hoepman, J. H., & Garcia, F. D. (2008). A practical attack on the MIFARE classic. In G. Grimaud & F.-X. Standaert (Eds.), *Proceedings of the 8th IFIP WG 8.8/11.2 International Conference on Smart Card Research and Advanced Applications* (LNCS 5189, pp. 267-282).

Hancke, G., & Kuhn, M. (2005). An RFID distance bounding protocol. In *Proceedings of the 1st International Conference on Security and Privacy for Emerging Areas in Communications Networks* (pp. 67-73). Washington, DC: IEEE Computer Society.

International Organization for Standardization. (1994). *ISO/IEC 9797: Information technology - security techniques - data integrity mechanisms using a cryptographic check function employing a block cipher algorithm.* Retrieved from http://www.iso.org/iso/iso_catalogue/catalogue_tc/catalogue_detail.htm?csnumber=22053

International Organization for Standardization. (2005). *ISO/IEC 9798-6: Information technology - security techniques - entity authentication - part 6: Mechanisms using manual data transfer.* Retrieved from http://www.iso.org/iso/iso_catalogue/catalogue_tc/catalogue_detail.htm?csnumber=39721

Kim, C., Avoine, G., Koeune, F., Standaert, F., & Pereira, O. (2009). The Swiss-knife RFID distance bounding protocol. In *Proceedings of Information Security and Cryptology* (pp. 98-115).

Laboratories, R. S. A. (2002). *PKCS #1 v2.1: RSA cryptography standard.* Retrieved from ftp://ftp.rsa.com/pub/pkcs/pkcs-1/pkcs-1v2-1d2.pdf

Peeters, R., Kohlweiss, M., & Preneel, B. (2009a). Threshold things that think: Authorisation for resharing. In J. Camenisch & D. Kesdogan (Eds.), *Proceedings of the IFIP WG 11.4 International Workshop on Open Research Problems in Network Security* (LNCS 309, pp. 111-124).

Peeters, R., Sulmon, N., Kohlweiss, M., & Preneel, B. (2009b). Threshold things that think: Usable authorisation for resharing. In *Proceedings of the 5th Symposium on Usable Privacy and Security* (p. 18). New York, NY: ACM Press.

Rabin, T. (1998). A simplified approach to threshold and proactive RSA. In H. Krawczyk (Ed.), *Proceedings of the 18th International Cryptology Conference on Advances in Cryptology* (LNCS 1462, pp. 89-104).

Rasmussen, K., & Capkun, S. (2009). Location privacy of distance bounding protocols. In *Proceedings of the ACM Conference on Computer and Communications Security* (pp. 149-160). New York, NY: ACM Press.

Rasmussen, K., & Capkun, S. (2010). Realization of RF distance bounding. In *Proceedings of the 19th Usenix Conference on Security* (p. 25).

Rivest, R., Shamir, A., & Adleman, L. (1978). A method for obtaining digital signatures and public-key cryptosystems. *Communications of the ACM, 210*(2), 120–126. doi:10.1145/359340.359342

Schnorr, C. P. (1989). Efficient identification and signatures for smart cards. In G. Brassard (Ed.), *Proceedings of Advances in Cryptology* (LNCS 435, pp. 239-252).

Shamir, A. (1979). How to share a secret. *Communications of the ACM, 220*(11), 612–613. doi:10.1145/359168.359176

Shoup, V. (2000). Practical threshold signatures. In B. Preneel (Ed.), *Proceedings of the International Conference on the Theory and Application of Cryptographic Techniques* (LNCS 1807, pp. 207-220).

Singelée, D., & Preneel, B. (2007). Distance bounding in noisy environments. In F. Stajano, C. Meadows, S. Capkun, & T. Moore (Eds.), *Proceedings of the 4th European Workshop on Security and Privacy in Ad Hoc and Sensor Networks* (LNCS 4572, pp. 101-115).

Wong, T. M., Wang, C., & Wing, J. M. (2002). *Verifiable secret redistribution for threshold sharing schemes* (Tech. Rep. No. CMU-CS-02-114). Pittsburgh, PA: Carnegie Mellon University.

ENDNOTES

[1] One only needs a threshold number of cooperating devices to reconstruct the secret and hence evaluate the cryptographic function.

[2] The reduction function RED reduces the bit length of the input to a fixed bit length $2n$.

This work was previously published in the International Journal of Handheld Computing Research (IJHCR), Volume 2, Issue 3, edited by Wen-Chen Hu, pp. 22-37, copyright 2011 by IGI Publishing (an imprint of IGI Global).

Chapter 3
Survivability Enhancing Techniques for RFID Systems

Yanjun Zuo
University of North Dakota, USA

ABSTRACT

Radio Frequency Identification (RFID) has been applied in various high security and high integrity settings. As an important ubiquitous technique, RFID offers opportunities for real-time item tracking, object identification, and inventory management. However, due to the high distribution and vulnerability of its components, an RFID system is subject to various threats which could affect the system's abilities to provide essential services to users. Although there have been intensive studies on RFID security and privacy, there is still no complete solution to RFID survivability. In this paper, the authors classify the RFID security techniques that could be used to enhance an RFID system's survivability from three aspects, i.e., resilience, robustness and fault tolerance, damage assessment and recovery. A threat model is presented, which can help users identify devastating attacks on an RFID system. An RFID system must be empowered with strong protection to withstand those attacks and provide essential functions to users.

INTRODUCTION

Radio Frequency Identification (RFID) is a wireless technology for automatic item identification and data capture. It uses radio signals to identify a product, an animal or a person. Given its technical and economic advantages, RFID has been applied to various fields. Many retailers and wholesalers use RFID systems to manage product shipments and inventory tracking. RFID has also been used in critical information systems in military, healthcare, and crisis management. For instance, the US Food and Drug Administration proposed attaching RFID tags to prescription drug bottles as a pedigree. The Department of Defense is moving towards RFID-based logistic control.

The structure of a typical RFID system is shown in Figure 1 and such a system includes the following two major subsystems (Karygiannis, Eydt, Bunn, & Phillips, 2007; Zuo, Pimple, & Lande, 2009):

DOI: 10.4018/978-1-4666-2785-7.ch003

Figure 1. Architecture of an enterprise RFID system

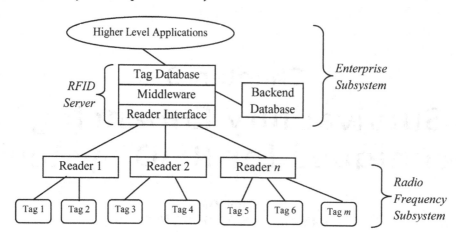

1. **Radio Frequency (RF) Subsystem:** It consists of components which perform tag identification and wireless communications and transactions between RFID readers and tags. A tag is physically attached to an item with a unique identification. A reader is a device that can recognize the presence of RFID tags and read the information supplied by them. It can be a PDA, a mobile phone or any kind of devices capable of communicating with an RFID tag. To obtain data from a tag, a reader first queries the tag and then forwards the received identity information to a backend server, which maintains a database of tag entries (see below). After being authorized, the reader can obtain more detailed information about the tag and the tagged item.

2. **Enterprise Subsystem:** It consists of two components: a backend database and an RFID server.

   a. **Backend Database:** The backend database contains an entry for every valid tag in the system. Each entry corresponding to a tag contains at least the following information: the tag identification number, the secret key shared between the database and the tag, and optional item descriptions of the tagged item. An RFID reader and the backend server communicate through a secure channel.

   b. **RFID Server:** It consists of systems and applications which communicate with the RFID readers and process data acquired from the RF subsystem. Three components constitute an RFID server:

   i. **Reader Interface:** An interface to communicate with the readers.

   ii. **Middleware:** An intermediate layer between the lower level RF subsystem and the higher level database and a set of user applications. The main purpose of the middleware is to pre-process the data collected from the RF subsystem and provide cleaned data to the tag database to be used by the higher level business applications. The raw data collected from the RF subsystem must be pre-processed to remove redundancy and/or transformed to appropriate formats before they can be used. Middleware also contains security mechanisms to ensure data confidentiality and integrity.

iii. **Tag Database:** Cleaned tag data is stored in this database to be used by the high level applications. This database may also contain high level events retrieved from the lower level tag data.

Although RFID systems provide numerous benefits arising from automatically identifying object items in a wide range of applications, they imply security concerns. Military RFID tags could be attacked by enemy forces. Supply chain RFID tags could be scanned by competitors for sensitive logistic information. RFID enabled passports may release personal data if not appropriately protected. Unavailability of RFID tags implanted into human bodies due to malicious attacks could result in life losses in emergency situations. RFID systems are attractive targets for computer criminals. RFID may have a financial character, and it is important for applications used for national security (i.e., military logistics).

Various mechanisms have been developed to enhance the security of RFID systems. However, no solution can guarantee that an RFID system is not hackable. In particular, the high mobility of RFID components, the open nature of an RFID system, and the increasingly sophistication of attacks, all make an RFID system vulnerable to various security threats. RFID systems are being used in various applications and technologies which rely on dependable, robust and highly secure properties. There is a pressing need to improve an RFID system's ability so that it continues to function to support mission-critical services despite malicious attacks and system failures. In this paper we address the issues of RFID survivability and classify and discuss the existing RFID security techniques in the literature that could greatly improve the survivability of an RFID system. We call such techniques survivability enhancing techniques.

The major challenge for the survivability of an RFID system is the limited resources (e.g., memory, computing power, and area space) of RFID tags available for security purposes. This is particularly true for those low-cost tags deployed in a massive scale in a decentralized system. There are several restrictions for those low to middle cost RFID tags due to their inherent physical features. For instance, the current consumption of security architecture of implementation on RFID tags must not exceed 15 μA (Feldhofer, Dominikus, & Wolkerstorfer, 2004). The measure of gate-equivalents (GE) allows a technical-neutral estimation of the required physical space. There is an oft-quoted consensus that out of 1000-10,000 GE on a restricted device such as an RFID tag, around 200-3,000 GE might be available for security (Juels & Weis, 2005; Peris-Lopez, Hernandex-Castro, Estevez-Tapiador, & Ribagorda, 2006). For the low to mid cost tags, the number of gates for security cannot exceed 2,500 – 5,000 gates.

In this paper, we focus on the survivability enhancing techniques for low to middle cost RFID tags, which have limited computing and memory resources for security. Since the backend server and the readers can be secured using standard security mechanisms (e.g., public key cryptography), we assume that they are secure and trustworthy. Furthermore, RFID survivability is a multi-layered problem crossing application, communications and physical layers of a RF subsystem. We are mainly concerned with survivability enhancing techniques at the application layer. But, we will address some issues at the communication layer.

In the following discussions, we first explain system survivability in general and RFID survivability in particular. Then, according to the unique features of RFID systems and the threat model, the desired survivability characteristics for RFID systems are specified. Next, the survivability enhancing techniques in the literature are classified and discussed. Those techniques are based on the identified desired survivability characteristics of an RFID system. Finally, we conclude this paper with discussions on future research in this area.

RFID SURVIVABILITY

The Concept of Survivability

Survivability refers to a system's abilities to withstand malicious attacks and tolerate system failures in such a way that the system can continue support mission-critical services. A survivable system may operate at a degrading model but the essential system functions must be guaranteed in spite of attacks and system faults. Survivability is different from the traditional security since it makes no assumptions that the system cannot be compromised. Rather, a survivable system assumes that some parts of the system could be damaged but the system as a whole can manage to mask the damage in order to support the essential services.

System survivability has been studied from different application areas and based on different abilities that a critical system should have. Various definitions (Deutsch & Willis, 1988; Knight, Strunk, & Sullivan, 2003; Hiltunen, Schlichting, Ugarte, & Wong, 2000) about system survivability have been proposed and they share some common understandings about the nature of survivability. For instance, survivability is widely understood as a system property, relating the level of services provided to the level of damage present in the system and operating environment; a survivable system must support the system's mission; operating in a hostile environment, a survivable system may offer degraded (but acceptable to users) services to users and have the ability to recover when the environment improves. Tarvainen (2004) identifies a set of key properties that a survivable system should have: (1) a survivable system delivers essential services and maintains essential properties of those essential services, e.g., specified levels of integrity, confidentiality, performance and availability; (2) requirements of survivability are often expressed in terms of maintaining a balance among multiple attributes such as security, reliability, and modifiability; and

(3) it is crucial to identify the essential services, and the essential properties that support them, within an operational environment.

RFID Survivability Threats

In this paper, we discuss survivability issues in the context of RFID systems. An RFID system differs from other systems in the following ways: a) it is composed of a large number of highly mobile components (e.g., tags or even readers); b) it uses wireless communications among various components, such as between a tag and a reader, or between a reader and the backend RFID server. The nature of wireless communications opens various opportunities for adversaries to intercept, eavesdrop and tamper the messages transferred; c) If an adversary can physically possess RFID components (e.g., tags), he/she can clone, swap, reprogram and corrupt the components; d) The low cost tags are cheap devices which have limited memory and computational power. They can only carry out simple functions like one-way hashing and simple bitwise operations. Because of this limitation, the RFID systems are often vulnerable to security breach.

In our discussions, RFID survivability refers to the ability of an RFID-enabled system to provide trustworthy, reliable, and timely information to support business functions. In the threat model, we assume that an attacker attempts to compromise the survivability of a system by attacking the system through various components and access points. In the following discussions, we follow the procedure used in (Zuo, Pimple, & Lande, 2009) for RFID threat analysis. In that model, an important procedure is to go through the systems assets and review a list of possible attacks for each asset (e.g., tag, reader, RFID server). The model applies attack trees (Schneier, 1999) to identify the critical threats related to an asset. In such a tree structure, the root represents the asset under threat and each node represents a type of threat. For each threat, the key P and I are used to describe whether it is possible or impossible for the threat to occur given other restrictions under consideration. Figure 2 shows an attack tree for an RFID tag as an important asset of the RFID system. The limited resources (both memory and processing power) of an RFID tag and the nature of wireless communications between the tag and a reader determines that an RFID system is vulnerable to various malicious attacks. Figure 2 shows an adversary can attack an RFID tag in one of the following ways (for a more complete list of attacks on RFID tags, please refer to Mitrokotsa, Rieback, & Tanenbaum, 2009):

1. **Permanently Disabling Tags:** Possible ways of rendering an RFID tag permanently inoperable are tag removal, tag destruction or using the KILL command. Tag removal and tag destruction are physical ways of destroying or damaging the tag. While a KILL command is a valid operation performed by the enterprise to permanently disable a tag after the tag is no longer needed, an adversary can exploit this method to sabotage RFID communication. However, since most tags have pin-protection features

built in, utilizing KILL commands to disable a tag can be very difficult. The identifier 'I' in the attack tree in Figure 2 denotes that this kind of attack is impossible given the embedded security mechanisms of a tag.

2. **Temporarily Disabling Tags:** A prospective thief can use an aluminum foil-lined bag (a simple Faraday Cage) to shield it from electromagnetic waves (such as those of the checkout reader) and steal any product undisturbed. RFID tags also run the risk of unintentional temporary disablement caused by environmental conditions (example: unintentional temporary disablement of a tag can be caused by covering it with ice/snow).

3. **Relay Attacks:** In a relay attack an adversary acts as a man-in-the-middle. An adversarial device is placed surreptitiously between a legitimate RFID tag and a reader. This device is able to intercept and modify the radio signal between the legitimate tag and reader. The legitimate tag and reader are fooled into thinking that they are communicating directly with each other.

4. **Attacks on Tag-Reader Communications:** An attacker can capture keys or other tag

Figure 2. Attack tree for tags as assets of RFID system

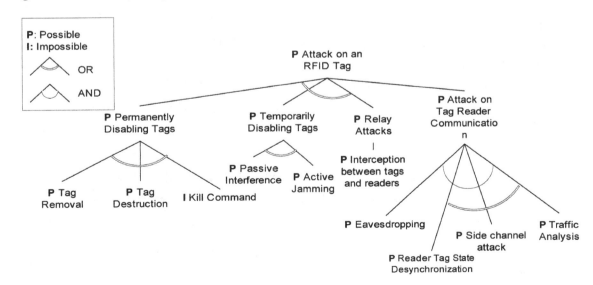

data communicated between a reader and a tag that allow for impersonation. An attacker inserts or modified messages exchanged by legitimate principles (such as tags and readers) without being detected. Then he/she can conduct an active eavesdropping in which the attack makes independent connections with the a reader and a tag and relays (and possibly modifies) messages between them, making them believe that they are talking directly to each other over a private connection when in fact the entire conversation is controlled by the attacker.

Similarly, threats can be identified for other important assets and entry points such as the RFID readers, backend database, RFID server, and inter-enterprise communications. An individual attack tree should be developed for each component. Due to page limitation, we will not show the detailed attack tree for each component. Rather, the major threats for each of those assets are listed in Table 1 (Zuo, Pimple, & Lande, 2009).

Desired Characteristics of Survivable RFID Systems and Measurements

The general survivability requirement for an RFID system is to provide trustworthy, reliable, and timely information to support higher-level applications in asset tracking and inventory management. Depending on individual application, a survivable RFID system should satisfy the general

Table 1. Threats related to assets

Assets	Major Threats
Reader	Impersonation, Eavesdropping, Relay Attack, Spoofing, Passive/Active Interference, Unauthorized Tag Reading
Backend Database	Denial of Service Attack, Buffer Overflow Attack
RFID Server	Malicious Code Injection

requirement of data confidentiality, integrity, and availability. In addition, an RFID system must have the ability to resist various attacks, maintain data privacy, recognize intrusions quickly, and recover from damage as soon as possible. Together, those requirements ensure that a survivable RFID system is capable of withstanding devastating attacks and supporting the system's mission. Survivability requirements for an RFID system can be specified from two perspectives: (1) specific functional requirements (items 1-3 in Table 2); and (2) survivability property requirements (items 4-6 in Table 2). In order to provide quantitative measurements of each survivability requirement, we abstract an RFID system as below and the related symbols are used in the survivability measurements in Table 2.

If an RFID system has most of the survivability properties listed in Table 2, we say the RFID system is survivable. RFID survivability must be quantified based on different applications (e.g., assigning different weights to the above categories). For instance, for RFID-enabled military logistic system, resistance and recognition are more important than recovery since military materials must be supplied in a timely manner. Any delay due to system recovery may not be acceptable in a critical war time. On the other hand, a commercial supply chain may put more weights on RFID recovery property since the major concern is the cost and it may be very expensive to build highly resistant and intrusion recognizable RFID systems. Rather, the option of reliable recovery is acceptable for many commercial applications as compared with the critical military applications.

RFID SURVIVABILITY ENHANCING TECHNIQUES

As discussed earlier, RFID survivability enhancing techniques refer to technical means and plans of actions that can improve the ability of an RFID

Table 2. Survivability characteristics and measurements

	Requirements	Quantitative Measurements	Symbols/Explanations
1	Completeness and Soundness Requirement	$\Pr(Accept(t) \mid !\exists(t, d) \in D) < \alpha$ $\Pr(Denial(t') \mid \exists(t', d') \in D) < \alpha'$	The possibility of acceptance of a tag t not belonging to the system is negligible (α & α' are thresholds) – soundness; The possibility of denying a legitimate tag t' is negligible - completeness.
2	Information Access Requirements	$Max(AccessTime(t_1, t_2, ..., t_n)) < \beta$ $Ave(AccessTime(t_1, t_2, ..., t_n)) < \beta'$	The maximum and average time to access rquired information stored in a set of tags $t_1, t_2 ... t_3$ to support a service should be no long than two defined threshold β and β', respectively.
3	Privacy Requirements	Given: (1) t_0 & t_1 s.t. $\exists(t_0, d) \in D$ & $\exists(t_1, d') \in D)$; and (2) t_b where random $b \in \{0, 1\}$. Verify: $\Pr(CorrectGuess(b=0 \text{ or } b=1))$ $< \frac{1}{2} + 1/poly(s)$	For any given two tags t_0 and t_1 in the system and randomly chosen one of them (denoted as t_b), the possibility for an attacker to correctly guess the right one (t_0 or t_1 depending on the actual one chosen) is negligibly larger than ½. *poly(s)* represents an arbitrary polynomials and s represents a security parameter.
4	Recognition Property	$Max(DetectionTime(A)) < \partial$ $Ave(AssessmentTime(\textstyle\sum)) < \partial'$ $Max(ResponseTime(\textstyle\sum)) < \partial''$	The ability of an RFID system $\sum$ to detect attacks A promptly (measured by the time for $\sum$ to recognize A), evaluate the extent of the damage (measured by the time for $\sum$ to perform survivability analysis and damage assessment), and adjust its behaviors (measured by the time for $\sum$ to respond). $\sum$ meets the survivability requirements if those parameters are below certain levels.
5	Resistance Property	$\dfrac{(N - N_C)}{N} > \chi$	The ability of an RFID system $\sum$ to withstand devastating attacks. It can measured as the ratio of the difference between the total number of tags N and the number of compromised tags Nc divided by N at a given moment. $\sum$ is resistant deadly attacks if and only if this ratio is larger than a threshold χ (the required minimum number of tags to support critical services is matinatained).
6	Recovery Property	$\dfrac{(N_{Rec})}{N_C} > \gamma$	The ability of an RFID system $\sum$ to recover damage. It is measured as the ratio of the number of recovered tags N_{Rec} over the number of compromised tags N_C in a given period of time. A RFID system meets the survivability requirement for recovery if and only if this ratio is larger than a threshold γ.

system to withstand malicious attacks and continuously support the essential service of the system. The existing techniques can be classified in two general categories: (1) software based (better protocols suitable for RFID tags, readers, and RFID sever); and (2) hardware based (more powerful and efficient hardware to support security and survivability software). Due to the cost pressure of massively deployed RFID tags, however, it is not possible in the near future to produce low-cost RFID tags with sufficient functionalities to support standard security methods (e.g., public-key cryptography and other multi-party security protocols). In the following sections, we focus on the software based approach for RFID survivability. We discuss the survivability enhancing techniques based on the desired characteristics of a survivable RFID system, i.e., resilience, robustness and fault tolerance, and damage assessment and recovery.

Resilience Techniques

This category includes the techniques that empower an RFID system's ability to resist attacks by using novel security mechanisms, environment measures, and operational protection means. In the literature, RFID techniques in this category include RFID system component authentication (e.g., between readers and tags), tag protection, tag disabling (temporarily or permanently), reader delegation, and tag physical security.

RFID Component Authentication

There is an inherent need for strong and reliable authentication protocols between tag, reader, RFID server and backend database. There have been intensive studies on RFID tag-reader authentication protocols. Authentication is important since a malicious attacker can change the state of a tag (e.g., secrets) so that the reader would not recognize the tag or put the tag into "sleep" or "kill", thus rending the tag useless. The major challenges are to develop protocols which are suitable for low-cost, low-complexity RFID tags. Most existing protocols use pseudonyms to thwart eavesdropping or privacy attacks. Many protocols also require that the tag and reader update their shared secret keys after each authorization process in order to avoid the tag tracking by attackers. We next discuss some authentication protocols in the literature.

Tsudik (Tsudik, 2006) proposed an RFID authentication protocol where a reader, R, shares a key, x_i, with a tag, T. T also stores an internal timestamp t_i which records the last time that the tag was interrogated by R. To query a tag, R sends the current time, t_R. The tag compares t_R with t_i. If $t_R \leq t_i$, then the tag outputs a random response. Otherwise, the tag outputs $Hx_i(t_R)$, where Hx_i represents the HMAC computed with secret key x_i. Finally, the timestamp maintained by T is also updated. To validate a response, R checks whether the received $Hx_i(t_R)=Hx_j(t_R)$ for any secret key, x_j, in its database. If so, then R can be certain that the response comes from a valid tag, T_j. Weis, et al.,(Weis, Sarma, Rivest, & Engels, 2003) proposed "hash lock" for private mutual authentication and tag access control. Each tag has two possible states: locked and unlocked states. In a locked state, the tag responds to all queries with only its meta-ID and offers no more functions; in an unlocked state, it performs privileged operations related to security and configuration. Their scheme ensures that a tag enters the unlocked state only if it receives an appropriate command

from a legitimate reader. Lim and Kwon (2006) proposed a strong and robust RFID security protocol with both forward and backward intractability. Their protocol uses a forward hash chain for tag identification and a backward hash chain for reader identification. The protocol allows up to m authentication failures between two valid sessions. A nice feature of the protocol is that it allows easy ownership transfer, i.e., transfer of authorization to read the tag to a different person. Ownership transfer must guarantee that the old owners should not be able to read the tag any more. The Hopper and Blum (HP) protocol (Hopper & Blum, 2001) and an augment version, HB++ (Juels & Weis, 2005), have been developed to secure against passive and active adversaries under the Learning Parity with Noise hardness assumption. Both protocols are secure under parallel and concurrent executions (Katz & Shin, 2006). There are also solutions for tag-reader mutual authentication using non-cryptographic primitives (e.g., Vajda & Buttyan, 2003; Juels, 2004).

RFID Component Protections and Physical Security

RFID tags are read-only, write-once, read-many, or fully rewriteable. To control access to tag data, password protection has been incorporated into tags to eliminate unauthorized user access. The RFID system shares a specific password (or PIN) with each tag. Writing of tag data must be password-protected. Furthermore, to protect the integrity of the tag data, digital watermark has been applied to RFID tags. Tags could be equipped with resources to support strong cryptographic primitives, tamper resistance packaging and other security enhancing features at a higher cost of $0.5-1 per tag (Vajda & Buttyan, 2003). The RFID system should deploy RFID tags such that physical access or close contact of the tags by the adversary is almost impossible. Policy and rules should regulate to control and monitor physical premier of RFID infrastructure. For instance, illegal scans of tags by unknown readers should be detected and prevented.

RFID tag killing has been used by many RFID systems. When a tag is no longer needed (e.g., change of owner or no need to be tracked) or the operating environment is getting too challenging, the tag can be disabled permanently by authorized readers. A tag specific kill PIN is implanted into a tag and the tag can simply disable itself upon receiving an authorized command. In addition, to prevent tags disabling by unauthorized entities, the tags can be designed to "screams" or "yells" when killed or removed.

Instead of disabling a tag permanently, an unauthorized reader may put the tag to "sleep," i.e., temporarily render it inactive. To "wake up" the tag when necessary, the reader must transmits another tag specific PIN to authenticate itself. Hence, an RFID tag can be in one of two states: sleep or awake. A physical trigger, like direct touch of a reader probe, might serve as an alternative means of waking up a tag (Stajano & Anderson, 1999).

Shield and blocker of RFID enabled products such as e-passports and credit cards are active techniques to protect tags from unauthorized interrogations. Juels, Rivest, and Szydlo (2003) propose a special form of RFID tag called a blocker. The blocking depends on the incorporation into tags of a modifiable bit called a privacy bit. A "0" privacy bit marks a tag as subject to unrestricted public scanning; a "1" bit marks a tag as "private." The scheme refers to the space of identifiers with leading "1" bits as a privacy zone. A blocker tag prevents unwanted scanning of tags mapped into the privacy zone.

Reader delegation is a way to reduce the exposure of an adversary breaking into an RFID reader. In this scheme, the system shares a set of tag secrets with a reader. When a tag is detected, the reader can use its local secrets to identify the tag instead of contacting the RFID server (e.g., the backend database) to get the tag identity. There are two main advantages of reader delegation: (1) it reduces the exposure in case that an adversary

breaks into the reader – instead of losing the secrets for all the tags, the system may lose only what was delegated to that reader (Tan, Sheng, & Li, 2006), and (2) it tolerates poor quality network connections between the reader and the backend server since a reader does not need to communicate with the server for every tag authorization. This can minimize the denial of service attack where readers cannot access the RFID server.

Physical security is crucial for tag anti-cloning and anti-counterfeiting. Generally, there are two research directions towards preventions of physical attacks. One is hardware approach, applying tamper resistance property to the system; the other is software approach, focusing on cryptographic ways to solve the problem. These two approaches can be used jointly. For instance, physically unclonable function (PUF) (Tuyls & Batina, 2006) is a technique applied to resist physical attacks on low-cost RFID tags. In order to thwart tag cloning attack, Tuyls and Batina (2006) proposes a PUF structure to store secret key materials in a tag. In order to make an item unclonable, two components are needed: (1) Physical protection, which uses unclonable physical structure embedded in the package (removal of the structure leads to its destruction). One or more unique fingerprints derived from the physical structure will be printed on the product for the verification of the authenticity of the product; and (2) Cryptography protection, which provides digital signature to detect and prevent tampering with the fingerprints derived from a physical object. It also provides secure identification protocols to identify a product.

A physical unclonable function maps challenges to responses and it is embodied in a physical object (Tuyls & Batina, 2006). It satisfies the following properties: (1) easy to evaluate: the physical object can be evaluated in a short amount of time; and (2) hard to characterize: from a number of measurements performed in polynomial time, an attacker who no longer has the device and who only has a limited (polynomial) amount of resources can only obtain a negligible amount of knowledge about the response to a challenge that is chosen uniformly at random.

Robustness and Fault Tolerance Techniques

This category of the techniques focuses on improving an RFID system's ability to respond to incidents through masking damage and tolerating faults. Those approaches can effectively enhance an RFID system's ability to adapt to the changing environment and respond to attacks correspondingly. The techniques in this category include tag monitoring, tag blocking, fault tolerance, and proxy-based protection.

Forward and Backward Security

Fault tolerance is an important aspect of enhancing RFID system survivability. In case of individual tag failure or tag compromise by attackers, inference reasoning can be applied to deduce individual tag information by using statistic analysis, time stream of correlative tag data, and tag grouping rules. Placement of tags should also consider such factors as fault tolerance since high correlation of RFID tags make it possible for tag data inference to effectively recover or mask damage of individual tags.

At the protocol level, an important approach to achieve fault tolerance is through forward and backward security in RFID tag-reader authentication, particularly when the ownerships of the tags need to be changed (Song, 2008). Forward security refers to the situation that the old owner, say S_{old}, will not be able to identify and control the tag, say, T, after it passes the ownership of T to a new owner, say S_{new}. Forward security can be achieved by simply requiring that S_{new} applies the mutual authentication protocol and updates the currently shared key k' to a new key k (by incorporating some secret materials only known to S_{new}). Since S_{old} cannot figure out k, it has no way to identify any transactions related to T conducted after the

key update. Backward security refers to the situation that the new owner will not be able to use any information it has on the tag to back track the past interactions conducted related to tag T. Once again, one simple key update is sufficient to achieve this goal. Before S_{old} transfers the ownership of T to S_{new}, it updates the key k'' to k'. Then S_{old} passes the secret k' to S_{new} via a secure channel. Since S_{new} has no way to calculate k'' based on k' (a one-way function such as a hash function can ensure this), it cannot identify any transactions related to T conducted before the ownership transfer. The major challenge of those methods is, however, to securely deliver the secure key from the current owner to the new owner. In the literature, key delivery is often conducted using an off-band secure channel.

Proxy Service

Proxy service provides context-aware access control to RFID tags. As a much more powerful device, a proxy server senses the environments and adjusts the access control policies in order to protect the RFID tags. Compared with the low-cost RFID tags, a proxy server is more capable of performing complicated computations and communications to implement standard security mechanisms such as public-key cryptography.

Dimitriou (2008) presented a proxy framework for protecting the privacy of users carrying RFID products. This approach uses a mobile phone (or any other similar device) as a proxy for interacting with readers on behalf of the user carrying tagged item. The user can specify when and where information will be released. Essentially, the proxy acts as a mediator for tag access to ensure the tag can withstand malicious attacks. The major operations are summarized in Figure 3, where the symbol "||" stands for message concatenation; *CID* stands for the current ID of the tag; *NewID* stands for the new ID of the tag; N_R represents a random nonce generated by the proxy; N_T represents another random nonce generated by the tag; $F_{cid}(.)$ stands

Figure 3. The proxy framework

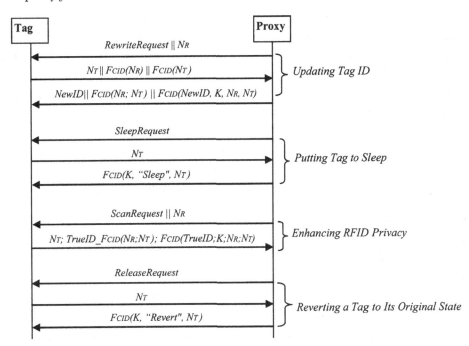

for an encryption function using key *cid*; and *K* represents the authentication key shared between the tag and the proxy.

Guardian (Rieback, Crispo, & Tanenbaum, 2005) is a device that acts as an intermediary between tags and readers and must always be alert in protecting tag responses from unauthorized read attempts. It allows reader queries, appropriately re-issues queries in encrypted form, or actively blocks tag answers. The RFID Guardian offers granular access control - coordination of security primitives, context-awareness, and tag-reader mediation. For instance, Guardian can control which RFID readers can query which RFID tags under which conditions. Also Guardian can act as a "man-in-the-middle", mediating interactions between RFID readers and RFID tags. Mediation can take either a constructive or destructive form. For a constructive mediation, an untrusted RFID reader first passes a request for a desired query to the RFID Guardian. Upon the successful completion of a possibly complex security negotiation, the Guardian re-issues the query in encrypted form and forwards the cryptographically-protected queries to RFID tags. The Guardian then receives the encrypted tag response, decrypts it, and forwards the response to the RFID reader that requested it. Selective RFID jamming is an example of destructive mediation where the RFID Guardian blocks unauthorized RFID queries on the behalf of RFID tags. By filtering RFID queries, Selective jamming provides off-tag access control (Figure 3).

Fault Tolerance for Communication Channels

Noisy tag (Castelluccia & Avoine, 2006) was not designed to tolerate the noisy communication channels between an RFID reader and its tags, but it can be used for this purpose. A noisy tag is owned by the reader's manager and set out within the reader's field. It is a regular tag that generates noise on the public channel between the reader and the queried tag, such that an eavesdropper cannot differentiate the messages sent by the queried tag from the ones sent by the noisy tag. There are several versions of the protocol in Castelluccia and Avoine (2006) to describe noisy tag operations and we only show the basic one as below. The protocol is to establish a secret key between a reader and a tag so that a secure channel can be set up between them. Here, we assume that a tag T sends a sequence of secret bits b to the reader R with help a noisy tag NT. R and NT shares a secret K.

1. R broadcasts a random nonce N.
2. Both NT and T reply simultaneously with one bit (one bit per time slot) until R halts the protocol. The ith bit sent by NT is the ith bit of a hash value of K and N. The ith bit sent by T is random.
3. Since R can predict the sequence of bits sent by NT, it can easily filter them out, and recovery the bits sent by T.
4. If both NT and T send the same bit, then an eavesdropper is able to figure out the secret bit (For example, let's assume that bit 1 is implemented by a pulse of xmV, bit 0 is implemented by pulse of 0mV. If both T and NT reply with the same bit, say 1, then a pulse of 2.xmV will be generated on the channel. In that case, an adversary knows that both T and NT sent the bit 1. Similarly if both T and NT send the bit 0, a pulse of 0mV will be generated on the channel and the adversary knows that both T and NT sent the bit 0). To avoid this situation, R sends to T the relevant time slots' numbers. T uses this information to recover the secret bits that should be used to compute the shared secret.

Damage Assessment and Recovery Techniques

A survivable system must be able to recover from damage quickly. We summarize the techniques in this category for RFID systems in two major thrusts: tag restoration and tag-read state synchronization as well as tag search. The former highlights the requirement for the mutual authentication protocols between RFID system components to be robust against various anomalies such as tag-reader desynchronization and side channel attacks. The latter indicates that it is important for an RFID system to automatically assess the current status of current tags in the system and detection of any missing tags, tag fraud or tag theft, given the large-scale of an RFID system.

Tag Restoration and Tag-Reader State Resynchronization

Tag restoration is to reset the secret of a tag as shared with the reader/backend server whenever it is necessary. It can be achieved by an explicit key reset channel via PIN matching, manual intervention (e.g., physical contact of a RFID tag to trigger tag key reset), or simply normal scan of the tag. Since RFID tag and reader maintain a synchronized state and update their shared secret keys in a coordinated fashion, reading the tag one or two times in a secure environment effectively serves as a tag reset method. Even when an attack has already possessed the tag secret key, reading the tag can trigger a key update using some new materials supplied by the reader. After such secure reading, the attack effect has been wiped out. In addition, a special PIN can be built into a tag, and knowledge of the PIN will allow the reader to change the tag secret(s).

Another technique in lieu of RFID recovery is the system's ability to resynchronize the state of tag and reader if its state is desynchronized possibly by malicious attackers. Desynchronization is a dangerous denial of service attack. If a tag is permanently desynchronized with its owner or the backend server, then the tag is considered completely compromised. Since an attacker has access to the communication channel between a tag and a reader, the attack could block some messages and make the states of the tag and reader out of pace. For instance, an attacker could block the key update message from reaching the tag. Therefore, the backend server updates the key but the tag does not. So, the tag and the backend server will not be able to authenticate each other since their keys are out of synchronization.

Henrici, *et. al.*, (Henrici & Muller, 2004) proposed to solve the synchronization problem by having a tag emit the difference between its current transaction number *TID* and the last successful transaction number *LST*, i.e., *ΔTID=TID-LST*. So, the reader will be able to determine the current state of the tag. The reader also maintains two entries with each tag, one is for the "should-be" values and the other is for the last successful authentication. A row in the database is never overwritten until the other entry has been addressed by the tag proving that one being currently valid and the one to be overwritten being obsolete. The protocol is summarized below (also see Figure 4):

- Reader *R* sends a HELLO message to a tag *T* (here we do not consider tag singleton).
- T increments its current transaction number TID by one and then responds with message *A=h(ID)*, *h(ID||TID)*, *ΔTID*, where *h(.)* is a hash function and ID represents the ID of tag *T*.
- *R* receives A and forwards it to the backend server *S* for tag data.
- S selects the entry with *HID=h(ID)*. The stored last successful transaction number *LST* and the received *ΔTID* are added to obtain the current transaction ID of tag *T*, i.e., *TID'*. Then *h(ID||TID')* is calculated

to verify the received *h(ID||TID)*. If they don't match, the transaction is terminated.

- If the *TID'* is not higher than the *TID*, then a replay attack is in process and the message is discarded.
- If the above is verified, *S* replaces the stored *TID* with *TID'*. In this way, *S* and *T* synchronize their states.
- Next, *S* creates a new *ID' = ID||RND*, where *RND* is a random nonce. *S* creates a reply *h(RND||TID'||ID)* and send it (together with *RND*) to *R*, which forwards the message to *T*.
- *T* checks *h(RND||TID'||ID)* and update its ID. In the mean time, *T* sets its last successful transaction number to the current *TID* value.

Tag Search

We propose a secure and private tag search protocol for a reader *R* to search for tag t_i (based on its ID id_i) from a set of tags in its range (e.g., t_j, ..., t_j ... t_k). Tag search is important to facilitate automatic assessment of the current status of certain tags in the system and detection of any missing tags, tag fraud or tag theft. The proposed protocol is shown in Figure 5 and the protocol explanations are given below.

In step 1, *R* broadcasts a message to every tag in its range: $f_{ki}(id_i \oplus H(n_1)) \| f_{ki}^N(id_i \oplus H(n_1)) \| n_1$, where $f_{ki}(.)$: a pseudorandom function with a seed k_i, $\|$ represents a message concatenation operator, $\oplus$ represents a bit-wise mapping operator (i.e., XOR), *H(.)* represents an one-way hash function,

Figure 4. A desynchronization-resistance tag-reader authentication protocol

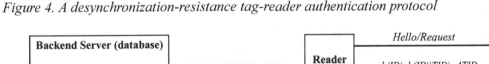
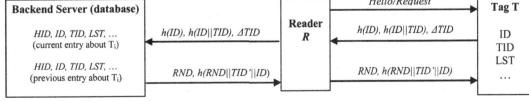

Figure 5. Secure and private tag search protocol

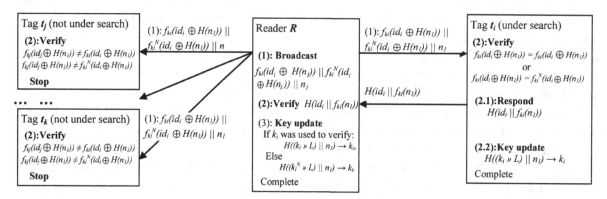

and n_l represents a random nonce generated by R. In our protocol, R (or the backend database) maintains both the current key shared with a tag, t_i, i.e., k_i, and the next "should-be" secret key, denoted as k_i^N (k_i^N is initially calculated as $H((k_i \gg L) \| n_l)$). Using both the current and next predicted key to generate search message allows a RFID system to tolerate fault and mitigate the desynchronization attack where an attacker causes t_i to update the shared key in the last successful authentication, but R did not. In general, our protocol is more robust.

In step 2, each tag uses its own ID, the key shared with R and the received n_l to verify both $f_{ki}(id_i \oplus H(n_1))$ and $f_{ki}^N(id_i \oplus H(n_1))$. For those tags such as t_j, ..., t_k, since they are not under search, such verifications should fail. Then they keep silent and stop the search process. However, for the tag t_i under search, after plugging in the corresponding parameters, either $f_{ki}(id_i \oplus H(n_1))$ or $f_{ki}^N(id_i \oplus H(n_1))$ should be true. Then tag t_i knows R is searching for itself. Thus, t_i responds by sending a message $H(id_i \| f_{ki}(n_l))$ to R. This message allows R to authenticate t_i since no other tag except t_i is supposed to know the secret k_i shared with R. In the mean time, t_i updates the shared key $H((k_i \gg L) \| n_l) \to k_i$, where $\gg$ represents a right cyclic rotating operator and L represents a predefined length value used in bit rotation.

In step 3, R verifies the received $H(id_i \| f_{ki}(n_l))$ by applying id_i, either k_i or k_i^N, and n_l. If any cal-

culation matches the received message, t_i has been successfully searched. Otherwise, R believes that either the response is not valid or t_i is not in its range. Next, R correspondingly updates the shared key with t_i in order to keep a synchronized state. If k_i was used to verify the message, then update the key by $H((k_i \gg L) \| n_l) \to k_i$. Otherwise, the key is updated by $H((k_i^N \gg L) \| n_l) \to k_i$.

The protocol ensures reader-tag communication privacy since a secret key is used to mask any message transmitted, and no attacker is able to interpret both the search message and the response from a tag. The protocol also ensures tag identity privacy. An attacker, A, is unable to figure out the identity of the tag which responds to a search query, since she does not pose the secret key used to encrypt the messages. The protocol also resists tag impersonation attacks. In order for A to impersonate a valid tag, t_i, A must compute the valid response $H(id_i \| f_{ki}(n_l))$ sent to the reader. But, it is impossible to construct such a valid response without knowledge of k_i and id_i. A cannot also launch interleaving and reflection attacks, because different values of nonce n_l is used in different rounds of searches. R maintains such n_l value for each search session and different n_l in different search sessions are not linkable. The protocol also resists tag-reader state desynchronization since R recognizes both the current k_i and the next "should-be" k_i of each tag. Any desynchronization between the tag and the reader will be resynchronized after successful protocol run.

Next, we give performance evaluation of the protocol. The storage requirement is l_1+l_2 for the tag and $(2l_1+l_2)*n$ for the server, where l_1 represents the bit length of key k_1, l_2 represents the bit length of tag ID id_1, and n represents the number of tags in the range of R. The average computation is $6f$ for the tag under search, $4f$ for other tags not under search and $7f$ for the server, where f represents the computational costs of a complex function (e.g., a pseudorandom function or a hash function). The communication requirement is C_R+C_S for the tag under search, C_R for any other tag not under search, and C_R+C_S* for the server, where C_R represents the cost of sending a message, C_R represents the cost of receiving a message and C_S* represents the cost of broadcasting a message. Those performance metrics indicate that the protocol is suitable for low to middle cost RFID tags.

CONCLUSION AND RESEARCH DIRECTION

RFID survivability is a less studied area although RFID systems have been applied to various fields including some high security and high setting applications. In this paper, we discussed existing RFID security and privacy techniques which could potentially improve the survivability of an RFID system. We classified those techniques from three aspects based on the desired characteristics of a survivable RFID system: resilience, robustness and fault tolerance, and damage assessment and recovery. Although those techniques may not resist all the attacks on the RFID systems given their open nature and large scale of distribution, they provide the baseline to further design and develop survivable RFID systems. Possible future research on RFID survivability includes (1) development of tamper-proof hardware and software capable of

foiling physical attacks; (2) Trustworthy computing platform suitable for RFID readers and servers; (3) advance hardware techniques to enable more computationally capable RFID tags with a reasonable level of cost; (4) advanced lightweight cryptographic algorithms and mechanisms suitable for low-complexity RFID tags; (5) scalable, flexible, and reliable RFID infrastructure to supply large scale distributed RFID enterprise systems; (6) advanced techniques to improve the security of wireless communications among various components of an RFID system; (6) advanced techniques for the RFID system to detect rogue and cloned readers by the mechanisms such as behavior observation, data reading analysis and non-self detection; and (7) algorithms and protocols for RFID system adaptation to changing environments, system repair and self-healing.

REFERENCES

Castelluccia, C., & Avoine, G. (2006). Noisy Tags: A Pretty Good Key Exchange Protocol for RFID Tags. In *Proceedings of the 7th IFIP WG 8.8/11.2 International Conference*, Tarragona, Spain.

Deutsch, M., & Willis, R. (1988). *Software Quality Engineering: A Total Technical and Management Approach*. Upper Saddle River, NJ: Prentice-Hall.

Dimitriou, T. (2008). Proxy Framework for Enhanced RFID Security and Privacy. In *Proceedings of the Consumer Communications and Networking Conference*, Las Vegas, NV.

Feldhofer, M., Dominikus, S., & Wolkerstorfer, J. (2004). Strong Authentication for RFID Systems Using the AES Algorithm. In M. Joye & J. Quisquater (Eds.), *Cryptographic Hardware and Embedded Systems* (LNCS 3156, pp. 357-370). New York: Springer.

Henrici, D., & Muller, P. (2004). Hash-based Enhancement of Location Privacy for Radio-frequency Identification Devices using Varying Identifiers. In *Proceedings of the Second IEEE Annual Conference on Pervasive Computing and Communications Workshops*, Orlando, FL.

Hiltunen, M., Schlichting, R., Ugarte, C., & Wong, G. (2000). Survivability through Customization and Adaptability: The Cactus Approach. In *Proceedings of the DARPA Information Survivability Conference and Exposition* (pp. 294-307).

Hopper, N., & Blum, M. (2001). Secure Human Identification Protocols. In C. Boyd (Ed.), *Advances in Cryptology – ASIA CRYPT 2001* (LNCS 2248, pp. 52-66). Berlin: Springer Verlag.

Juels, A. (2004). Minimalist Cryptography for Low-cost RFID Tags. In *Proceedings of the Fourth Conference on Security in Communication Networks*, Amalfi, Italy (pp. 149-153).

Juels, A., Rivest, R., & Szydlo, M. (2003). The Blocker Tag: Selective Blocking of RFID Tags for Consumer Privacy. In *Proceedings of the ACM Conference on Computer and Communication Security* (pp. 103-111).

Juels, A., & Weis, S. (2005). Authenticating Pervasive Devices with Human Protocols. In V. Shoup (Ed.), *Advances in Cryptography – Crypto 05* (LNCS 3126, pp. 198-293). Berlin: Springer Verlag.

Karygiannis, T., Eydt, B., Bunn, L., & Phillips, T. (2007). *Guidelines for Securing Radio Frequency Identification (RFID) Systems*. National Institute of Standard and Technology.

Katz, J., & Shin, J. (2006). Parallel and Concurrent Security of the HB and HB++ Protocols. In *Proceedings of the Advances in Cryptology (EURO CRYPT 2006)* (LNCS 4004, pp. 73-87). New York: Springer.

Knight, J., Strunk, E., & Sullivan, K. (2003). Towards a Rigorous Definition of Information System Survivability. In *Proceedings of the DARPA Information Survivability Conference and Exposition*, Washington, DC.

Lim, C., & Kwon, T. (2006). Strong and Robust RFID Authentication Enabling Perfect Ownership Transfer. In *Proceedings of the 8th Conference on Information and Communications Security*, Raleigh, NC.

Mitrokotsa, A., Rieback, M., & Tanenbaum, A. (2009). Classification of RFID Attacks. *Information System Frontiers: A Journal for Innovation and Research*.

Peris-Lopez, P., Hernandex-Castro, C., Estevez-Tapiador, J., & Ribagorda, A. (2006). RFID Systems: A Survey on Security Threats and Proposed Solutions. In *Proceedings of the International Conference on Personal Wireless Communications* (LNCS 4217).

Rieback, M., Crispo, B., & Tanenbaum, A. (2005). RFID Guardian: A Battery-powered Mobile Device for RFID Privacy Management. In. *Proceedings of the Australian Conference on Information Security and Privacy*, 3574, 184–194. doi:10.1007/11506157_16

Schneier, B. (1999). Attack Trees. *Dr. Dobb's Journal of Software Tools*, 24, 12–29.

Song, B. (2008). RFID Tag Ownership Transfer. In *Proceedings of the 4th Workshop on RFID Security*, Budapest, Hungary.

Stajano, F., & Anderson, R. (1999). The Resurrecting Duckling: Security Issues for Ad-hoc Wireless Networks. In *Proceedings of the 7th International Workshop on Security Protocols* (LNCS 1796, pp. 172-194). Berlin: Springer Verlag.

Tan, C., Sheng, B., & Li, Q. (2006). Secure and Serverless RFID Authentication and Search Protocol. *IEEE Transactions on Wireless Communications, 7*(3).

Tarvainen, P. (2004). Survey of the Survivability of IT Systems. In *Proceedings of the 9th Nordic Workshop on Secure IT-systems*, Helsinki, Finland.

Tsudik, G. (2006). YA-TRAP: Yet Another Trivial RFID Authentication Protocol. In *Proceedings of the 4th Annual IEEE International Conference on Pervasive Computing and Communications*, Pisa, Italy.

Tuyls, P., & Batina, L. (2006). RFID-Tags for Anti-Counterfeiting. In *Proceedings of the Cryptographer's Track at the RSA Conference*, San Jose, CA.

Vajda, I., & Buttyan, L. (2003). Lightweight Authentication Protocols for Low-cost RFID Tags. In *Proceedings of the Second Workshop on Security in Ubiquitous Computing*, Seattle, WA.

Weis, S., Sarma, S., Rivest, R., & Engels, D. (2003). Security and Privacy Aspects of Low-cost Radio Frequency Identification Systems. In *Proceedings of the 1st International Conference on Security in Pervasive Computing*, Boppard, Germany.

Zuo, Y., Pimple, M., & Lande, S. (2009). A Framework for RFID Survivability Requirement Analysis and Specification. In *Proceedings of International Joint Conference on Computing, Information and Systems Sciences and Engineering*, Bridgeport, CT.

This work was previously published in the International Journal of Handheld Computing Research (IJHCR), Volume 2, Issue 1, edited by Wen-Chen Hu, pp. 25-40, copyright 2011 by IGI Publishing (an imprint of IGI Global).

Chapter 4

Mobile Agent Based Network Defense System in Enterprise Network

Yu Cai
Michigan Technological University, USA

ABSTRACT

Security has become the Achilles' heel of many organizations in today's computer-dominated society. In this paper, a configurable intrusion detection and response framework named Mobile Agents based Distributed (MAD) security system was proposed for enterprise network consisting of a large number of mobile and handheld devices. The key idea of MAD is to use autonomous mobile agents as lightweight entities to provide unified interfaces for intrusion detection, intrusion response, information fusion, and dynamic reconfiguration. These lightweight agents can be easily installed and managed on mobile and handheld devices. The MAD framework includes a family of autonomous agents, servers and software modules. An Object-based intrusion modeling language (mLanguage) is proposed to allow easy data sharing and system control. A data fusion engine (mEngine) is used to provide fused results for traffic classification and intrusion identification. To ensure Quality-of-Service (QoS) requirements for end users, adaptive resource allocation scheme is also presented. It is hoped that this project will advance the understanding of complex, interactive, and collaborative distributed systems.

INTRODUCTION

Security has become one of the most critical issues in today's computer-dominated society. Security threats have increased in sophistication, frequency and complexity in the past couple of decades. Despite the continuous efforts from the security community, organizations are being attacked at an alarming rate nowadays. Particularly for large scale enterprise network consisting of a number of mobile and handheld devices, there is a mismatch between the level of protection that current security measures are providing and the level needed to address their actual degree

DOI: 10.4018/978-1-4666-2785-7.ch004

of risks. The characteristics of the mobile and handheld devices make incorporating security very challenge. The constraints on mobile and handheld devices make the design and operation different from the contemporary wired networks. Security has become the Achilles' heel of networks of all sizes.

The monitoring and surveillance of security threats in network systems are mostly done by Intrusion Detection System (IDS) (Douligeris & Serpanos, 2007). IDSs are based on the principle that attacks on computer systems and networks will be noticeably different from normal activities. The job of IDS is to detect these abnormal patterns by analyzing information from different sensing sources in the network.

IDSs may be classified into host-based IDSs, network based IDSs and distributed IDSs, according to the source of audit information. Host-based IDSs get data from host audit trails; network-based IDSs collect network traffic as the data source; distributed IDSs gather audit data from multiple hosts and the network. There has been a shift from a centralized and monolithic IDS framework to a distributed one (Balasubramaniyan & Fernandez, 1998). Distributed IDSs usually include multiple sensors or agents for intrusion detection, and information fusion modules for data correlation.

However, when deploying IDSs in enterprise network with a large number of mobile and handheld devices, the IDS systems usually suffer from a number of limitations.

- **Configurability, Controllability, and Manageability:** Today's networks are dynamic with mobile and handheld devices. IDSs need to support flexible on-demand reconfiguration and dynamic deployment of new nodes. For example, tasks like loading attack signatures at run-time, creating new detection sensors, being adaptive to changes in the environment, and detecting new attacks should be supported.

- **Interoperability:** Today's networks are heterogeneous with mobile devices from different vendors. Many IDSs are developed and operated in specific domains and environments. It is a complicated and error-prone task to integrate and coordinate multiple IDSs. Mechanisms need to be designed to support the effective integration, cooperation and collaboration of heterogeneous IDSs.

- **Scalability, Extensibility, and Robustness:** IDSs should be scalable to monitor large scale networks with limited overhead imposed. IDSs need to be lightweight to run on mobile and handheld devices. IDSs need to protect themselves from attacks. They should also be able to recover quickly from system crashes or network failure.

- **Effectiveness:** IDSs suffer from the problem of high false alarm rate, including false positive and false negative. Algorithms need to be designed to produce real time, high-confidence detection results by fusing information from multiple data sources.

- **Global Coordination:** Network intrusion detection and prevention should utilize both local surveillance and global coordination. Local surveillance secures a protection domain by proactively identifying and thwarting attacks. Global coordination integrates the information from different parts of the network and coordinates collective countermeasures.

Therefore, IDS involving mobile and handheld devices is similar to a standard, wired IDS, but has additional deployment requirements as well as some unique features specific to wireless network.

In this paper, we proposes a highly-configurable, well-integrated intrusion detection & response framework named Mobile Agents-based Distributed (MAD) security system for enterprise network with mobile and handheld devices. *The key idea of MAD is to use mobile autonomous agents (hereby referred to as AA) as lightweight, independently-running entities to provide unified interfaces for intrusion detection, intrusion response, information fusion and dynamic reconfiguration.* All these AAs are mobile agents which can travel among network hosts and take actions at target spots. The reason of using mobile AA is that they can be sent to security hot spots or blank spots to install new AAs or perform other tasks with great flexibility. Lightweight agents reduce the running overhead on mobile and handheld devices. This feature is designed to fit dynamic nature of enterprise network with mobile and handheld devices.

Multiple AAs are coordinated to obtain a global view of the security state and take collective countermeasures against attacks. AAs are designed to carry out tasks in a flexible, adaptive and intelligent manner that is responsive to changes in the environment.

The MAD framework supports dynamic system generation, adaptive reconfiguration and real-time responses. All these features are critical to network with mobile and handheld devices. The MAD system provides enhanced configurability, controllability, interoperability, and scalability with simplified management interfaces.

SYSTEM ARCHITECTURE

The key features of MAD are listed below:

1. MAD supports a hybrid, integrated and flexible intrusion detection model. The intrusion detection network consists of a family of AAs: active AAs which are regular intrusion detection agents, hibernative AAs which are usually in hibernation but can turn active upon requests and auxiliary AAs which provide interfaces between MAD and existing IDSs from other vendors.

2. MAD is designed as an open framework using modular structure. AAs can dynamically and intellectually download and install appropriate modules, signatures and policy files from the central servers. New intrusion detection techniques and capabilities can be easily integrated into the MAD framework. This allows different organizations and individuals to contribute to the development of MAD.

3. A data fusion and event analysis engine (mEngine) and an object-based intrusion modeling language (mLanguage) are designed. Both mEngine and mLanguage are domain-independent. In MAD, confidence-level based data fusion techniques are used. Instead of 0 or 1 (legitimate or illegitimate), the data fusion output is a value from 0 - 1 indicating how likely the traffic is legitimate (or illegitimate). This enables IDS systems to be flexible in raising alerts and enables network systems to be flexible in intrusion response.

4. Quality-of-Service (QoS)-based intrusion responses on routers and backend servers are designed. Based on the data fusion result, incoming traffic is classified and placed into different queues with different QoS requirements. By adaptively managing resources allocated to different traffic queues, legitimate clients will be ensured with sufficient resources.

Figure 1 shows the architecture of MAD framework. The sample network is divided into three autonomous zones, which can be based on subnets. Each autonomous zone consists of multiple intrusion detection agents, data fusion agent(s) and control agent(s). These agents can be installed on mobile/handheld devices or on the base stations. Multiple autonomous zones are coordinated by the central servers to obtain global view and take global actions.

Figure 1. The architecture of the MAD framework

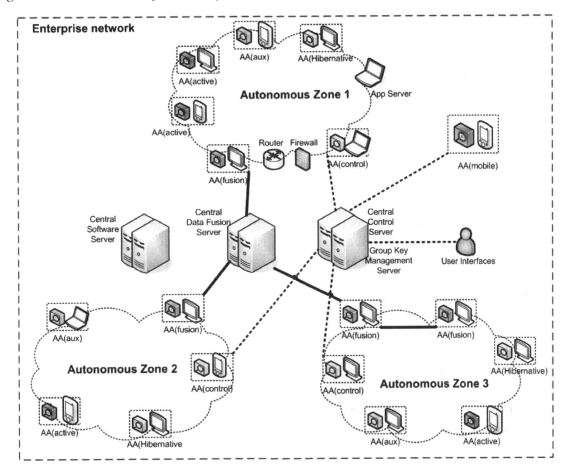

A group key management system is used to provide secure and scalable group communication and group management for heterogeneous entities in MAD. The group key system can ensure the confidentiality, authenticity, and integrity of messages delivered between group members. In MAD, all participating nodes are authenticated, and all control messages and data are encrypted. This is a necessary measure against insertion and evasion attacks on distrusted IDS itself.

MAD is organized in a hierarchical structure which improves system scalability. Intrusion detection AAs are at the bottom of hierarchy. Multiple layers of data fusion AAs and control AAs can be used to correlate intrusion detection

data and take intrusion responses. The central servers are at the top of hierarchy, which provide global intrusion data fusion, intrusion responses and system management.

There are of five types of AAs and three types of central servers in MAD architecture. The family of AAs is explained below:

- **Active AA:** Active AAs monitor network traffic and transfer collected intrusion detection information to data fusion agents. The intrusion detection can be host-based, network-based or hybrid. Active AAs can load intrusion detection signatures, policies and additional modules at run-time without restarting the whole MAD system. Active AAs can also take local intrusion response to improve system responsiveness.

- **Hibernative AA:** Hibernative AAs usually stay in hibernation state and impose almost no overhead on local hosts and the network. Upon receiving "wake-up" command from control agents, they turn into active AAs. The rationale behind this is as follows. Excessive intrusion detection agents and data traffic will impose unacceptable overhead on network systems. Therefore, if no attacks, some AAs should hibernate; if there is an alarm for suspicious activities, the hibernating AAs in the affected domain should turn into active mode to watch the intrusion closely.

- **Auxiliary AA:** This type of AA is used to integrate MAD and the existing IDSs from other vendors, like the proprietary IDSs or Snort. Auxiliary AAs provide interfaces between multiple IDSs so they can share audit information and take coordinated intrusion responses. Different interface modules can be plugged into auxiliary AAs based on operational requests. The reason to have auxiliary AAs is to take advantage of the existing intrusion detection resources.

- **Data Fusion AA:** This type of AA runs as information fusion agent, which collects, correlates and fuses information from multiple data sources, and generates intrusion detection analytical results with a value of 0-1 indicating confidence-level. Different data fusion and event analysis modules, like mEngine, neural network or Bayesian belief network, can be plugged into the system.

- **Control AA:** This type of AA gets intrusion detection analytical results from the data fusion agent. Based on intrusion response policies, control AA makes deci-

sions on appropriate actions and notifies corresponding entities. The control AA also keeps track of all agents in its domain and maintains an agent information database.

AAs play a central role in the MAD framework. AAs can receive high-level control commands and take predefined actions. Based on operational requirements, different modules can be dynamically plugged in, loaded and unloaded on agents at run time. AAs greatly simplify the deployment, configuration and management of distributed IDSs.

The family of central servers in MAD is listed as follows:

- **Central Data Fusion Server:** It collects refined intrusion detection information from multiple data fusion AAs and performs information fusion for the whole network. Different levels of reporting from downstream agents are defined.

- **Central Control Server:** It gets global intrusion detection results and makes decision on global intrusion responses, including QoS-based response on router and end server, firewall-based response, and IDS dynamic reconfiguration. Three end-user interfaces are provided for system administration: web-based, command-line-based and script-based.

- **Central Software Server:** It contains a software module repository, an intrusion signature database, and a collection of response policies. Two software installation modes are supported. In pull mode, data is pulled by local agents from the central server. In push model, data is pushed by central server to selected local agents.

The data flow diagram in MAD is shown in Figure 2. Local intrusion detection AAs are not

Figure 2. The data flow diagram of MAD

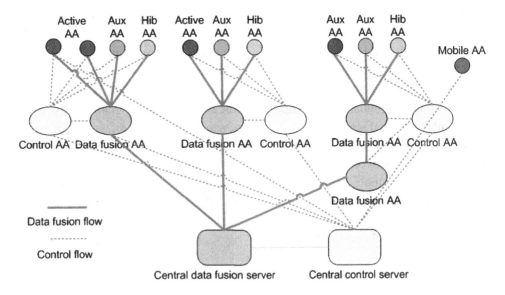

designed to communicate with each other. They can communicate with up-stream data fusion AAs, control AAs and central servers. On the other hand, control AAs and data fusion AAs can only communicate with agents within their authorized domains. Central control server can send commands to all agents in the network and override commands from local control AAs. This enables global coordination and collaboration.

AUTONOMOUS AGENTS AND CENTRAL SERVERS

Figure 3 illustrates the process of agent installation in MAD. First, a java installer is executed manually or by mobile AA for automatic installation. Second, the installer program will contact the central software server(s), download a C agent module based on current platform information, and install the module on local host. Third, the newly-installed C agent program takes control from the Java installer. The software agent sup-

Figure 3. Agent installation

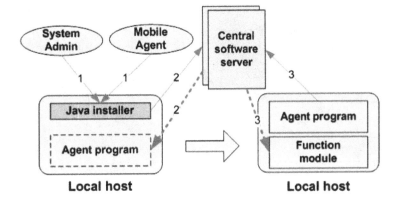

ports dynamic modules. It can intellectually scan current systems and the network to decide what additional modules are needed. It then downloads and installs the modules. It can load and unload modules at run-time without restarting itself. It also supports network socket communication to receive high-level control commands from upstream control agents or central servers.

The architecture of mobile agent system is shown in Figure 4. It consists of one central mobile agent server and multiple mobile agent interfaces in the network. A small code fragment (mobile agent) can move to the network devices where it is allowed to invoke other codes to com-

plete the service. By dispersing centralized network management tasks to subnet hosts, mobile agents help to conserve network bandwidth and improves management efficiency.

Key management protocols are the core of the secure communications between network hosts. Recently, many dynamic key management schemes for the wireless sensor network have been proposed (Zhang, 2008). The architecture of a group key system in MAD is shown in Figure 5. Users are authenticated through the use of digital certificate. Group key are issued when members are joined or leaves to ensure the security policy. All control messages and data ex-

Figure 4. A mobile agent-based system

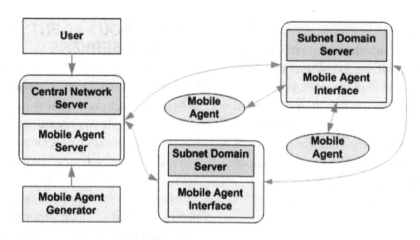

Figure 5. A secure groupware system

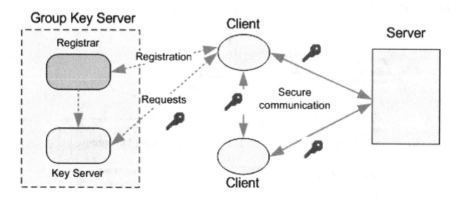

changed are encrypted. A group key may be an encryption key, a signing key, a security-association in IPSec, etc.

The central servers may become a single-point of failure. To improve performance and robustness, server clusters will be used, e.g., Linux Virtual Server (LVS) and Windows clusters. Redundant upstream servers can be configured for fail-over protection, which is similar to the use of redundant name servers in Domain Name System (DNS).

DATA FUSION AND EVENT ANALYSIS ENGINE

To reduce high false alarm rate and identify new attacks, we propose a multi-stage data fusion and event analysis engine (mEngine) in MAD. The architecture of mEngine is illustrated in Figure 6. It is designed to be a domain-independent architecture. The mEngine starts with a preprocessing module which exam incoming data to ensure validity like data source, data format and required attributes. The noise filtering module filters out unwanted background noise. Kalman filter (Anderson & Moore, 1979) is a recursive filtering algorithm

for stochastic dynamic systems, which is robust to background noise existing in the monitored data. The duplicate reduction module removes duplicated information reported by multiple agents.

Then data moves from data processing stage to information analysis stage. The instance clustering module is responsible for clustering data into attack instances. The host/network integration module associates network-based data with host-based data that are related to the same attack instance. The hotspot identification module identifies suspicious activity instances and security hotspots in the network.

Next, data moves into knowledge analysis stage. The multi-step analysis module is responsible for identifying attacks involving multiple steps. A connection-history based anomaly detection algorithm based on (Toth & Kruegel, 2002) will be designed and implemented. The idea here is to detect the trend, not the burst. Therefore, this module is useful to identify attacks and worms at early stage. Based on the analytical result, precocious response actions will be taken, for example, waking up hibernative AAs, relocating mobile AAs and classifying traffic with lower priority.

The distribution analysis module identifies attacks based on traffic distribution analysis.

Figure 6. The multi-stage data fusion and event analysis engine (mEngine)

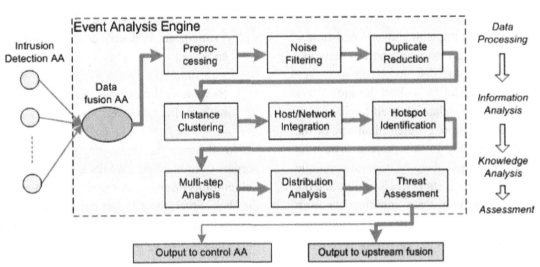

61

The basic approach is to derive traffic distribution features from the normal network traffic and use them as a baseline for comparison, e.g., byte frequency distribution (Kruegel & Vigna, 2003) and position-aware distribution signature (Tang & Chen, 2005). This can identify previously unknown and new attacks.

The last element in mEngine is the threat assessment module. It is used to evaluate the impact of attacks, determine the effectiveness of analytical results, and prioritize the output results. The analytical results of mEngine are values between 0-1 indicating confidence level. The control AA will determine response actions based on these values and response policies. Also, the refined intrusion detection dataset is outputted to upstream data fusion entities.

We also implemented a two-phase data fusion approach based on neural network. The openness of MAD framework makes it easy to incorporate new techniques and capabilities into the system.

Neural network is a well explored area with various types of applications and usages (Hall, 2001). Offline trained neural network models are robust to environmental noise but have restricted use as there may be a prediction error when the dynamics of the environment changes beyond a certain bound. Instead, simpler neural networks can be trained online during the traffic monitoring process. But training a neural network to model the entire traffic situation involves large computations, hence, may not be implementable in real-time.

As an alternative to using online networks alone, it is observed that switching to a well trained offline model for a short duration while the online model is adapting itself to changes in the dynamics, is more efficient than any of the individual networks. In this project, a training scheme based on Levenberg-Marquardt technique for online networks is studied. The online and the offline neural networks are combined to form a multi-network structure. By having multiple networks and a dynamic switching technique,

more accurate data correlation hopefully will be achieved.

INTRUSION MODELING LANGUAGE

Intrusion modeling languages have received attention from the intrusion detection community. In (Cheung, Lindqvist), the authors proposed intrusion modeling languages named CAML and STATL respectively. In MAD, an object-based (OB) intrusion modeling language (mLanguage) is designed. OB has become especially popular in scripting languages with abstraction, encapsulation, reusability, and ease of use being the most commonly cited reasons. The other benefit of OB is to be compatible and inter-operable with XML (Extensible Markup Language) and Web Services. The mLanguage uses C/C++ alike syntax to reduce the learning curve.

There are six predefined classes in mLanguage. Users may define new classes based on needs.

- **Event:** It describes network "event," or "activity," for example, a scan action.
- **Source:** It describes the source of attack, like IP address, port number and running services.
- **Target:** It describes the target of attack.
- **State:** It describes the state and status of attacks and the network.
- **Transition:** It describes the transition of state and event. This is useful for multi-step attacks.
- **Response:** It describes intrusion responses with confidence level.

To illustrate the mLanguage, let's look at a scenario. An attacker exploits a buffer overflow vulnerability on an e-Commerce web server. From the Web server, the attacker tries to mount a file system to access some sensitive data. This attack may be observed by multiple AAs. For instance, a signature-based network IDS may detect the buffer overflow attack, and an anomaly detection component may detect the unusual file access. The mEngine is responsible to correlate and fuse different pieces to get a whole picture of the attack. A piece of mLanguage code describing the attack is shown in Algorithm 1.

QOS-BASED INTRUSION RESPONSE

MAD provides flexible intrusion response. First, global intrusion response over the network and local intrusion response on local host or subnet are combined together. The local intrusion response is performed by local intrusion detection agent based on local response policy. It reduces the response time and communication overhead. The local intrusion response is also useful to detect and defense against internal attacks.

Second, in addition to classic intrusion responses on firewall, i.e., packet filtering and rate limiting, new QoS-based intrusion responses on front-end router and back-end server are designed. On router, traffic classification is used to classify the incoming traffic based on its confidence level from information fusion. On backend application server, QoS resource management mechanism is

Algorithm 1. A piece of mLanguage code describing buffer overflow and remote execution attack

```
1: Module Buffer-Overflow-Attack {
2: Set source = CreateObject (" mLang.Source ")
3: Set target = CreateObject (" mLang.Target ")
4: source.list = (141.218.2.22, 141.218.2.11)
5: target.list = (128.121.82.88: port 443)
6: Event event1 {
7: .type = bufferOverflow
8: .timestamp = 04092006121311
9: .target = target.list[0] }
10: Event event2 {
11: .type = exploitRemoteExec
12: .timestamp = 04092006141751
13: .target = target.list[0] }
14: if event2.timestamp > event1.timestamp then
15: state.current = attackSucceed
16: state.timestamp = event2.timestamp
17: response.type = block
18: response.confidenceLevel = 1
19: else
20: transition.action = exploitRemoteExec
21: transition.timestamp = event2.timestamp
22: transition.location = event2.target
23: response.type = block
24: response.confidenceLevel = computeConfidenceLevel()
25: end if
26: }
```

designed to provide differentiate services to each traffic class.

Figure 7 shows the QoS-based intrusion response. The rationale of QoS-based intrusion response is as follows. QoS is usually the target

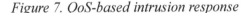
Figure 7. QoS-based intrusion response

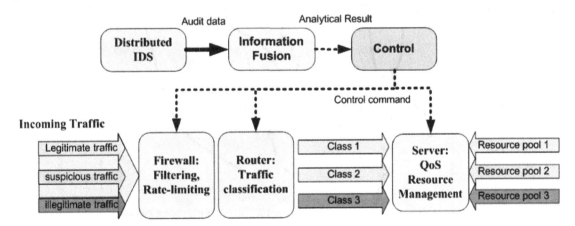

of attacks; but it can also be used to fight against attacks. Most current IDSs suffer from the high false alarm rate problem. Instead of classifying traffic into legitimate (0) and malicious (1), we can use a confidence level (0-1) generated by mEngine to measure the legitimacy of the incoming traffic. Traffic with less possibility of being malicious will be allocated with more system resources, and vice verse. This type of intrusion tolerance scheme seems more realistic in some scenarios. For example, an e-commerce site may not want to completely expel suspicious customers. Instead, the goal should be to reserve enough resources to serve legitimate customers.

Several QoS differentiation and regulation techniques were developed to support the QoS based intrusion response. We designed and implemented a dynamic process allocation approach, integrated with a proportional integral derivation (PID) feedback controller, on an Apache web server for proportional response time differentiation (Zhou, 2004).

AN ATTACK SCENARIO

To illustrate how MAD works, let's look at an attack scenario shown in Figure 8.

1. Attackers launch attacks against the large scale handheld device network.
2. Intrusion detection agents (two active AAs and one auxiliary AA) monitor the network traffic and host audit trails. The intrusion detection information is sent to data fusion AAs.
3a. Based on local response policy, active AA may take local intrusion response, like notifying firewalls on local hosts or subnets.
3b. Data fusion AA fuses intrusion information from multiple sources and sends analytical result to control AA.
3c. Data fusion AA sends refined intrusion information to central data fusion server.
4a. Control AA makes decision on whether to wake up hibernative AAs.

Figure 8. An example showing attacks and defense

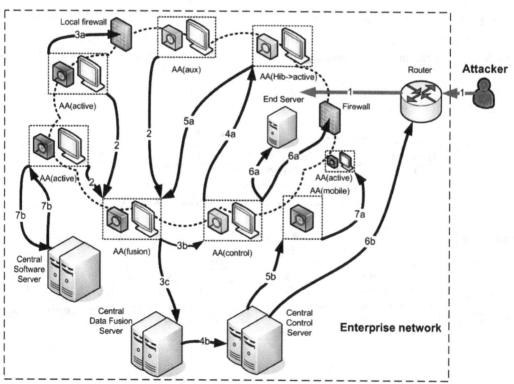

4b. Central data fusion server analyzes the intrusion information and notifies the central control server.

5a. Hibernative AAs wake up, collect intrusion information and report to data fusion AA.

5b. Central control server decides whether to relocate mobile AAs to target spots.

6a. Control AA make decision on intrusion responses like notifying router, firewall and end-server, reconfiguring the IDS nodes, and reloading new signatures, policies and modules.

6b. Central control server makes global intrusion responses over the whole network.

6c. Mobile AAs move to the specified destination.

7a. Mobile AAs deploys new active agent based upon control messages.

7b. Agents download and install new modules, policies and signatures at run time from central software server based upon control messages.

EXPERIMENTAL RESULT

Prototype Implementation

The MAD prototype was implemented on the Linux (kernel 2.4 and 2.6) systems, the Windows Server 2003 systems and iphone devices. Migrating MAD to other systems will be investigated later.

A kernel patch was written to provide necessary kernel environment information and user interface to other MAD modules. To reduce maintenance overhead, the kernel patch was restricted to several kernel files with less than 10 lines of code changes.

The MAD system was designed into several independent modules based on functionality. MAD is a loosely-coupled system which allows modules to be loaded, unloaded and maintained dynamically upon operational request. This design greatly enhances system manageability, usability and flexibility.

The interface between MAD modules and end users was through /proc file system. End users can input parameters (i.e., resource allocation ratio, control message) and adjust system behavior at runtime. The secure communication within MAD was built on secure socket connection with OpenSSL.

Overhead Analysis

One of the design goals in MAD is to provide real-time intrusion detection and response. Experimental results show that the latency in MAD primarily comes from two sources.

1. The overhead of secure group communication between entities, particularly over the wireless connection. Experimental results indicate that communication overhead may be up to 30%. This is the price to pay for secure communication between network hosts. A possible solution to reduce secure communication overhead is to use faster authentication and encryption / decryption methods, such as (Ferguson, Whiting, & Schneier, 2003).

2. The latency of mEngine when conducting data fusion and event analysis, particularly when multiple stages involved. A solution is to use local response and signature-base detection. Other possible solutions include generating intermediate results during analysis in mEngine.

QoS and Resource Management

We would like to achieve QoS based intrusion mitigation in the MAD system. For example, in a web server system, the incoming traffic is classified into two classes. One class has higher priority, such as VIP clients; the other class has lower priority, such as suspicious traffic. We would like to maintain the ratio of the response time of the two classes to a predefined number through resource management. Figure 9 depicts the achieved response time ratio with its 95% confidence interval of two classes of traffic. The pre-specified differentiation weight ratio of two classes was set to be 1:2. The results have demonstrated the feasibility of providing predictable QoS differentiation by adaptive resource management.

RELATED WORK

Recent intrusion detection research has been heading towards a distributed framework on monitors that do local detection and provide information for global detection of intrusions (Balasubramaniyan & Fernandez, 1998). They rely on some predefined hierarchical organization and most of them perform centralized intrusion analysis. Gopalakrishna et al. present a framework for doing distributed intrusion detection with no centralized analysis component. Ning et al. (2001) present a decentralized method for autonomous but cooperative component systems to detect distributed attacks specified by signatures. The GrIDS project (Chen, Cheung, & Crawford, 1996) employs distributed modules to report information to engines that build a graph representation of activity to detect pos-

sible intrusions. Cooperative Security Managers (White, Fisch, & Pooch, 1996) perform distributed intrusion detection and do not need a hierarchical organization or a central coordinator. The EMERALD project (Porras & Neumann, 1997) proposes a distributed architecture for intrusion detection that employs service monitors to perform monitoring functions. In Balasubramaniyan and Fernandez (1998), the authors proposed a distributed IDS framework named AAFID using autonomous agent and central processing servers. The Hi-DRA project (Kemmerer & Giovanni, 2005) proposes a framework for the modular development of intrusion detection sensors in heterogeneous, high-speed environments.

Work has been done to develop Wireless IDS (Li, 2008). Zhang and Lee (2000) proposed a Wireless IDS architecture in which all nodes act as independent IDS sensor which are able to act independently and cooperatively. Karim (2006) used linear decision-making pool means to enhance the efficiency and reduced fault. Game theory was applied in Ad hoc network intrusion detection system to improve the efficiency of judgment and reduce the power consumption (Liu, 2006).

Figure 9. QoS-based intrusion mitigation

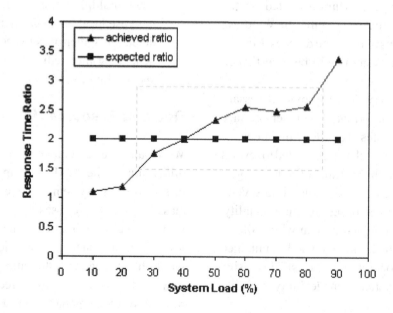

Multi-sensor data fusion is a challenging issue in distributed IDS. The effectiveness of distributed IDS relies heavily on data fusion and event analysis. Related works include (Bass, 2000).

In Julisch (2004), authors suggest that alarms should be handled by identifying and removing the most predominant and persistent root causes. In Ning, Cui, Reeves, and Xu (2004), techniques for constructing attack scenarios by correlating alerts on the basis of prerequisites and consequences of attacks are presented. Spice (Staniford, Hoagland, & McAlerney, 2002), probabilistic alert correlation (Valdes & Skinner, 2001), and alert clustering methods (Cuppens, 2001) correlates alerts based on the similarities between alert attributes.

CONCLUSION

This paper presents a highly-configurable, well-integrated intrusion detection and response framework named Mobile Agents-based Distributed (MAD) security system. The MAD framework is designed in a flexible, dynamic and intelligent manner that is adaptive to changes in the environment. By using lightweight, autonomous mobile agents, we will be able to installed and managed security mechanisms on mobile and handheld devices. With Object-based intrusion modeling language (mLanguage) and data fusion engine (mEngine), security information from different sources is analyzed for intrusion identification and traffic classification. Adaptive resource allocation scheme is used to ensure Quality-of-Service (QoS) requirements for end users.

We plan to release the source code of MAD prototype to general public over the Internet. http://sourceforge.net/ is one of many options.

This will greatly increase the visibility of MAD and get the open-source community involved in the development of MAD. Since the design goal of MAD is to build an open framework for intrusion detection, enormous modules and interfaces can be developed under GNU General Public License.

It is our hope that this project will advance our understanding of the complex, interactive, and collaborative behaviors among distributed autonomous agents. The improved IDS based on our work will enable the end clients to better protect their valuable enterprise information assets.

REFERENCES

Anderson, B. D. O., & Moore, J. (1979). *Optimal filtering*. Upper Saddle River, NJ: Prentice Hall Publishing.

Balasubramaniyan, J., & Fernandez, G. (1998). An architecture for intrusion detection using autonomous agents. In *Proceedings of the annual computer security applications conference (ACSAC)*.

Bass, T. (2000). Intrusion detection systems and multisensor data fusion. *Communications of the ACM, 43*(4), 99–105. doi:10.1145/332051.332079

Cheung, S., Lindqvist, U., & Fong, M. W. (2003). Modeling multistep cyber attacks for scenario recognition. In *Proceedings of the third DARPA information survivability conference and exposition*.

Cuppens, F. (2001). Managing alerts in a multi-intrusion detection environment. In *Proceedings of the 17th annual computer security applications conference*.

Douligeris, C., & Serpanos, D. N. (2007). *Network security: Current status and future directions.* New York: Wiley.

Ferguson, N., Whiting, D., & Schneier, B. (2003). *Helix: Fast encryption and authentication in a single cryptographic primitive* (LNCS 2887, pp. 330-346). New York: Springer.

Gopalakrishna, R., & Spafford, E. (2004). A framework for distributed intrusion detection using interest-driven cooperating agents. In *Proceedings of international symposium on recent advances in intrusion detection.*

Hall, D. L. (2001). *Handbook of Multisensor Data Fusion* (1st ed.). Boca Raton, FL: CRC Publishing.

Julisch, K. (2004). Clustering intrusion detection alarms to support root cause analysis. *ACM Transactions on Information and System Security,* 6(4), 443–471. doi:10.1145/950191.950192

Karim, R. (2006). An Efficient Collaborative Intrusion Detection System for MANET Using Bayesian Approach. In *Proceedings of the MSWiM.*

Kemmerer, R. A., & Giovanni, V. (2005). Hi-DRA: intrusion detection for internet security. *Proceedings of the IEEE,* 93(10), 1848–1857. doi:10.1109/JPROC.2005.853547

Li, H., Xu, M., & Li, Y. (2008). 802.11-based Wireless Mesh Networks. In *Proceedings of Complex, Intelligent and Software Intensive Systems conference.* The Research of Frame and Key Technologies for Intrusion Detection System in IEEE. doi:10.1109/CISIS.2008.42

Liu, Y., Comaniciu, C., & Man, H. (2006). A Bayesian Game Approach for Intrusion Detection in Wireless Ad Hoc Networks. In *Proceedings of the GameNets conference.*

Ning, P., Cui, Y., Reeves, D. S., & Xu, D. (2004). Techniques and tools for analyzing intrusion alerts. *ACM Transactions on Information and System Security,* 7(2), 274–318. doi:10.1145/996943.996947

Ning, P., Jajodia, S., & Wang, S. (2001). Abstraction-based intrusion detection in distributed environments. *ACM Transactions on Information and System Security (TISSEC),* 4(9), 407–452. doi:10.1145/503339.503342

Porras, P., & Neumann, P. (1997). EMERALD: Event monitoring enabling responses to anomalous live disturbances. *In Proceedings of the 20th NIS security conference.*

Staniford, S., Hoagland, J., & McAlerney, J. (2002). Practical automated detection of stealthy portscans. *Journal of Computer Security,* 1(10), 105–136.

Tang, Y., & Chen, S. (2005). Defending against internet worms: A signature-based approach. In *Proceedings of the IEEE Infocom conference.*

Toth, T., & Kruegel, C. (2002). Connection-history based anomaly detection. In *Proceedings of the IEEE workshop on information assurance and security.*

Valdes, A., & Skinner, K. (2001). Probabilistic alert correlation. *In Proceedings of the 4th international symposium on recent advances in intrusion detection.*

White, G. B., Fisch, E. A., & Pooch, U. W. (1996). Cooperating security managers: A peer-based intrusion detection system. *IEEE Network,* 5(3), 20–23. doi:10.1109/65.484228

Zhang, J., & Varadharajan, V. A. (2008). New Security Scheme for Wireless Sensor Networks. In *Proceedings of the IEEE Global Telecommunications Conference.*

Zhang, Y., & Lee, W. (2000). Intrusion Detection in Wireless Ad-Hoc Networks. In *Proceedings of the Sixth Annual International Conference on Mobile Computing and Networking.*

Zhou, X., Cai, Y., Godavari, G. K., & Chow, C. E. (2004). An adaptive process allocation strategy for proportional responsiveness differentiation on Web servers. In *Proceedings IEEE international conference on web services.*

This work was previously published in the International Journal of Handheld Computing Research (IJHCR), Volume 2, Issue 1, edited by Wen-Chen Hu, pp. 41-54, copyright 2011 by IGI Publishing (an imprint of IGI Global).

Chapter 5

Security Assurance Evaluation and IT Systems' Context of Use Security Criticality

Moussa Ouedraogo
Public Research Center Henri Tudor, Luxembourg

Haralambos Mouratidis
University of East London, England

Eric Dubois
Public Research Center Henri Tudor, Luxembourg

Djamel Khadraoui
Public Research Center Henri Tudor, Luxembourg

ABSTRACT

Today's IT systems are ubiquitous and take the form of small portable devices, to the convenience of the users. However, the reliance on this technology is increasing faster than the ability to deal with the simultaneously increasing threats to information security. This paper proposes metrics and a methodology for the evaluation of operational systems security assurance that take into account the measurement of security correctness of a safeguarding measure and the analysis of the security criticality of the context in which the system is operating (i.e., where is the system used and/or what for?). In that perspective, the paper also proposes a novel classification scheme for elucidating the security criticality level of an IT system. The advantage of this approach lies in the fact that the assurance level fluctuation based on the correctness of deployed security measures and the criticality of the context of use of the IT system or device, could provide guidance to users without security background on what activities they may or may not perform under certain circumstances. This work is illustrated with an application based on the case study of a Domain Name Server (DNS).

DOI: 10.4018/978-1-4666-2785-7.ch005

INTRODUCTION

Evolution is an inherent characteristic of IT systems. IT systems' models are made to evolve depending on the context, either because of new business or users requirements or owing to changes in the system operating environment (new threats for instance). However, as it is well known, different contexts may harbor different risks and eventually call for different security requirements.

The list of recent high profile security breaches is daunting; headlines have exposed major leaks among the largest organizations, resulting in loss of customer trust, potential fines and lawsuits (Le Grand, 2005). Vulnerable systems pose a serious risk to successful business operations, so managing that risk is therefore a necessary board-level and executive-level concern. Executives must ensure appropriate steps are being taken to audit and address IT flaws that may leave critical systems open to attack (Le Grand, 2005). As revealed by a study conducted by Wool (2004) on firewall configurations, a common but sometimes overlooked source of IT risks for large distributed and open IS is improper deployment of security measures after a risk assessment has been completed. The term security measures within the paper refer to security controls. In fact, risk countermeasures may be properly elucidated at risk assessment but their actual deployment may be less impressive or unidentified hazards in the system environment may render them less effective. How good, for instance, is a fortified door if the owner, inadvertently, leaves it unlocked? Or considering a more technical example, how relevant is a firewall for a critical system linked to the Internet if it is configured to allow any incoming connections? Therefore, monitoring and reporting on the security status or posture of IT systems can be carried out to determine compliance with security requirements (Jansen, 2009) and to get assurance as to their ability to adequately protect system assets. This remains one of the funda-

mental tasks of security assurance, which is here defined as *the ground for confidence on deployed security measures to meet their objectives.* To that extent, our understanding of security assurance is in line with the Common Criteria's definition of assurance (Common Criteria Sponsoring Organizations, 2006). Unfortunately most of what has been written so far about security assurance is definitional. Published literatures either aim at providing guidelines for identifying metrics example (Vaughn et al., 2002; Seddigh et al., 2004; Savola, 2007), without providing indications on how to combine them into quantitative or qualitative indicators that are important for a meaningful understanding of the security posture of an IT component; target end products (example of the Common criteria Common Criteria Sponsoring Organizations, 2006) or the software development stage (example of assurance cases Strunk & Knight, 2006; UMLSec, Jürjens, 2005; Secure Tropos, Mouratidis & Giorgini, 2007).

Our approach. We argue that evaluation of an operational system's security assurance only make sense when placed within a risk management context. To reflect this, our method literally takes place after the risk assessment has been completed and the countermeasures deployed. Figure 1 shows the security assurance evaluation model and how it relates to the risk assessment stage, which concepts are depicted in bold. The security requirements identified for the risks mitigation could come either on the form of security functions deployed on the system or on the form of guidelines for security relevant parameters i.e., those parameters that are not directly linked to security but when altered could induce a security issue. According to the NIST special publication 800-33 (Stoneburner, 2009), the assurance that the security objectives (integrity, availability, confidentiality, and accountability) will be adequately met by a specific implementation depends on whether required security functionality is present and correctly implemented, and effective. Heeding

Figure1. Security assurance evaluation model

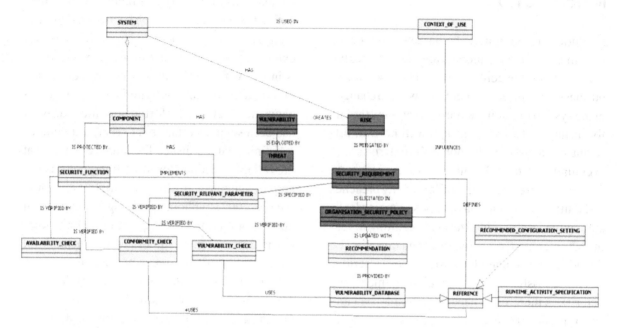

that call, our approach to evaluating the security assurance of a security measure is founded on key verifications that aim to:

1. Ensure that any security measures identified as necessary during the risk assessment stage have been implemented and is running (availability check).
2. Ensure the correctness posture of the security measures at any time: this includes deployment correctness (conformance of the security measure configuration) and activity correctness (concerns the security measure compliance to other procedural policies such as frequency of updating it, etc.).
3. The security level (or security effectiveness) required for a System context of use, which is dependent on the security criticality of the context, defined as the magnitude of the impact of an eventual security breach for an organization/individual in a specific context of use of the system.

These results are integrated in a security assurance function to yield a value of security assurance. Users may elect to use a system with a set of predefined security measures for its protection. However, once the system is deployed, previously unknown errors or vulnerabilities may surface for a given security entity or, environmental assumptions may need to be revised. Furthermore, the effectiveness of most security measures is limited in time. Today's state of the art protection may be by-passed with relative ease tomorrow as attackers' techniques are getting more and more sophisticated. As a result of operation, feedback could be given that would require the operator to correct the system security model or redefine its security requirements or environmental assumptions in view of strengthening the security of the system. To handle that eventuality, the vulnerability check (refer to Figure 1), which is associated to each evaluated security entity, uses a known vulnerability database such as the National Vulnerability Database (NVD, http://nvd.nist.gov) to verify whether any vulnerability has been identified for a specific version of an evaluated protection measure or security relevant parameter. Recommendations on how to overcome such matter are then taken into account and will help constitute the new reference against which any a posteriori conformity evaluation of the protection measure will be undertaken. This ensures that the system's security requirements is permanently updated and henceforth presents enough quality to face up to potential threats to the system. One of the main drawbacks of traditional risk management is that it is often a one-shot activity, or at best it is performed at regular but distant intervals of time (often every six months or so). To that extent, the continuous vulnerability check adds a hint of "dynamic risk management" to our approach.

Outline. The remainder of the paper is organized as followed: First a discussion on related work is presented. Then, a classification scheme for measuring a system's security criticality is provided. This is followed by the presentation of the actually security assurance evaluation methodology and a discussion on the choice of architecture for the approach. Finally, an illustration of its applicability with the aid of an application based on Domain Name Server (DNS) is presented along with some conclusions and directions for future work.

RELATED WORK

Considerable efforts have been made across computer science disciplines to address the ever-growing issue of security. Information System engineering, for instance, has recently called for the systematic integration of security in the development process, to help developers in producing more secure systems. In that perspective, modeling methodologies such as Secure Tropos (www.securetropos.org) (Mouratidis & Girogini, 2007) and UMLsec (Jürjens, 2005) have been proposed in the literature. The rationale is that without a rigorous and effective way of dealing with security at system development process, the end product cannot be secure. While this is true, the emphasis on design and process evidence versus actual product software largely overshadows practical security concerns involving the implementation and deployment of operational systems (Jansen, 2009).

The Common Criteria (Common Criteria Sponsoring Organizations, 2006) defines security assurance evaluation requirements for the development and design phases. The seven security assurance rating of the standard can be used as a platform for customers to compare the security product of different vendors. However, CC is not directly applicable for the evaluation of the security assurance of a system in operation. In fact, it defines how the system must be developed, but not how to maintain it in the "correct" (i.e., intended) state.

Defining metric taxonomy is also a topic where extensive works have been realized. The security metrics guide for Information Technology (Swanson et al., 2003) of the NIST provides guidance on how an organization, through the use of metrics, identifies the adequacy of in-place security measures, policies, and procedures. It describes the metric development and implementation process and how it can be used to adequately justify security measures investments. Amongst the taxonomies for the evaluation of security are the one proposed by Vaughn et al. (2002), Seddigh et al. (2004) and Savola (2007). However, these contributions only provide means to find the metrics and do not indicate how to combine them into quantitative or qualitative indicators more meaningful for appreciating the security posture of an IT component.

With respect to operational systems, some initiatives have been taken towards developing operational methodologies for the evaluation of IT infrastructure security assurance. Penetration tests (Klevinsky et al., 2002), for instance, have been routinely used for evaluating the security of a computer systems or networks by simulating an attack from a malicious source. The process involves an active analysis of the system for any potential vulnerabilities that may result from poor or improper system configuration, known and/or unknown hardware or software flaws, or operational weaknesses in process or technical countermeasures. Unfortunately, penetration testing is often conducted prior to system deployment in its operating environment or is used as a one off system audit. Bugyo (Bulut et al., 2007) can be cited as a pioneer methodology and tool for the evaluation of continuous security assurance. Inspired by the Common Criteria, security assurance in the context of Bugyo is understood as the ground for confidence that an entity meets its security objectives. Unlike the CC, Bugyo defines five levels of assurance and the methodology consists of five steps that are: Modelling, metric specification, assurance evaluation, aggregation, and finally, display and monitor. Pham et al. (2008) introduce an attack graph based security assurance assessment system based on multi-agents. In their approach, the authors use an attack graph approach to compute an attackability metric value (the likelihood that an entity will be successfully attacked) and define other metrics for anomaly detection to assess both the static and dynamic visions of the system under study. Unfortunately, the two above mentioned contributions only consider security correctness metrics and moreover, they do not account for the security criticality of the context of use of the system. Furthermore, these initiatives have targeted only security experts. However, although common users are not interested in the detailed of how security solutions are implemented, they still require assurance that the security measures are working as intended since most often, they are at forefront of impacts caused by security failures.

Based on the above analysis, the main contributions of this paper can be summarized as: (a) the proposal of metrics for appraising security assurance depending on the system context of use security criticality. Our metrics integrates information related to the correctness posture of the security measure at a given time and uses the information on the context of use security criticality to determine the assurance (confidence) on the security measure to meet its objective; (b) a methodology for the evaluation of security assurance of operational IT systems that takes into account emerging requirements for a security measure. With respect to the first contribution, we have linked a system security criticality level to the security level required for its protection and subsequently used probability theory, namely normal law of probability, to quantitatively evaluate the confidence that the security measure can adequately protect system assets in a specific context of use. As for the second contribution, we have extended Bugyo by making it more adaptable to changes occurring within the system security requirements through the integration of the vulnerability check dimension, which ensures that the conformance check is performed against an up to date security requirement specifications.

Our approach exhibits the following advantages: end-users' usage of IT systems is multipurpose. Therefore, providing them with continuous monitoring so to get some assurance on the security measures meeting their objective in a specific context of use, would help determine which activities can be performed (at a given time) with confidence, and those not to undertake or to be performed with more caveats. Moreover, providing assurance indicators on the correctness posture of security measures could assist security managers in the management of their systems security since a drop in security assurance level would imply that a component security posture is no longer compliant with the security requirements specifications. The consequence being that, ill-intentioned individuals might exploit such vulnerability to inflict damage. Our methodology and tool may therefore help practitioners identify areas of the security measures which need attention and address them before it is too late.

CATEGORIZATION AND DETERMINATION OF THE SECURITY CRITICALITY OF IT SYSTEMS' CONTEXT OF USE

IT systems and devices may be used in different environments for different purpose. Different environments will require different level of security and so will different purposes of use. Although a system security criticality is taken into account when defining the security measures, the actual evaluation of the confidence level in those measures has so far been conducted without considering the criticality of the context in which the system is running. In fact, far from being static, the security criticality of a system may change overtime (depending on several factors including new or changed business objectives, new or updated regulations, new threats and so forth) while the in place security measures remain unchanged. Consider the following example: Alice may feel very confident in using an unsecured wireless connection for simple Internet browsing but that confidence would drop considerably if Alice was to use it for Internet banking. We could state that purpose 1 (web browsing) requires low security or that its *context security criticality level* (α) is low (mainly because any potential risk impact for that context will be relatively low for the user); whereas purpose 2 has a higher *context security criticality level*.

For a systematic determination of a system context security criticality, a classification scheme with clear security criticality levels is necessary. The Federal Information Processing Standards (FIPS) in its publication FIPS PUB 199 (Evans et al., 2004) establishes security criticality categories for both information and Information System (IS). According to the FIPS, the security criticality levels are based on the potential impact on an organization, should certain events occur that jeopardize the information and IS needed by the organization to accomplish its assigned mission,

protect its assets, fulfill its legal responsibilities, maintain its day-to-day functions, and protect individuals (Evans et al., 2004). It is worth mentioning that Criticality Assessment is an activity of Risk Assessment (RA) dealing with the impact of incidents. In fact most risk assessment/ management methodologies such as OCTAVE (Alberts & Dorofee, 2001), CRAMM (http://www.cramm.com/), steps will begin with the identification of the assets for a given organization or system. However, the major difference between the approach for criticality assessment presented in this paper and the one conducted by existing risk assessment/ management methodologies is that the latter focus on the asset while the security criticality valuation the paper purports to undertake is based on the context of use of the system. In fact context of use security criticality, as we purpose to evaluate, does not seek to determine what need to be protected but rather emphasize more on valuating " what if protection fails in a specific situation?" and examine its consequence on different socio-economical and technological aspects. An asset criticality evaluation is about valuating the importance of that entity for the continuity of an organization or individual's activities. In light of that, an asset may present different criticality depending on the context of use and subsequently, the level of security provided should match the need of each context.

Based on the definition of security criticality provided by the FIPS, the following categories for impact evaluation have been deduced:

- **Health and Safety:** Does the occurrence of an Information Security incident for a given context lead to death or injury?
- **Revenue or Finance:** Does the occurrence of incident for a given context lead to financial losses?
- **Organization or Individuals' Intangible Assets:** Does the occurrence of an incident in a particular context lead to mate-

rial loss or to intangible assets such as an individual or an organization's reputation or competitiveness?

- **Organization or Individuals' Activity Performance:** Does the occurrence of an incident lead to the degradation of an individual or organization's activity performance?

Each of these categories can then be reviewed and a security criticality level, as defined by the FIPS assigned. The possible levels for security criticality are as follows:

- **Low:** The loss of confidentiality, integrity, or availability (CIA) in a particular context of use could be expected to have *no* or a *negligible* adverse effect on organizational operations, professional obligations, individuals and other assets.
- **Moderate or Medium:** The loss of CIA could be expected to have a *limited* adverse effect on organizational operations, professional obligations, individuals and other assets.
- **High:** The loss of CIA could be expected to have a *serious* adverse effect on organizational operations, professional obligations, individuals and other assets.
- **Very High:** The loss of CIA could be expected to have a *severe or catastrophic* adverse effect on organizational operations, professional obligations, individuals and other assets.

We aim at providing quantitative rather than qualitative security assurance level. Thus, the value of the system security criticality (we note α) could be in the range [0,1] or expressed in terms of percentage. In that perspective, the above defined qualitative security criticality level can be quantitatively classified as follows: *Low: $0 \leq \alpha < 0.25$; Moderate: $0.25 \leq \alpha < 0.5$; High: $0.5 \leq \alpha < 0.75$;*

Very high: $0.75 \leq \alpha \leq 1$. For a given context of use, there may be several impacts for the organization or the individual that may be considered as qualitatively the same (moderate, for instance), but with different relevance. Thus, associating security criticality levels to a range of values could serve such purpose. Importantly, we consider that a system with a *context security criticality level α* means that the security measures deployed for its protection should have an effectiveness level greater or equal to α or that that the system security level should be at least equal to α; provided the same classification granularity is used for security level and security criticality. Such idea is substantiated by the work of Holstein (2009). This implies that, for instance, in a context where $\alpha=0$, the security measures are required to have an effectiveness level ($\alpha >= 0$) i.e., the security measures working or not is irrelevant since the system is supposedly in a context where no IT security risks exist.

The determination of a system context security criticality is based on answering valuable questions for the individual or the organization. The questions should cover all the four categories previously reviewed. Rather than adopting a weighting system of the possible categories to account for their relevance for the individual or the organization, *the highest security criticality of the four categories will determine the security criticality of the system context.* Tables 1, 2, 3, and 4 provide detailed questionnaires proposed for the determination of the security criticality for each category. The questions have been identified based on the definition of security criticality and level of criticality provided by the FIPS. In addition, specific questions can be inserted by individuals or organizations to better suit their case, as it can be seen in Tables 1 through 4. For each category, the process of determining the security criticality level is as follows: When a "Yes" is given to a question (starting from up to down), the corresponding level should be noted and the

Table 1. Health and safety-related questions

Question	Classification Level
Does the occurrence of an Information Security incident for the particular context lead to death?	Very high
Organization/individual specific question here.	Very high
Does the occurrence of an Information Security incident for the particular context lead to permanent disability?	High
Organization/individual specific question here	High
Does the occurrence of an Information Security incident for the particular context lead to minor injuries?	Moderate
Organization/individual specific question here	Moderate

Table 2. Finance and revenue-related questions

Question	Classification Level
Does the occurrence of an Information Security incident for the particular context lead to severe financial losses?	Very high
Organization/individual specific question here	Very high
Does the occurrence of an Information Security incident for the particular context lead to considerable financial losses?	High
Organization/individual specific question here	High
Does the occurrence of an Information Security incident for the particular context lead to some minor financial losses?	Moderate
Organization/individual specific question here	Moderate

Table 3. Intangible assets-related questions

Question	Classification Level
Does the context involve the production or handling of date considered very sensitive for the individual or the organization?	Very high
Organization/individual specific question here	Very high
Does the context involve the production or handling of data that may affect the individual's reputation, privacy or competitiveness in case of security incident?	High
Organization/individual specific question here	High
Does the context involve data that may affect the company or individual's reputation, privacy or competitiveness locally in case of a security incident?	Moderate
Organization/individual specific question here	Moderate

Table 4. Activity performance-related questions

Question	Classification Level
Will the organization/individual's activity be halted in case of a security incident for that particular context?	Very high
Organization/individual specific question here	Very high
Will the organization/individual's activity (volume/quality) immediately be affected as a result of an IT security incident in that particular context of use?	High
Organization/individual specific question here	High
Will the organization/individual's activity (volume/quality) be affected over time as a result of an IT security incident in that particular context of use?	Moderate
Organization/individual specific question here	Moderate

next category of questions should be reviewed. There is no need to answer the remaining questions within the same category. Otherwise, if no "Yes" answer is recorded at the end of a category the corresponding security criticality level will be considered as "Low". A key assumption on the determination of the security criticality level using the questionnaires in Tables 1 through 4 is that there is a process for reaching a consensus in case of disagreement amongst those involve in answering the questions about the potential impact of a security incident. In that respect, the answer provided for each question is the result of such consensus.

It is important to note that the overall approach to determine security criticality based on the category and questionnaires has been primarily inspired by the OLF guideline (OLF, 2009), which was developed by professional and government experts. A key reason as to why it was adopted lies on the level of details it provides and the alignment of its classification scheme with the definition of the FIPS.

The ultimate aim of the OLF guideline was to provide guidelines to operators and suppliers within the oil and gas industry on the Norwegian Continental Shelf on how to perform classification of Process Control, Safety and support ICT Systems based on the systems criticality.

STEPS OF THE SECURITY ASSURANCE EVALUATION METHODOLOGY

The security assurance evaluation methodology was mainly inspired by the Bugyo methodology to evaluate, maintain and document the security assurance of telecom service (Bulut et al., 2007). However, it also distinguishes itself from Bugyo by considering assurance level as dependent on the context of use a system. The security assurance evaluation methodology consists of the following steps: Assurance components modelling, Specification of the security verification requirements; Verification of the component security posture; Assurance level aggregation; Display assurance level and monitoring.

Assurance Components Modeling

Modeling consists of determining and modeling assurance relevant components. This means that the model does not reflect the whole system

but only critical elements that are important and need to be security assured towards a system's well functioning continuity. An efficient way of identifying those critical components is an a priori use of a RA methodology. Weights can be assigned to each component to account for their respective impact on the overall system that is decomposed in hierarchical way.

Specification of the Security Verification Requirements

This step involves mainly specifying, for each assurance relevant component, the probe to conduct the verification of the correctness of a component and the frequency of the verification. A probe may be an existing security tool (such as firewall tester, IDS, system vulnerability scanner and so forth) or self developed programs to audit the operational system. The quality level (QL) of those probes affects the value of the security assurance. In fact a correlation exists between the quality of the probe used and one's confidence (assurance) in the check result achieved. Highly qualitative probes will provide higher confidence in the verification outcome. Consider for instance the following: Imagine you are using two different anti viruses AV1 and AV2 for checking whether your system is currently immune of any type of malware. Assuming you know from previous use of AV1 that its effectiveness to detect a malware is somehow dubious, whereas AV2 has proven to be very reliable in terms of malware detection precision. If you were to use AV1 and then AV2, and the verification outcome is "no virus found", you will certainly be relieved. However, your confidence in the actual result will differ depending on whether you have used AV1 or AV2. In fact, one's assurance (confidence) in the system security posture after using AV1 will be lower compare to the use of AV2. In view of addressing such eventuality; we developed a verification process quality metric taxonomy, based on the Common Criteria ATE:

test class and the Systems Security Engineering Capability Maturity Model (SSE-CMM), to be reflected in the correctness verification result.

The SSE-CMM-like tailored levels are used to define the different quality levels possible for probes evaluating a security measure while some of the Common Criteria EAL families serve as the requirements for being assured at some level of quality. The underlying concept of this process is that high assurance can be consistently attained if a process exists to continuously measure and improve the quality of the security evaluation process. The Common Criteria (CC) philosophy of assurance helps in identifying some of the quality requirements pertinent to assurance. As a matter of fact, the CC philosophy asserts that greater assurance results from the application of greater evaluation effort, and that the goal is to apply the minimum effort required to provide the necessary level of assurance. The increasing level of effort is based upon: *Scope or coverage*: The effort is greater because a larger portion of the IT system is included in the verification; *Depth*: The effort is greater because it is deployed to a finer level of design and implementation detail; *Rigor*: The effort is greater because the verification is applied in a more structured, formal manner. We use the five capability maturity levels of the SSE-CMM, which we tailored to represent the verification probe quality levels and some of the CC families (Scope, Rigor, Depth and Independent verification) as the minimum requirements to fulfill to be at some level of quality. Table 5 reviews the quality levels and their associated description. The matrix shown in Table 6 expresses the minimum requirements to achieve certain quality level.

Table 5. Probe quality level and description

Level 0: Not Performed	The quality of the verification process is unknown.
Level 1: Performed Informally	The verification is not rigorously undertaken nor planned and tracked. A human expert who relies on his own knowledge of the security measure may perform it.
Level 2: structurally performed	The evaluation process conforms to some specified standards or procedure and requirements with provision of appropriate tools to perform the process.
Level 3: Structured and Independent verification	Verifications are performed according to a well-defined process using approved standard or tools provided by third party.
Level 4: Semi complete verification	The verification follows a well-defined process with a usage of software tools that cover most of the relevant part of the security measure.
Level 5: Complete verification	The maturity of the verification is such that all known relevant part of the security measure are investigated appropriately in depth as well as in breadth.

Table 6. Probe quality metric taxonomy

Class	Family and Meaning	Quality Level: QL				
		1	2	3	4	5
QAM: Probe Quality Metric	QAM_COV: Coverage (Larger coverage of the verified security measure provides more confidence on the results about its status)	1	2	2	2	3
	QAM_DPT: Depth (A detailed verification of the security measure will decrease the likelihood of undiscovered errors.)	1	2	2	3	4
	QAM_RIG: Rigor (The more structured the evaluation of the deployed security measure, the more reliable the outcome of the verification)	1	2	2	3	3
	QAM_IND: Independent Verification (verification performed by a third party evaluator or software tool provides more assurance)	1	1	2	2	3

A probe satisfies quality level 3 (QL3), for instance, if at least the following requirements are met: QAM_COV.2 (Only some of the key areas of the security measure, known to be relevant for its well functioning are verified in the process.), QAM_DPT.2 (High level verification of the security measure through its interface.), QAM_RIG.2 (Semi-structure verification: A clear verification procedure exists for the verification.), QAM_IND.2 (Partial verification with independent means).

A detailed specification of the verification quality taxonomy has been submitted to Springer Journal of Software Quality (Ouedraogo et al., 2010). Nonetheless, interested readers may refer to the earlier version of the taxonomy (Ouedraogo et al., 2009) for further information.

Verification of the Components Security Posture

Once the metrics have been specified, the next step consists in performing the verification of the security measures through probes available in the system. Those probes will carry out the metric specifications or base measures, which provide information on the nature of the verification, the targeted component, the frequency of the verification and so on. The verification process of a security measure with respect to conformity check is depicted in Figure 2. It shows that verification probes verify implemented security measures by performing some base measures in accordance with *ISO/IEC 27004:2009* terminology (International Organization for Standardization, 2009). An interpretation of the results of the base measures is then performed i.e., compared to a reference which can be a mandatory or recommended configuration file, a best practices database, and the like; to inform on whether a security measure has been

Figure 2. Security measure conformity verification process

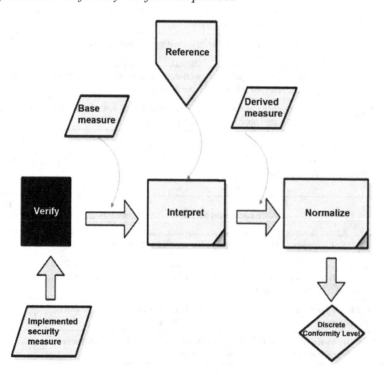

properly deployed. The result of the interpretation is a derived measure that will be normalized (associated a discrete value) to inform on the posture of the security measure. Expert knowledge is required to associate to each possible conformity mismatch a discrete number ($C \in [0, QL]$) which is correlated to the gravity of the mismatch for the system security. 0 will mean that nothing is conformed to the reference or that the mismatch is of a high risk and QL (probe quality will be used when there is a maximum match or mismatch with the lowest risk for the system security.).

As for Availability checks, the normalized value is Boolean: "0" to signify security measure not present and "1" for security measure present and working.

Assurance Level Aggregation

As previously mentioned, the confidence in a security-enforcing mechanism to adequately protect system assets, in a specific context of use, is dependent on whether the concerned security measure is present and properly implemented. For the sake of verifying each of the deployed security measure, dedicated software probes are used. As shown in Figure 3, probe agents collect results from probes and proceed to a first aggregation aiming at determining the overall indicator value for each of the assurance parameters (availability, conformity). Possible aggregation functions (Agf) are described at the end of this section. Such an aggregation is particularly necessary since there may be need to verify more than one parameter (with different importance) of the security measure to account for its status with respect to each category of checks. Once that aggregation is completed, the actual value of the security measure security assurance is determined using a mathematical function. As a matter of fact, the security assurance of a security measure *(S)* with a conformity level *c* with respect to the specified security requirement, working in a context with a security criticality level α is a function *SAL: $R^3 \rightarrow [0,1]$* taking as input a 3-tuples of values on that are: *Availability (A), Conformity (c) of (S) and the Context Security Criticality α, and* produces a real value within the range [0,1]. Some important elements to take into account in defining the assurance function are: (1) No assurance can be gained

Figure 3. Security assurance aggregation process

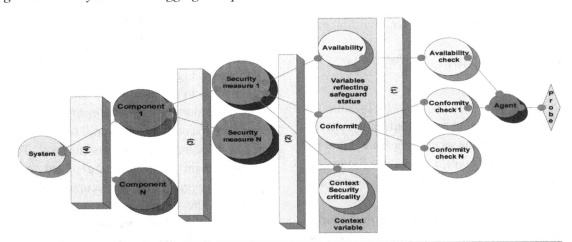

(1) : Obtain overall values for Availability and Conformity using an appropriate aggregation fucntion Agf
(2): Determine safeguard security assurance using the function   A*(1-α^c)
(3): Determine a critical component security assurance by aggregating the values of assurance for its associated safeguards
(4): Use an appropriate aggregation function (Agf) to determine the overall security assurance for the system

if the evaluated security measure is not "available" i.e., present and working *(A=0→SAL=0)*; (2) A complete lack of conformity of the security measure with respect to the reference provided by the security requirements will result in no confidence (c=0→ SAL=0); (3) In the event that the system is working in a context with a security criticality level α= 100% or 1, no assurance can be gained since no security measure in practice can guarantee over 100% security.

An interesting mathematical characterization closer to the requirements above specified can be derived from probability theory, namely from normal law of probability. Let α be the estimated security criticality of a system context of use and C be the value of conformity between deployed security measure and the security policy specification. The assurance that the security measure will exhibit an effectiveness level at least equal to α given its current conformity level C is given by the following formula:

$$1 - á^c \tag{1}$$

Given that the non-deployment or non availability of a security measure would result in no assurance at all since the security measure is not prompt to advert security risks, we extend (1) by adding the availability parameter.

$$\boldsymbol{SAL(S)} = A*(1 - á^c) \tag{2}$$

with $\alpha \in [0,1]$, $c_i \in [0,QL]$, $A=\{0,1\}$ and c= Agf$\{c_i\}_1$ <=ci<= number of conformity checks performed within the security measure. For a fix value of α (same context criticality- Figure 4a) the higher the conformity level, the higher the confidence in the security measure.

One obvious way of increasing the security assurance level through the value of c, would be to use a probe with a high quality level. When one switches to a context with higher security criticality (Figure 4b), the confidence level is still

positively correlated to the conformity level, but drops sharply. The value of the security assurance for a value of c=QL (probe quality level) is called the achievable security assurance level. It represents the highest assurance level one could expect while using a given probe. The value of the security assurance for c=QL also reflects the well functioning of the security measure. Therefore it should be displayed along with the actual value of the security assurance to help the operator appreciate the gap between the two values and subsequently take appropriate actions to get closer to the achievable assurance level.

A system in general, a critical component in particular is normally associated with several security measures protecting it against specific threats. The security assurance of such component or system requires therefore the consideration of the security assurance levels of all its associated security measures. An aggregation function (Agf) is therefore necessary. Equation (2) can be generalized as follows:

$$SAL(Component) = Agf \{SAL (SM_i)\} \tag{3}$$

with 1<=i<= number of associated security measures, and where SAL (SM$_i$) represents the security assurance level of an associated security measure. Subsequently, the security assurance level of a critical system or service can be obtained by aggregating the assurance level of its critical components. Figure 4 shows the sequence of aggregation up to a system level. The aggregation process is undertaken, using a linear or non-linear algorithm, to compute the overall security assurance at infrastructures and system level. There are three main algorithms (Agf) used for the operational aggregation: the recursive minimum algorithm, the recursive maximum algorithm and the recursive weighted sum algorithm. The recursive minimum algorithm, applied to systems with several critical points, implies that the overall assurance of a service is represented by the lowest level of its components. One potential

Figure 4. Assurance fluctuation (a) with conformity level; (b) with the context

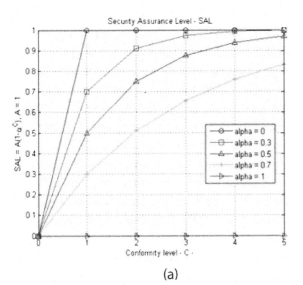

(a)   (b)

usage of this model can occur when the security assurance evaluation system is deployed at the first time or when operators want to deploy assurance metrics but without any strong idea on relative importance of each component taking part in the system or service. The recursive maximum algorithm, however, commands that the assurance of a system or its component be the highest of its components metrics. The recursive average weighted sum algorithm is used for systems composed of security functions that contribute in different, non-critical manner (the failure of one security function does not lead to the complete failure of the system security) to service security implemented by independent observed system parts. It uses weight properties model to calculate the assurance level value at each level of the assurance model of a service and reflects more precisely the assurance needs described in the model. The recursive weighted average sum algorithm, although more complex, can be considered to be the most optimal as it does not require powerful computation and can monitor any minor change in the infrastructure.

Display and Monitoring of Assurance Level

Once the overall value of the security assurance level has been estimated, an almost real time display of security assurance of the system is performed. Furthermore, a comparison between the current value of the security assurance level and the achievable security assurance level will result in an appropriate message being displayed in case of mismatch between the two. This can help the security manager identify causes of security assurance deviation and also assist him/her in making decisions.

Security Measure Vulnerability Check

As underlined in the introductory section of the paper, the ultimate aim of vulnerability monitoring is to ensure best practices in terms of deployment and implementation of a security measure are permanently applied. The vulnerability monitoring may impact directly on the system security model either through recommendation of new topology for the system security model; removal, addition or upgrade of a component within the current model. The impact could also be on step two of the methodology i.e., specification of the security verification requirements, in a sense that new guidelines on how to tackle a given security risk vulnerability will result on the consideration of a new reference against which the verification of the security measure posture must be performed. Those guidelines and principles are obtained from the vulnerability databases such as the NVD.

ARCHITECTURAL CHOICE FOR THE SECURITY ASSURANCE EVALUATION METHODOLOGY

Given the highly distributed nature of most current systems, verification of the security measures is more challenging due to issues such as concurrency, fault tolerance, security and interoperability. Multi-agent systems (MAS) (Wooldridge, 2002) offer interesting features for verifying the security of such systems. In our work, we consider an agent as an encapsulated computer system that is situated in some environment and that is capable of flexible, autonomous action in that environment in order to meet its design objectives (Jennings, 1999). As agents have control over their own behavior, they must cooperate and negotiate with each other to achieve their goals (Jennings, 1999). The convergence of these agents' properties and distributed systems behavior makes MAS architecture an appropriate mechanism to evaluate the security assurance of critical infrastructures run by distributed systems. In our framework, the security assurance evaluation approach has been implemented using Jade (http://jade.tilab.com) with an agent organization involving a hierarchy of three types of agents: server agents (embedded in the server), multiplexer agents, or MUX-agent (For huge and multi domains systems with dedicated firewalls, crossing each firewall every time a check is needed is not recommendable. Thus,

MUX agents can be defined, for each sub-domain, at firewall level to relay the information to probe agents); and probe agents (agents triggering a probe or collecting information from probe during assurance evaluation). Importantly, although an XML based format is used for the message between server agent and MUX agents, the message format between probe agents and probe is specific to the probe. (Bulut et al., 2007)

Thus the message from the probe agent has to be transformed to a format understandable by the probe. The following scenario and Figure 5 (Bulut et al., 2007) describes how the agents are hierarchically organized and interact when conducting a security assurance measurement: The Server-Agent receives a request including the list of verifications to perform or base measures (BM1, BM2) with the targeted network elements (1). It consults the roles directory for determining the sub-domains to which the targeted network elements belong and also the MUX-Agent (2). Then, it sends the request to the concerned MUX-Agent (3). The MUX-Agent dispatches the base measures request (4, 5) after determining which Probe-Agents can perform these base measures. The Probe-Agents receive the request intended for them, launch the measurement at the probe level (6), get the result and format it in a well defined agent-messaging format (7) and send it to the MUX-Agent (8). The MUX-Agent collects all the measurement results and sends them to the Server-Agent (9). The Server-Agent collects all the measurement coming from the MUX-Agent or generally from all the MUX-Agents, aggregates them before generating the response (10).

APPLICATION TEST-BED

For the purpose of illustrating the security assurance evaluation approach, let us consider the following scenario, brief summary of a larger case study used in a technical report:

Figure 5. Agents interaction scenario

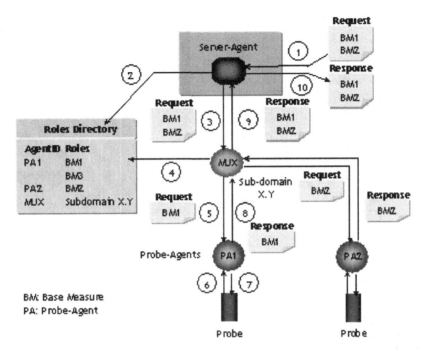

Bob is an employee who frequently uses his office IT networks for other activities such as online banking, personal emails check and so on. A critical component part of the network and relevant to Bob is the DNS server. In fact, an illicit modification of the DNS configuration can cause the transmission of malicious information and affecting the security of the DNS dependent elements. This may result in him being tricked while visiting his bank website and subsequently give his banking details on a replica of the website controlled by fraudsters.

We propose to evaluate the security assurance of the DNS server taking into account the context of use of the system by Bob. For that, we consider a scenario where errors have been injected in the address resolution file in order to corrupt the DNS bind9. An illicit modification of the DNS configuration can cause the transmission of malicious information and affect the security of the DNS dependent elements. In this scenario, the Samhain (http://www.la-samhain.de/samhain), an open source host-based intrusion detection system using cryptographic checksums of files to detect modifications, is used as probe for the verification. An effective functioning of that probe helps detect the address resolution files integrity being corrupted as a result of a malicious attack. Furthermore, self-developed scripts have been used to verify whether the DNS file was well constructed. Following the steps of the methodology discussed in the previous section, the following analysis is performed.

Step 1: Assurance Components Modeling

Figure 6 provides an overview of the system model and the assurance relevant component, particularly the DNS.

Step 2: Specification of the Metrics

Based on information obtained from the Samhain documentation and by considering the quality metric taxonomy in Table 6, we derived the following conclusions:

Figure 6. A simple representation of Bob's workplace network

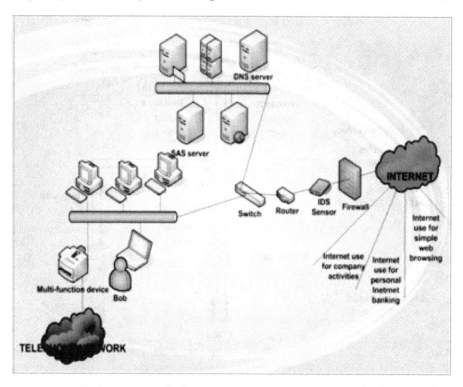

- **Coverage:** The coverage of the Samhain measurements satisfies QAM_COV.2. In fact the measures only represent a static behavior of the service and not a dynamic network view (in and outgoing flows from and to the DNS server).
- **Depth:** The measures undertaken by the Samhain target the address resolution file. This is good because a missing address resolution file or a bad content is relevant for the correct DNS behavior. This satisfies at least QAM_DPT.3.
- **Rigor:** Samhain is a dedicated open source integrity check software tool (QAM_RIG.3). A recent version (v2.4.5) was used with a continuous evolution of the dictionary.

The main weaknesses of the DNS controlled Independence of Verification (QAM_IND.3); the Samhain is a third party tool and although scripts

were also used for the verification they were not developed by the person who set up the DNS configuration.

The values of the Samhain (S) capabilities do not explicitly correspond to any of quality level of Table 1. Nonetheless, all the parameters of the Samhain (Coverage: 2, Depth: 3, Rigor: 3 and Independent verification: 3) are greater or equal to those of quality level 4, while some are lower than those of quality level 5. We can here conclude that the Samhain probe corresponds to quality level 4.

Step 3: Verification of the Components Security Posture

For undertaking the security assurance evaluation of the DNS, two probe agents have been launched for verifying the DNS both the DNS conversion file integrity and the conversion file construction. The verification frequency was set to 1 hour during which the agents were dispatched in the network to perform the evaluation. Table 7 describes the requirements for the DNS conformity verification.

Step 4: Assurance Level Aggregation- Determining the Security Assurance Level of the DNS Based on Bob's Experience

Availability value. Availability check only concerns security measures which have been implemented or installed, in the case of the DNS verification for instance, it is rather a case of security relevant parameter where altering some elements of its configuration could result in a security loophole. In such case the value of availability is by default set to 1.

Conformity value. Taking into account the quality level of the Samhain (QL=4) and the possible results obtain from the Samhain, the overall conformity level for the DNS (depending on the gravity of the security breach in the expert view) can be summarized as follows. If the address resolution files integrity is compromised:

- **Corrupted Files:** An evil-minded modification then the conformity level is 0 and a message is then triggered.
- **Configuration Errors:** the conformity level is 1 and an appropriate message is display.
- **Otherwise:** If everything is fine the conformity level is 4.

The above classification of the mismatch assumes that a configuration error is less serious than a malicious modification.

Determining the security criticality of the contexts of use for Bob: For the purpose of illustrating the security criticality determination process, we consider Bob's daily usage of his workplace system for his personal matters and provide estimates of the security criticality level of each of the context of use (Internet banking, emails checks and web browsing) using the questionnaire and the categorization of security criticality previously presented, the context of use security criticality analysis depicted in Tables 8, 9, and 10 led to the conclusion that the security criticality were as follows: Internet banking: *Very high,* Email check: *High and* Web browsing: *Moderate.* Determining the qualitative security criticality level is relatively easier compare to picking up a value within the different range of possible values assigned to each qualitative level. In fact except the Revenue/ finance, and activity performance categories where one could more or less accurately estimate the impact of a security incident in terms of percentage of losses, objectively deciding on a value for categories such as Intangible assets, and health and safety can be challenging.

As a matter of fact, we could objectively assert that should a security incident occur in the case of the Internet banking, Bob may incur a financial loss of up to 80% of his personal wealth (very high security criticality) and that his company's activity performance will not be directly affected 0% (Low security criticality). However, it would be difficult choosing a value to express the extent

Table 7. Requirements for the DNS conformity verification

Reference	Base Measures	Required Probes	Frequency
DNS conversion files integrity	Check the integrity of address resolution file.	Samhain	1 hour
DNS conversion files construction.	Files corrupted: Check if the resolution files content is well constructed.	Scripts	1 hour

Table 8. Determining the security criticality for internet banking

Context Description		Aggregated Security Criticality	
Internet banking		Very high	
Category	Security Criticality	Rationale	
Health and safety	Low	The context is not directly related to health and safety	
Revenue/finance	Very high	An IT security incident may result in financial losses for Bob. The severity of the loss may depend on Bob's financial capability and the time that he takes to realize something irregular is happening with his account. We here assume that the impact may be the highest i.e., severe or catastrophic for Bob.	
Intangible assets (Reputation, privacy, competitiveness,...)	Moderate	An IT security incident may result in the disclosure of Bob's banking information, which is a breach to his privacy.	
Activity performance	Low	The context has no direct impact Bob's other activity and especially on his professional activities.	

Table 9. Determining the security criticality for email checks

Context Description		Aggregated Security Criticality	
Email checks		High	
Category	Security Criticality	Rationale	
Health and safety	Low	The context is not directly related to health and safety	
Revenue/finance	Low	The context is not directly related to revenue and finances.	
Intangible assets (reputation, privacy, competitiveness, etc.)	High	An IT security incident may result in Bob's email password being snooped leading to personal message being viewed by third party, spams, and so on. The implication in terms of privacy is high.	
Activity performance	Low	The context has no direct impact on Bob's other activities and especially on his professional activities.	

Table 10. Determining the security criticality for web browsing

Context Description		Aggregated Security Criticality	
Web browsing		Moderate	
Category	Security Criticality	Rationale	
Health and safety	Low	The context is not directly related to health and safety	
Revenue/finance	Low	The context is not directly related to revenue and finances.	
Intangible assets (reputation, privacy, competitiveness, etc.)	Moderate	An IT security incident may result in the monitoring of the websites visited by Bob, which is a privacy issue. The privacy implication is here moderate compare to the email check case.	
Activity performance	Low	The context has no direct impact Bob's other activity and especially on his professional activities.	

to which his reputation, privacy will be affected. In the absence of expert judgments or agreement to help determine a value within the range of values for each level of security criticality, we recommend considering the median value for the interval. For example, the email checks context which has a "high" security criticality $(0.5 \leq <0.75)$ will be assigned a default value of $(0.5+0.75)/2 = 0.625$ or 62.5% while the web browsing context with a "moderate" security criticality $(0.25 \leq <0.5)$ will be assigned the default value of $(0.5+0.25)/2 = 0.375$ or 37.5%. In terms of security level, Internet banking will require at least a 0.8(since the required security level should at least equal to the security criticality), email checks will require at least 0.625 while web browsing will need a least 0.375, as effectiveness level for the in place security measures

Security assurance level. An application of the assurance function to Bob cases is provided in Table 11. The figures on the table show that the achievable assurance level (i.e., no errors

Table 11. Security assurance value for the DNS

DNS security assurance values SAL=A(1-α^c) with A=1 and depending on the context	Conformity Levels		
	0	1	4
Internet banking (α= 0.8)	0	0.2	0.59
Personal email checks (α=0.625)	0	0.375	0.84
Simple web browsing (α=0.375)	0	0.625	0.98

detected) for Internet banking is 0.59. In case of security loophole leading to a conformity level c=0 (corrupted files, evil-minded modification) the displayed assurance level is 0. Presence of configuration errors (c=1) gives an assurance level of 0.2.

Step 5: Display and Monitoring

Once the security assurance level has been evaluated, the value is displayed on the assurance evaluation system as shown in Figure 7. The current security assurance level is then compared to the

Figure 7. Security assurance monitoring cockpit

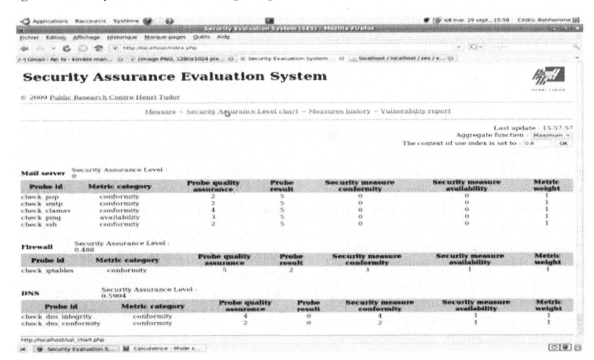

previous evaluated value. The operator can then use the assurance measurements history to find out whether the assurance level has experienced any drop, increased or is just static.

CONCLUSION AND FUTURE WORK

This paper has presented a methodology for the evaluation of security assurance of an IT system given the criticality of the context in which it is operating. A classification scheme that aims at helping to evaluate system context security criticality has also been provided. Our approach exhibit the following advantages: On one hand, by giving security assurance information that is dependent on the context of use, our metrics and methodology can help them in knowing which activities can be performed with more confidence and which one is advisable to refrain from doing based on the security assurance values. On the other hand, it could help security experts manage better deployed security measures by identifying areas of the system that need attention before a security loophole is exploited by malicious individuals. Nonetheless, the categories for criticality assessment proposed still require some investigation so to ensure they are complete enough to be use for most systems.

During the application of the methodology, we realized that a sound knowledge of the system components is imperative for a better specification of the metrics which in turn will guarantee realistic values for security assurance.

We currently envisage a wider application of the approach to a variety of system in view of its improvement. Moreover, we plan to integrate the approach with network management tools and risk assessment software and to simulate real world complexity. Another future direction, would be the development of self-learning agents (for performing the measurement) capable of reasoning and capable of dynamically adapting to network different network context.

ACKNOWLEDGMENT

This work has been supported by the TITAN project, financed by the national fund of research of the Grand Duchy of Luxembourg under contract C08/iS/21.

REFERENCES

Alberts, C. J., & Dorofee, A. J. (2001). *OCTAVE criteria, version 2.0* (Tech. Rep. No. CMU/SEI-2001-TR-016). Pittsburgh, PA: Carnegie Mellon University.

Bulut, E., Khadraoui, D., & Marquet, B. (2007). Multi-agent based security assurance monitoring system for telecommunication infrastructures. In *Proceedings of the Fourth IASTED International Conference on Communication, Network and Information Security* (pp. 90-95).

Common Criteria. (2006). *Common criteria for information technology, part 1: Introduction and general model version 3.1*. Retrieved from http://www.commoncriteriaportal.org/files/ccfiles/CCPART1V3.1R1.pdf

Evans, D. L., Bond, P. J., & Bement, A. L. (2004). *Standards for security categorization of federal information and information systems*. Gaithersburg, MD: NIST.

Holstein, D. K. (2009). A systems dynamics view of security assurance issues: The curse of complexity and avoiding chaos. In *Proceedings of the 42nd Hawaii International Conference on System Sciences* (pp. 1-9).

International Organization for Standardization. (2009). *ISO/IEC 27004: Information technology - Security techniques - Information security management measurements*. Geneva, Switzerland: International Organization for Standardization.

Jansen, W. (2009). *Directions in security metrics research* (Tech. Rep. No. NISTIR7564). Gaithersburg, MD: National Institute of Standards and Technology.

Jennings, N. R. (1999). An agent-based software engineering. In *Proceedings of the 9th European Workshop on Modelling Autonomous Agents in a Multi-Agent World.*

Klevinsky, T. J., Laliberte, S., & Gupta, A. (2002). *Hack I.T.—Security through penetration testing.* Reading, MA: Addison-Wesley.

Le Grand, C. H. (2005). *Software security assurance: A framework for software vulnerability management and audit.* Longwood, FL: CHL Global Associates and Ounce Labs, Inc.

Mouratidis, H., & Giorgini, P. (2007). Secure Tropos: A security-oriented extension of the tropos methodology. *International Journal of Software Engineering and Knowledge Engineering, 17*(2), 285–309. doi:10.1142/S0218194007003240

OLF. (2009). *OLF Guideline No 123: Classification of process control, safety and support ICT systems based on criticality.* Retrieved from http://www.olf.no/Documents/Retningslinjer/100-127/123%20-%20Classification%20of%20process%20control,%20safety%20and%20support.pdf?epslanguage=no

Ouedraogo, M., Mouratidis, H., Khadraoui, D., & Dubois, E. (2009). A probe capability metric taxonomy for assurance evaluation. In *Proceedings of the UEL's AC&T Conference.*

Ouedraogo, M., Savola, R., Mouratidis, H., Preston, D., Khadraoui, D., & Dubois, E. (2010). Taxonomy of quality metrics for security verification process. *Journal of Software Quality.*

Savola, R. M. (2007). Towards a taxonomy for information security metrics. In *Proceedings of the International Conference on Software Engineering Advances,* Cap Esterel, France.

Seddigh, N., Pieda, P., Matrawy, A., Nandy, B., Lambadaris, L., & Hatfield, A. (2004). Current trends and advances in information assurance metrics. In *Proceedings of the Conference on Privacy, Trust Management and Security* (pp. 197-205).

Stoneburner, G. (2001). *Underlying technical models for information technology security.* Gaithersburg, MD: National Institute of Standards and Technology.

Strunk, E. A., & Knight, J. C. (2006, May 23). The essential synthesis of problem frames and assurance cases. In *Proceedings of the Second International Workshop on Applications and Advances in Problem Frames.*

Swanson, M., Nadya, B., Sabato, J., Hash, J., & Graffo, L. (2003). *Security metrics guide for information technology systems* (Tech. Rep. No. NIST-800-55). Gaithersburg, MD: National Institute of Standards and Technology.

Vaughn, R. B., Henning, R., & Siraj, A. (2002). Information assurance measures and metrics – state of practice and proposed taxonomy. In *Proceedings of the IEEE International Hawaii Conference on System Sciences* (p. 331.3).

Wool, A. (2004). A quantitative study of firewall configuration errors. *Computer, 37*(6), 62–67. doi:10.1109/MC.2004.2

Wooldridge, M. (2002). *An introduction to multi-agent systems.* New York, NY: John Wiley & Sons.

This work was previously published in the International Journal of Handheld Computing Research (IJHCR), Volume 2, Issue 4, edited by Wen-Chen Hu, pp. 59-81, copyright 2011 by IGI Publishing (an imprint of IGI Global).

Section 2
Mobile Evaluations and Analyses

Chapter 6
Modeling and Analyzing User Contexts for Mobile Advertising

Nan Jing
Bloomberg L. P., USA

Yong Yao
IBM Silicon Valley Lab, USA

Yanbo Ru
Business.com Inc., USA

ABSTRACT

Context-aware advertising is one of the most critical components in the Internet ecosystem today because most WWW publisher revenue highly depends on the relevance of the displayed advertisement to the context of the user interaction. Existing research work focuses on analyzing either the content of the web page or the keywords of the user search. However, there are limitations of these works when being extended into mobile computing domain, where mobile devices can provide versatile contexts, such as locations, weather, device capability, and user activities. These contexts should be well categorized and utilized for online advertising to gain better user experience and reaction. This paper examines the aforementioned limitations of the existing works in context-aware advertising when being applied for mobile platforms. A mobile advertising system is proposed, using location tracking and context awareness to provide targeted and meaningful advertisement to the customers on mobile devices. The three main modules of this comprehensive mobile advertising system are discussed, including advisement selection, advertisement presentation, and user context databases. A software prototype that is developed to conduct the case studies and validate this approach is presented.

DOI: 10.4018/978-1-4666-2785-7.ch006

1. INTRODUCTION AND MOTIVATION

Online advertising constitutes a large portion in the financial ecosystem of web sites nowadays, including search engines, commercials, blogs, news, reviews etc. Driven by recent Internet revolution and the tremendous increases in on-line traffic, a huge growth in spending on online advertising is seen in last few years. eMarketer in 2007 reported a total Internet advertising spending of nearly 20 billion US dollars, just in 2007. This number supports the World Wide Web (WWW) to be amongst the top 3 advertisement medium, along with TV and print media. In these online advertisements, contextual advertising is a main category that we have identified in providing the advertising content matching the keywords of the user searches or the content of the web pages where the advertising content will be placed. How to optimize the advertising content in this method is always an important research topic with the dual goals of increasing revenue of both publisher and advertising business. However, if we check most of the current advertisements, the majority of these advertisements are either serendipitous or solely dependent on the keywords of users' searches, both forms of advertising where all the users who perform the same search are deemed identical and are thus shown the same advertisements. These approaches of selecting advertisement mostly result in the advertisement banners that we have seen on various web sites and are trying to sell the products that we will never buy or even look for details. A superior online advertising approach should always provide the customers with information that match their contexts and interests as much as possible (Chatterjee, 2006). Every possible factor that contributes to understand the interests and contexts of the existing and potential customers should be appropriately captured and carefully investigated.

Meanwhile, mobile computing technologies have profoundly transformed the way that people communicate and receive information from various media including WWW. With mobile devices becoming more powerful and affordable, the user base has expanded from the early business elites to ordinary people. Consequently, mobile information access is gaining widespread prominence with improving connection speed and access technologies leading to richer content explosion and user experience. The addition of mobility has opened up new prospects as devices are expected to be with users at all times providing reliable information on user interests and contexts. In turn, the next generation of mobile computing applications should leverage mobility of devices combined with advanced approaches to provide more customized information and, at meantime, more targeted advertisement. Mobile advertising, which is a cross area of mobile computing and online advertising, is a form of advertising that targets users of handheld wireless devices such as mobile phones and PDAs. In comparison with traditional advertising, mobile advertising can be more location-dependent, user-targeted, and device-customized. Mobile advertising can reach the target customers anywhere, anytime; customers can be aware of mobile advertisements and services while they are walking. Location tracking in our opinion is most important in such scenarios and so is worthy of being studied distinctively before other contexts. Using demographic information collected by mobile devices in the current locations of customers, much targeted advertising can be done. Such advertisements can also inform a user about ongoing specials (such as restaurants, hotels and malls) in surround areas. The messages can be sent to all the users located in a certain area or to the users who previously subscribe to certain advertising messages. Furthermore, combing with the mobile user context, including user profile (such as a special preference on certain kinds of products) and user's previous activities (such as clicks on the advertisement belonging to one particular seller), advertising companies can provide the customers more targeted and acceptable advertisements at the exact time the customers may need these advertisements and at the exact location they may find it close to use the contact information in the advertisement, not just "spam" them with advertisements they will never buy.

As the user profile may be obtained properly from the registration record and user location can be provided by the GPS unit in the mobile device or the cell station of the carrier network, it is understandably difficult for a computer to judge the preference of a user and predict what advertisement may be of her favor at a specific time in a specific location. Recent mobile computing research is investigating how to collect and analyze the contexts of user activities in order to predict users' preferences (Bardram, 2004). Recently an interesting trend has emerged in building recommendation systems that leverage the opinions of other users to make predictions for the user, by utilizing a methodology, namely collaborative filtering. It is a technology that has emerged in e-Commerce applications to produce personalized recommendations for users (Schafer, 1999). It is based on the assumption that people who like the same things are likely to feel similarly towards other things. This has turned out to be a very effective way of identifying new products for customers. It works by combining the opinions of people who have previously expressed inclinations similar to one user to make a prediction on what may be of interest to the user now. So far, collaborative filtering has mostly been applied to online applications, where the user contexts are rarely available as discussed earlier. On mobile devices, a user's decision can be influenced by many things in the surrounding context, such as, locations, weather, and device capabilities. The user preferences are undoubtedly bound with then context. Some context, like location, may be much more important than others, in determining the similarity of user activities and preferences. There are existing research regarding context-aware collaborative filtering, however, they either failed

at utilizing context information in the calculating the user similarity (Lawrence, 2001), or did not clearly specify what contexts should be included and how to specially deal with most important context, such as location, in the mobile computing applications (Chen, 2005). Therefore, an improved approach based on collaborative filtering is needed to provide well designed and illustrated solutions to select and present targeted advertisements on mobile devices by modelling and analysing the user contexts.

This paper describes a context-aware mobile advertising approach, which engages a comprehensive and operational process, based on context awareness and collaborative filtering, to select and present targeted advertisement from the raw contents provided by advertising providers. It also defines the types of mobile user contexts which should be included and investigated in the aforementioned selection and presentation process. This paper presents a prototype system developed based on the aforementioned approach and discusses the case studies that were undertaken using this prototype to validate this approach. As such, the rest of this paper structures as follows: Section 2 reviews a few schools of studies which are relevant to this work. The context-aware mobile advertising approach is described in Section 3. Section 4 presents the software prototype and case studies. Finally Section 5 concludes this paper and outlines the open issues that are to be addressed to extend and improve this approach.

2. BACKGROUND REVIEW

As mentioned in Section 1, our challenge is to develop a context-aware mobile advertising approach using the collaborative filtering approach. To put this discussion in perspective, this section reviews a variety of studies in relation to context awareness, Internet-based advertising, and collaborative filtering.

2.1. Context and Context Awareness

Context

In general context means situational information. One of its popular definitions (Dey, 2001) is "any information that can be used to characterize the situation of an entity. An entity is a person, place or object that is considered relevant to the interaction between a user and an application, including the user and applications themselves." In the studies we have reviewed, there are mainly two ways of using contexts in software applications. First, applications can optimize their outputs according to the contexts. Major search engines using the keywords and web page content to provide more targeted advertising content fall in this category. Second, the context information can be used to create new types of applications, such as location-based applications. In these studies, context is often categorized into physical context representing the environment of the activity and logical context representing more abstract information about the stakeholder and the application. Physical context properties are at a very low level of abstraction. They are continuously updated to take into account the fact that the state of the stakeholder and the application continuously changes, such as spatial and temporal information. Logical context information is needed to enrich the semantics of physical context information (e.g., stakeholder's preferences) thus making it meaningful for high-level purposes (e.g., stakeholder's visits to certain locations) (Kappel, 2002). Theoretically any information available in the course of an interaction can be used as context information, such as time of the interaction, user identity, application status. In our research, the focus is the context information that is useful and critical to determine the context-aware advertising content in the mobile applications.

We acknowledge that context has no standard definition, since every school of study can give their understanding about context to a valid pur-

pose. However, in a particular area such as mobile platform, the target of using context is to better serve users by providing needed information to these users on mobile devices. Classification of contexts should embody mobile-user-centric essence and, particularly, in our research, it should be directly helpful for us to generate and select more targeted and purposeful advertising content for the users. In our study, we also recognize that in mobile computing, the notion of context is often equated simply with key words and contents in PC web or just locations of mobile devices. Actually the mobile context is more complex than that. Mobile application usage can vary continuously because of changing circumstances and differing user needs. To fit into these circumstances and satisfying these needs, manufacturers and developers have built numerous devices, databases, and communities to model these circumstances and capture the needs in order to better serve the users. The information that they have modeled and captured, which is usually open to the public, is very helpful and should not be ignored in generating and selecting context-aware advertising content.

Context Awareness

In fact, context awareness is not a new topic. It has been pioneered by Mark Weiser around fifteen years ago who then focused on the context-aware computing area under the vision of ubiquitous computing (a.k.a. pervasive computing or ambient intelligence). Ubiquitous computing is a method devised to make distributed computing available by multiple computers throughout the physical environment and make them transparent to the stakeholders (Weiser, 1991). Context awareness as a scientific term was first introduced by Schilit (1994) in ubiquitous computing. In his research, context is put into three categories: computing context, user context, and physical context. By these categories, Schmidt further defined context as knowledge of the user's and IT device's state, including surroundings, situation and locations

(Schmidt, 1999). Recently researchers have paid long due attention to context acquisition and utilization in various mobile platforms. Khedr et al. (2005) apply agent-based approaches for building mobile context-aware platform using the network-level context. Biegel and Cahill (2004) described a framework of utilizing environmental observance for context aware application development in ubiquitous computing. Gu et al. (2004) described context models using ontology in mobile intelligent environments. Major mobile organizations such as Open Mobile Alliance (OMA), W3C and IETF (Internet Engineering Task Force) have worked on standardization that has greatly influenced the research on mobile platforms. However, there is still lack of effective ways to utilize contexts for delivering more targeted and purposeful content. W3C's Cascading Style Sheet media queries determine a specific style sheet based on the type of media that is accessing the web page, such as PC, PDA, etc. Another standard, Synchronized Multimedia Integration Language (SMIL) also supports checking the characteristics of the system whose dynamics are governed by the runtime mobile environment. The User Agent Profile (UAProf) by OMA is commonly used by mobile researchers and developers to identify device characteristics using a pre-defined vocabulary over RDF. WURFL is another popular resource description mechanism used by the mobile platform. One limitation of this mechanism is in that it needs the developers to constantly solicit information from client devices and update the database that holds all the resource information. More importantly, there is no well-established approach or procedure to apply the device resource information described by WURFL or such mechanism for optimizing the information provided to the users, not to mention effectively utilizing this information to select advertising content on mobile web.

2.2. Internet-Based Context-Aware Advertising

As an emerging research topic, Internet-based advertising has very few publications, even less for context-aware advertising. Wang et al. (2002) stated that the advertising contents must be relevant to the user's interest to match with the user's experience and promote the chances of later interactions. Ribeiro-Neto et al. (2005) worked on a ground-breaking report from the information retrieval perspective in which they examined a number of strategies to match pages to ads based on search keywords. More recently, the fast-growing popularity of sponsored search in online advertising, such as major search engines, has motivated more researchers from multiple disciplines, such as information retrieval, query optimization, and database management, to study various topics. Dean and Ghemawat (2004) presented their approach of extracting keywords from web pages to match with advertising contents. Broder et al. (2007) proposed a framework for matching ads using a large taxonomy including both semantic and syntactic features. Ribeiro-Neto et al. (2005) tried to use additional pages using a Bayesian model to overcome the difference between the vocabularies of Web pages and ads. Yih et al. (2006) presented an original approach for context-aware advertising in reducing it to the problem of sponsored search advertising by extracting phrases from the page and matching them with the bid phrase of the ads. They used various features to determine the importance of page phrases for advertising purposes. Another school of study tries to estimate the click through rate of ads using data analysis tools such as clustering analysis for keyword matching and classification (Regelson, 2006). In this work, the ads are clustered by their bid phrases. The click-through rate is averaged over each cluster.

In a summary of all the reviewed works, they have provided valuable references and solid grounds for building the frameworks and approaches to match user context information with the advertising content. However, most of them have only associated user context with either the content of web pages, or the keywords of user searches, or just location information of mobile devices, and therefore, even with solid-grounded matching approaches, their works cannot be extend to the complex information given by context-aware advertising challenges. A new framework is needed to utilize the contexts on mobile platforms to generate and select targeted and purposeful advertising content for mobile users.

2.3. Collaborative Filtering Process

Collaborative filtering is used to predict what items a user will like, given a set of feedback made by like-minded users. Generally the input to a collaborative filtering system is a user with a list of items and the system returns a list of predicted preferences of the user for these items. The algorithms used for building a collaborative filtering system include neighbor-based methods, Bayesian networks, singular value decomposition, and inductive rule learning. The essence of these methods is mostly the same, i.e., choosing a set of users based on their similarity to the active user and then using the weighted aggregate of their ratings to generate predictions for the concerned user. These methods also have similar sequence of steps. Choosing one of the most popular collaborative filtering algorithms, the key steps are briefly introduced here to provide a basic understand and point of reference for the audience. The first step of a collaborative filtering process is to build the user profile from her feedback made on a set of items. Second step is the key of collaborative filtering, to locate other users with profiles similar to that of the concerned user, i.e. the neighbors. In collaborative filtering approaches, this similarity is usually calculated based on the similarity of the ratings that these users give to the same items. A commonly used method for this calculation is the Pearson correlation coefficient, which measures

the degree of a linear relationship between two variables (Breese, 2008). Last, all the neighbors' ratings are combined into one prediction by computing a weighted (i.e. correlations) average of the ratings. With the basic steps of a collaborative filtering process introduced, in the next section we will discuss how the context awareness, such as tracking location and analysing user profile and activities, can work with collaborative filtering to build a comprehensive and operational mobile advertising system.

3. A CONTEXT-AWARE MOBILE ADVERTISING SYSTEM

The aim of our approach is to design a system that provides mobile users with targeted and customized advertisements by modeling and analyzing the user contexts. Figure 1 shows the high-level view of our system.

As shown in the figure, when the users send a query to the web server from her mobile device, the web server returns the results according to the user query and, meanwhile, sends a request to the advertising content provider. This request is normally composed based on the query keywords. After receiving the request, the advertising content provider will respond with the advertisements accordingly. This is a common practice undertaken by most of the existing web servers. However, the advertisements returned from the provider are often roughly selected and not targeted at the needs of the users, which certainly does not help to achieve the original purpose of providing such advertisements. The advantage of our system is to scan through all these advertisements and then only keep the targeted and meaningful ones for the specific user. To achieve that, this system first checks each advertisement to see if it is location dependent or independent. The location-independent advertisement will be directly sent to an advertisement selection module. The location-dependent advertisement will be first matched

Figure 1. Context-aware mobile advertising system

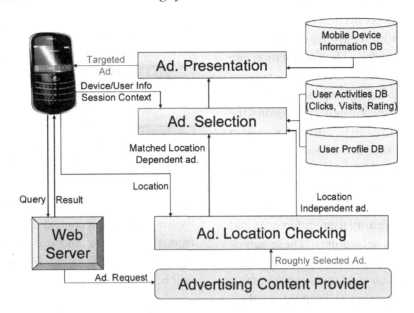

with the location of the mobile device, only matched ones being sent to the advertisement selection module. The advertisement selection module uses the information of user profiles and user activities to select targeted advertisement for the particular user, with a ranking of these advertisements based on the level of their match with the user information. Selected advertisements will be sent to an advertisement presentation module, which determines how these advertisements can be properly displayed on the mobile device based on the device capabilities, such as the screen resolution, orientation and touch-screen support.

This mobile advertising system has a dual goal of improving mobile advertising relevance and not sacrificing the user's overall experience in using the mobile application. The procedure taken to select and present mobile advertisements must be carefully evaluated for mobile devices which have very limited resources compared to Desktop computers. The system accomplishes this goal by considering the user locations and contexts in various steps. Consequently, the advertisement selection and presentation modules interact with the mobile device, which has the user location and

user session information, and backend databases, which store user contexts including profile and activities, to decide what advertisements to be presented and how to present them. In the rest of this section, we will describe the advertisement selection, advertisement presentation, and backend database in great details. To clearly present the workflow and how they interact with each other, we will first describe the backend databases and then discuss how to use their data in the advertisement selection and presentation modules.

User Context Databases

Based on the understanding and recognition we have discussed in Section 2.1, the user contexts in our research, in a high level view, should be divided into three categories: mobile device, user profile and user activity. They are represented by three backend databases: mobile device information database, user profile database, and user activity database. Mobile device information database stores the device features and capabilities, including device manufacturer, model, touchscreen support, screen resolution, equipped keyboard, portrait/landscape, memory size, multimedia rendering, and Bluetooth connection. Device features and capabilities differ vastly amongst various manufactures and models. This information can be identified by the web server when accepting a request and analyzing the request header from the mobile device. User profile database stores the profile information that can be usually obtained in a user registration, such as user locations, email addresses, phone numbers, age group, newsletter preference, technological interests, etc. The third database stores the user's activities that are important for us to identify user preferences and select advertisement accordingly. These activities include users' clicks on the advertisements, users' visits to the web resources, and the rating scores that the users give as feedback to their visited business. This user activity also stores user session contexts associated with each activity, such as location, time and weather. When a new request is being processed, the user session context is collected from the device to indicate the context information of the current session. The session context object is updated and maintained continuously during the session, similar to the user profile context but more accurate and recent. It can also be used by the advertisement selection module. For example, when a user searches for restaurant information at lunch time in a rainy day, our system will present the advertisements for the restaurants, which are open for lunch, geographically close to the user and offer in-door dining. When the session is over, the information in this session context will be written back to the user activity database when the session is over. Table 1 describes the detailed information stored in these three databases.

After the three databases are established, they can be constantly updated when the web server process the new user requests. Upon accepting a new request, the server can detect the user identity by checking the login authority, cookie values, or a previously assigned special link to the user. Once the user identity is determined, the web

Table 1. Mobile user context databases

Databases	Context Details
Mobile device information database	Manufacturer, model, touchscreen, resolution, keyboard, portrait/ landscape, memory, stylus, multimedia, Bluetooth, etc.
User profile database	Home address, age group, newsletter registration, dining interests, etc.
User activity database	Clicks, visits, ratings, session contexts

server can retrieve the user profile, previous activities and session context of these activities, all from the database, and use that information to select the advertisements. And this new request will be used to update the user activity database.

Advertisement Selection Module

Online advertisements are generally implemented as a quality-based bidding approach. For instance, Google and Yahoo! search marketing rank sponsored search advertisement by the bid price on matching key words plus a quality score evaluated by the advertisement's click-through-rate, keyword relevancy with landing page, and site quality (Bernard, 2008). However, when being applied in mobile advertising domain, this approach has not considered the contextual factors such as location, user profile, and user activities. These factors are not available in PC computing world, but can be found in the mobile devices. They are very critical factors in selecting advertisements for mobile users.

On another front, collaborative filtering (CF) has turned out to be a very effective way of identifying new services for customers. Generally it works by combining the opinions of people who have expressed inclinations similar to the user in the past to make a prediction on what may be of interest to the user now. Accordingly, we have applied and refined CF based on two assumptions: first, people who have the similar contexts (e.g. user profile and session context) likely feel similarly towards the same business service; second, the activities of people, who have more similar contexts (with the user) than others, are more valuable in predicting the user's interest. Therefore, in our system, it works by investigating the activities of people who have similar contexts to a user to make a prediction on what advertisement may be of interest to that user. It will first analyze the similarity between the context of the user and the contexts of other users, and then use the previous activities of other users with similar contexts to recommend more targeted advertisement. More details about this refined CF method are presented below.

Collecting User Information

Traditional collaborative filtering (CF) systems only consider the users' feedback (i.e. rating in user activity) when building the user profile. With the richer contexts we can get from mobile devices, we should capture all these contexts when collecting user information, including user profile contexts from user profile database, such as home addresses and dining interests, and user session context from user activity database, such as location and weather. Feedback, which is commonly used in existing CF systems, are included in the user activity database, together with users' clicks on and visits to the web resources (in our case, online advertisements).

Measuring the Contextual Similarity of Users

In this step, our system evaluates the profile context and session context of the current user against the profile contexts and session contexts of other users in the system and calculates the similarity between these contexts. For one user, there may be two kinds of 'neighbors': first kind has the similar profile contexts, such as they are in the same age groups and similar dinning interests. The

second kind has the similar session contexts, such as location and weather. Obviously, the activities (clicks, visits and ratings) that were conducted by this kind of 'neighbor' are great references when selecting the advertisement for the current user.

Based on this observation of neighbors, in order to calculate the similarity of the two contexts, we first define the context as a tuple of different context types from either user profile context or session context:

$$C = (C_1, C_2, ..., C_n)$$

in which the exact profile or session contexts included in measuring similarity may differ case to case, e.g. dining interest in user profile context and location in session context definitely matter when the current user is looking for a restaurant.

Context types can vary widely and it may be difficult to define a universal similarity function for each context type. Nevertheless, for each context that is publicly used, there should be a commonly accepted function to measure the closeness or distance quantitatively between two values, such as using altitude and longitude for address, use Fahrenheit or Centigrade for weather. In the rare case there is not any function to quantitatively measure two values in one context type, we will match the value in two contexts to yield a 0 or 1 value, 1 indicating an exact match and 0 otherwise.

We use the Pearson's correlation coefficient to measure the similarity between the contexts of two different users, u_o and u_q in the equation:

$$col(u_o, u_q) = \frac{\Sigma_{j=1}^{n} \left(c_o, j - c_q, j \right)}{\sigma \left(u_o \right) \sigma \left(u_q \right)} \quad (1)$$

where $c_{o,i}$ and $c_{q,i}$ are the value of the context i of user u_o and user u_q, and $\sigma(uo)$ and $\sigma(u_q)$ are standard deviation of context of u_o and u_q, assuming all the contexts have normalized value (Breese, 2008).

Recommending Advertisements

After sort out the contextual similarity between two users, our system in this step will recommend advertisements for a user by combining the previous activities of other users, who have similar context, in a weighted-average method. This method uses contextual similarities of other users' as the weights. Generally a collaborative filtering (CF) system uses Equation 2 to calculate a prediction for the user u_o, where $P_{o,i}$ is defined as the predicated rating of u_o on advertisement i, n is the number of u_o' neighbors, $r_{q,i}$ is the rating that user u_q gives to item i, r_o is the average rating of u_o gives to every item, r_q is the average rating of u_q gives to every item and k is the normalized factor of the averaging rating of u_q.

$$P_{o,i,c} = \overline{r_o} + k\sum_{q=1}^{n}(r_{q,i} - \overline{r_q}) * w_{o,q} \qquad (2)$$

In our system, Equation 1 has been refined to infer the $w_{o,q}$ in Equation 2, by calculating the contextual similarity and considering all the related activities, such as clicks, visits and ratings to the business resources, of the current mobile user and other mobile users, instead of just ratings.

In order to represent user activities in our consideration, we define a value for the previous activities of a user u_o as:

$$a_{o,i} = k_1 * n_{clicks,i} + k_2 * n_{visits,i} + k_3 * r_{o,i} \qquad (3)$$

where k_1, k_2, k_3 are the normalized factors for the number of clicks, number of visits, and the rating to the advertisement i.

Based on the activity value definition, the Equation 2 is refined to predicate the preference of u_o:

$$P_{o,i,c} = \overline{a_o} + k\sum_{q=1}^{n}\left(a_{q,i} - \overline{a_q}\right) * w_{o,q} \qquad (4)$$

where, compared with Equation 1, $\overline{a_o}$ is the sample means of previous activities (clicks, visits, and ratings) u_o has taken on advertisements, $\overline{a_q}$ is the sample means of the previous activities (clicks, visits, and ratings) u_q has taken on advertisements, $a_{q,i}$ is the activity value for user u_q on advertisement i, and $w_{o,q}$ is the weight of u_q's activity value in recommending an advertisement for u_o, denoted by the contextual similarity between the u_o and u_q.

Taken the Equation 2 into Equation 4, we have the Equation 5 to combine all the weighted activity value, with respect to contextual similarity of all the neighbor users, to give an comprehensive recommendation for the current user on the advertisement i.

$$P_{o,i,c} = \overline{a_o} + k\sum_{q=1}^{n}\left(a_{q,i} - \overline{a_q}\right) * \frac{\sum_{j=1}^{n}\left(c_{o,j} - c_{q,j}\right)}{\sigma\left(u_o\right)\sigma\left(u_q\right)} \qquad (5)$$

Advertisement Presentation Module

Mobile advertisements can be displayed in alternative formats on a mobile device as simple text links, colorful images, or animated images. The size of a mobile web page should be much smaller compared to an Internet web page in order to reduce the download time and to fit the page to the small screen of mobile devices. An image advertisement is more eye-catching than a text link, but also takes longer to download and occupies a bigger part of the screen. An animated image advertisement can be more attractive to some mobile users than a simple text link advertisement, but also requires heavy processing power of the mobile device.

Based on all such considerations, advertisement presentation module uses mobile device information, user profile and user session context to define heuristics to determine which kind of advertisements presentation are more appropriate for the current user on the mobile device. The most popular online advertising types include sponsored advertising, text taglines, and animated image, etc. Based on these types, some sample heuristics can be proposed below.

First, device information is a key factor in this module because the capabilities of mobile devices vary significantly. The new generation of mobile devices is usually equipped with a big touch screen, and thus can show more advertisements on the same page without interrupting the user. This is compared to old devices with a smaller screen, and the user can only scroll the page by repeatedly pressing navigational keys. In particular, it is acceptable to show several advertisements on a mobile phone with a screen size of 480*320, but it would take almost half of the screen displaying the same number of advertisements on an old phone with a resolution of only 160*128.

Second, user profile is also important to select the presentation type of advertisement. For example, mobile Web Banner Ad is a popular way to present advertisements on mobile web pages, which composes a still or animated image and optional text Taglines. The aspect ratio and the size of the banner image not only need to be adjusted to the user's mobile device, but also have to be adjusted to the users' preferences. If the users are unfamiliar with image banners, many don't realize the image banners can be navigated to and clicked on, while a Text Tagline can be added to generate a higher click rates. If the information in the user profile suggests that the user is familiar to image banners, then the Text Tagline can be removed to improve user browsing experience.

Last but not least, user session context can also be used in pre-defined heuristics to present the selected advertisements. Imagine if the user is browsing a category of restaurants or searching for the closest gas station, a sponsored search advertisement is more relevant to the user session. Similarly, if the user is reading blogs on his mobile device, then the text taglines is a better format to be presented to the user at the bottom of the blog page.

4. PROTOTYPE IMPLEMENTATION AND CASE STUDY

In order to provide an exemplary application and conduct appropriate case studies to validate our approach, we have implemented a prototype system for the proof of concept, namely Skyhelper, based on the approach described in Section 3 and the advanced mobile computing technologies. Skyhelper has two versions, one is a mobile native application using Java™ micro edition and the other is a mobile web site rendered in XHTML and AJAX. The two applications have the same functionality, which is to allow users to search for the information of local restaurants and then, by using our approach, to return appropriate results with targeted advertisement based on the user locations, profile, previous activities and session contexts. Mobile native application is for the users that use high-end PDA with decent processing power. Mobile web site is for other users, who have traditional cell phones with too limited capabilities to support the native application. Based on our approach, Skyhelper consists of three main layers: client, web server and database management system (DBMS), as shown in Figure 2.

- **Client:** The client can be any browser-equipped mobile devices, from the traditional cell phones to the state-of-the-art smart PDAs such as Blackberry and Google phone. For J2ME customers, we will both provide advertisements when the customers send queries and, if customers prefer, use J2ME MIDP 2.0 Push Registry to push advertising promotions. This push registry is part of the device application management system (AMS), which is responsible for each application's life-cycle (installation, activation, execution, and removal). For WAP customers, with their preferences, Skyhelper pushes both the selected mobile advertisements and advertising promotions to them through WAP push

Figure 2. System architecture of Skyhelper

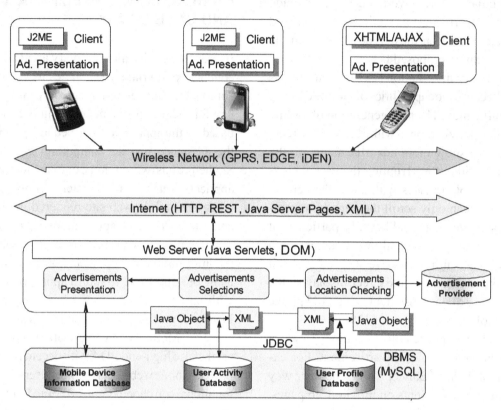

technology. The web server communicates with the WAP Push Proxy Gateway using the Push Access Protocol (PAP) over-the-air (OTA) to deliver the advertisements to the customers.

- **Web Server:** We use a Tomcat web server as the container for restaurant-search and advertising services. The request from the client applications, including information query, device location and user session contexts, come through the wireless network (such as GPRS, EDGE, iDEN) and arrive at the web server in HTTP. The web server is implemented in Java Server Pages that handles the HTTP communication between the client application and the server via REST, which is a light-weight and XML-based communication protocol. The server divides the functionality of advertisement location checking, advertise-

ment selection and advertisement presentation into three subsystems. As specified in our approach, the location checking subsystem first intercepts the raw advertisement information from the advertising provider and matches the location of the advertisement with the location of the device. Then the matched advertisements, or location-independent advertisements, will be sent to the selection subsystem for further processing based on the user contexts. The presentation subsystem will take the selected advertisements and send them to the device in an appropriate format, either an XML message for J2ME application or an XHTML page as a mobile web page. Each subsystem will use XML to transfer static data or use serialized Java object for dynamic data and control.

- **Database Management System (DBMS):**
  We use MySQL™ 5.0 to store user profiles, users' previous activities, and mobile device information in three databases on clustering servers to decent scalability and short response time. The web server can access the information in these databases via the interface of Java™ Database Connection (JDBC). As specified in Section 3, the information from user profile database and user activity database is used by advertisement selection subsystem. The mobile device information database serves the advertisement presentation subsystem.

Based on this software prototype Skyhelper, two case studies have been taken in our work, one, one with the full support of our approach (i.e. advertisement location checking, selection and presentation subsystems are all running properly), and the other only with the advertisement presentation subsystem and not the other two, i.e., the raw advertisements from the providers are shown to the customers.

First, we configured a few emulators for the devices with different capabilities (screen resolution, GPS support, etc) and selected a small group of five users with different profiles and various activities conducted previously. Second, in both case studies, the users are suggested with the same set of queries that they can submit for searching for restaurants in their daily activities. Third, we provided the users with the query results and also the selected (or not, in the second case) advertisements, such as restaurants, movies, and tourist attractions, and then tracked the clicks of these users on the advertisements. Finally, the comparison between the users' clicks on the advertisements in both studies gave us a clear idea of the level of improvement that our approach has made in the match between the presented advertisements and users' interests.

The user feedback data is shown in Figure 3, where the average user clicks for the first set of the search results is compared with the average clicks for the second set. According to the data we have obtained, more user clicks (3 more clicks out of 10, i.e. 30% improvement) have been observed for the second set of the search results, i.e. on Skyhelper with the support of our approach. In addition, most test users think that the advertisements returned with the full support are more suited to their need than from the one without. Therefore, based on this preliminary analysis and with certain limitations caused by the nature of case studies (e.g. limited user profiles and case selections), it still can be clearly seen that this approach well fulfills its objectives.

5. CONCLUSION AND FUTURE WORK

In this paper, we have proposed a mobile advertising system that provides targeted and customized advertisements based on context awareness and location tracking. We classified mobile user contexts into three categories according to their characteristics and discussed the benefits and challenges of adapting these contexts to the mobile advertisement selection and presentation

Figure 3. Comparison of users' clicks

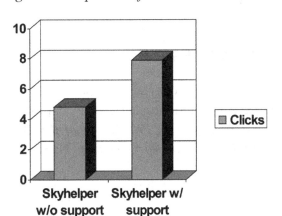

process. We used collaborative filtering to automatically predict the customer's preference for advertisements based on both the past activities of like-minded users and contextual similarity between these like-minded users and the current user. We have discussed the three main modules of this comprehensive mobile advertising system, including advisement selection, advertisement presentation and user context databases. We also presented a software prototype which is developed based on this approach and then discussed the case studies that were undertaken using this prototype to validate our approach.

It is our next step to further evaluate the approach and prototype system. We will need much more real data from user participations in Skyhelper to validate whether the preference predictions match the user's actual decision. In addition, Skyhelper needs to consider the privacy issue when storing and investigating the preferences and activities information of the mobile users, since this information may potentially lead to the users' privacy. On the theoretical side, we will concentrate on mining context information in mobile application and develop more intelligent context-aware advertisement selection algorithms. It would be interesting to look specifically into different types of contexts, including social (e.g. relatives and community) and technical (e.g. device), and consider the impact these contexts may bring to interaction amongst the user, application and advertisement.

REFERENCES

Bardram, J. E. (2004). Applications of context-aware computing in hospital work: Examples and design principles. In *Proceedings of the ACM Symposium on Applied Computing*. New York, NY: ACM Press.

Bernard, J. J., & Tracy, M. (2008). Sponsored search: An overview of the concept, history, and technology. *International Journal of Electronic Business*, *6*(2), 114–131. doi:10.1504/IJEB.2008.018068

Biegel, G., & Cahill, V. (2004). A framework for developing mobile, context-aware applications. In *Proceedings of the Second IEEE Annual Conference on Pervasive Computing and Communications* (p. 361). Washington, DC: IEEE Computer Society.

Breese, J. S., Heckerman, D., & Kadie, C. (2008). Empirical analysis of predictive algorithms for collaborative filtering. In *Proceedings of the Fourteenth Conference on Uncertainty in Artificial Intelligence* (pp. 43-52).

Broder, A., Fontoura, M., Josifovski, V., & Riedel, L. (2007). A semantic approach to contextual advertising. In *Proceedings of the 30th Annual International ACM SIGIR Conference on Research and Development in Information Retrieval* (pp. 559-566). New York, NY: ACM Press.

Chatterjee, P., Hoffman, D. L., & Novak, T. P. (2006). Modeling the clickstream: Implications for web-based advertising efforts. *Marketing Science*, *22*(4), 520–541. doi:10.1287/mksc.22.4.520.24906

Chen, A. (2005). Context-aware collaborative filtering system: Predicting the user's preference in the ubiquitous computing environment. In T. Strang & C. Linnhoff-Popien (Eds.), *Proceedings of the International Workshop Location- and Context-Awareness* (LNCS 3479, pp. 244-253).

Dean, J., & Ghemawat, S. (2004). MapReduce: Simplified data processing on large clusters. In *Proceedings of the Sixth Symposium on Operating System Design and Implementation* (pp. 137-150).

Dey, A. (2001). Understanding and using context. *Personal and Ubiquitous Computing*, *5*(1), 4–7. doi:10.1007/s007790170019

Gu, T., Wang, X. H., Pung, H. K., & Zhang, D. Q. (2004). An ontology-based context model in intelligent environments. In *Proceedings of Communication Networks and Distributed Systems Modeling and Simulation* (pp. 270-275).

Kappel, G., Retschitzegger, W., Kimmerstorfer, E., Pröll, B., Schwinger, W., & Hofer, T. (2002). Towards a generic customisation model for ubiquitous Web applications. In *Proceedings of the 2nd International Workshop on Web Oriented Software Technology*, Málaga, Spain.

Khedr, M., & Karmouch, A. (2005). ACAI: Agent-based context-aware infrastructure for spontaneous applications. *Journal of Network and Computer Applications*, 19–44. doi:10.1016/j.jnca.2004.04.002

Lawrence, R. D., Almasi, G. S., Kotlyar, V., Viveros, M. S., & Duri, S. S. (2001). Personalization of supermarket product recommendations. *Data Mining and Knowledge Discovery*, 5, 11–32. doi:10.1023/A:1009835726774

Regelson, M., & Fain, D. (2006). Predicting click-through rate using keyword clusters. In *Proceedings of the Second Workshop on Sponsored Search Auctions*.

Ribeiro-Neto, B., Cristo, M., Golgher, P. B., & de Moura, E. S. (2005). Impedance coupling in content-targeted advertising. In *Proceedings of the 28th Annual International ACM SIGIR Conference on Research and Development in Information Retrieval* (pp. 496-503). New York, NY: ACM Press.

Schafer, J. B., Konstan, J., & Riedl, J. (1999). Recommender systems in e-commerce. In *Proceedings of the 1st ACM Conference on Electronic Commerce* (pp. 158-166). New York, NY: ACM Press.

Schilit, B., Adams, N., & Want, R. (1994). Context aware computing applications. In *Proceedings of the IEEE Workshop on Mobile Computing Systems and Applications*, Santa Cruz, CA (pp. 85-90). Washington, DC: IEEE Computer Society.

Schmidt, A., Aidoo, K. A., Takaluoma, A., Tuomela, U., Laerhoven, K. V., & de Velde, W. V. (1999). Advanced interaction in context. In *Proceedings of the First International Symposium on Handheld and Ubiquitous Computing*, Karlsruhe, Germany (pp. 89-101).

Wang, C., Zhang, P., Choi, R., & Eredita, M. (2002). Understanding consumers attitude toward advertising. In *Proceedings of the Eighth Americas Conference on Information System* (pp. 1143-1148).

Weiser, M. (1991). The computer for the 21st century. *Scientific American*, 94–104. doi:10.1038/scientificamerican0991-94

Yih, W., Goodman, J., & Carvalho, V. R. (2006). Finding advertising keywords on web pages. In *Proceedings of the 15th International Conference on World Wide Web* (pp. 213-222). New York, NY: ACM Press.

This work was previously published in the International Journal of Handheld Computing Research (IJHCR), Volume 2, Issue 3, edited by Wen-Chen Hu, pp. 38-52, copyright 2011 by IGI Publishing (an imprint of IGI Global).

Chapter 7

Effect of Personal Innovativeness, Attachment Motivation, and Social Norms on the Acceptance of Camera Mobile Phones:
An Empirical Study in an Arab Country

Kamel Rouibah
Kuwait University, Kuwait

T. Abbas H.
Kuwait University, Kuwait

ABSTRACT

This study develops a model to assess the consumer acceptance of Camera Mobile Phone (CMP) technology for social interaction. While there has been considerable research on technology adoption in the workplace, far fewer studies have been done to understand the motives of technology acceptance for social use. To fill in this gap, this study develops a model that is based on the following theories: the technology acceptance model, the theory of reasoned action, the attachment motivation theory, innovation diffusion theory, and the theory of flow. The first research method used was a qualitative field study that identified variables that most drive CMP acceptance and build the research model using a sample of 83 consumers. The second method was a quantitative field study. Data was collected from a sample of 240 consumers in Kuwait and used to test the proposed model. The results reveal two types of use: "social use" and "use before shopping," explaining 32.3% and 30% of the variance respectively. Most importantly, the study reveals that personal innovativeness, attachment motivation and social norms have an important effect on CMP acceptance. The implications of this study are highly important for both researchers and practitioners.

DOI: 10.4018/978-1-4666-2785-7.ch007

INTRODUCTION

The mobile phone industry is quickly growing to offer new services such as Internet access, games, digital camera capabilities, text messaging and mobile commerce (m-commerce). With the increasing number of mobile phones around the world (4 billion users in 2009; see The Mobile World, 2009), there is no doubt that the mobile services industry is aware of the factors leading potential consumers to adopt such services. The number of studies focusing on the effects and adoption of mobile devices over different cultures has increased over the last decade: m-commerce (Venkatesh & Ramesh, 2003; Lin & Wang, 2003; Hung et al., 2003; Lin & Wang, 2003; Cheong & Park, 2005; Nysveen et al., 2005a,b; Lin & Shih 2008), mobile services in Norway (Nysveen et al., 2005a; Nysveen et al., 2005b), mobile phone use in Africa (Meso et al., 2005), use of mobile phones in academic libraries in Malaysia (Abdul Karim et al., 2006), m-learning in Korea (Kim & Ong, 2005), portable device assistant (PDA) adoption in the USA (Sarker & Wells, 2003; Yi et al., 2006), text message services in China (Yan et al., 2006), mobile Internet use in Japan (Okazaki, 2006), mobile payment (Dahlberg et al., 2008), mobile wireless technology adoption in South Korea (Kim and Garrison, 2009); mobile advertisements in Taiwan (Yang, 2007), and mobile entertainment (Ha et al., 2007).

Although there are some studies that focus on the motives for using mobile devices and services, there are several unexplored dimensions related to our understanding of mobile services, such as CMP usage. The authors observed a lack of studies pertaining to CMP usage, with the exception of Yao and Flanagin (2006). Although Yao and Flanagin (2006) focused on the social aspect, they did not use a well known theoretical model as their base.

This paper seeks to understand the motivational factors for the adoption of CMP in an Arab country for three main reasons. First, the Arab world has not received much research attention (Rouibah,

2008) and so Western researchers lack an understanding of how technologies are adopted in the Arab world. Second, the Arab world belongs to a collectivist culture where the motives to use technologies are primarily different than those in individualism cultures (i.e., most Western cultures) (Hofstede, 2009). Third, unlike past studies on technology acceptance, this paper will focus on actual usage of the CMP service for social use instead of the intentions to use in the workplace because very few studies have focused solely on social use instead of the use of technologies in the workplace (Nysveen et al., 2005a, b; Li et al., 2005; Rouibah, 2008; Rouibah & Hamdy, 2009; Lu et al., 2009; Fang et al., 2009).

The remainder of the paper is structured as follows: a literature review is given, followed by the discussion of the research model and hypotheses. Next, the authors describe the research methodology. An analysis is presented as well as results pointing out the managerial implications and future research directions. Finally, a conclusion is given.

Literature Review

Past studies on wireless mobile device adoption have investigated technology adoption from different theory perspectives including: the Theory of Reasoned Action (TRA) (Fishbein & Ajzen, 1975), the Innovation Diffusion Theory (IDT) (Rogers, 1983), the Technology Acceptance Model (TAM) (Davis, 1989) and its variations (e.g., Venkatesh & Davis, 2000; Venkatesh et al., 2003; Yi et al., 2006; Rouibah, 2008), the theory of flow (Csikszentmihalyi, 1990) and combined theories (e.g., Kim & Garrison, 2009). Still, efforts are being undertaken to propose new models or extend previous ones in order to fit the unique features of different technologies. In line with this observation, several variations of TAM were proposed to increase the parsimony of the model by combining different existing TAMs (Venkatesh & Davis, 2000; King & He, 2006; Rouibah, 2008;

Kim & Garrison, 2009; Liao et al., 2009; Mallat et al., 2009). This study is, therefore, oriented to fill in the gap and contribute to direct marketing and industrial efforts to enhance the comprehension pertaining to the requirements of potential adopters.

This paper focused on a new model because most past acceptance models were developed to support technology in the workplace environment and, consequently, these models would seem inappropriate, even after modification, for social purposes. The use of CMP is distinct from technology use in the workplace environment context in several aspects. First, CMP and other mobile devices differ from office systems in their notion of ubiquity reflecting the fact that mobile phones are portable, and, therefore, useful in maintaining a high attachment to family members. This is why the proposed model in this study includes the attachment motivation variable, derived from the *attachment theory* of Bowlby (1969). Attachment motivation is a personality attribute that reflects an individual's desire for social interaction and a sense for communication with others and is thus an intrinsic motivation variable. While this variable has been well studied in different disciplines such as social psychology (Baumeister & Leary, 1995), organizational behavior (Mowday et al., 1982), and marketing (Morgan & Hunt, 1994), it is surprising that only one study examined it in the IS field (Li et al., 2005). Second, the organizational context is another criterion that distinguishes between the two technologies. Mobile phones differ in being personal devices and are used for social purposes besides being valued by the users as desktop computers are. Mobile phones also differ from other technology devices in workplaces due to the emphasis on using them for entertainment during work hours. Finally, technology usage in the workplace is mandatory, and it used to achieve a number of objectives such as increased productivity and enhanced performance. However, mobile phones have voluntary usage, and some cultures can exert pressure on people to use them.

Additionally, those people who show signs of innovativeness have a tendency to use them more often than others. Derived from IDT, perceived innovativeness has been proposed (Agarwal & Prasad, 1998) to denote the risk-taking propensity that is higher in certain individuals who are more willing to take a risk by trying out an innovation than others who are hesitant to change their practices. While a number of empirical studies revealed the impact of personal innovativeness (PI) on different technologies (Agarwal & Karahanna, 2000; Lewis et al., 2003; Hung & Chang, 2005; Lassar et al., 2005; Lu et al., 2005; Yi et al., 2006; Thompson et al., 2006; Lian & Lin, 2008), relatively few studies have been carried out regarding the potential effects of innovativeness differences in the context of technology use for social use.

Another important variable is related to the role of social norm, derived from TRA, in the Arab world. An analysis of Arab culture and the state of IT/ICT in the Arab world demonstrate a number of different (Rouibah & Hamdy, 2009). First, Arab culture is highly social and family-oriented with a high context and collectivist dimension (Al-Khatib et al., 2008). Arabs are more affected by normative and social values and they are requested to adopt attitudes and behaviors that comply with the established norms that govern their daily life (Kabasakal & Bodur, 2002). Thus, they may use CMP to overcome these barriers. Second, Arab culture promotes high socialization and interaction amongst relatives and friends. Arabs have a strong tendency for face-to-face interactions. We thus expect that new technologies such as CMP could be used to maintain close relationships with friends and important people (Rouibah, 2008). While many studies included social norms in technology acceptance, see literature review done by Schepers and Wetzels (2007), only two of 51 studies focused on the Arab world (Al-Khaldi & Al-Jabri, 1998; Selim, 2003). Moreover, two other studies, not listed in Schepers and Wetzels (2007) also studied the effect of culture on technology acceptance (Loch et al., 2003; Rouibah, 2008). Past studies on social usage of instant messaging have found that Kuwaiti adults use instant messaging for social norms pressure and for perceived enjoyment (Rouibah, 2008; Rouibah & Hamdy, 2009), with high frequency (Rouibah & Hamdy, 2009), and also for Internet usage (Loch et al., 2003). Other studies about social usage outside the Arab region focused on instant messaging and found that it is used for attachment motivation, perceived enjoyment and pressures from friends and social groups (Li et al., 2005).

Taking into account the characteristics of a collectivist culture, it is unknown what impact attachment motivation has on actual use compared to social norm. Another unexplored dimension consists of assessing the relative effect of attachment motivation compared to other important factors (usefulness and ease of use) highlighted by the TAM theory and enjoyment from the flow theory. In seeking to understand the determinant of CMP use in Kuwait, an Arab and collectivist culture, this study raises the following question: *How does attachment motivation (AM), personal innovativeness (PI) and social norms (SN) relate to perceived ease of use (PEOU), perceived usefulness (PU), perceived enjoyment (PE), and level of CMP usage in a collectivist culture?*

The research could be further refined to discuss whether CMP adoption depends on instrumental and cognitive complexities (usefulness and ease of use), effects of social pressure (subjective norm), and flow (perceived enjoyment). This paper seeks answers to these questions from the perspective of an Arab country.

Kuwait is the focus of this study due to its high IT/ICT penetration achieved amongst the Gulf Cooperation Countries (GCC) countries (see Table 1) in terms of PC, Internet, and mobile phone penetration. In addition, the GDP per person in Kuwait is the second highest ($31860) after Qatar in the GCC. Kuwait was also ranked third in the GCC after the UAE and Qatar in terms of utilizing ICT. Recently, Kuwait's economic growth has been remarkably strong with an annual real GDP growth rate of 8.2% after UAE (8.5%) and Qatar (8.4%), and this growth rate is predicted to continue increase by 5.4% (ICT Qatar 2009). Additionally, other basic ICT indicators in the GCC countries are reported in Table 1.

Finally, this study is useful for both Arab and non Arab researcher since there is a lack of knowledge about how new technologies are perceived in the Arab world. Additionally, except for very few studies, the literature review indicates that most empirical evidence was collected outside

Table 1. Basic ICT indicators in the GCC countries (World Development Indicator, 2009)

Indicator	Year	Kuwait	Bahrain	Oman	Qatar	Saudi Arabia	UAE
Population (thousands)	2000	2190	672.0	2442.0	606.4	20660.7	3247
	2005	2535.4	726.6	2566.9	812.8	23118.9	4533.1
GDP per capita ($)	2000	17222	11861	8135.	29290	9120	21740
	2005	31860.6	17773	9460	52239	13399	28611
PC penetration rate	2000	11.41	14.13	3.30	14.85	6.29	12.31
	2005	23.66	16.65	4.60	16.36	36.66	18.75
Mobile penetration rate	2000	21.73	30.61	6.63	19.93	6.65	43.98
	2005	93.86	100.00	51.93	88.17	57.53	100.00
Internet penetration rate	2000	6.85	5.95	3.68	4.95	2.23	23.5
	2005	27.6	21.3	11.10	26.9	6.86	30.82

Figure 1. Research model

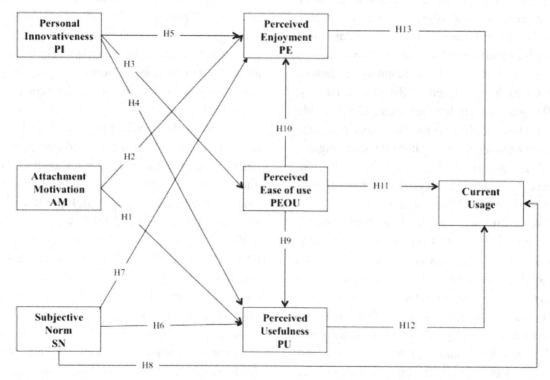

the Arab world (see a summary of Arab literature review Rouibah et al., 2009). This is why Rouibah and Hamdy (2009), after investigating instant messaging acceptance in Kuwait, called for additional research to shed light on other new technologies. This study will reveal why and how Arab people use CMPs for social use.

Conceptual Model and Hypotheses Development

This research seeks to develop an extended TAM by including other constructs from four theories (IDT, TRA, flow, and the attachment motivation theory). The extended model examines the relationship between seven variables taken from past studies (Figure 1).

External Variables

Attachment Motivation (AM)

Attachment, according to the attachment motivation theory (Bowlby, 1969), is an essential and central concept for building and maintaining friendship. AM is defined as the tendency of human beings to have a pervasive drive to form and maintain interpersonal friendships (Li et al., 2005). And it is used to keep frequent and pleasant contacts with important people such as family members, spouses and close friends. Those who are motivated to keep an attachment with others feel safe with each other, and are very active in seeking support from their social networks (Li et al., 2005). Unlike individuals who are highly motivated to build and maintain relationships people with less attachment motivation feel negative towards and are anxious about social interaction with others. To keep frequent contact, high motivated people are constantly seeking new communication tools such as CMP (Li et al., 2005). Therefore, in the context of using this tool (CMP), people with high levels of attachment motivation may perceive this technology to be very useful, and enjoy a high level of happiness with their partners. This is supported by Li et al. (2005) who found that AM was a significant determinant of perceived usefulness during instant messaging usage. Since camera mobile phone usage implies also communication with multimedia data, thus, we hypothesize:

H1: Attachment motivation is positively associated with perceived usefulness.

Compared with other technologies, a CMP has the advantage of instantly taking pictures and sending to people which are important for an individual to experience enjoyment, ensuring reciprocal effects, supports and responsiveness between friends and family members. In line with Li et al. (2005), we hypothesize:

H2: Attachment motivation is positively associated with perceived enjoyment.

Personal Innovativeness (PI)

Several recent studies have largely supported the notion that perceived usefulness (PU) and perceived ease of use (PEOU) fully mediate the influence of the individual difference variables on usage behavior. PI is originated from IDT (Agarwal & Prasad, 1998). Agarwal and Prasad (1998) adapted this construct in the domain of IT and defined as the "willingness of an individual to try out any new IT." Agarwal and Prasad (1998) proposed a new metric for the measurement of domain-specific individual innovativeness. While several authors studied the direct effect of PI on current behavior (e.g., Lu et al., 2005; Hung & Chang, 2005; Lian & Lin, 2008; Thompson et al., 2006; Fang et al., 2009), this study will not consider

such an effect. PI is an individual characteristic that exerts an impact through mediating variables (PEOU, PU or enjoyment).

Past studies found that PI was a significant determinant of PEOU (Agarwal & Karahanna, 2000; Lewis et al., 2003; Hung et al., 2003; Lu et al., 2005; Yi et al., 2006) and PU (Agarwal & Karahanna, 2000; Lewis et al., 2003; Lu et al., 2005; Yi et al., 2006). In addition, while Yi et al. (2006) found a significant relationship between and subjective norm and PI, this paper will not consider such a link as Yi et al. (2006) did, since this study considers the subjective norm to be an independent external variable (Venkatesh & Davis, 2000). Thus, we hypothesize:

H3: Personal innovativeness is positively associated with perceived ease of use.

H4: Personal innovativeness is positively associated with perceived usefulness.

Because this study aims to explore the effect of PI on behavior in Arab (i.e., non-Western) culture, we hypothesize for the first time a link between PI and perceived enjoyment. The rational is that the more a user shows signs of innovativeness to use new technologies and loves everything new, the more he will enjoy its use. We posit the following hypothesis:

H5: Personal innovativeness is positively associated with perceived enjoyment.

Subjective Norm (SN)

The subjective norm is derived from TRA as a determinant of behavioral intention, and it was later included in TAM2 (Venkatesh & Davis, 2000). Subjective norm refers to a person's perception of what people important to him think he should or should not perform in accordance to a behavior in question. Through the direct effect of subjective norm, users may adopt a behavior that complies with the group which they identify themselves with. The direct compliance effect of subjective norm is the case when an individual perceives that a social actor wants him to perform a certain behavior, and the social actor has the ability to reward the behavior or punish it in case of its absence (Venkatesh &Davis, 2000).

Past studies have found a positive link between subjective norm and perceived usefulness (Venkatesh & Davis, 2000; Yi et al., 2006; Schepers & Wetzels, 2007; Rouibah, 2009), with perceived enjoyment (Rouibah, 2008); and with actual behavior (Loch et al., 2003; Hung et al., 2003; Nysveen et al., 2005b; Schepers & Wetzels, 2007; Rouibah, 2008; Rouibah et al., 2009). Thus, we hypothesize:

H6: Subjective norm is positively associated with perceived usefulness.

H7: Subjective norm is positively associated with perceived enjoyment.

H8: Subjective norm is positively associated with the individual level of CMP usage.

Mediating Variables

Perceived Usefulness (PU) and Perceived Ease of Use (PEOU)

TAM assumes that beliefs about perceived usefulness (PU) and perceived ease of use (PEOU) are always the primary determinants of current behavior (Davis, 1989; Venkatesh & Davis, 2000). PU refers to the extent to which a person believes that using a technology would have a value for his tasks. PEOU, on the other hand, refers to the extent to which a person believes that by using a technology they will be free from any mental effort. There is extensive empirical evidence accumulated over the last two decades that finds a link between PEOU-PU, PEOU-usage and PU-usage (see recent findings in Rouibah, 2008; Rouibah et al., 2009). One of the important aspects of CMPs

is its digital nature. This is important due to the fact that images captured in this format can be effortlessly copied and easily distributed to a large number of people important to an individual in a short period of time.

With regard to a possible relationship between PEOU and perceived enjoyment (PE), the literature review indicates two possible relationships, either enjoyment influences PEOU (Yi & Hwang, 2003; Sun & Zhang, 2006), or PEOU influences enjoyment and PU (Heijden, 2004; Nysveen et al., 2005b; Li et al., 2005; Sun & Zhang, 2006; Rouibah, 2008). The first stream posits that the effects of PE increase over time as users gain more experience with the technology. This study's view is in line with the second stream, which posits that PEOU is an antecedent of PE, which directly affects current behavior. The rationale behind this is that the more a user feels the technology is easy to use, the more likely he will enjoy using it and the more likely he will spend more time using this technology. Thus, we hypothesize:

H9: PEOU is positively associated with perceived usefulness.

H10: PEOU is positively associated with perceived enjoyment.

H11: PEOU is positively associated with individual level of CMP usage.

Since the current study focuses on CMP use for social and recreational purposes, we argue that users will not use the technology primarily due to of its usefulness and value, but rather for other motives such as perceived enjoyment and fun (Heijden, 2004; Li et al., 2005; Rouibah, 2008). This is why we hypothesize the absence of an association between PU and current behavior, which is in line with past studies on social use of hedonic technologies (Rouibah, 2008).

H12: PU will not affect individual level of CMP usage.

Perceived Enjoyment (PE)

Originated from the theory of flow (Csikszent-mihalyi, 1990), PE is defined as the extent to which the activity of using a new technology is perceived to be enjoyable in its own right, apart from any performance consequences that may be anticipated (Davis et al., 1989). Individuals, who experience pleasure and fun from using the CMP, are likely to use it more extensively than others. Several recent studies have found a positive link between perceived enjoyment and behavioral usage of several technologies (Lee et al., 2003; Loch et al., 2003; Li et al., 2005; Sun & Zhang, 2006; Rouibah, 2008) including mobile devices (Nysveen et al., 2005a, b). Thus, we hypothesize:

H13: Perceived enjoyment is positively associated with individual level of CMP usage.

RESEARCH METHODOLOGY

This study is relatively new and therefore requires rigorous justification so as to the choice of employed research methods. In many traditional studies on IS, either quantitative or, to some lesser extent, qualitative methods were used, but not both (Loch et al., 2003). To complete this study, two research methods were used (Rouibah & Abbas, 2006). The qualitative method was used to identify the basic motives and guide the design of the instrument while the quantitative research method was used to validate the research model.

Qualitative Data Collection

As argued, a qualitative approach was used for the first stage of this research to collect initial sets of motives for the adoption of CMPs by a sample of students at a leading College of Business Administration in Kuwait University. Several brainstorming sessions were held with 84 students to highlight why they use CMPs. Respondents were encouraged to list, in free format comments, their motives behind using CMPs. The answers were gathered, compiled, and summarized in Table 2, which provides a summary list of encouraging factors.

A literature review of past technology acceptance models was conducted, and a list of motives (constructs) was identified. A comparison between the above table and motives from past studied was done. The identified constructs are those depicted in the research model.

Quantitative Data Collection

Based on the results of the qualitative findings, the data collection instrument was developed and checked by two faculty members at College of Business Administration (CBA) of the Kuwait University who have experience in technology adoption. In order to examine questionnaire validity, a control check was performed over completeness, readability, and comprehensibility. After this, 270 questionnaires were distributed randomly among students at the CBA. The survey consisted of two sections: the first one focused on demographic data related to respondents; and the second section detailed the respondents' behavior with regards to CMP, representing the components of the research model. The questionnaire was circulated to students of nine sections (each with 30 students) soliciting their beliefs and attitudes regarding CMP use. Students were chosen because of two reasons. First, it was an appropriate sample to test and CMP acceptance where subjective norm plays important role; and second, students are the heaviest CMP users in the local mobile market. Among the 270 questionnaires, only 240 were fully completed, and the data was analyzed using SPSS software.

Table 2. Motive factors for using CMPs in Kuwait

No.	Construct	Why is a camera mobile phone (CMP) used?	%
1	Enjoyment	To take family pictures and save them for later.	61
2	Enjoyment	To share funny pictures and video clips with peers.	51
1	Ease of use	Because it is easy to use while traveling.	51
2	Ease of use	Because it is easy to take pictures anywhere and at anytime.	50
3	Usefulness	Because it is easy to take and exchange pictures and video clips between mobile devices and/or PCs.	49
4	Usefulness	It can be used as a scanner, such as photographing the office hours of instructors.	48
5	Usefulness	Because it is useful to take exceptional pictures or unusual events.	46
6	Usefulness	Because it is useful to take pictures of famous people such as actors, singers and political figures.	46
7	Ease of use	Because it is easier to use and carry than normal (digital) cameras.	45
1	Innovativness	It has many services in one (mobile, camera, radio, movie player, etc.).	45
8	Enjoyment	I love and am interested in taking pictures.	44
1	Usage	I use CMPs to send pictures and video clips to people not seen in a long time.	43
2	Ease of use	Because it is easy to take pictures and share them with others.	41
2	Usage	I can take a picture of everything new and attractive.	41
3	Attachment Motivation	Because it is useful to take nice photos when traveling and send them via SMS service to people (family and friends) important to me.	40
4	Innovativness	To use the CMP as a mirror to view oneself.	39
3	Attachment Motivation	I use it to instantly share pictures and video clips with my family members and friends when I am in distant places.	38
4	Innovativeness	I like to use new technology.	37
5	Innovativeness	I like to see and send multimedia (MMS).	36
6	Usefulness	It is useful in recording evidence of crime or accidents (e.g. car accidents).	34
7	Usefulness	To take and capture opinions of important people (family members and friends) before shopping.	33
9	Enjoyment	To use for enjoyment, fun, and entertainment.	32
8	Social Norms	To keep up with and follow updates in the the field of new technologies	31
9	Attachment Motivation	To send pictures to my family members and friends instead of visiting them.	30
10	Social Norms	Because it offers easy communication with friends and family via MMS instead of actually visiting.	28
11	Social Norms	Because my family uses it.	25
12	Social Norms	To show off in public using the latest technology.	23
13	Social Norms	Because I feel pressured to use CMPs as it is in accordance with the social norm (peers and family members).	20
14	Social Norms	Because using a CMP will enhance my appearance to my peers and friends.	18
15	Social Norms	To enhance my social status.	17
16	Social Norms	I use a CMP because my peers and friends use it too.	15
17	Usefulness	Because it offers use for indecent purposes such as cheating on exams.	14
18	Innovativeness	To capture different styles to imitate the latest fashion.	7
19	Innovativeness	It is a convincible approach for disseminating rumors.	7
10	Ease of use	To associate phone numbers with their respected pictures.	4
11	Ease of use	Because it is useful to capture pictures for friends.	3

Constructs Measurement

Table 3 shows the measures used. They had been validated in prior research, but their wording was modified in order to fit the context of the CMP. Actual usage was operated using self reporting measure as was the case of many previous studies (Rouibah, 2008; Rouibah & Hamdy, 2009). Respondents were also asked to indicate what kinds of pictures they most often take choosing from a list of options: family, friends, famous people, self picturing, fashion designs, decorations, recent models, hairstyles, house designs, clothes, strange events, funny and beautiful pictures, proof of crime, urgent events, immoral pictures, and cheating. Since this was the first study on CMPs.

RESULTS AND DISCUSSION

Demographic Data and Behavior of Respondents with CMP Usage

The demographic profile of the survey respondents (Table 4) shows that 66.8% of the respondents were female and 33.2% male. The most represented age group was ages 18-25 (94.2%). Most participants were not married (94.2%) and are doing well at university since 63.5% had a GPA above 3.0. The average frequency of use of the respondents was 3. With regard to CMP usage, 68.5% use their CMP to transfer pictures to other media storage, 67.2% used it to listen and watch video clips, and 66% shared pictures with their family members, friends, or social groups. In addition, 48% use their CMP to share their opinions about items to purchase before effective shopping.

Factor Analysis

Factor analysis was conducted to confirm the existence and relevance of the existing variables. A total of eight factors instead of seven with eigen values greater than 1.0 were identified and factors

loading was greater than 0.50 (Hair et al., 1998). These factors explained 71.708 of the total variance of the research model. The measurement scales showed high reliability; with Cronbach's alpha coefficients for the all variables exceeding 0.88 (see Table 5). Construct validity was strongly supported by the factor analysis, which indicates that the instrument is valid.

The construct "usage" was split into two distinct factors; the first one is related to social interaction with people, which was decided to be called "social use"; while the second one is related to CMP usage before shopping, which was decided to be called "use before shopping." All the variables had mean values higher than 3.02, indicating that on the average most respondents agreed to the items set in the questionnaire.

Evaluation of Research Model and Hypotheses Testing

After checking absence of collinearities, five regression analyses were conducted to estimate the relationship between the constructs of the research model. Same approach has been adopted by past studies (Venkatesh & Davis 2000; Venkatesh et al., 2003). And significant paths are those with t-value greater or equal to 1.96.

Table 5 shows that the usage is split into two types: *social use* and *use before shopping*. Thus, testing hypotheses H8, H11, H12, and H13 needed to be refined into two additional hypotheses that were named "a" and "b" (e.g. H8a and H8b).

In Table 6, 32.3% of the variance of social use is explained by two variables: perceived enjoyment and social norms. 30% of CMP for *use before shopping* is explained by three variables: perceived enjoyment, social norms and PEOU. Accordingly, three hypotheses H8 (H8a and H8b) (SN-Usage), H12 (H12a and H12b) (PU-Usage) and H13 (H13a and H13b) (PE-Usage) are accepted, while one hypothesis H11 (PEOU-Usage) is partially supported (H11b is accepted and H11a is rejected) (Figure 2).

Table 3. List of measures

Construct	Measure	Source
Attachment Motivation (AM)		
AM1	CMP helps me be close to others, watch them more, and help me as if relating to them on a one-to-one level.	Li et al., (2005)
AM2	CMP helps me be around others and find out about them and that is one of the most interesting things I can think of using the CMP.	
AM3	I feel like I have really accomplished something valuable when I am able to get close to someone through the technology of the CMP.	
AM4	I like watching people and using CMPs helps me to this and gives me enjoyment.	
Personal Innovativeness (PI)		
PI1	Amongst my friends, I am usually the first to explore new IT.	Agarwal and Prasad (1998)
PI2	I like to experiment with new CMPs.	
PI3	I like to try new CMPs.	
Social Norm		
SN1	People who influence my behavior (e.g. friends) think that I should use a CMP.	Rouibah (2008)
SN2	People who are important to me (e.g. family) think that I should use a CMP.	
SN3	The traditions and customs influence me to use a CMP.	
Perceived Enjoyment		
PE1	I use a CMP when I feel I am bored.	Rouibah (2008)
PE2	I believe using a CMP is enjoyable.	
PE3	The actual process of using a CMP is pleasant.	
PE4	I have fun using a CMP.	
Perceived Usefulness		
PU1	I use a CMP to reduce loneliness and anxiety.	Li et al. (2005)
PU2	I use a CMP for security and well being.	
PU3	I use a CMP to maintain interpersonal relationships with others.	
Perceived Ease of Use		
PEOU1	Interacting with a CMP does not require a lot of my mental effort.	Davis (1989)
PEOU	I find a CMP easy to use.	
PEOU3	I find a CMP flexible to interact with.	
Usage (U)		
U1	I take family pictures to socialize with my family members.	Variety of use, Rouibah (2008)
U2	I take pictures of friends to socialize with them.	
U3	I take pictures of famous people (artists, singers, celebrities) and exchange them with my friends.	
U4	I take pictures of fashion designs and get opinions of important people about them before shopping.	
U5	I take pictures of recent models and get opinions of important people about them before shopping.	
U6	I take pictures of clothes of well dressed people and buy similar items.	

Table 4. Demographic profile of respondents

Gender	Number	Percent
Male	80	33.2
Female	161	66.8
Age		
13-17	10	4.2
18-25	227	94.2
25-50	4	1.7
GPA		
Less than 1.0	2	0.8
Between 2.0 and 3.0	86	35.7
Above 3.0	153	63.5
Marital Status		
Single	227	94.2
Married	14	5.8
Frequency of Use		
About one picture each week	79	32.45
Several pictures each a week	93	38.55
About one picture each day	28	11.6
Several pictures each day	41	17.0

In Table 6, three variables among four contribute positively to explain 98.3% in the variance of PU. These are attachment motivation, social norms and PEOU. Thus, hypotheses H1 (AM-PU), H6 (SN-PU) and H9 (PEOU-PU) are supported while H4 (PI-PU) is rejected.

In Table 6, three variables among five contribute positively to explain 35.7% in the variance of PE. These are personal innovativness, attachment motivation and ease of use. Thus, hypotheses H2 (AM-PE), H5 (PI-PE) and H10 (PEOU-PE) are accepted and H7 (SN-PE) is rejected. Personal innovativness contributes to explain 9.7% in the variance of ease of use, leading to accept H3.

Table 5. Descriptive statistics and psychometric properties of measures

Constructs	Loading	Mean	SD	Cronbach's Alpha
Attachment Motivation (AM)				
AM1	0.774	3.200	1.461	
AM2	0.801	3.630	1.464	.899
AM3	0.809	3.020	1.442	
AM4	0.693	3.000	1.462	
Personal Innovativeness (PI)				
PI1	0.791	3.040	1.387	
PI2	0.854	3.560	1.198	.888
PI3	0.749	3.240	1.277	
Social Norm (SN)				
SN1	0.845	3.040	1.315	
SN2	0.783	3.800	1.281	.893
SN3	0.750	3.300	1.268	
Perceived Enjoyment (PE)				
PE1	0.665	3.500	1.300	
PE2	0.759	3.540	1.250	.890
PE3	0.822	3.780	1.109	
PE4	0.774	3.380	1.272	
Perceived Usefulness (PU)				
PU1	0.711	3.280	1.354	
PU2	0.846	3.880	1.394	.885
PU3	0.853	3.150	1.385	
Perceived Ease of Use (PEOU)				
PEOU1	0.804	4.230	.910	
PEOU2	0.873	4.380	.771	.894
PEOU3	0.818	4.240	.892	
Social Usage (SU)				
SU1	0.641	4.701	.970	
SU2	0.699	4.020	.930	0.857
SU3	0.682	4.620	.900	
Usage Before Shopping (UBS)				
UBS1	0.781	3.700		
UBS2	0.718	4.012		0.826
UBS3	0.805	3.788		

Table 6. Regression results explaining dependent variables of the model

Regressions	Std. Error	Beta	T	Sig.	R² (%)	F
Regression of PEOU						
H3: PI-PEOU	0.041	0.260***	4.799	0.000	9.700	23.020
Regression of PU						
H4: PI-PU	0.010	0.040	0.404	0.687	98.300	F= 3744
H2: AM-PU	0.007	0.492***	55.369	0.000		
H6: SN-PU	0.011	0.201***	22.322	0.000		
H9: PEOU-PU	0.031	0.798***	96.936	0.000		
Regression of Enjoyment						
H5: PI-PE	.072	.339***	5.809	.000	35.700	F=32.70
H1: AM-PE	.049	.329***	5.741	.000		
H7: SN-PE	.077	0.072	1.232	0.219		
H10: PEOU-PE	.221	.109*	2.048	.042		
Regression of Social Usage (SU)						
H11a: PEOU- SU	.063	.040	.628	.531	32.300	F=30.28
H12a: PU- SU	.064	.097	1.501	.135		
H13a: PE- SU	.064	.144*	2.229	.027		
H8a: SN- SU	.020	.168**	2.666	.008		
Regression of Regression of Use Before Shopping (SBS)						
H11b: PEOU- SBS	.074	0.159*	2.574	.011	30.000	F=32.49
H12b: PU- SBS	.076	.017	.277	.782		
H13b: PE- SBS	.076	.241***	3.821	.000		
H8b: SN- SBS	.024	.151*	2.452	.015		

* Significant path (p<0.05); ** significant path (p<0.01) and *** significant path (p<0.001)

DISCUSSION

The current study shows that attachment motivation, considered an intrinsic motivation, affects both PU and enjoyment. Such a result is partially consistent with Li et al. (2005) who only found a positive association between attachment motivation (AM) and enjoyment. Moreover, results of this study reveal its indirect effect on current behavior through the mediation of perceived enjoyment. With regard to its effect, attachment motivation exerts the second strongest effect on enjoyment after personal innovativeness. Thus, people with high attachment motivation, perceived more enjoyment from CMPs and use it to maintain social friends with important people

such as family members, relatives and friends. In addition, those with high level of motivation use CMPs for requesting opinions of important people about items or products before shopping.

This study shows two types of behavioral use of CMPs: for *social use* and *use before shopping*, each with different antecedents. Such a result is similar to Rouibah and Hamdy (2009) who found two different types of usage for social use of instant messaging in Kuwait.

This study also shows the importance of personal innovativeness as an individual characteristic that affects both types of CMP uses. Results of this study partially support findings of previous studies that studied the effect of innovativeness. In particular, our findings partially

Figure 2. Results of the CMP usage in Kuwait

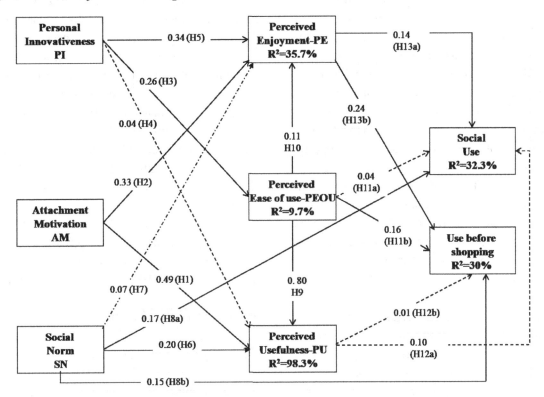

support the finding of Agarwal and Karahanna (2000); Lewis et al. (2003); and Yi et al. (2006). While Yi et al. (2006) found a positive association between PI-PEOU, and PI-PU, our results did not. A plausible explanation could be explained by the nature of the two studies: Previous studies focused on technology in the workplace and the current study focused on social use. Unlike previous studies that focused on innovativeness, this study extends results of previous studies (Agarwal & Karahanna, 2000; Lewis et al., 2003; Yi et al., 2006) and succeeds to show for the first time a positive association between PI and PE. PI affects CMP use (for social and before shopping) through the mediation of enjoyment and ease of use. Moreover, the effect of PI on current CMP behavior is stronger through the mediation of enjoyment (ß = 0.34) than that of ease of use (ß = 0.26). Such a result reveals that people with a high level of innovativeness tend to perceive more

enjoyment when using CMP for both social and before shopping purposes.

Results also reveal that social norm (ß=0.17) exerts the strongest effect on CMP social use, followed by PE (ß=0.14). Such findings are in line with findings of three studies in collectivist cultures: Internet usage in Egypt (Loch et al., 2003) social usage of instant messaging by adults in Kuwait (Rouibah, 2008), and e-banking in Malaysia (Rouibah et al., 2009). Our findings are also consistent with the study of Li et al. (2005) about social use of instant messaging outside the Arab world. Surprising, our results are not consistent with those of Nysveen et al. (2005b), who studied cell phone-based IM usage for socializing among Norwegian users and found the effect of subjective norm to exert the weakest effect. Differences in culture may be one cause of the different results, since Arab culture is marked by very high levels of collectivism, compared to northern European countries (e.g., Norway), which are more individualistic. This result confirms the strong effect of social norms on technology acceptance for social use in the Arab world.

With regard to the effect of perceived enjoyment on usage, the results indicate that this variable affects only CMP use before shopping. Unlike its effect on social use, perceived enjoyment exerts more effect (ß=0.24) on CMP use before shopping than does social norm (ß=0.15) and this effect almost 1.5 times. This study highlights a new result that none past study did mention it: the effect of social norm and perceived enjoyment differ based on the nature of user behavior whether it is for social use or for shopping purposes.

The relationship between PEOU and PU is supported by a large body of findings from past TAM studies (e.g., Rouibah et al., 2009). The lack of association between PU and current behavior contrasts with most findings about technology acceptance where PU emerged as the major determinant of technology acceptance in the workplace in particular with TAM1 (e.g., Davis et al., 1989), and TAM2 (Davis et al., 1989; Venkatesh & Davis, 2000) or outside the workplace (Yi et al., 2006). Unlike previous studies in the workplace, our findings are consistent with those of recent studies on technology use for social use outside the workplace (Teo et al., 1999; Nysveen et al., 2005b; Rouibah, 2008).

With regard to the relative effect of PU and PE, this study shows that enjoyment plays more of an effect on behavior than does PU. This is consistent with previous studies (See Lee et al., 2003; Li et al., 2005; Nysveen et al., 2005b; Rouibah, 2008) which found that PE exerts a stronger indirect effect on intention to use than does PU. For example, Lee et al. (2003) differentiated in their proposed model between intrinsic and extrinsic variables. According to their findings, an intrinsic variable in terms of PE is the most significant construct and the key factor affecting a student's adoption of an Internet-based learning medium. This result is in line with Teo et al. (1999) who

found that perceived enjoyment has a strong impact on Internet usage. The relative impact of the enjoyment result confirms one observation that is valid for technology based social usage. That is when perceived enjoyment and usefulness are integrated in social use, usefulness losses its effect compared to intrinsic internal motivation in terms of perceived enjoyment. Such a result however requires and calls for additional future studies.

During the result analysis, this study has shown differences with previous studies. These contradictions could be explained by four motives; setting of the technology used in each study (workplace vs. communication technology), usage context (workplace vs. social use), culture, and the uniqueness of a CMP as an entertainment device.

Managerial Implications

This study has several practical implications. Results of the study could be used to promote CMPs before shopping for m-commerce. We advocate that two motives push towards the use of CMPs before shopping, i.e., high speed of life and effect of social norms in the society. With regard to the first motive, family members are spending less time together, and even dedicate a smaller amount of time shopping together. With the increase in multi-services of mobile phones, it is possible that one family member shops alone and may request the opinion of another family member for effective shopping using the functionalities of CMPs. With regard to the second motive, CMP use before shopping is appropriate and helps to overcome cultural barriers. Indeed, Kuwaiti society is a conservative society where females have problems sharing opinions with others when shopping and trying out outfits.

This study also encourages the increased use of CMPs in the workplace but with caution. News agencies and hospitals are promising settings as appropriate and potential applications. In educational institutions, although CMPs may be encouraged to be used during lectures and trainings, it also should be prohibited during exams, since students may use it to cheat. CMPs should also be prohibited in female sports locations and women changing and fitting rooms. Since Kuwait is an Arab country that is highly influenced by tribal and Islamic beliefs, CMP usage may violate women's privacy and thus contribute to create tension between families. In such extreme situations, the case may be exaggerated and might lead to legal cases. Mobile development companies are encouraged to design a new CMP that may make a small sound during recordings. This may alert people of the recording especially where it is prohibited (e.g., museums, female changing rooms). Because of its portability, CMPs can also be used evilly to capture forbidden documents or scenes such as marriage events which are permitted either for group of people, ages, or sexes. Moreover, the usage of CMPs should also be restricted in some workplaces such as ministries to prevent competitive intelligence usage by enemies, and competitors.

Finally, this study encourages mobile phone providers, online shopping, social networking and wireless devices manufacturers to focus on how potential consumers may use their CMP not only before shopping but also during shopping for increased effectiveness. While this study does not provide sufficient understanding from the standpoint of system designers and requirements to design such a system, it has the merit to raise the ideas and leave the door open for additional creative ideas about its implementation. The basic idea consists to enable a potential shopper: (i) to take picture of the product he wants to buy; (ii) exchange his viewpoints with people important to him, (iii) take a picture of the product ID and send it to the e-shopping; then the shopping will process and fulfill his order. This idea goes beyond the social shopping (e.g., iVillage.com) which combines social networking with shopping and allow users to exchange shopping tips and display favorite purchases.

Limitation and Suggestion for Future Research

The above findings should be interpreted in light of three limitations.

The first limitation is related to the usage of students as respondents of the study. The findings and implications of our research are based on a single method of study involving one particular group (i.e., students) whom we noticed is a major group among CMP users. However, CMP technology is widely acceptable and known to other sectors of the Kuwaiti population, who are also considered to be heavy users and can be appropriate targets for future research. Having other groups of CMP users is expected to enrich the research and enhance the types and alterations of this technology utilization. Moreover, while a group of students is an appropriate sample, most past studies on TAM also used students (e.g., Schepers & Wetzels, 2007), which was found to be just as valid and reliable as when other groups (e.g., adults) where used (McCoy et al., 2007). While the use of a student sample limits the generalization of our research, for example to employees who are driven by other norms, this group does represent well-educated and CMP–skilled, and well-being consumers and their most popular reason for CMP usage is enjoyment and social orientation.

The second limitation is related to the current focus of this study. Opposite to most studies on technology acceptance which focused on intention to use new technology, our study concentrates on CMP current usage, which has an element of enjoyment outside the workplace. The study would have been more useful if it were performed in business environments.

The third limitation is related to time. This study focuses on current use of CMP instead of continuous usage. The data gathering process was carried out at a sample point in time and did not examine either short or long term usage (Venkatesh & Morris, 2000) or continuous usage (Liao et al., 2009).

From a research perspective, this study suggests several research directions.

First, the variance extracted from CMP use (32.3% [for social use] and 30% [for before shopping]) is acceptable. However, a large percentage remains unknown (67.7% and 70% successively for the two types of use) which calls for additional research to discover other motives. CMP use involves taking and exchanging pictures. This use may raise the violation of individual privacy, which calls for the need to include privacy construct in future studies. Despite the large number of studies that included this construct and were developed specifically for ecommerce (e.g., Malhotra et al., 2004; Vijayasarathy, 2004), it is surprising to note the lack of a commonly agreed upon scale for privacy, as is the case of PEOU, PU, and enjoyment. This study encourages future studies to develop a scale for CMP privacy or to customize exiting ones, e.g. the one of Malhotra et al. (2004), and to investigate its potential contribution to behavior.

Second, there is a need to extend the external validity of the model. Since this study only concentrated on one group, it is advisable that subsequent studies focus on other groups such as employees in the workplace. Two potential settings that merit attention are the use of CMP in news agencies and in hospitals by doctors (Howard et al., 2006). While students base their behavioral decisions on enjoyment, innovativeness and PEOU, employees may focus more on usefulness or other concrete motives (e.g., mandatory usage endorsed by top managers). Such an extension could shed more light on the motives behind CMP adoption.

Third, this study also encourages to conduct a cross Arab study to test and enrich standard technology adoption theories. The Arab world includes 22 countries which share a similar culture, values, language, history and geographic location. It spans from the west coast of Africa through the northern part of Africa to the Arabian Gulf, and from Sudan to the Middle East GCC states, which include Kuwait, Bahrain, Oman, Qatar, the United Arab Emirates, and Saudi Arabia. While the Arab world is large in terms of population (335,827,330 in 2009), very few studies have focused on IT/ICT in this part of world compared to those in North America and Europe (see Rouibah & Hamdy, 2009). It is true that Arab countries share a similar culture, but they differ in terms of wealth. Therefore, access to IT/ICT usage is different and it is also expected that motivational factors for technology adoption are also different. This is important if the perspective is to validate the model across different Arab countries. An interesting and promising research direction is to extend the study to users from different Arab countries and integrating perceived user resources (Meso et al., 2005). The objective is to shed light on the interactions between socio-economic development and technology adoption in the Arab world. Answers to how such a construct will affect PEOU, PU, PE and current usage, and how the effect is different with regard to users from different Arab countries would serve to build a more generalized set of insights for international technology transference to the Arab world.

Fourth, by extending this study to workplace and social use in Kuwait and other Arab countries future studies are encouraged to study the effect of gender on CMP acceptance because past studies of technology acceptance found that gender is an important factor (Venkatesk & Morris, 2000; Nysveen et al., 2005a; Sanchez-Franco et al., 2009). For example, Nysveen et al. (2005a) found that enjoyment is the most significant factor for females while usefulness is considered most significant factor for males when using mobile device services (payment, text messaging, games and contact).

CONCLUSION

The study developed a model for voluntary usage of CMPs for social purposes and maintaining friendships in Kuwait, which is known to have a collectivist culture. The study contributes to a better theoretical understanding of the factors that promote technology adoption in an Arab country and reveals two types of usage, each with its own antecedents. This study achieves several contributions that have potential implications on both research and practice.

ACKNOWLEDGMENT

This research was funded by Kuwait University, Research Grant IQ 06-05.

REFERENCES

Abdul Karim, N. S., Darus, S. H., & Hussin, R. (2006). Mobile phone applications in academic library services: a students' feedback survey. *Campus-Wide Information Systems, 23*(1), 35–51. doi:10.1108/10650740610639723

Agarwal, R., & Karahanna. (2000). Time flies when you are having fun: Cognitive absorption and beliefs about information technology usage. *Management Information Systems Quarterly, 24*(4), 665–694. doi:10.2307/3250951

Agarwal, R., & Prasad, J. (1998). A conceptual and operational definition of personal innovativeness in the domain of information technology. *Information Systems Research, 9*(2), 204–216. doi:10.1287/isre.9.2.204

Al-Khaldi, M. A., & Al-Jabri, I. M. (1998). The relationship of attitudes to computer utilization: new evidence from a developing nation. *Computers in Human Behavior, 14*(1), 23–42. doi:10.1016/S0747-5632(97)00030-7

Al-Khatib, J., & Malshe, A., & AbdulKader, M. (2008). Perception of unethical negotiation tactics: A comparative study of US and Saudi managers. *International Business Review, 17*(1), 78–102. doi:10.1016/j.ibusrev.2007.12.004

Baumeister, R. F., & Leary, M. R. (1995). The Need to belong: Desire for interpersonal attachment as a fundamental human motivation. *Psychological Bulletin, 117*(3), 497–529. doi:10.1037/0033-2909.117.3.497

Bowlby, J. (1969). *Attachment and loss. Vol. 1: Attachment.* New York: Basic Books.

Cheong, J. H., & Park, M. (2005). Mobile internet acceptance in Korea. *Internet Research, 15*(2), 125–140. doi:10.1108/10662240510590324

Csikszentmihalyi, M. (1990). *Flow: The Psychology of Optimal Experience.* New York: Harper and Row.

Dahlberg, T., Mallat, N., Ondrus, J., & Zmijewska, A. (2008). Past, present and future of mobile payments research: A literature review. *Electronic Commerce Research and Applications, 7*(2), 165–181. doi:10.1016/j.elerap.2007.02.001

Davis, F. D. (1989). Perceived usefulness, perceived ease of use, and user acceptance of information technology. *Management Information Systems Quarterly, 13*(3), 319–340. doi:10.2307/249008

Davis, F. D., Bagozzi, R. P., & Warshaw, P. R. (1989). User acceptance of computer technology: A comparison of two theoretical models. *Management Science, 35*, 982–1003. doi:10.1287/mnsc.35.8.982

Fang, J., Shao, P., & Lan, G. (2009). Effects of innovativeness and trust on web survey participation. *Computers in Human Behavior, 25*, 144–152. doi:10.1016/j.chb.2008.08.002

Fishbein, M., & Ajzen, I. (1975). *Belief, Attitude, Intention and Behavior: An Introduction to Theory and Research.* Reading, MA: Addison-Wesley.

Ha, I., Yoon, Y., & Choi, M. (2007). Determinants of adoption of mobile games under mobile broadband wireless access environment. *Information & Management, 44*(3), 276–286. doi:10.1016/j.im.2007.01.001

Hair, J. F., Anderson, R. E., Tatham, R. L., & Black, W. C. (1998). *Multivariate Data Analysis* (5th ed.). Upper Saddle River, NJ: Prentice-Hall.

Heijden, D. H. V. (2004). User acceptance of hedonic information systems. *Management Information Systems Quarterly, 28*(4), 695–704.

Hofstede, G. (2009). *Geert Hofstede cultural dimensions*. Retrieved June 20, 2009, from http://www.clearlycultural.com/geert-hofstede-cultural-dimensions

Howard, A., Hafeez-Baig, A., Howard, S., & Gururajan, R. (2006). A framework for the adoption of wireless technology in healthcare: An indian study. In *Proceedings of the 17th Australasian Conference on Information Systems*, Adelaide.

Hung, S., Ku, C., & Chang, C. (2003). Critical factors of WAP services adoption: An empirical study. *Electronic Commerce Research and Applications, 2*, 42–62. doi:10.1016/S1567-4223(03)00008-5

Hung, S. Y., & Chang, C. M. (2005). User acceptance of wap services: Test of competing theories. *Computer Standards & Interfaces, 28*, 359–370. doi:10.1016/j.csi.2004.10.004

Kabasakal, H., & Bodur, M. (2002). Arabic cluster: A bridge between east and west. *Journal of World Business, 37*(1), 40–54. doi:10.1016/S1090-9516(01)00073-6

Kim, G., & Ong, S. M. (2005). An exploratory study of factors influencing m-learning success. *Journal of Computer Information Systems*, 92–97.

Kim, S., & Garrison, G. (2009). Investigating mobile wireless technology adoption: An extension of the technology acceptance model. *Information Systems Frontiers, 11*(3), 323–333. doi:10.1007/s10796-008-9073-8

King, W. R., & He, J. (2006). A meta-analysis of the technology acceptance model. *Information & Management, 43*, 740–755. doi:10.1016/j.im.2006.05.003

Lassar, W. M., Manolis, C., & Lassar, S. S. (2005). The relationship between consumer innovativeness, personal characteristics, and online banking adoption. *International Journal of Bank Marketing, 23*(2), 176–199. doi:10.1108/02652320510584403

Lee, M., Cheung, K. O., Christy, M. K., & Chen, Z. (2003). Acceptance of internet-based learning medium: The role of extrinsic and intrinsic motivation. *Information & Management, 42*, 1094–1104.

Legris, P., Ingham, J., & Collerette, P. (2003). Why do people use information technology? A critical review of the technology acceptance model. *Information & Management, 40*, 191–204. doi:10.1016/S0378-7206(01)00143-4

Lewis, W., Agarwal, R., & Sambamurthy, V. (2003). Sources of influence on beliefs about information technology use: An empirical study of knowledge workers. *Management Information Systems Quarterly, 27*(4), 657–678.

Li, D., Chau, P. Y. K., & Lou, H. (2005). Understanding individual adoption of instant messaging: An empirical investigation. *Journal of the Association for Information Systems, 6*(4), 102–126.

Lian, J., & Lin, T. (2008). Effects of consumer characteristics on their acceptance of online shopping: Comparisons among different product types. *Computers in Human Behavior, 24*(1), 48–65. doi:10.1016/j.chb.2007.01.002

Liao, C., Palvia, P., & Chen, J. (2009). Information technology adoption behavior life cycle: Toward a Technology Continuance Theory (TCT). *International Journal of Information Management, 29,* 309–320. doi:10.1016/j.ijinfomgt.2009.03.004

Lin, H., & Wang, Y. (2003). An examination of the determinants of customer loyalty in mobile commerce contexts. *Information & Management, 43,* 271–282. doi:10.1016/j.im.2005.08.001

Lin, Y.-M., & Shih, D.-H. (2008). Deconstructing mobile commerce service with continuance intention. *International Journal of Mobile Communications, 6*(1), 67–87. doi:10.1504/IJMC.2008.016000

Loch, K. D., Straub, D. W., & Kamel, S. (2003). Diffusing the internet in the Arab world: The role of social norms and technological culturation. *IEEE Transactions on Engineering Management, 50*(1), 45–63. doi:10.1109/TEM.2002.808257

Lu, J., Yao, J. E., & Yu, C.-S. (2005). Personal innovativeness, social influences and adoption of wireless internet services via mobile technology. *The Journal of Strategic Information Systems, 14,* 245–268. doi:10.1016/j.jsis.2005.07.003

Lu, Y., Zhou, T., & Wang, B. (2009). Exploring Chinese users' acceptance of instant messaging using the theory of planned behavior, the technology acceptance model, and the flow theory. *Computers in Human Behavior, 25,* 29–39. doi:10.1016/j.chb.2008.06.002

Malhotra, N. K., Kim, S. S., & Agarwal, J. (2004). Internet users' information privacy concerns (IUIPC): The construct, the scale, and a causal model. *Information Systems Research, 15*(4), 336–355. doi:10.1287/isre.1040.0032

Mallat, N., Rossi, M., Tuunainen, K., & Öörni, A. (2009). The impact of use context on mobile services acceptance: The case of mobile ticketing. *Information & Management, 46,* 190–195. doi:10.1016/j.im.2008.11.008

McCoy, S., Galletta, D. F., & King, W. R. (2007). Applying TAM across cultures: the need for caution. *European Journal of Information Systems, 16,* 81–90. doi:10.1057/palgrave.ejis.3000659

Meso, P., Musa, P., & Mbarika, V. (2005). Towards a model of consumer use of mobile information and communication technology in LDCs: The case of Sub-Saharan Africa. *Information Systems Journal, 15*(2), 119–146. doi:10.1111/j.1365-2575.2005.00190.x

Morgan, R. M., & Hunt, S. D. (1994). The commitment-trust theory of relationship marketing. *Journal of Marketing, 58*(3), 20–38. doi:10.2307/1252308

Mowday, R. T., Porter, L. W., & Steers, R. M. (1982). *Employees-Organization linkage: The psychology of commitment, absenteeism, and turnover.* New York: Academic Press.

Nysveen, H., Pedersen, P. E., & Thorbornsen, H. (2005a). Explaining intention to use mobile chat services: Moderating effects of gender. *Journal of Consumer Marketing, 22*(5), 247–256. doi:10.1108/07363760510611671

Nysveen, H., Pedersen, P. E., & Thorbornsen, H. (2005b). Intention to use mobile services: antecedents and cross-service comparison. *Journal of the Academy of Marketing Science, 33*(3), 330–346. doi:10.1177/0092070305276149

Okazaki, S. (2006). What do we know about mobile Internet adopters? A cluster analysis. *Information & Management, 43,* 127–141. doi:10.1016/j.im.2005.05.001

Qatar, I. C. T. (2009). *Launch of New Telecommunications Licenses for Fixed and Mobile Services.* Retrieved June 6, 2009, from www.ict.gov.qa/files/marketoverview.pdf

Rogers, E. M. (1983). *Diffusion of Innovations.* New York: Free Press.

Rouibah, K. (2008). Social Usage of Instant Messaging by individuals outside the workplace in Kuwait: A structural Equation Model. *IT & People*, *21*(1), 34–68. doi:10.1108/09593840810860324

Rouibah, K., & Abbas, H. (2006). *Modified Technology Acceptance Model for Camera Mobile Phone Adoption: Development and validation.* Paper Presented for the 17th Australasian Conference on Information System Web. Retrieved from http://www.acis2006.unisa.edu.au/

Rouibah, K., & Hamdy, H. (2009). Factors Affecting Information Communication Technologies Usage and Satisfaction: Perspective From Instant Messaging in Kuwait. *Journal of Global Information Management*, *17*(2), 1–29.

Rouibah, K., Ramayah, T., & May, O. S. (2009). User acceptance of internet banking in Malaysia: Test of three acceptance models. *International Journal of E-Adoption*, *1*(1), 1–19.

Sanchez-Franco, M. J., Ramos, A. F. V., & Velicia, F. A. M. (2009). The moderating effect of gender on relationship quality and loyalty toward Internet service providers. *Information & Management*, *46*, 196–202. doi:10.1016/j.im.2009.02.001

Sarker, S., & Wells, J. D. (2003). Understanding mobile handheld device use and adoption. *Communications of the ACM*, *46*(12), 35–40. doi:10.1145/953460.953484

Schepers, J., & Wetzels, M. (2007). A meta-analysis of the technology acceptance model: Investigating subjective norm and moderation effects. *Information & Management*, *44*, 90–103. doi:10.1016/j.im.2006.10.007

Schlosser, A. E., Shavitt, S., & Kanfer, A. (1999). Survey of Internet users' attitudes toward Internet advertising. *Journal of Interactive Marketing*, *13*(3), 34–54. doi:10.1002/(SICI)1520-6653(199922)13:3<34::AID-DIR3>3.0.CO;2-R

Selim, H. M. (2003). An empirical investigation of student acceptance of course websites. *Computers & Education*, *40*(4), 343–360. doi:10.1016/S0360-1315(02)00142-2

Sun, H., & Zhang, P. (2006). Causal relationships between perceived enjoyment and perceived ease of use: An alternative approach. *Journal of the Association for Information Systems*, *7*(9), 618–645.

Teo, T. S. H., Lim, V. K. G., & Lai, R. Y. C. (1999). Intrinsic and extrinsic motivation in Internet usage. *Omega*, *27*(1), 25–37. doi:10.1016/S0305-0483(98)00028-0

The Mobile World. (2009). Retrieved from www.themobileworld.com

Thompson, R., Compeau, D., & Higgins, C. (2006). Intentions to use information technologies: An integrative model. *Journal of Organizational and End User Computing*, *18*(3), 25–46.

Venkatesh, V., & Ramesh. (2003). Understanding usability in mobile commerce. *Communications of the ACM*, *46*(12), 53–56. doi:10.1145/953460.953488

Venkatesh, V., & Davis, F. D. (2000). A theoretical extension of the technology acceptance model: Four longitudinal field. *Management Science*, *46*, 186–204. doi:10.1287/mnsc.46.2.186.11926

Venkatesh, V., Morris, M. G., & Ackerman, P. L. (2000). A longitudinal field study of gender differences in individual technology adoption decision making processes. *Organizational Behavior and Human Decision Processes*, *83*, 33–60. doi:10.1006/obhd.2000.2896

Venkatesh, V., Morris, M. G., Davis, G. B., & Davis, F. D. (2003). User acceptance of information technology: Toward a unified view. *Management Information Systems Quarterly*, *27*(3), 425–478.

World development Indicator. (2009). *Internet based database.*

Yan, X., Gong, M., & Thong, Y. L. (2006). Tow tales of one service: User acceptance of short message service (SMS) in Hong Kong and China. *Info*, *8*(1), 16–28. doi:10.1108/14636690610643258

Yang, K. (2007). Exploring factors affecting consumer intention to use mobile advertising in Taiwan. *Journal of International Consumer Marketing*, *20*(1), 33–49. doi:10.1300/J046v20n01_04

Yao, M. Z., & Flanagin, A. J. (2006). A self-awareness approach to computer-mediated communication. *Computers in Human Behavior*, *22*, 518–544. doi:10.1016/j.chb.2004.10.008

Yi, M. Y., Jackson, J. D., Park, J. S., & Probst, J. C. (2006). Understanding information technology acceptance by individual professionals: Toward an integrative view. *Information & Management*, *43*, 350–363. doi:10.1016/j.im.2005.08.006

This work was previously published in the International Journal of Handheld Computing Research (IJHCR), Volume 2, Issue 1, edited by Wen-Chen Hu, pp. 72-93, copyright 2011 by IGI Publishing (an imprint of IGI Global).

Chapter 8
A Framework for the Quality Evaluation of B2C M-Commerce Services

John Garofalakis
University of Patras, Greece

Antonia Stefani
University of Patras, Greece

Vassilios Stefanis
University of Patras, Greece

ABSTRACT

Business to consumer m-commerce services are here to stay. Their specifics, as software artifacts, indicate that they are primarily and most importantly user-driven; as such user perceived quality assessment should be an integral part of their design process. Mobile design processes still lack a formal and systematic quality control method. This paper explores m-commerce quality attributes using the external quality characteristics of the ISO9126 software quality standard. The goal is to provide a quality map of a B2C m-commerce system in order to facilitate more accurate and detailed quality evaluation. The result is a new evaluation framework based on decomposition of m-commerce services to three distinct user-software interaction patterns and mapping to ISO9126 quality characteristics.

INTRODUCTION

Mobile services are now a reality that are seamlessly and cost-effectively experienced by a large corpus of users. The migration of services from the WWW to the mobile arena was spearheaded by e-commerce vendors which took the opportunity of using a new medium to promote their goods.

There is an enthusiasm in business, academia and users for mobile services, and this enthusiasm is the impetus for not only the research of the novel but for the adaptation of the old (Jahns, 2009; Bouwman et al., 2008; Büyüközkan, 2009).

E-commerce, in the form of Business to Consumer transactions is one of the primary business successes of the WWW. It is only natural that

DOI: 10.4018/978-1-4666-2785-7.ch008

enterprises sought to increase their market share by moving to the mobile Web as well. Mobile commerce (m-commerce) systems are developed at an increasing rate in recent years. As a business process, m-commerce is viewed as particular type of e-commerce (Coursaris, 2002) and refers to a transaction with monetary value that is conducted via a mobile network. When users conduct m-commerce such as e-banking or purchase products, they do not need to use a personal computer system. Indeed, they can simply use some mobile handheld devices such as Personal Digital Assistants (PDA) and mobile phones to conduct various e-commerce activities. In the past, these mobile devices or technologies were regarded as a kind of luxury for individuals. However, this situation has changed. Technology has driven the growth of the mobile services industry thus creating a new opportunity for the growth of m-commerce (Ngai, 2007; Huang et al., 2007; Gunasekaran & McGaughey, 2009). Location-based services are also attracting the attention of the business world (Junglas, 2007).

Focusing on B2C services (Business to Consumer services), this uniqueness is both a blessing and a curse. Being user-intensive, it is absolutely imperative that the software satisfies mobile user needs; mobile commerce user needs are, in many perspectives different than Internet-based e-commerce user needs mainly because the access medium is different. Thus, the quality of the software itself, which is the satisfaction of implied and non-implied user needs, is of primary importance. To date, most research efforts focus on Quality of Service that deals mostly with low-level network attributes (Ghinea & Angelides, 2004; Lu et al., 2009). The literature also includes an ever-increasing number of research efforts that analyze specific technical (Chen & Nath, 2008), socio-economic (Li & McQueen, 2008) and cultural (Dai & Palvi, 2009; Constantiou et al., 2009) issues involved in m-commerce adoption and use.

The research on the quality of B2C m-commerce systems is a new and challenging task; especially the quality of mobile commerce systems as it is perceived by the end-user is only now becoming a research issue. However, providers of mobile services and mobile hardware have always paid attention to ergonomics and usability. Google's Android platform (http://www.android.com) is an approach that aims to attract the novice user and actually increase the total target group of advanced mobile services by creating new users. Usability is not the only dimension of software quality. According to ISO standards, many dimensions to software quality need to be satisfied. A user perspective, rather than a developer perspective, of quality is important (Hong et al., 2008).

The quality of software is a principle concern to end-users and developers as well. It is increasingly difficult to evaluate diverse software such as m-commerce. The latter provides a wealth of different services, different in the sense that different technologies and user-service interaction patterns are used. By identifying these differences in the level of basic services it becomes easier to apply different evaluation methods that are suitable for each case. Such a method would permit a detailed quality evaluation with an increased practical impact. After all, different software artifacts should be evaluated with methods focusing on their uniqueness. Having these in mind, one of the main questions posed is how to identify these differences and how to cluster the services according to them. Another problem is that a formal evaluation method should be used in order to provide a concise solution. It is with the above observation that this paper examines the quality attributes of m-commerce systems adopting the ISO9126 software quality standard (ISO, 2001). ISO9126 is a general standard for software quality that is user-driven. Because of its generality, it can be applied to any kind of software. In order for it to be practical however it must be seen in the light of a specific application domain. Adopting and adapting ISO 9126 for specific domains is not new and not foreign to the standard itself (Losavio, 2004; Cote, 2005). A usual approach is to enhance the hierarchical and (by design) open scheme to include more attributes suitable for a domain (Stefani & Xenos, 2008).

Building on ideas initially presented in (Garofalakis et al., 2007), this paper explores B2C m-commerce quality attributes using the external quality characteristics of ISO9126 of Functionality, Usability, Efficiency and Reliability. A new evaluation framework is proposed based on decomposition of m-commerce services to three distinct user-software interaction patterns and mapping to ISO9126 quality characteristics. The contribution of this work is the m-commerce specificity of the proposed technique, a technique that is flexible and extendable.

Software Quality Evaluation

M-commerce as an application area deals with a diversity of characteristics organized in three main categories (Tarasewich, 2003): environment, participants and activities. Environment relates physical properties, location and orientation issues, availability and quality of devices and communications. The Participant category focuses on personal properties (age, education, preferences, etc), personalization aspects and expectations. Finally, the Activity category includes tasks and goals of the mobile user combined with events of the environment. Environment, participants and activities interact in order to achieve the success of the m-commerce system that is the consumer-centered quality satisfaction.

In this paper, ISO9126 quality standard is proposed in order to map end user interaction with the mobile software m-commerce applications. ISO attacks the problem of defining software quality by decomposing it to several sub-problems and by questioning about what are the different behavioral patterns of software as it interacts with the hardware, the users or other systems. As a result, many different standards were defined, creating a lot of confusion. After a significant effort to reduce the numbers of standards, the ISO9126 standard

release 2004 was defined and is considered to date the main software quality standard of ISO (International Organization for Standardization, 2004). It includes guidelines of how the software should behave internally and externally in order to be of good quality; it provides tangible tools called metrics as practical measures of quality. The standard, by definition does not provide guidelines on how to build quality software but guidelines on the characteristics of good quality software. For this it has received some criticism about its practicality, especially compared to relevant W3C initiatives. We consider them complementary as they have different goals.

According to ISO9126, quality is defined as a set of features and characteristics of a product or service that bear on its ability to satisfy stated or implied needs. In order to provide a developer view of software, besides the end user's view and guidelines for overall assessment of quality (Cote et al., 2005) the latest revision of four-part ISO9126 software quality standard has been proposed.

ISO9126: Part 1 defines the quality model for software products (Figure 1). The other three parts discuss the metrics that are used to evaluate the quality characteristics defined in Part 1 which are internal metrics, external metrics, and quality in use metrics. The quality model is subdivided into two parts: the quality model for internal quality characteristics and external quality characteristics, and the quality model for quality in use. A quality characteristic defines a property of the software product that enables the user to describe and appraise some product quality aspect. A characteristic can be detailed into multiple quality sub-characteristics.

External quality characteristics are observed when software products are used, that is, they are measured and appraised when the products are tested, resulting in a dynamic view of the software. Evaluation of internal quality characteristics is accomplished by verifying the software project and source code, resulting in a static view. The quality model for internal and external character-

Figure 1. ISO9126 software quality standard: part 1

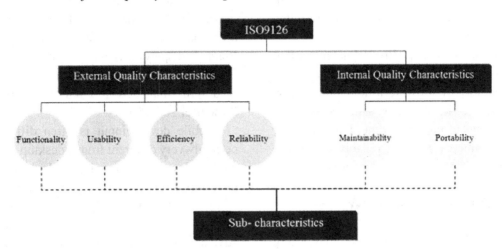

istics categorizes quality attributes into six characteristics: functionality, usability, efficiency, reliability, maintainability, and portability. Each of these characteristics is subdivided into quality sub-characteristics. These quality characteristics can be used as goals to be reached in development, selection and acquisition of components and also as factors in predicting properties of component-based applications.

The external quality characteristics of ISO9126 quality model may be used as basis for m-commerce quality evaluation but further analysis and mapping of its characteristics is required. The main issue is how m-commerce system's quality can be analyzed using this standard. In this work, we use the following external quality characteristics of ISO 9126 to evaluate m-commerce systems: Functionality, Usability, Efficiency and Reliability. Each of the above mentioned characteristics provide the quality framework (actually the baseline) on which an m-commerce system may be built, taking into account end-users requirements. The external quality characteristics of ISO9126 are defined as follows:

- **Functionality:** A set of functions and specified properties that satisfy stated or implied needs. The meaning of Functionality

is to provide integrative and interactive functions in order to ensure end-user convenience. Especially for m-commerce systems Functionality refers to the existence of these functions and services that support end user's interaction via the mobile system.

- **Usability:** A set of attributes that bear on the effort needed for the use of a product or service, based on the individual assessment of such use by a stated or implied set of users. Usability is an important quality characteristic as all functions of an m-commerce system are usually developed in a way that seeks to facilitate the end-user by simplifying end-user's actions; this fact can however affect negatively the system in certain cases.

- **Efficiency:** A complex concept that entails both conceptual challenges as well as implementation difficulties. Efficiency is defined as the capability of the system to provide appropriate performance, relative to the amount of resources used, under stated conditions. It refers to a state where system functions are both usable and successful, i.e. they achieve their aim, the reason for their existence. One of the main criteria of efficiency of an m-commerce system is the quality relating to time and resource behavior.

- **Reliability:** The quality characteristic that refers to a set of attributes that bear on the capability of software to maintain its performance level under stated conditions for a stated period of time. Especially for m-commerce systems, reliability refers to systems tolerance on end users actions.

Mobile-Readiness of WWW

The World Wide Web is not mobile-ready. Many Web pages are laid out for presentation on desktop size displays exploiting capabilities of desktop browsing software (Burigat et al., 2008). Accessing such a Web page on a mobile device often results in a poor experience. The main factor resulting in this negativity is page size and layout. Because of the limited screen size and the limited amount of material that is visible to the user, context and overview are often lost. A page may require considerable (vertical) scrolling to be visible, especially if the top of the page contains many images and/or navigation links. Layout patterns such as dense text and chunks of hyperlinks are also discouraging user from continuing their on-line experience. A few of the parameters that affect mobile browsing in general include page layout, input devices used, network speed and device ergonomics with respect to software handling.

A psychological rule for successful browsing is to facilitate, as soon as possible, the creation of a mental picture of the site a user chooses to visit. This is a seamless process for most web sites. This is not however the case when a mobile device is used. Disorientation, difficulty to decode the structure of a web page, that is, no immediate feedback as to whether information needs are fulfilled may result to increased drop-out rate (the user leaves the web site with a high probability of not visiting it again). Consistency is becoming

a vital factor to success. Dense text, numerous hyperlinks, large images, lengthy forms and tables are negatively affecting the browsing experience. Figure 2 displays a web page with dense text. It is obvious that so much information cannot be read in the small screen of a mobile phone even if the user zooms in.

Mobile device input is often difficult and certainly very different from a desktop computer equipped with a keyboard. Mobile devices often have only a very limited keypad, with small keys, and sometimes with no pointing device. Latest releases include track balls and touch screens, an advance that significantly facilitates user input. Lengthy URLs and those that contain a lot of

Figure 2. A classic B2C site with dense hyperlink structure as seen by a mobile browser

punctuation are particularly difficult to type correctly. Because of the limitations of screen and input, forms are hard to fill in as well. This is because the navigation between fields may not occur in the expected order and because of the difficulty in typing into the fields. While many modern devices provide back buttons, others do not, and in some cases, where back functionality exists, users may not know how to invoke it. This means that it is often very hard to recover from browsing errors.

Mobile networks can be slow compared to fixed data connections and still have a measurably higher latency. This can lead to long retrieval times, especially for lengthy content and for content that requires a lot of navigation between pages. Mobile data transfer costs money. The fact that mobile devices frequently support only limited types of content means that a user may follow a link and retrieve information that is unusable on their device. Even if the content type can be interpreted by its device there is often an issue with the experience not being satisfactory - for example, larger images may only be viewable in small pieces and require considerable scrolling. Web pages may contain content that the user has not specifically requested for - especially advertising-related images or large images. In the mobile world this data contributes to poor usability and may add considerably to the cost of the retrieval. Cost is an issue if the user is charged by the kilobyte.

Mobile users typically have different interests compared to users of fixed or desktop devices. They are likely to have more immediate and goal-directed intentions than desktop Web users. Their intentions are often to find out specific pieces of information that are relevant to their context. An example of such a goal-directed application might be a user requiring specific information about schedules for a journey he/she is currently undertaking. Mobile users are typically less interested in lengthy documents or in browsing lengthy pages. The ergonomics of the device are frequently unsuitable for reading lengthy documents, and users will often only access such information from mobile devices only when more convenient access is not available.

Developers of commercial Web sites should have in mind that different commercial models are often at work when the Web is accessed from mobile devices as compared with desktop devices. For example, some mechanisms that are commonly used for presentation of advertising material (such as pop-ups and large banners) do not work well on small devices.

As noted above, the restrictions imposed by the keyboard and the screen typically require a different approach to page design than for desktop devices. Various other limitations may apply and these have an impact on the usability of the Web from a mobile device. Mobile browsers usually do not support scripting or plug-ins, which means that the range of content that they support is limited. In many cases, the user has no choice of browser and upgrading is not possible. Some activities associated with rendering Web pages are computationally intensive - for example re-flowing pages, laying out tables, processing unnecessarily long and complex style sheets and handling invalid markup. Mobile devices typically have quite limited processing power, which means that page rendering may take a noticeable time to complete. As well as introducing a noticeable delay, such processing uses more power as does communication with the server. Many devices have limited memory available for pages and images, and exceeding their memory limitations results in incomplete display and can cause other problems.

The above-mentioned limitations apply in e-commerce sites as well. In fact, e-commerce users are much more demanding than a regular Internet user with a general interest in information browsing. Frequent on-line buyers, having used to high quality e-commerce services in the WWW, (a level of quality, which was reached after some years of maturing both technologically and ergonomically), are more demanding (and less forgiving) from m-commerce sites. Penalty for low quality (probably) affects both the device used and the site visited; and the penalty for poor services is the slow death of on-line commerce: a shrinking number of visits and a resulting reduced income.

Mobile Web Best Practices

The limitations presented briefly in the previous section were noticed early on, and significant efforts, especially by the W3C were initiated in order to overcome them. The W3C mobile web best practices were born as a result (W3C, 2008; W3C, 2010). They can be considered the first step towards increasing the usability and partially the efficiency of web sites when accessed from a mobile browser. Their advantage is that they are practical however, they do not embrace quality as a whole; at least quality as it is addressed by ISO.

Mobile web best practices and mobile ok basic tests are the result of two different working groups of W3C Mobile Web Initiative (MWI). The Mobile Web Initiative is led by worldwide key players in the mobile production chain, including authoring tool vendors, content providers, adaptation providers, handset manufacturers, browser vendors, and mobile operators. There are nineteen MWI Sponsors: Ericsson, France Telecom, HP, Nokia, NTT DoCoMo, TIM Italia, Vodafone Group Services Limited, Afilias, Bango, Jataayu Software, Mobileaware Ltd., Opera Software, Segala, Sevenval AG, Rulespace, and Volantis Systems Ltd.

The mobile web best practices document specifies best practices for delivering Web content to mobile devices. The principal objective is to improve the user experience of the Web when accessed from such devices. The recommendations refer to delivered content and not to the processes by which it is created, nor to the devices or user agents to which it is delivered. In other words, mobile web best practices refer to how the web content should be presented to the end user, independently to his/her device or to the adaptation mechanisms the network may use (e.g. content

adaptation proxies). There is no proposition yet specifically from m-commerce.

The sixty best practice statements are grouped in five categories:

1. **Overall Behavior:** General principles that underlie delivery to mobile devices.
2. **Navigation and Links:** Because of the limitations in display and of input mechanisms, the possible absence of a pointing device and other constraints of mobile devices, care should be exercised in defining the structure and the navigation model of a Web site.
3. **Page Layout and Content:** This category refers to the user's perception of the delivered content. It concentrates on design, the language used in its text and the spatial relationship between constituent components. It does not address the technical aspects of how the delivered content is constructed.
4. **Page Definition**
5. **User Input:** This section contains statements relating to user input. This is typically more restrictive on mobile devices than on desktop computers and often a lot more restrictive.

In order to allow content providers to share a consistent view of a default mobile experience, the W3C has defined the Default Delivery Context, a simple and largely hypothetical mobile user agent. This allows providers to create appropriate experiences in the absence of adaptation and provides a baseline experience where adaptation is used. The Default Delivery Context (DDC) has been determined by the W3C as being the minimum delivery context specification necessary for a reasonable experience of the Web. It is recognized that devices that do not meet this specification can provide a reasonable experience of other non-Web services. The Default Delivery Context is presented in Table 1.

Note that many devices exceed the capabilities defined by the DDC. Content providers are encouraged not to diminish the user experience on

Table 1. Default delivery context

Characteristic	Value
Usable Screen Width	120 pixels, minimum
Markup Language Support	XHTML Basic 1.1 delivered with content type application/ xhtml+xml
Character Encoding	UTF-8
Image Format Support	JPEG and GIF 89a
Maximum Total Page Weight	20 kilobytes
Colors	256 Colors, minimum
Style Sheet Support	CSS Level 1. In addition, CSS Level 2@media rule together with the handheld and all media types
HTTP	HTTP ver1.0 or more recent HTTP ver1.1
Script	No support for client side scripting

those devices by developing only to the DDC specification, and are encouraged to adapt their content, where appropriate, to exploit the capabilities of the device used. Web applications should adapt to known or discoverable properties of the Delivery Context by adjusting the content, navigation or page flow, with a view to offering a good user experience on as broad a range of devices as possible. In order to discover device's capabilities server side or client side techniques (e.g. JavaScript) it may be used. From the other hand, if a large number of devices are being targeted, or if the mobile web application is sensitive to the permutations of a large number of configuration properties, the number of application variants required might quickly become unmanageable. To combat this, classification of target devices is proposed (W3C, 2010). Then the developer may build a single application variant for each class.

B2C M-Commerce System Quality Patterns

The overall idea of modeling a B2C m-commerce system is that software artefacts that exhibit different behavior when invoked require a different evaluation approach. One cannot usually evaluate with the same method different thinks and expect to get precise measurements. Thus, by recognizing the different service categories that need to be handled differently by the user, either as a process or as a user-software interface and then by grouping the provided functions to these categories we create distinct function evaluation clusters. By mapping the functions to ISO9126 external sub-characteristics we provide a focus for the evaluation. There is strength for each relation between a function and the sub-characteristics. We consider this strength mostly user-perceived and so we have contacted a survey to record it. Binding these two steps together we answer the question of how to evaluate which function.

In order to model the interaction among the end user and the m-commerce system we consider four different interaction patterns: Presentation, Navigation, Purchasing and Location-based. Presentation describes how a product or service is presented to the end user. For example, a book may be presented using an image-snapshot of its content and an electronic device by a 3D animation. Navigation describes the various mechanisms provided to the end user for accessing information and services of the m-commerce system. Site structure, menus, shortcuts and all those means that facilitate the browsing process are included here. Purchasing refers to the facilities provided for the commercial transaction per se. These interaction patterns are usually applied through a browser, just as in web e-commerce. Mobile device however take into account user location. Either push or pull m-commerce services are available. In the first category, the user's location triggers software proximity switches and adds or offers from nearby points of sale may appear. The user may choose to enable or disable such services or even make a list of the products or providers of interest. Information may come in the form of an SMS or a comment-like banner on a map. Pull services are invoked by the user usually through a query mechanism. Location-based pull services provide m-commerce information based on a proximity or a geographic area query. For example, queries such as "show me the electronic retails shops that sell iPhone near my current position" or "which electronic retail shops are offering discounts in an area of about 1km around my location" are common. This type of service is not the classic B2C commerce that we know, but it definitely requires an on-line presence since information, either pull or push must be available to mobile users. This is a type of m-marketing or m-recommender mechanism similar to the classic recommender mechanisms of a B2C system. It is however location-based driven. The most frequent medium to access such information are maps provided either through a browser or through a map service.

Applying the above steps to m-commerce requires an adjustment to the attributes that the system presents because of its wireless communication character. In the following paragraphs we present the functions (we call them attributes because they include both services and systems characteristics) of mobile systems that constitute end user purchasing process.

The aim of this paper is not to describe all existing B2C m-commerce attributes or fully present their use but rather to offer a quality evaluation of these attributes and to present a quality framework for m-commerce systems. The patterns are discussed in the following sub-sections while their categorization and mapping is presented later. One could say that an attribute is mapped to all external quality sub-characteristics of ISO9126 so why should there be a need for mapping? The hypothesis is correct however not all relations of the type attribute-quality sub-characteristic

are of the same importance to the quality evaluation process. In fact, some of the relations are stronger than others. On the other hand none are so weak as to be considered negligible. We call the strength of the relation, weight. We consider the weight mostly user-dependable. This means that the quality performance of an attribute is evaluated by the user and thus the user contributes to the forming of a strong or a weak relation between the attribute and the expectations he/she has of the software. Expectations are closely linked to needs and thus to quality. It is difficult to measure the exact weight of a relation even when expert evaluators participate in a survey to determine them. Although exact measurements are not feasible within the scope of this research, a crude measure of the weight for each relation is calculated through a user survey.

The Presentation Pattern

Presentation is supported basically by text and images because mobile devices present limitation such as screen size and resolution, number of supported colors, computation power, memory size, rate of data transfer and energy required for proper functionality. Color usage is also important. Using colors obviously gives a pleasant and friendly interface, but a too colored screen confuses. All the pages of the m-commerce system must have the same colors so the user can feel that he/she is navigating in the same environment. By removing background images, background colors and text colors we increase the readability of the content. The use of images in Internet applications is common. Nevertheless, using images in mobile web applications significantly increases download and response time and thus, usage cost. See Figure 3.

Presentation issues are also related with thematic consistency and the default delivery context which intends to provide an acceptable mobile environment for any end user from different mobile devices. The clarity of the text presented with meaningful, short and simple words and the pre-

Figure 3. Presentation of a book using image, text and hyperlinks. A limited snapshot of the contents is also available.

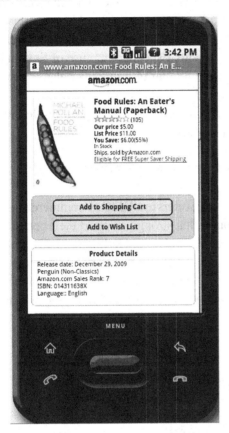

Figure 4. Limited use of image links and a handful of only the most important shortcuts in this Amazon web page

sentation of the central meaning at the first page of each mobile device contributes attributes that an m-commerce system should provide to the end user for an accessible mobile environment. Additionally providing a descriptive title for the page allows easy identification of the content and by keeping the title short reduces page weight, and bear in mind that it may be truncated.

The Navigation Pattern

The navigability of an m-commerce system is a critical factor for its success. Navigation is an important design element, allowing users to acquire more of the information they are seeking and making that information easier to find. See Figure 4.

Navigation issues support m-commerce systems quality by taking into account the quality of components such as indexes, navigation bars, site maps and quick links. The availability of these components facilitates access of information and services and enables users to locate efficiently the information they need, while avoiding usability bottlenecks. Additionally, navigation concerns the facilities for accessing information and the connectivity of the above systems.

Navigation refers at attributes that support the navigability of the m-commerce systems. These refer to navigation bars, which according to W3C Mobile Web Best Practices 1.0, should be placed on the top of the page. Any other secondary navigational element may be placed at the bottom of the page if really needed. It is important the users should be able to see page content once the page has loaded without scrolling. M-commerce systems, as e-commerce systems provide simple metaphors such as shopping cart where the end user can insert the products that intend to buy. Mobile devices present limitations on text inputting so an m-commerce system will be enabled by attributes such as access keys (keyboard short cuts), by providing defaults at any function that the user should select an action and by avoiding free text with minimum text inputting. The navigability of the mobile system is supported from search services that are also related with device capabilities and context presentation as well. Search with simple text inputting in an AND/OR operator format enables the user to find the information needed without navigating to several mobile pages. Search attributes can reduce the cost of mobile browsing and prevent navigability difficulties.

Additionally, because of the limitations in display and of input mechanisms, the possible absence of a pointing device and other constraints of mobile devices, care should be exercised in defining the structure and the navigation model of a Web site. Especially the use of links should be limited aiming to provide a balance between having a large number of navigation links on a page and the need to navigate multiple links to reach content.

The Purchasing Pattern

Purchasing refers to all B2C m-commerce systems attributes that strongly support their commercial character of web systems (Figure 5). In particular, it refers to attributes that support the interaction with the m-commerce system. These attributes are also related to the navigability of the system but

Figure 5. Two steps of the purchasing process: shipping details and credit cards information entry

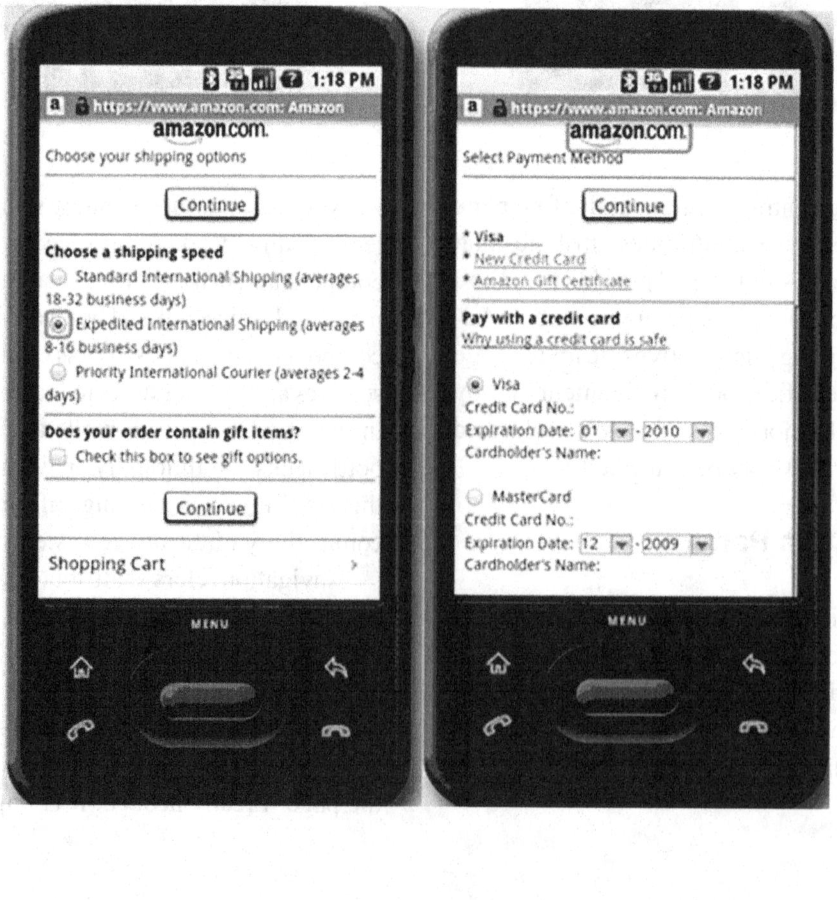

they are categorized differently because of their significant contribution to the purchasing process.

Purchasing process success is also related to the stability of the process via the m-commerce system and issues like error tolerance and error recovery at this crucial procedure. M-commerce systems success and trustworthiness is based on the system's tolerance on the above issues.

Authentication and personalization attributes support an m-commerce system where the end user can provide private information (i.e. Credit Card Number).

The Location-Based Pattern

Localization services can enable the presentation of the products and service because the m-commerce system can recommend the best selection based on end user's positioning (Figure 6). User's positions may be acquired automatically through GPS or may be entered manually from the user. Additionally notification services provide great advantage to m-commerce systems because they can also be combined with localized information. Alternative payment methods either support a complete transaction via the m-commerce system or otherwise combined with localized information can allow the mobile user to conclude a transaction to the closest sales point.

The main functions that make use of basic context information (e.g. the current location of the user) could be categorized as follows:

- **View:** They are generally four available view: map, traffic map, satellite and enriched map. The plain map depicts the roads and blocks of a city without any other information of interest to the user. Additional information is depicted only when the user performs a query. Traffic maps are an extension, provide traffic information. They are usefully mainly in Business to Business service like fleet management or goods monitoring. Satellite maps provide

Figure 6. Location of bookstores near the user's location. Most of them exist on-line as well.

mainly terrain information and are useful for special applications that make use of geospatial services. An enhanced map contains points of interest (POI). The type of the POI is defined by the user. In m-commerce, it may include sale-points, hotels, bookstores etc.

- **Navigation Functions:** In LBS the user browses information which is located in a map. Just as in the case of a browser, there are standards and special functions that facilitate navigation. They include options for free moving over the map, zoom, Search for POIs in the map, directions (from starting point to finishing point), history, back and forth buttons, browsing over POIs, listing of POIs information. These functions belong to the Navigation facet of the system.

- **Context-Awareness:** Attributes that make explicit use of the positioning mechanism include calculation of current location and appearance on the map, triggered messages, location-based billing (mobile vouchers), direction (from or to current location), local information services and POIs near current location.

Although the above-mentioned attributes do not directly constitute m-commerce functions they are often used as supportive functions. For example, the user query "show me in which stores near my location I can pick-up the book I bought on-line?" involves the pinpointing of the user's current location (localization) and a search for specific POIs in a region of interest.

Location-based services are either pull or push. Pull services are activated by the user (e.g. a query) and push services by the service provider (e.g. sales' offers near current location). In m-commerce notification services are the most commonly used. Location information can be mixed with time-dependent information especially as a support to mobile ticketing services.

New devices that make use of a wealth of sensors will be able to support more supporting functions that pull data depending not only on the location but on other parameters as well (e.g. orientation, speed etc.) (Wright, 2009).

QUALITY EVALUATION OF B2C M-COMMERCE

In this survey approach, three expert quality evaluators were selected in a heuristic evaluation method (Nielsen, 1990). Heuristic evaluation is done by looking at an interface and trying to come up with an opinion about what is good and bad about the interface. Ideally people would conduct such evaluations according to certain rules, such as those listed in typical guidelines documents. The evaluators for this method are IT experts

with experience in quality evaluation and mobile systems as well. The nature of the presented evaluation method demands the use of expert evaluators because of its technical character.

There was a two-step evaluation process. First, the evaluators were asked to proceed a complete purchase using a mobile phone and two different emulators from their PC. For the evaluation process we have used the Nokia N70 mobile phone. The N70 has a screen with resolution 176x208 pixels and supports 262.144 colors. The phone can also connect to 3G networks for high rate data transfers using the Opera Mobile 8.51 browser. In order to avoid operability issues for the Nokia N70, help about the functionalities of the device was provided during the evaluation process. The emulators were Google's Android and OpenWave.

The three evaluators have browsed in three popular m-commerce systems according to Google Search in order to have a recent m-commerce experience. Each evaluator was asked to assess specific m-commerce attributes and evaluates each one by assigning one value of relevance (r_{ij}). Relevance defines the correlation among the m-commerce systems attribute i (presented in Table 2) and software quality characteristic j ordered as they presented in the paper (i.e. j=1 for Functionality, j=2 for Usability, j=3 for Efficiency and j=4 for Reliability) using a five-grade Liker-type scale. The evaluator may select from the Liker-type scale assigning one different value for each quality characteristic.

$$r_{i,j} \begin{cases} 1, no \\ 2, weak \\ 3, strong\ corellation \\ 4, very\ strong \\ 5, critical \end{cases}$$

This provides a qualitative representation of m-commerce systems quality and especially gives emphasis on external quality characteristics.

Table 2. Mapping of attributes to quality characteristics per pattern and weights of the relations

M-Commerce Attributes	Quality Characteristics			
	F	U	E	R
Presentation				
Product's description	3	5	3	3
Still images	3	5	3	1
Use of Text	4	5	3	2
Use of Colors	3	5	4	2
Use of Graphics	4	5	3	2
Clarity	3	5	4	2
Content Theme	3	5	4	1
Text inputting	4	5	4	1
Thematic consistency	2	5	4	2
Provide defaults	3	5	4	2
Navigation				
Navigation mechanism	4	4	4	3
Uploading Time	3	3	5	4
Access keys	4	5	4	2
Use of Links	4	4	3	3
Help	5	5	3	3
Feedback	3	5	4	3
Undo functions	5	3	3	5
User oriented hierarchy	2	5	4	3
Redirection	5	3	4	3
navigation bar	5	5	3	1
Scrolling	3	5	4	2
Search response time	2	4	5	4
Search results processing	3	4	5	3
Purchasing				
Shopping cart –Metaphor	4	5	4	2
Security mechanism	3	2	4	5
Pricing Mechanism	3	4	3	3
Alternative payment methods	4	4	3	4
Authentication	5	2	3	5
Personalization	4	5	4	2
Transaction recourses behavior	3	3	5	4
Error recovery	3	3	3	5
Errors tolerance	4	3	4	4
Stability	4	3	3	5

continued in following column

Table 2. Continued

M-Commerce Attributes	Quality Characteristics			
	F	U	E	R
Location-Based				
Mobile ticketing	5	5	2	3
Mobile vouchers	4	4	3	2
P2P information service	4	4	2	3
Localization	4	5	3	1
Notification service	3	5	3	4

Quality evaluation of m-commerce systems attributes provides a quantitative representation of e-commerce systems' quality. Table 2 provides the evaluation results for the m-commerce attributes presented in the previous section. Especially presents the values of function relevance (r) for each attribute. These values are the average values of all evaluators approximated in monad.

Based on the evaluation results, quality of B2C m-commerce systems can be modeled in external quality characteristics and attributes. Providing a value for each attribute an ordered list for each external quality characteristic is provided. These values provide a first impression of end users preferences and perquisites about m-commerce systems' attributes.

The categorization of these attributes provides important feedback for m-commerce systems' assessment which is in an initial stage. By evaluating the attributes that an m-commerce system provides to the end user we also offer an end user perception of quality. End user's experience is a critical determinate of success in mobile web applications. If end users, who are also the customers, cannot find what they are searching for, they will not buy it; a site that buries key information impairs business decision making. Poorly designed interfaces increase user errors, which can be costly. A user-centered evaluation approach supports all the tasks users need to accomplish using different m-commerce systems' attributes. The above evaluation process provides measurement results which can be also be defined as metrics for a quantitative representation of m-commerce systems' quality.

In order to evaluate m-commerce systems features a new metric that summarizes the relevance of each attribute is introduced. This metric is called Mobile Attributes Weight (MAW) and it provides an evaluation weight with respect to the four quality characteristics. It is calculated by the following formula:

$$MAW = normalized \sum_{i=1}^{4} r_{ij} \in [0,1]$$

where r_{ij} is the relevance for every listed m-commerce system attribute. The value for MAW provides a numerical value for every m-commerce system attribute and an ordered list about end user preference based on external quality characteristics. MAW actually represents attributes importance for the end user and can be used at the development phase in order to define end user preferences. The values for *MAW* need to be further specified, probably with experience testing in future work and the use of different end users' groups.

The evaluation process provides also interesting results about the quality characteristics. In an up and down processing of values r_{ij} the WF=0,24, WU=0,30, WE=0,26, WR=0,20 values have been defined as the normalized average values for each quality characteristic. From these values arises that m-commerce end users gives great emphasis to Usability and Efficiency issues and less on Functionality and Reliability. These values differ from e-commerce systems where Usability and Functionality have equally great importance (Stefani & Xenos, 2008). In e-commerce systems the end users expects different and usable functions/services, but in m-commerce systems the end user desires the basic functions with increased

efficiency as far as time and resource behavior are concerned.

CONCLUSION

In this paper, we presented a quality evaluation for selected attributes of m-commerce systems and particularly B2C m-commerce systems. This evaluation provides an extendable framework useful for mobile system developers. We believe that this is a step towards more effective measurement of m-commerce systems' quality. We acknowledge that our attributes does not include a complete set and may not cover every aspect of m-commerce systems. The above evaluation results provide an initial research for m-commerce systems' quality.

In this paper a new method has been introduced which measures the value of relevance for each m-commerce system attribute. The theoretical framework for this metric is also presented. The validity of the presented measures should further examine with different user groups in alternative evaluation cases and it is included in future work. It should be mentioned that the values presented are not strictly defined as numerical results but present the correlation among m-commerce systems attributes and external quality characteristics.

Practical application of the evaluation is always an issue. That is, providing tangible information to developers on how to design and develop quality m-commerce applications. A valuable tool to address this need are metrics, the bottom level of the ISO916 model. Metrics are measures of quality. While quality attributes provide a somewhat generic view of quality and for this reason they have attracted criticism for their practicality, metrics provide more information to the mobile application developer/designer. W3C mobile OK tests use such metrics for evaluating the appropriateness of web content for presentation through mobile devices. For example, the existence of long vertical scroll bars in a web site deteriorates its representation in a mobile phone where the screen is of a limited size. This metrics has two values, yes for a need for vertical scrolling and no otherwise. Although this is a somewhat rough approach to quality (i.e. there is no information on how much the vertical scrolling is, if it is existent), it provides an insight on what developers and designers expect from quality evaluation techniques: tangible information upon which design decisions can be relied. Note that metrics do not make the use of ISO characteristics obsolete. They are actually the fine-grained level of the ISO9126 quality pyramid. ISO has recognized the usefulness and has included several metrics for software evaluation in the latest release of ISO9126. However, these metrics are too general to be applied in m-commerce in terms of practical impact. There is a need to produce a new set of mobile-specific web metrics, perhaps beginning with the existing corpus of web metrics and fine-tuning or alter were necessary.

There is a wealth of works that present, analyze, or evaluate the use of web metrics; the majority focusing on web usability. There are no specific e-commerce metrics that could be considered the parental link to m-commerce metrics. Usability is of course an issue. But m-commerce quality is much more than that: it includes the process itself, the functionality, reliability and all the external characteristics defined in ISO9126.

Location-based services pose a new challenge. For example proximity post-sales services (e.g. special offers to clients approaching a sales point) could prove vital to a business engaged in m-commerce. M-commerce functions of the Purchasing facet may be mixed with time and location information services. So where does one start to present useful metrics for m-commerce. Using the patterns as a starting point and the existing corpus of web metrics as a basis, a categorization is possible.

Location-based will turn into context-aware in the near future. Sensors such as magnetic compasses and accelerators are already standard equipment in new mobile/smartphones. New challenges will arise when merging of personal and context data will be made available for processing by the AI-capable mobile devices of the near feature.

M-commerce is an intriguing research area with high dynamicity. New software and hardware create the opportunities for a large future user base. Increased user diversity and the provision of advanced functions to novice users requires software of high quality. Building such software is difficult and the fine-tuning of existing quality evaluation methods would help towards easing the burden of designers and programmers. User driven standards such as ISO9126, when suitably enhanced, are able complement practical initiatives as the ones of W3C. Although practicality will always remain an issue, insight on how to offer quality mobile services is feasible. The work presented in this paper is a step towards this direction.

REFERENCES

Bouwman, H., De Vos, H., & Haaker, T. (Eds.). (2008). *Mobile service innovation and business models*. New York, NY: Springer. doi:10.1007/978-3-540-79238-3

Burigat, S., Chittaro, L., & Gabrielli, S. (2008). Navigation techniques for small-screen devices: An evaluation on maps and web pages. *International Journal of Human-Computer Studies, 66*(2), 78–97. doi:10.1016/j.ijhcs.2007.08.006

Büyüközkan, G. (2009). Determining the mobile commerce user requirements using an analytic approach. *Computer Standards & Interfaces, 31*(1), 144–152. doi:10.1016/j.csi.2007.11.006

Chen, L., & Nath, R. (2008). A socio-technical perspective of mobile work. *Information Knowledge Systems Management, 7*(1-2), 41–60.

Clarke, I. (2001). Emerging value propositions for m-commerce. *The Journal of Business Strategy, 18*(2), 133–148.

Constantiou, I. D., Papazafeiropoulou, A., & Vendelø, M. T. (2009). Does culture affect the adoption of advanced mobile services? A comparative study of young adults' perceptions in Denmark and the UK. *SIGMIS Database, 40*(4), 132–147. doi:10.1145/1644953.1644962

Cote, M., Suryn, W., Laporte, C., & Martin, R. (2005). The evolution path for industrial software quality evaluation methods applying ISO/IEC 9126:2001 quality model: Example of MITRE's SQAE method. *Software Quality Journal, 13*(1), 17–30. doi:10.1007/s11219-004-5259-6

Coursaris, C., & Hassanein, K. (2002). Understanding m-commerce. *Quarterly Journal of Electronic Commerce, 3*(3), 247–271.

Dai, H., & Palvi, P. C. (2009). Mobile commerce adoption in China and the United States: A cross-cultural study. *SIGMIS Database, 40*(4), 43–61. doi:10.1145/1644953.1644958

Garofalakis, J., Stefani, A., Stefanis, V., & Xenos, M. (2007). Quality attributes of consumer-based m-commerce systems. In *Proceedings of the ICETE-Business Conference* (pp. 130-136).

Ghinea, G., & Angelides, M. C. (2004). A user perspective of quality of service in m-commerce. *Multimedia Tools and Applications, 22*(2), 187–206. doi:10.1023/B:MTAP.0000011934.59111.b5

Gunasekaran, A., & McGaughey, R. E. (2009). Mobile commerce: Issues and obstacles. *International Journal of Business Information Systems, 4*(2), 245–261. doi:10.1504/IJBIS.2009.022826

Holzinger, A. (2005). Usability engineering methods for software developers. *Communications of the ACM, 48*(1), 71–74. doi:10.1145/1039539.1039541

Hong, S. J., & Lerch, F. J. (2002). A laboratory study of customers' preferences and purchasing behavior with regards to software components. *The Data Base for Advances in Information Systems, 33*(3), 23–37.

Hong, S. J., Thong, J. Y., Moon, J., & Tam, K. (2008). Understanding the behavior of mobile data services consumers. *Information Systems Frontiers, 10*(4), 431–445. doi:10.1007/s10796-008-9096-1

Huang, W. W., Wang, Y., & Day, J. (2007). *Global mobile commerce: Strategies, implementation and case studies*. Hershey, PA: IGI Global. doi:10.4018/978-1-59904-558-0

International Organization for Standardization. (2004). *ISO/IEC 9126: Software product evaluation –quality characteristics and guidelines for the user*. Geneva, Switzerland: International Organization for Standardization.

Jahns, V. (2009). Mobile computing and urban systems: A literature review. In *Proceedings of the Conference on Techniques and Applications for Mobile Commerce* (pp. 17-26).

Junglas, I. (2007). On the usefulness and ease of use of location-based services: Insights into the information system innovator's dilemma. *International Journal of Mobile Communications, 5*(4), 389–408. doi:10.1504/IJMC.2007.012787

Kwon, O. B., & Sadeh, N. (2004). Applying case-based reasoning and multi-agent intelligent system to context-aware comparative shopping. *Decision Support Systems, 37*(2), 199–213.

Li, W., & McQueen, R. J. (2008). Barriers to mobile commerce adoption: An analysis framework for a country-level perspective. *International Journal of Mobile Communications, 6*(2), 231–257. doi:10.1504/IJMC.2008.016579

Losavio, F., Chirinos, L., Matteo, A., Levy, N., & Ramdane, A. (2004). ISO quality standards for measuring architectures. *Journal of Systems and Software*, *72*, 209–223. doi:10.1016/S0164-1212(03)00114-6

Lu, Y., Zhang, L., & Wang, B. (2009). A multi-dimensional and hierarchical model of mobile service quality. *Electronic Commerce Research and Applications*, *8*(5), 228–240. doi:10.1016/j.elerap.2009.04.002

Ngai, E. W. T., & Gunasekaran, A. (2007). A review for mobile commerce research and applications. *Decision Support Systems*, *43*, 3–15. doi:10.1016/j.dss.2005.05.003

Nielsen, J., & Molich, R. (1990). Heuristic evaluation of users interfaces. In *Proceedings of the SIGCHI Conference on Human Factors in Computing Systems* (pp. 249-256).

Saunders, S., Ross, M., Staples, G., & Wellington, S. (2006). The software quality challenges of service oriented architectures in e-commerce. *Software Quality Journal*, *14*, 65–75. doi:10.1007/s11219-006-6002-2

Stefani, A., & Xenos, M. (2008). E-commerce system quality assessment using a model based on ISO 9126 and belief networks. *Software Quality Control*, *16*(1), 107–129.

Tarasewich, P. (2003). Designing mobile commerce applications. *Communications of the ACM*, *46*(12), 57–60. doi:10.1145/953460.953489

W3C. (2008). *Mobile Web best practices 1.0*. Retrieved from http://www.w3.org/TR/mobile-bp/

W3C. (2010). *Mobile Web application best practices*. Retrieved from http://www.w3.org/TR/2010/CR-mwabp-20100211/

Wright, A. (2009). Get smart. *Communications of the ACM*, *52*(1), 15–16. doi:10.1145/1435417.1435423

This work was previously published in the International Journal of Handheld Computing Research (IJHCR), Volume 2, Issue 3, edited by Wen-Chen Hu, pp. 73-91, copyright 2011 by IGI Publishing (an imprint of IGI Global).

Section 3
Mobile Applications

Chapter 9
MICA:
A Mobile Support System for Warehouse Workers

Christian R. Prause
Fraunhofer FIT, Germany

Marc Jentsch
Fraunhofer FIT, Germany

Markus Eisenhauer
Fraunhofer FIT, Germany

ABSTRACT

Thousands of small and medium-sized companies world-wide have non-automated warehouses. Picking orders are manually processed by blue-collar workers; however, this process is highly error-prone. There are various kinds of picking errors that can occur, which cause immense costs and aggravate customers. Even experienced workers are not immune to this problem. In turn, this puts a high pressure on the warehouse personnel. In this paper, the authors present a mobile assistance system for warehouse workers that realize the new Interaction-by-Doing principle. MICA unobtrusively navigates the worker through the warehouse and effectively prevents picking errors using RFID. In a pilot project at a medium-sized enterprise the authors evaluate the usability, efficiency, and sales potential of MICA. Findings show that MICA effectively reduces picking times and error rates. Consequentially, job training periods are shortened, while at the same time pressure put on the individual worker is reduced. This leads to lower costs for warehouse operators and an increased customer satisfaction.

DOI: 10.4018/978-1-4666-2785-7.ch009

INTRODUCTION

The four fundamental processes of a warehouse are (Tompkins & Smith, 1998):

1. To receive incoming goods for storing.
2. To store goods until they are required.
3. To prepare requested goods for shipping (picking).
4. To ship the picked goods (sometimes called packing).

Among all the processes of logistics, picking is the most problematic one because it is highly error-prone (Miller, 2004). Many different types of errors are known (Lolling, 2003): picking of wrong types or quantities of articles, complete omission of a type, and insufficient quality of delivered articles (see Figure 1). All these errors cause high costs for manufacturers and warehouse operators, either because extra shipments and returns are necessary, or, in the worst case, because contract penalties have to be paid.

In today's lean production, where only small resource reserves are kept at the manufacturing site, the resources necessary for production are usually delivered to a customer just when he needs them. The orders are possibly known to the ware-

Figure 1. Picking errors

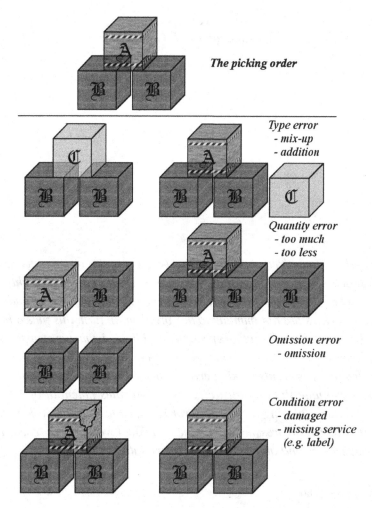

house weeks before but delivery is expected exactly at the specified date. If an important item from the order is missing, this can mean that the whole production has to stop, incurring extra costs for the warehouse for courier delivery, and the customer who then lets the warehouse pay for the financial damage of the production halt. Besides causing huge costs, this certainly has potential to annoy customers. Accordingly, the primary goal for warehouses is to eliminate or at least reduce the number of errors.

Especially warehouses with human workers are confronted with returns caused by incorrect delivery of items. But although humans constitute the soft spot in this process, completely automated solutions are not an option for most warehouses because human workers are much more flexible (see Figure 2).

During economic peak times, warehouses are forced to employ unskilled workers in order to cope with the increased workload. These unexperienced workers are not familiar with the structure and organization of the warehouse, yet have to be operational in a short time. They do not have the time to learn from experienced workers where an ordered article can be found, what the fastest routes through the warehouse are, what the exact processes are, or what a certain article looks like.

Nevertheless, work has to be completed without errors and under the same high time pressure that also skilled workers face. Picking errors and time pressure constitute the major problems for unskilled and skilled workers. Hence, there is a need for an intelligent assistance system that supports the workers. By preventing errors, such system also reduces the pressure put on each single warehouse worker. An assistance system for employees needs to support untrained workers as well as experienced workers in their usual way of working and not force them to change habits.

Based on an initial requirement analysis, we propose the Interaction-by-Doing paradigm, which was realized in a first MICA prototype. Its

Figure 2. Worker in a non-automated warehouse

success led to the development of the second MICA pilot for field testing in a productive environment. In a field test we evaluated MICA's usability, its effect on worker efficiency and estimated its world-wide sales potential. Finally, we present related work and conclude.

REQUIREMENTS AND CONCEPTS

In collaboration with two medium-sized non-automated warehouses, we initially collected requirements for our envisioned MICA system. Such system would reduce the pressure put on workers by assisting them with their tasks, and by preventing errors and the costly effects thereof.

For end-users and designers, design proposals can be understood as design probes to explore the characteristics and usefulness of a proposed system. When a prototype is available, end-users can try it and gain personal experience with it. The active involvement of users and a clear understanding of their tasks is the key for a successful system development.

The ISO13407 "Human-centred design processes for interactive systems" standard does not prescribe specific methods for how to achieve these goals; they are to be chosen according to what is state of the art and what is appropriate under the respective project circumstances. Based on practical experiences from other projects, we have devised a scenario-based approach, combined with user interviews, participatory observation and expert analysis, based on the structure proposed by Robertson and Robertson (1999) for mastering requirements.

Their Volere process ensures that all important aspects of requirements are carefully addressed and that the methods applied have proven their value in practical work. The Volere process makes a distinction between global constraints affecting the project, functional requirements and non-functional requirements. Associated with this process is the Volere template. The template makes fine-grained distinctions between different types of requirements and requires that they are assessed in various categorizations. It also captures the rationale for each requirement as well as fit criteria. These criteria are used to evaluate customer satisfaction and dissatisfaction if a requirement is implemented or not. Hence, the Volere process ensures that all important aspects of requirements are carefully addressed.

Requirements

The requirements gathering process was directed by interviews with all stakeholders and focus groups at the two involved warehouses. These provided valuable insight into the work at the warehouse and contributed to a worthwhile understanding of the current and immanent problems, limits, tasks and benefits of a future support system. But much more important than the interviews and focus groups was the participative observation of the work in the warehouses for several days. Initially, the warehouse workers were quite cautious to work exactly to rule but soon they completely forgot the presence of the observers and reverted to their old habits.

Quite remarkably this brought to light several important issues that would never have been discovered with interviews or focus groups: we could observe workers that were deliberately committing errors. For example, when an item was missing or broken they scanned in the correct bar code from the shelf and put another (wrong) item into the order box, thus producing two severe errors: an order with a wrong item (mix-up error) and removing another item from the warehouse that would leave the warehouse inventory in an inconsistent state, causing a missing article in a subsequent order.

The reason for this observed behavior was the cumbersome process that had to be followed when an item of an order was missing or broken: the warehouse worker had to stop his picking and take the warehouse manager in. First he needed to find the warehouse manager in his office – usually on the other side of the warehouse – then inform him that an item was broken or missing. This could lead to some awkward situations due to the short-tempered nature of the warehouse manager. After that both had to go back to the location in the warehouse, so that the warehouse manager could investigate the situation. Then, mostly not without a new critique, the warehouse manager had to go back to his office, enter the error into the system and restart the picking order. Only now the worker could continue picking. This easily could consume more than 30 minutes.

All in all we collected over 50 requirements that were prioritized by our own observation and by experienced workers and warehouse managers. The most important requirements can be summarized as follows:

1. Reduce error-rates. To be effective, MICA must reduce error-rates compared to existing assistance systems. Only this alleviates the workers' fear of picking errors, while at the same time justifying the use of MICA from an economic point of view.
2. Provide the opportunity to process several picking orders at the same time. As experienced workers tend to process several orders at the same time to reduce walking distance.
3. Support trained and untrained workers. MICA must support trained as well as untrained workers, because at peak times the permanent staff is reinforced with unskilled workers. Both groups have inherently different needs. Unskilled workers need a certain time to acquire the knowledge where to find an article. Navigation assistance should help to avoid making detours and to find articles on the picking list. At the same time, MICA should not interfere with habits of experienced workers.
4. Unobtrusive guidance. A novel technology gets rejected if people have to change their way of working or if they feel patronized.

To achieve acceptance, a guidance system may not require much attention but should work in the background. It should only draw attention to prevent a picking error.

5. Usability. Interfaces should be intuitive, provide a good overview and be easy to learn because user training is expensive. Keeping the system responsive enables seamless user interaction and avoids idle time for workers.
6. Working hands-free. Simultaneous handling of mouse or keyboard would disrupt picking. Additionally, workers are usually unfamiliar with computer user interfaces.
7. Environmental conditions. Different environmental conditions play an important role for MICA. For example, the display should remain readable under unfavorable light conditions, while sound should remain hearable despite ambient noise.

Interaction-By-Doing

From above requirements, we derive MICA's Interaction-by-Doing concept that builds on multi-modal and implicit interaction with a calm computing system that provides pro-active help to users with different experience levels. The different concepts' implications are described in the sections below. Interaction-by-Doing is an enhancement of Interaction-by-Movement (Lorenz, Zimmermann, & Eisenhauer, 2005). Interaction-by-Movement means that moving towards a location is recognized by the system, which reacts with a proactive help. In Interaction-by-Doing, interaction is not reduced to movement only but to multiple kinds of behavior. By this, no explicit interaction is necessary.

A practical example is the picking process: when picking with support of a scanner, the worker passes articles by his scanner. The scanner confirms with an acoustical response. Not till then may the worker continue picking. In contrast, when the process is extended with an Interaction-by-Doing system, the worker just puts an article into the box. The Interaction-by-Doing system identifies the article in the background and interrupts the worker (using an appropriate modality) only in case of an error. Interaction-by-Doing eliminates the additional control process in picking and automatically helps with solving the problem by telling the user to remove the wrong article.

Multi-Modality

People's interaction in a completely engineered environment relies on different modalities because a single modality may fail in a certain situation. For example, people in motion cannot focus their full attention on the interaction with a touch screen (Brewster, 2002). In contrast, noisy environments make speech less reasonable than other modalities.

Implicit Interaction

In explicit forms of interaction the user assertively communicates his wishes to the system via specific IT devices like mouse, keyboard or touch screen, and well-known interaction rules like clicking, dragging or typing. Other explicit interaction forms like gesture recognition can be socially obtrusive if they involve sweeping gestures. When such gestures are required for interaction, the user might feel embarrassed because it looks strange to casual bystanders. Still, even if more subtle gestures (like a small tilting) are used, the user wants to explicitly and consciously communicate something to the machine.

Implicit interaction is a passive form of formulating wishes. The idea is to analyze natural movement of the user and to derive reactions by the system. To be able to extract interaction information the system needs to know the tasks and intentions of the user in the specific context. Implicit interactions are unobtrusive but are also less reliable.

Yet the success of a system is highly dependent on its usability because simple and intuitive interaction increases user acceptance. Natural interac-

tion means multi-modal interaction with explicit and implicit interaction modalities. Therefore, the idea of MICA is to offer a suitable interaction modality to every particular situation.

Calm Computing

Weiser and Brown came up with the term "calm computing." They demanded that technology engages both the center and the periphery of our attention, and in fact moves back and forth between the two (Weiser & Brown, 1997). In MICA, the normal picking process remains the center of the worker's attention. MICA stays in the periphery where additional information about the current task can be retrieved. It only moves to the center if an error occurs. A system based on the Interaction-by-Doing principle beholds the workers' actions with various sensors to detect problems.

Proactive Help

Proactive applications want to give adequate help when the user is expected to need it (Kaufmann et al., 2007). A computer system offers proactive help if it deems that this could reduce a worker's stress or if he is about to make an error.

The intention is not to point out that the worker errs but to avoid errors. This saves time and reduces worker frustration. Because of that, help is presented as a non-binding offer. This is necessary so that the worker does not feel patronized. Furthermore, wrong reactions of the system – i.e., the worker's intention was not predicted correctly and hence the system reacts in an undesired way – do not cause unnecessary disturbance as the worker is never explicitly interrupted in his work.

One example is well-known from navigation systems: if the user leaves the proposed path, the route is silently recalculated accounting for the deviation. Hence, the proactive help assures that the optimal path is always displayed without interrupting the user with error messages or commandments. Similarly, MICA does not interrupt the current workflow when it recognizes a deviation. Instead, it just highlights the help button, so that appropriate help can be obtained with a single click.

Concrete situations or problems are identified by observing the worker's actions. Systems' reactions rely on the analysis of the movement history. Hence, movement sensors and a positioning technology are necessary.

Different Experience Levels

Calm computing makes it possible that experienced users are not interrupted by system alerts addressed to unskilled workers. They can work in the efficient ways they are familiar with. As Interaction-by-Doing facilitates implicit interaction, less explicit interaction forms have to be learned by users. This supports both, skilled and unskilled users. Proactive help is designed as a non-binding offer so that experienced workers are not disturbed. At the same time, unexperienced workers are thankful for the help when in need.

DEVELOPMENT METHODOLOGY OF MICA

The first MICA project started in 2004 running until 2006. The goal of the project was to prototypically explore the possibilities of supporting humans in their every-day working environment using technologies like context-awareness and user modeling with mobile devices. For this exploration it did not matter what working environment that was.

The prototype served three purposes: firstly, it allowed us to gather first experiences with a variety of hardware and software technologies used in MICA. Secondly, it shows the technical feasibility of the concept. And thirdly, early feedback from potential users and customers can be collected. For example, it is possible to run user tests under laboratory conditions, or to show the prototype on trade fairs. This helps to review the general concept and shows opportunities for future development.

Depending on success and reception of the prototype, a second project can be set up. For this successor the prototype is reworked, and a more mature pilot version is then tested in the field, i.e., in a productive environment.

First Prototype

As a first setting for MICA, a warehouse scenario was chosen, where MICA would assist warehouse workers in the picking process. However, the MICA architecture was not meant to be constrained to this specific setting but to be a platform for realizing very different scenarios. The major challenges of the first MICA prototype are:

- Workers need hands-free support.
- Interaction modalities for system input must fit the current situation and task.
- Volume levels of audio output must automatically adapt to environment noise levels.
- A display must react to rapidly changing lighting conditions.
- The system must be highly responsive, never leaving the worker to wait for it.

MICA has to provide a combination of explicit and implicit interaction methods in blue-collar environments. It faces situations in which the spatial relations of objects change dynamically. Hence, the worker's environment has to be monitored and interpreted in real time. On the one hand, this enables MICA to identify a worker's need for help and to react on implicit interaction clues like stumbling or search behavior. On the other hand, workers interact explicitly with MICA, for example by pressing the "OK" button to confirm the execution of a proposed action after a warning message. In particular the combination of implicit and explicit interaction on various modalities leads to natural blended interaction (Eisenhauer, Lorenz, Zimmermann, Duong, & James, 2005).

MICA guides the worker through the warehouse, keeps track of articles picked (see Figure 7b) and pro-actively offers help. The following subsections give short descriptions of the different components of the original MICA system. At the same time, this section serves as a baseline for explaining the modifications of the MICA pilot in the next section. Both sections are therefore structured similarly, describing the necessary ingredients for Interaction-by-Doing like the MICA trolley, navigation, article identification, and the software design. This is followed by summaries of the realized concepts and the interaction experience. As a case study, the MICA prototype provided valuable lessons learned that were used to improve its successor, the second MICA (pilot).

Trolley

The multi-modal interaction in MICA relies on a wide range of different sensors for retrieving implicit interaction information, an input device for providing explicit interaction possibilities and devices to give feedback to the worker; among these is pen-touch display, RFID readers and antennas, a WLAN data connection, and a high-performance CPU for coordinating all interaction devices in the worker's direct vicinity. Besides being heavy and bulky by themselves, the devices also need a power supply large enough to keep the system running during a complete workday.

A device that workers use throughout the entire picking process is the trolley on which order boxes are placed. The trolley therefore became the heart of MICA hosting all the devices required, as one essential requirement is not to strain the worker in his work with heavy equipment.

Article Identification

MICA receives picking orders from a server and lets workers select their next order. Workers – while always being connected to MICA via the picking trolley – are presented a picking list, which is automatically synchronized with the articles already picked.

For this, the MICA trolley is equipped with RFID technology which continuously monitors the items that are already picked. Articles and boxes (into which the articles are picked) are tagged with passive RFID tags. The floor of the MICA trolley is made up of a 4x4 array of RFID readers that read articles placed above them (see Figure 3). When an article or a box is placed on the trolley, MICA can determine if the article is correct and if it is placed in the right box. Hence, with MICA the worker is not constrained to work on one order at a time, but he may choose several orders and work on them simultaneously, while MICA makes sure that no article is wrong, forgotten or picked into a wrong shipping box.

Our RFID readers operate at High-frequency (approx. 14MHz, HFID) electromagnetic wave band, which has a limited reading range, but reveals good characteristics when used with conduc-

Figure 3. RFID antenna array for identifying articles and their location on the MICA trolley

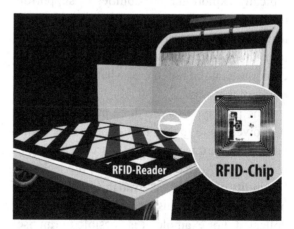

tive materials or liquids. This is important because the prototype is intended for office items (e.g., hole puncher) which occasionally contain metal or liquids. Diverting from the original plan, the MICA prototype only has a 1x2 array of RFID readers at the bottom. The reason is that the shape and diameter of an antenna affects the reach of the reader's field above the antenna. So their large size enables an elongated vertical read area up to 30cm above the trolley loading area, while limiting the maximum number of simultaneous orders that can be processed at a time to two.

In practice, it turns out that reading articles on a second layer above ground layer is unreliable. Articles randomly disappear or are attributed to the wrong box if an article is placed near the boundary of two adjacent RFID areas.

Navigation

MICA realizes an indoor navigation system to guide the worker to the next article on the picking list. Such navigation system essentially consists of two parts: A hardware part for determining the physical position of the worker, and a software part calculating and presenting a route.

Drafts for tracking mechanisms in MICA combined Ekahau - a low precision WLAN tracking - with fine grained ultra-wide band (UWB) tracking (Eisenhauer, Lorenz, Zimmermann, Duong, & James, 2005). However, the actual realization of the MICA prototype completely relied on WLAN tracking leaving aside other tracking technologies because of the high cost associated with equipping large areas with UWB tracking systems.

Given a list of articles the worker has to pick for his orders and storage locations of individual articles, the navigation system determines an optimal route for the worker through the warehouse. For this computation, the route calculation determines the distances between each pair of articles along valid paths in the warehouse with the A* algorithm (Hart et al., 1968). In the next step locations are ordered so that the round trip

that visits all locations has an optimal length. This sorting – known as the Traveling Salesman Problem (TSP) (Menger, 1932) – is solved with a brute-force algorithm that finds the optimal solution. This is the most expensive computation step with a complexity of $O(n!)$.

Software

MICA uses a bus concept for input and output data. The buses are based on a Jabber/XMPP (see Saint-Andre, 2004) instant messaging infrastructure. Collected data – for example, raw sensor data from pointing gestures – are refined into messages with MICA data – e.g., pointing direction and angle – and posted to a chat room, and thus forwarded to subscribed receivers in this room. Here, user modeling and dialog management servers pick up that data. Also connected to the bus are server components that provide databases, host navigation processing, manage users and trigger pro-active help. Having higher memory and processor requirements, these components are placed on a stationary machine providing enough resources for the software. Responses from server components are then posted on the output bus for rendering on user interfaces (Schneider, Lorenz, Zimmermann, & Eisenhauer, 2006). A known problem with Jabber is that the amount of network traffic grows exponentially with the number of participants in a room. The common bus is therefore split into several rooms so that the number of participants and messages in each room is less than that in one common room for all.

The XMPP protocol was chosen because of its openness and interoperability. Protocol implementations exist for different programming languages and platforms; even resource restricted mobile devices. For this resource thriftiness, Lorenz and Zimmermann (2006) embedded ARFF (http://www.cs.waikato.ac.nz/~ml/weka/arff.html) messages (containing the data sent by MICA subsystems) into the XMPP messages. ARFF messages have their own header containing sender and receiver information, and data specific to the command.

All components of the MICA server are implemented in a stand-alone Java application. On the client side, the GUI is implemented in C#. Originally the client included several intelligent functions, but due to high processor load and resulting low responsiveness of the GUI several functionalities were moved to the server. A light-weight client receiving detailed instructions remained.

Proactive Help

Kaufmann et al. (2007) interviewed and accompanied several workers and managers during their work in two non-automated warehouses. The aim was to get to know the authentic processes in a warehouse in order to identify situations where help was needed. Besides helping workers to avoid picking errors by observing picked articles, and helping them to find articles quicker by guiding them to the storage location, other stressing or time consuming activities were identified. Resulting from these, eight implicit interaction principles were developed that indicate if a worker is in need

for help. These interaction principles are used as input to a finite state machine, which then triggers a means of pro-active help.

For example, a common situation was that an article was damaged or missing. When this happened the worker had to fetch the warehouse manager to show to him the situation. During this time, the picking process is effectively on hold for up to 30 minutes (as described in Section 2). For such situations, the MICA prototype includes a wireless headset and video camera so that the worker can remotely consult the warehouse manager without having to walk to her office (see Figure 4). One of the indicators is that a worker is looking about or moving his head horizontally in a noticeable way (i.e., scanning for a specific item). The video stream continuously generated by the helmet-mounted video conference camera was re-purposed to detect horizontal movements between individual video frames.

Other indicators used as input for the proactive help system included a worker standing still in the warehouse, the picking of a wrong item, or the worker walking too far away from his trolley in the wrong direction.

Figure 4. Warehouse worker with headset and a small video camera mounted on his hardhat

Multi-Modal Interaction

Besides allowing graphical interaction with the worker, the client outputs speech like "Put two hole punchers into box A." The voice output is composed from sampled partial sentences, hence providing high quality output at the cost of flexibility and a higher cost for hiring professional speakers to speak the partial sentences, quantities and article names. By offering several output modalities, the chance of misunderstanding due to inappropriate environmental conditions like loud noise is reduced.

The MICA software architecture is a mix of centralized and distributed components. Input from different kinds of sensors is collected by MICA clients, fused with other sensor data on the same client, and filtered through a recognition mechanism. The pre-processed data is then published to a data bus on the local network that is used by MICA components (like sensors, servers, etc.) for their communication.

Besides touchscreen interaction, Interaction-by-Doing is offered as input modality. Each of the modalities is chosen for the correct situation. In the normal case of picking, the worker is situated beside the trolley, so that it is difficult to type on the touchscreen. Thus, Interaction-by-Doing is used. In cases where the worker has to go to the screen anyway, e.g., when video chatting with the overseer, touchscreen input is expected.

Lessons Learned from the First Prototype

The biggest problems we encountered were of physical nature. The batteries for power supply of the mobile components (e.g., RFID readers) have to be well proportioned. Long before the batteries are depleted, the readers' reliability begins to degrade. This adds to an already existing problem with the article identification. The original assumption that every reader reads only the area above itself and not the area above another reader cannot be confirmed. The readers' electromagnetic fields overlap or even twist around each other when objects are brought into the field. Though some modern antennas' lobes have almost the desired shape in free space, it is impossible to prevent fields from getting distorted when there are obstacles. Similarly, the radio-based localization needed frequent recalibration in an ever-changing environment.

On the software side, unreliabilities in the WLAN connection caused problems with the XMPP-based communication bus infrastructure on. A disruption of connection meant that either messages in chat rooms were lost or that messages were delivered twice. This could mean that first a lot of outdated messages were delivered before current messages were received. This led to some confusion for the test persons.

Finally, there were some usability problems because the mobile PC used a pen input. Many people would have preferred to use their fingers and have larger buttons or active areas. However, much of the functionality in the prototype's interface was not needed.

Second Prototype

After the first prototype of MICA proved its feasibility, a follow-up pilot project started in 2008. The goal was to evaluate the road capability of MICA as well as its usability and market potential. A pilot allows testing a design and making adjustments in time. It also shows if anything is missing and provides quantitative proof that the system has potential to succeed on the larger (full) scale.

For MICA, the new requirement of being suitable for productive use brought new challenges so that many aspects of the first prototype were re-engineered. Also, the scenario where MICA would be used (warehouse) was now fixed from the beginning.

With Antriebs- und Regeltechnik GmbH (ART), a medium-sized enterprise for testing MICA was found. Founded in 1955, 700 employees in Germany, Romania and Poland develop machinery supplies for the manufacturing industry. MICA ran at the headquarters' central warehouse. ART faces the typical problems of non-automated warehouses, and at the same time is continuously looking for solutions that support their just-in-sequence and just-in-time services. ART actively supported and contributed to the pilot and assembled the new hand pallet truck at its workshop.

Compared to the warehouses of the first MICA prototype, the requirements at ART are a bit different. They produce many of their articles specifically for their customers in small quantities (sometimes only one single unit). So for them, even one error in tens of thousands of picking items is too much. Their warehouse is smaller with fewer workers, and a much friendlier atmosphere. Environmental conditions like lighting and noise levels are rather constant. Importantly, ART's products constitute a challenge for MICA's original article identification and positioning: most items mainly consist of metal parts, are stocked in metal shelves and are handled in metal baskets. This is a conceivably bad situation for RFID and WLAN tracking. The typical order includes several baskets that are interchangeable but have to be prepared in a fixed order.

These differences and the immanent productive use of MICA necessitated design changes that are discussed in the following sections.

The MICA Hand Pallet Truck

The working process at ART requires that two boxes stand side by side on a pallet, and can be stacked up to the fourth level at eye height. A finger touch enabled Tablet PC is mounted at common eye level. It is tiltable and turnable so that people can adjust it to their personal needs. Two batteries supply the Tablet PC and the RFID readers for article identification and positioning (see following sections). A standard power supply cable for charging ensures the easy recharge of

batteries. Full charging takes four hours so that it fits well between two working shifts. A fully charged battery supplies energy for ten hours of MICA-enhanced work so that it lasts an entire shift without having to recharge.

The MICA hand pallet truck is resistant against impact and scratches, electronic parts are protected. It is as easy to use as a normal hand pallet truck and fulfills safety at work guidelines. At the same time, hardware costs do not endanger profitability of the whole system.

Navigation

The positioning engine of the MICA pilot requires a higher accuracy. It needs to identify each single stockyard, situated no more than 30 cm apart from each other. Ekahau cannot assure this accuracy, especially in a metal flooded and constantly changing environment. Hence, we changed to RFID-based tracking: RFID tags, working on another frequency range as the article tags, are placed on the floor of the storehouse. A RFID reader mounted underneath the trolley scans these tags and gets the corresponding location from a database. Floor RFID tags are usually placed in holes in the floor. This is not possible at ART because of a special anti-static and expensive ground. Using special adhesive labels which are resistant against physical force, we avoided drilling holes.

As corridors are wider than the width of the hand pallet truck, we placed three tags in a row orthogonal to the corridor to guarantee reading a tag on any track the truck could move through the corridor. Each of these tags is associated with the same X/Y coordinates.

The worker's walking direction is derived from his current and previous positions. The new MICA system rotates the map so that viewing direction is always up as opposed to a fixed orientation with north always being up. Article locations are stored as their "true" location, which is not directly on the paths defined by corridors. Instead the articles are projected to the nearest path; the angle between projection vector and moving direction determines if an article is in the left or right hand shelf (see Figure 5). Additionally, now 3D map data may be annotated with individual path costs, therewith allowing to reduce traffic in selected corridors.

The brute-force path length optimization with NP-complete TSP from the first MICA performs unsatisfactorily when applied to a picking list with 20 or more different locations. But an approximation algorithm (the "nearest neighbor" algorithm) usually provides good results with an average length of 1,25 times the shortest possible route. In the worst case, the route is at most O(log d) longer than the optimal route (Rosenkrantz, Stearns, & Lewis, 1977). Its computational complexity at any given time is O(n) for sweeping through all the remaining article locations to find the article with the shortest distance to the current location. With a normal PC this means that planning for thousands of locations is possible without a noticeable delay.

Article Identification

As the RFID readers of the first prototype were integrated in the trolley's bottom they occasionally failed to read articles on the second level. In the pilot setting this detection ratio is even lower due to the fact that articles are metallic. This requested a change in RFID technology, from HF to Ultra High Frequency (UHF). UHF is prevalent in logistical applications: besides worldwide standardization in ISO 18000-6C (EPC global) (ISO 18000-6C), it features long reading ranges, bulk reading of several transponders at favorable prices, and compact chip design.

Additionally, we need to be 100% sure that no article passes the scanner without being recognized. Picked articles appearing and disappearing randomly are obviously not acceptable. Also cross checking after picking is no option as it would entail no enhancement compared to the situation without MICA. Therefore, the RFID readers are mounted on a frame that is put on top of the boxes. These readers do not scan the contents of boxes, but register articles that are moved in and out of a box. The reading area is carefully calibrated to exactly cover the whole box opening. It thus assures that every passing article is read but nothing else from nearby shelves. This preserves a natural way of working and preserves valuable energy on the mobile device.

In order to identify the direction of movement of a transponder through the box opening

Figure 5. Picking article from right hand shelf

we analyze the RSSI value (Receive Signal Strength Indication (IEEE, 802.11)) of the tag. The obtained value qualifies the received field strength of wireless communication applications and is transmitted by the reader for every reading event of a transponder. Mathematical calculations yield the information if movement is upwards or downwards when two readers are placed on top of each other.

For two boxes stand side by side on the pallet, two readers (for RSSI movement direction detection) cover each box opening. A frame carrying one pair of readers slides up and down the truck's mounting to rest on the rear stack of boxes. The second pair is attached to the front of the frame with its read area covering the front box opening (see Figure 6). The frame construction is quite heavy, but workers need to move it up to the fourth

level of boxes (height of the head). Therefore, a Bowden cable enables effortless moving.

Maintaining a continuous electromagnetic RFID field consumes much energy. This is too much for a battery powered device, particularly if there are five such readers (four for article identification and one for positioning). A way to reduce energy consumption is to turn off the field while no article is potentially in range. Built-in ultrasonic (US) sensors that consume negligible amounts of electricity are coupled with the readers and automatically trigger a reading process. The US sensors detect moving objects within a range of up to 70cm. Any object passing the sensor immediately turns on the reader that creates the electromagnetic read field within milliseconds. Every transponder within reading range is then read. Configuration and placement of the rear

Figure 6. Lifting truck with boxes and RFID mounting of the second MICA pilot

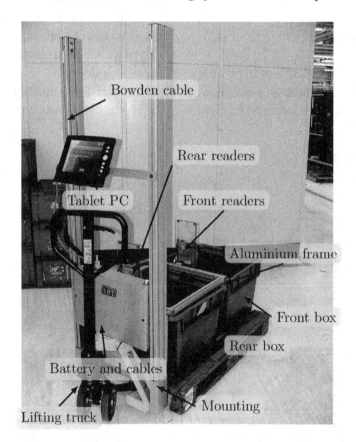

readers' triggers is carefully adjusted because US waves emitted by them would reflect from the back side of the front readers and trigger a read because the echo would mistakenly be interpreted as coming from an article. After a predefined period the reader switches back to standby. This method ensures that a reader only reads while an article is placed into a box or taken out of it. Thus the power consumption is reduced to a minimum, resulting in a higher operating time of the whole system.

Besides saving energy, the trigger has two additional advantages:

1. The already low probability of reading articles in the shelf is further reduced, because readers sleep most of the time.
2. It is possible to identify if an article is not detected by RFID, in case of a broken RFID tag. In that case, the MICA system prompts the worker to re-pick the article and initiates a multi-step error correction process.

Proactive Help

Several indicators that were used for the pro-active help of the first MICA are no longer available. For example, the video conferencing is no longer necessary due to the warehouse size, and therefore the head-mounted camera is not available. Similarly, persons are no longer tracked and it is therefore impossible to detect if they wander away from their trolley. As a consequence of this, we concentrated our efforts on the reliability of the remaining areas of pro-active help, navigation support, article identification through article pictures and picking error correction.

Multi-Modal Interaction

The constant environment in terms of lighting and noise within the warehouse makes a combination of auditive and visual output on the table PC most reasonable. Speech composition

from partial sentences is not feasible anymore because there is an unlimited and dynamically changing space of possible wordings. The MICA pilot rather generates audio output with a text-to-speech engine allowing synthesizing any sentence from its written form. As ART has a complicated naming scheme for articles where each name is composed of 15 numbers and letters it makes no sense to spell this name. By referring to an article's stockyard instead, there is no danger the text-to-speech engine generates incomprehensible article names. The visual interface has been re-factored to present less information at once.

Menus and other complex explicit interaction constructs in the GUI have been removed because most workers have little computer experience. Instead, buttons and other active areas have been enlarged. The main interaction method, Interaction-by-Doing, remains: movement and picking still make up for most of the interaction between worker and MICA.

Software Components

Building a system that is used in a productive environment requires higher diligence than building an experimental prototype serving only as a proof of concept. The first prototype was implemented with an architecture for a wide and diffuse application domain reflecting personal research interests. It was to be used under laboratory conditions. Implementation sprints with ad-hoc processes took place before trade fairs. Necessary cleaning-up and refactoring with reuse in mind would occur in the time in between, where deadlines were less troubled. This, however, did not happen. MICA was not suitable for field-testing. In preparation of the study, MICA's software sources experienced a general overhaul and large portions of MICA's code had to be rewritten from scratch by an all new team because scientists of the first prototype had changed to new projects.

Differently from its predecessor, the new MICA system is much more focused, and no longer a platform for studying multi-modal interaction and user modeling under different conditions. Instead it is fully aimed at the warehouse scenario. Usability and understandability to the common warehouse worker are the primary objectives. Also a higher degree of reliability is necessary to not disrupt daily work of the productive environment. However, one design goal is still to keep the MICA software open for using different hardware devices and software components with the system.

MICA's backbone - instant messaging based communication - is replaced by the Java Message Service (JMS), which provides enterprise-class reliability in message oriented architectures. JMS guarantees message delivery for the publish-subscribe model, thus replacing the chat room mechanism, where messages could get lost if a client temporarily disconnects from the chat room. Also messages are sent with a time to live, which prevents a flood of messages that could occur if a component reconnected to the message bus. To save bandwidth low-priority messages (like position updates) are discarded earlier in case of a transport bottleneck. As JMS is capable of transporting entire Java objects, the ARFF message encoding was mostly removed, leaving it only as an interface to the .NET GUI.

The original monolithic server part is broken up into distributable components that communicate with each other through JMS messages, thus facilitating the relocation of individual components between different physical machines. The bus architecture is modified according to the observer design pattern: Every component (e. g., a worker location sensor) publishes its messages, while other components (e. g., navigation and route computation) subscribe to the update of components that they are interested in.

THE EFFECTS OF MICA ON ART

The biggest change for ART resulted in the elimination of an explicit cross-check in their picking process: prior to the introduction of MICA, workers went through the repository with a shopping-cart-like trolley. They collected articles, brought them to the shipping area, cross checked the articles, and put them into a new box on a pallet. Afterwards, they repeated all steps until the order was completed.

The manual cross checking process has now been replaced by MICA's tag identification during the picking process. A box-by-box cross-check to keep an overview is no longer needed. The old MICA trolley is replaced by a hand pallet truck equipped with MICA technology (see Figure 6). Thus, picking onto a pallet on the hand pallet truck accomplished in one step. This safes time compared to the former multi-step process.

One of MICA's software modules connects it to the SAP Warehouse Management server of ART. This real-time data link keeps stock lists up to date. When picking starts, the items of the order are marked as "reserved," and are immediately removed from stock when picking finishes. Formerly, there was no "reserved" state and updates would take hours. It was possible that the last item of a kind was picked in a previous order, so that the next one could not complete although started. Stock outage is a frequent problem and requires that picking starts early with enough reserve time before delivery, so that the missing piece can become available again. Hence, MICA reduces the time reserved for picking and packing.

USABILITY EVALUATION

We put a considerable amount of effort into the design of the MICA GUI. It underwent several changes in the last years (see Figure 7a and Figure 7b): internal design reviews, customization to

Figure 7. Original design in English (a) and SAP design in German for ART (b)

a

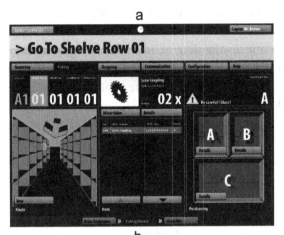

b

the SAP design guide, translation from English to German, and an adaptation to ART processes. Finally, a usability study was conducted with ART warehouse workers.

A usability evaluation measures the extent to which users in their specific context can achieve their goals effectively, efficient and satisfactorily (ISO 9241, 1998b). Effectiveness is put on level with usability. A product, system or software supports effective processing of a task, if it provides all functions needed by a user to completely achieve his goals. Beyond that it can be rated as efficient if its functions are operated accurately and effortless. Satisfaction with the product, system or software necessarily (but not sufficiently) results

from the perceived easiness and intuitiveness to operate the system. This unfolds in a model of stages, which stimulates the presentation of results. Effectiveness constitutes a measure of the usability potential. Results of a user test represent the efficiency, and satisfaction is measured with a questionnaire depending on the sample size of the usability study. Tests like the Software Usability Measurement Inventory (SUMI) (Kirakowski & Corbett, 1993) are recommended for sample sizes above twelve users. With lower sample sizes, like in our case, a shorter questionnaire provides a rough impression of the overall satisfaction. This, of course, is not statistically representative, but nevertheless delivers useful information for the usability assessment and suggestions for improvements.

We first conducted a participatory expert evaluation of the interface with three experts – which meant that all experts were accompanied by two testers – one taking notes and the other in dialog with the expert taking care that no action on the interface is left uncommented and all relevant aspects of the system have been covered. That means that the tests were conducted in close collaboration with experts with the "thinking aloud"-method (e.g., Nielsen, Clemmensen, & Yssing, 2002) and subsequent discussions. In this test we identified 51 violations of requirements for the design of dialogs (according to ISO 9241-10 (ISO 9241, 1998a)). Most problems were related to violations of conformity with user expectations, error tolerance, or self-descriptiveness. Additionally, we identified and eliminated screens, which had lost functions, like the registration screen.

The original plan arranged for testing eight workers. But due to time restrictions this evaluation had to be conducted prior to the full completion of the system. In particular tests with RFID-reader hardware and its interplay with the software were only partially completed. The evaluation of the MICA pilot was therefore conducted with a limited set of three more experienced workers. The main reason for this limitation was that the system still was not completely robust, and that there was a risk that it had to be rebooted.

The participants were tested with real picking orders. Each usability test consisted of a short introduction of the worker into the nature of the test and to the method of "thinking aloud". The tests were always conducted with three investigators: one that constantly stimulated the worker to comment each of his actions and two that took notes and pictures. All critical incidents were then collected and analyzed according to ISO 9241-10 and their specific violation of the dialog requirements. At the end of the test the workers were asked to complete a questionnaire measuring their satisfaction with the system. As expected this early testing revealed a potential for optimization. A series of critical incidents could be assigned to 69 violations of dialog requirements that limit the efficiency with respect to task completion (see Figure 8a). In particular a high number (20) of malfunctions occurred that limit the effectiveness of the MICA System. However, the tests also revealed many positive aspects of MICA where the participants distinctively emphasized an unqualified success in conformance with dialog principles, as well as eight malfunctions that were directly eliminated during the test (see Figure 8b).

The results of questionnaire and interview indicate a good satisfaction with MICA in spite of its malfunctions. Participants were confident that the system simplified their picking process, saved time and prevented errors. Specifically, they highlighted that search and identification of items in the picking process with the system was much easier, less error prone and much simpler than picking without MICA. The violations of dialog principles and identified malfunctions resulted in a revised version of MICA in the actual field-tests.

EVALUATION OF EFFICIENCY

In this section we present results of the MICA evaluation based on (Gillmann, 2008). An economic evaluation of investments must always be based on solid investment appraisals. This in turn requires well evaluated data, describing the logistical process. Key economical measures of logistic systems are time, quality and cost. To evaluate these data, the MICA pilot is implemented and tested in the ART warehouse.

To measure potential quality improvement and time reduction, as well as potential cost reduction, the paper based manual picking process is benchmarked against a MICA supported picking process. In particular, process times and failure rates are monitored. To evaluate the impact of MICA as a support system for (particularly unskilled) workers, both the manual picking process and the MICA supported picking process are tested with groups of skilled and unskilled workers. Skilled workers included all of the workers that worked in the respective ART warehouse (less than ten). A similar number of unskilled workers were brought into the warehouse either from other departments of the company or by us.

The workers in the evaluation worked on real orders that were finally shipped to customers. During the entire evaluation phase we accompanied workers in the warehouse, drawing maps of the paths they took and taking the picking times. In the MICA experimental groups, the warehouse's paper-based quality control procedures were carried out in parallel by an experienced worker in order to detect problems that MICA might not have detected. In all experimental groups, the orders were finally (after picking was completed) cross-checked by an experienced worker to ensure the quality of the delivery. The results of this were also recorded to check for quality improvements through MICA.

The duration of the evaluation phase was four months. The full time was sliced into multiple smaller time slots, each one to host one of the

Figure 8. Usability violations (a) and usability conformance (b)

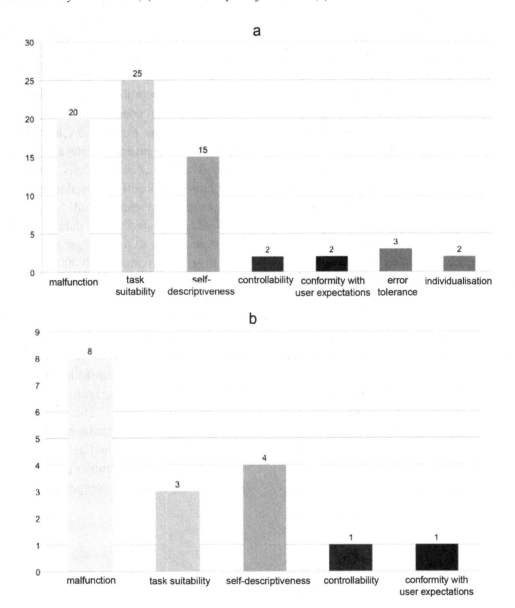

four variations (with MICA vs. without MICA, experienced vs. inexperienced).

One major customer of the warehouse in our study produces large industrial machines (several meters in size), who receives electrical components and cables in various sizes from ART. In our study we included all orders for one series of machines from this customer. Hence, this study is highly representative for such orders. Due to the high number of orders from this customer, it can in turn also be considered as representative for all of the warehouse's orders.

To accurately determine the efficiency of MICA, we additionally need to take into account the cost of processes implied by MICA (e.g., tagging of articles before they are stored), the monetary cost of MICA hardware like RFID, and potential maintenance costs due to hardware wearout. We use common investment appraisal methods (Tucker, 1963) to calculate economic sense and an investment risk for this specific case.

The overall picking process divides into four sub-processes: preparation, picking, inspection, and confirmation & hand over. Picking and inspection times depend on item quantities. Preparation and confirmation & hand over are independent of item quantities. Using MICA, the item dependent process time is reduced by 36%, when used by skilled workers, and by 75% when used by unskilled workers. A significant processing time reduction of 46%, for skilled workers and 61% for unskilled workers is demonstrated for the item independent process time (see Figure 9. Picking times are hidden due to data confidentiality).

The field trial focuses on two different error rates: error rate per order item (E_i) and error rate per picking order (E_o). E_i shows how many order items are picked mistakenly. E_o shows how many picking orders are processed mistakenly and in turn would result in a customer complaint. Without using MICA, the error rate per position is 0.48%/4% (skilled/unskilled workers). More than this, a subsequent final cross-check of articles picked cannot detect all picking errors. Thus the error rate per order turns out to be 14.3%/25%

(skilled/unskilled workers). In contrast, with using MICA, all of the human errors are detected, thus the error rate per item and per order drops to 0%, for skilled and unskilled workers.

The investment costs divide into one-time expenses and operating costs. One-time expenses are, for example, expenses for hardware, software, and infrastructure. Furthermore, assuming a good "confidence" in inventory data, according to correct and real time confirmations, dispatching can be organized more efficiently, thus "safety stock" can be reduced, yielding additional cost savings.

Based on experimental data, investment in MICA reaches a ROI of 38% at a yearly picking quantity of approximately 80,000 items. A typical amortization time of 18 months is reached at a yearly picking quantity of approximately 280,000 items. MICA demonstrates a substantial reduction of processing times, as well as a zero picking failure rate, both with a group of skilled and unskilled workers. Additionally, unskilled workers using MICA reached a picking performance almost as good as skilled workers. This shows very high potential of MICA for broad spread industrial use.

In summary, the evaluation shows that MICA successfully supports complex picking processes. Potential customers are companies with small to

Figure 9. Average picking process times

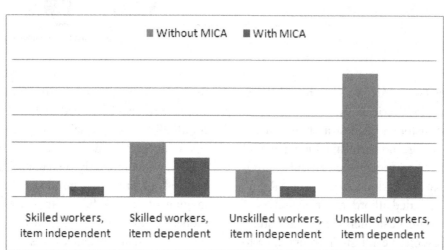

medium sized goods, each of medium to high value, with inhomogeneous order structures and with high overall picking quantities.

ANALYSIS OF SALES POTENTIAL

In order to find out the potential of MICA for the market, we finally present an analysis of sales potential based on (Brauner, 2009). The sales potential is defined as the portion of the market potential that a particular firm can reasonably expect to achieve (Lucas, 1975). With the analysis we address the question, how big the market is in the relevant industries.

As MICA is a new system, no reviews of existing similar systems can be taken into account. Instead, a questionnaire was distributed to 300 companies, 25 completed questionnaires were sent back. The amount of companies for which MICA can be useful is too large to contact them all. Because of that, a mix of random and quota selection was conducted (Schnell & Kreuter, 2003). As only particular employees of a company can answer the questions, we tried to contact experts via already existing contacts.

The questionnaire introduces the MICA systems and afterwards lets the probands reflect about their current picking system. Finally, the probands are asked if they consider buying a MICA-like system to determine the sales potential. About 17% consider buying a MICA-like system, so we assume the sales potential to be 17%.

According to the Bureau of Labor Statistics (2009) there are about 6,000 pickers in the US machinery industry. Hence, the sales potential for the US machinery industry is 17% of 6,000=1,020 MICA trolleys. We gathered the same information for the rest of the world and interpolated data where no data was available according to general economic performance data. In summary, we computed a sales potential of several ten thousand MICA-like trolleys for the whole world. Again

concrete number are subject to data confidentiality but can be obtained from SAP on request.

RELATED WORK

MICA combines in a single application context-awareness and RFID technology. On the one hand, its context-awareness is quite similar to that of systems like Oppermann and Specht (2000), Eckel (2001) or Long et al. (1996). These systems determine the physical context of the user from his location and orientation in order to provide him with useful information about his surrounding, i.e. exhibits in a museum or sights in tourism, respectively. In contrast, MICA is multi-modal, deals with a changing environment, and uses a set of different sensors to obtain context information. It does not only provide plain situational information but assists the worker with his current task. With respect to this, it is more like the context-aware assistants of Dey et al. (1999), Yan and Selker (2000) or Rhodes (1997). MICA, however, also serves a logistical assurance objective by guaranteeing that picking errors do not occur. MICA also has some aspects in common with smart shopping applications like that from Asthana et al. (1994) because it helps the user to locate items and guides him to their storage location.

On the other hand, RFID is attracting more and more attention in business environments as it provides an interface between the physical world and an enterprise's information infrastructure (Want, 20004). A major role in the business environment plays logistics, where the tracking of goods and their delivery status are traditional applications (Weinstein, 2005). Companies use huge RFID gates on the entrance and exit areas of their warehouses to automatically register incoming and outgoing goods while these are still loaded on trucks (Lefebvre, Lefebvre, Bendavid, Wamba, & Boeck, 2006). Historically, RFID technology and tags were too expensive to track individual items. As prices for tags have recently dropped significantly, even item level tracking of low-price goods becomes possible. Bendavid et al. (2006) present a study where a RFID-based single-item hand-over between customer and supplier is analyzed under laboratory conditions. Yet their system does not provide support or detect errors during the picking process. With this respect, RFID-enabled shopping carts are what comes closest to MICA's article tracking (e.g., Kourouthanassis, 2003). For RFID-based locating Luimula et al. (2010) use a similar approach but as MICA is operated by humans, obstacle tagging is not necessary.

Unlike other systems like the IBM MAGIC system, which uses gaze tracking for the prediction of cursor movement (Zhai, 1999), in MICA a combination of speech and pen input with user movement in a physical environment was favored. MICA, as opposed to many other multi-modal applications, has prominent physical parts that have original practical reasons: the trolley is already there so that the worker does not need to carry the items himself. Only article and box identification are added but they do not change the natural way of working. Likewise, the worker would be moving in physical space also without MICA. This observation is the basis for Interaction-by-Doing.

CONCLUSION

We presented a support system for warehouse workers: its requirements, the Interaction-by-Doing design principle, its evolution through prototype and pilot, and evaluations of usability, effectiveness and sales potential.

The MICA prototype is successful as a proof of concept. The reworked MICA pilot proves successful in field-tests, where it measurably supports warehouse workers in the picking process. Thanks to multi-modal interaction MICA is applicable to the changing environments of a warehouse. Calm computing principles and pro-active help

make it usable for unskilled workers as well as for experienced workers without having to interrupt the familiar way of working. Intuitive handling as well as support for every worker's experience level is assured by the Interaction-by-Doing concept. Pro-active help prevents errors before they occur.

During its evaluation, several parts of MICA were adapted and improved. The positioning system is enhanced from WLAN-based to RFID-based technology, thus increasing its accuracy when identifying a single stockyard. The new RFID-enabled hand pallet truck frame proves to scan 100% of the picked articles without interfering with the usual way of working. At the same time, no articles from the repository are accidentally read. The frame fulfills the picking process requirements of the ART warehouse and could be adapted to the needs of other warehouses.

The usability evaluation reveals that search and identification of articles is much improved with MICA. The results already indicate a good satisfaction with MICA in spite of still existing malfunctions. Participants are confident that the system would simplify their picking process, save time and prevent errors. It can be expected that the new introduced interaction concepts and an improved MICA system that overcomes the detected malfunctions and violations of dialog principles will fully satisfy users' needs.

MICA's potential to reduce the number of errors and to speed-up picking processes in non-automated warehouses is shown in the economic evaluation. Enterprises with high picking quantities are offered a high potential to reduce costs. During economic peak times MICA helps as well unskilled workers to reduce picking errors even under high time pressure. The analysis of sales potential shows that there is a market for MICA-like systems.

The future work consists of transforming the pilot into a productive system. For this, mainly stability must be enhanced and errors have to be found in high-pressure tests. Besides software errors, the reliability of the RFID reading has to be checked. Multi-modality can be advanced to more modalities as in the current status. However, MICA's future also depends on economic decisions of companies that traditionally have conservative customers.

ACKNOWLEDGMENT

Our thanks go to our designer Lars Zahl for the images, and to our former colleagues who helped realizing the MICA pilot: Oliver Kaufmann, Feras Nassaj, Rossen Rashev, Daniel Meckenstock and Kin Voong, as well as the teams of SAP Research, deister electronic and ART. Comments from the anonymous reviewers helped us very much to improve this paper.

REFERENCES

Asthana, A., Cravatts, M., & Krzyzanowski, P. (1994). An Indoor Wireless System for Personalized Shopping Assistance. In *Proceedings of the IEEE Workshop on Mobile Computing Systems and Applications*, Santa Cruz, CA (pp. 69-74).

Bendavid, Y., Wamba, S. F., & Lefebvre, L. A. (2006). Proof of concept of an RFID-enabled supply chain in a B2B e-commerce environment. In *Proceedings of the 8th international Conference on Electronic Commerce: the New E-Commerce: innovations For Conquering Current Barriers, Obstacles and Limitations to Conducting Successful Business on the internet (ICEC '06)* (Vol. 156, pp. 564-568). New York: ACM.

Brauner, S. *(2009)*. Investigation of the international market potential of an interactive, context-sensitive and multi-modal picking tool for SAP AG with final recommendation for further strategic proceeding. *Unpublished Master's Thesis*.

Brewster, S. (2002). Overcoming the lack of screen space on mobile computers. *Personal and Ubiquitous Computing, 6*(3), 188–205. doi:10.1007/s007790200019

Bureau of Labor Statistics. (2009). *Packers and packagers, hand*. National Employment Matrix.

Dey, A. K., Futakawa, M., Salber, D., & Abowd, G. D. (1999). The Conference Assistant: Combining Context-Awareness with Wearable Computing. In *Proceedings of the 3rd International Symposium on Wearable Computers (ISWC)*, San Francisco (pp. 21-28).

Eckel, G. (2001). LISTEN - augmenting everyday environments with interactive soundscapes. In *Proceedings of the I3 Spring Days Workshop Moving between the physical and the digital: exploring and developing new forms of mixed reality user experience*, Porto, Portugal.

Eisenhauer, M., Lorenz, A., Zimmermann, A., Duong, T., & James, F. (2005). Interaction by movement - one giant leap for natural interaction in mobile guides. In *Proceedings of the International Workshop on Artificial Intelligence in Mobile Systems*.

Gillmann, L. (2008). Wirtschaftliche Evaluierung eines innovativen multimodal interaktiven Systems zur Unterstützung von Kommissionierprozessen (Economic evaluation of an innovative multi-modal interactive system in support of picking processes). *Unpublished Master's Thesis*.

Hart, P. E., Nilsson, N. J., & Raphael, B. (1968). A formal basis for the heuristic determination of minimum cost paths. *IEEE Transactions on Systems Science and Cybernetics, 4*(2), 100–107. doi:10.1109/TSSC.1968.300136

IEEE 802.11. (2007). IEEE Standard for Information technology-Telecommunications and information exchange between systems-Local and metropolitan area networks-Specific requirements: Part 11: Wireless LAN Medium Access Control (MAC) and Physical Layer (PHY) Specifications.

ISO 13407. (1999). Human-centred design processes for interactive systems.

ISO 9241. (1998a). ISO 9241-10: Ergonomic requirements for the design of dialogs - part 10 guidance on usability.

ISO 9241. (1998b). ISO 9241-11: Ergonomic requirements for office work with visual display terminals - part 11 guidance on usability.

ISO/IEC 18000-6. (2004). Information technology - Radio frequency identification for item management - Part 6: Parameters for air interface communications at 860 MHz to 960 MHz.

Kaufmann, O., Lorenz, A., Oppermann, R., Schneider, A., Eisenhauer, M., & Zimmermann, A. (2007). Implicit interaction for pro-active assistance in a context-adaptive warehouse application. In *Proceedings of the 4th international conference on mobile technology, applications, and systems and the 1st international symposium on Computer human interaction in mobile technology*, Singapore.

Kirakowski, J., & Corbett, M. (1993). SUMI: the software usability measurement inventory. *British Journal of Educational Technology, 24*(3), 210–212. doi:10.1111/j.1467-8535.1993. tb00076.x

Kourouthanassis, P., & Roussos, G. (2003). Developing Consumer-Friendly Pervasive Retail Systems. *IEEE Pervasive Computing / IEEE Computer Society and IEEE Communications Society, 2*(2), 32–39. doi:10.1109/MPRV.2003.1203751

Lefebvre, L. A., Lefebvre, E., Bendavid, Y., Wamba, S. F., & Boeck, H. (2006). RFID as an Enabler of B-to-B e-Commerce and Its Impact on Business Processes: A Pilot Study of a Supply Chain in the Retail Industry. In *Proceedings of the 39th Annual Hawaii International Conference on System Sciences*.

Lolling, A. (2003). *Analyse der menschlichen Zuverlässigkeit bei Kommissioniertätigkeiten* (Analysis of human reliability in picking processes). Shaker.

Lorenz, A., & Zimmermann, A. (2006). User modelling in a distributed multi-modal application. In *Proceedings of the Workshop on Ubiquitous User Modeling*, Riva del Garda, Italy.

Lorenz, A., Zimmermann, A., & Eisenhauer, M. (2005). Enabling natural interaction by approaching objects. In *Proceedings of the Workshop on Adaptivity and User Modeling in Interactive Systems*. DFKI.

Lucas, H. C. Jr, Weinberg, C. B., & Clowes, K. W. (1975). Sales response as a function of territorial potential and sales representative workload. *JMR, Journal of Marketing Research, 12*, 298–305. doi:10.2307/3151228

Luimula, M., Sääskilahti, K., Partala, T., Pieskä, S., & Alaspää, J. (2010). Remote navigation of a mobile robot in an RFID-augmented environment. *Personal and Ubiquitous Computing, 14*(2), 125–136. doi:10.1007/s00779-009-0238-3

Menger, K. (1932). Botenproblem (Messenger Problem). *Ergebnisse eines Mathematischen Kolloquium, 2*, 11-12.

Miller, M. (2004). *Technology: Cost per error and return on investment*. Retrieved from http://www.vocollect.com/np/documents/CostPerErrorWhitePaper.pdf

Nielsen, J., Clemmensen, T., & Yssing, C. (2002). Getting access to what goes on in people's heads?: reflections on the think-aloud technique. In *Proceedings of the second Nordic conference on Human-computer interaction (NordiCHI '02)* (pp. 101-110). New York: ACM.

Oppermann, R., & Specht, M. (2000). A Context-sensitive Nomadic Information System as an Exhibition Guide. In *Proceedings of the Second Symposium on Handheld and Ubiquitous Computing* (LNCS, pp. 127-142). New York: Springer.

Rhodes, B. J. (1997). The Wearable Remembrance Agent: A System for Augmented Memory. *Personal Technologies Special Issue on Wearable Computing, 1*(1), 218–224.

Robertson, S., & Robertson, J. (1999). *Mastering the requirements process*. Reading, MA: Addison-Wesley.

Rosenkrantz, D. J., Stearns, R. E., & Lewis, P. M. (1977). An analysis of several heuristics for the traveling salesman problem. *Fundamental Problems in Computing*, 45-69.

Saint-Andre, P. (2004). *RFC 3920: Extensible messaging and presence protocol (xmpp): Core. Request for Comments, IETF*. Retrieved from http://tools.ietf.org/html/rfc3920

Schneider, A., Lorenz, A., Zimmermann, A., & Eisenhauer, M. (2006). Multimodal Interaction in Context-Adaptive Systems. In *Proceedings of the Second Workshop on Context Awareness for Proactive Systems* (p. 101).

Schnell, R., & Kreuter, F. (2003). *Separating interviewer and sampling-point effects. In UC Los Angeles: Department of Statistics, UCLA.* Retrieved from http://www.escholarship.org/uc/item/7d48q754

Tompkins, J. A., & Smith, J. D. (1998). *Warehouse Management Handbook* (2nd ed.). New York: Tompkins Associates.

Tucker, S. (1963). *The Break-Even System: A Tool for Profit Planning*. Upper Saddle River, NJ: Prentice Hall.

Want, R. (2004). Enabling ubiquitous sensing with RFID. *IEEE Computer, 37*(4), 84–86.

Weinstein, R. (2005). RFID: A Technical Overview and Its Application to the Enterprise. *IT Professional, 7*(3), 27–33. doi:10.1109/MITP.2005.69

Weiser, M., & Brown, J. (1997). The coming age of calm technology. *Beyond Calculation. The Next Fifty Years of Computing, 8*, 75–85.

Yan, H., & Selker, T. (2000). Context-Aware Office Assistant. In *Proceedings of the 5th Infernational Conference on Intelligent User Interfaces*, New Orleans, LA (pp. 276-279). New York: ACM.

Zhai, S., Morimoto, C., & Ihde, S. (1999). Manual and gaze input cascaded (MAGIC) pointing. In *Proceedings of the Conference on Human Factors in Computing Systems (CHI'99)* (pp. 246-253). New York: ACM.

This work was previously published in the International Journal of Handheld Computing Research (IJHCR), Volume 2, Issue 1, edited by Wen-Chen Hu, pp. 1-24, copyright 2011 by IGI Publishing (an imprint of IGI Global).

Chapter 10

Ontology-Based Personal Annotation Management on Semantic Peer Network to Facilitating Collaborations in E-Learning

Ching-Long Yeh
Tatung University, Taiwan

Chun-Fu Chang
Tatung University, Taiwan

Po-Shen Lin
Tatung University, Taiwan

ABSTRACT

The trend of services on the web is making use of resources on the web collaborative. Semantic Web technology is used to build an integrated infrastructure for the new services. This paper develops a distributed knowledge based system using the RDF/OWL technology on peer-to-peer networks to provide the basis of building personal social collaboration services for e-Learning. This paper extends the current tools accompanied with lecture content to become annotation sharable using the distributed knowledge base.

INTRODUCTION

Web sites provide effective intermediaries between instructors and learners to share lecture contents and exchange messages. From the course web site, learners choose to download lecture notes and slides used in class, lecture videos, etc., and study them offline either in printed form or on screen. During the course of reading, learners may want to take down notes, or annotations, about specific pieces of the contents as reminder that has been done. If learners have chance to share their annotations with each other, then it is possible to work out the collaborative intelligence according

DOI: 10.4018/978-1-4666-2785-7.ch010

to the topics of the lecture content (O'Reily, 2005). Instructors, on the other hand, may know learners' responses by collecting the annotations attached to lecture contents as an aid to future teaching. Yang, et al. summarize that this kind of personal annotation has the advantage in helping student in focusing and organizing learning content, and providing a place for discussion (Yang & Shao, 2004).

Web 2.0 services, for example, blogs, wikis, social websites, etc., provide a collaborative environment by sharing content and response among authors and readers (O'Reily, 2005). It has been investigated in the e-Learning domain that new services are helpful to enhance the interactions among instructors and learners (Owen, Grant, Sayers, & Facer, 2006; Anderson, 2007; Franklin, & van Harmelen, 2007; Gillet, Ngoc, & Rekik, 2005; Gillet, Helou, Yu, & Salzmann, 2008). Since the advent of Semantic Web, annotation mechanisms, for example, Annotea (Kahan, Koivunen, Prud'Hommeaux, & Swick, 2001), CREAM (Handschuh & Staab, 2003; Handschuh, Staab, & Ciravegna, 2002), have long been developed for as means of knowledge sharing among users. Through browsers, a user adds annotations to web pages and the resulting ontology-based metadata in RDF (Beckett, 2004) is either inserted in the web page document or stored in RDF server for others to use.

Semantic Web technology has been employed to capture author-created wikitext in formal way, for example, Semantic Mediawiki (Krötzsch, Vrandecic, & Völkel, 2006), Semantic Wikipedia (Krötzsch & Vrandecic, 2009) and Rhizome (Souzis, 2005). The wikitext inserted with metadata in RDF makes the knowledge management of wiki content more efficient. Though the introduction of Semantic Web technology improves the collaboration and reuse of knowledge in Web 2.0 services, they however are isolated from each other and hence form disconnected communities and the valuable knowledge acquired from public

effort, for example, Wikipedia, would be difficult to reuse (Bojars, Breslin, Peristeras, Tummarello, & Decker, 2008). For integrating the social web sites, the SIOC project develops an ontology describing entities found in Web 2.0 services as the infrastructure to build the integrated platform (Bojars, Breslin, Peristeras, Tummarello, & Decker, 2008; Bojars, Breslin, Finn, & Decker, 2008).

Yang, Chen, and Shao (2004) develop a personalized annotation management system as the basis to support collaborations in e-learning applications. Documents and the associated annotations are stored in two relational data models, respectively. Annotations on the objects of documents are clustered according to their semantic similarities. User's query is first computed to determine the semantically related clusters and then further refine the search to obtain match result from the potential clusters. In their succeeding work (Yang, 2006), ontology technology is used to construct the context-aware environment for ubiquitous learning and peer-to-peer technology is employed to develop collaborative learning, including learning content access, personalized annotation management and discussion group.

In Web 2.0, the mechanism is naturally embedded in various kinds of services to support the multi-directional information flow among authors and readers. In Semantic Web context, the collaboration is further employed to create new knowledge to be reuse based on the exchange of knowledge on the distributed knowledge-based environment (d'Aquin, Motta, Dzbor, Gridinoc, Heath, & Sabou, 2008). As mentioned previously, annotation taking and sharing is essential to facilitate knowledge management and collaborative learning. In this paper, we attempt to take advantages of both technologies, i.e., collaborative services in Web 2.0 and semantic integrated infrastructure in Semantic Web, to develop a platform for managing personal annotations. On this platform, user can annotate lecture contents and share the annotated results using Web 2.0

services, like wiki, blog, instant messaging, etc., or reading tools, for example, MS PowerPoint, Adobe Acrobat.

The infrastructure of the platform is a distributed RDF management system built on a peer-to-peer (P2P) network. The P2P network allows any device to join the network as a peer node to exchange messages with each other and collaborate independently of the underlying network topology. The distributed RDF management system is designed by extending the single ring in the DHT-based RDFPeer (Cai & Frank, 2004) to including additional level of ring for ontology schema. It performs two tasks, the DHT topology management and distributed RDF store management. The former is carried out by the Two-level Routing Module along with Intra- and Inter-finger tables. The latter is divided into management functions on distributed store and local RDF store. The distributed store consists of the triples of distributed RDF store dispatched to a node according to of the DHT topology. Services Layer provides an administration interface for user to boot and configure the peer itself and access both the local and distributed RDF stores. The annotation management service mentioned above is implemented in the Service Layer as well by accessing the distributed RDF store.

Wiki technology can adequately support knowledge sharing in students' collaboration (Stahl & Hesse, 2009; Larusson & Alterman, 2009). To show how the technology is integrated into the semantic peer network we develop, in this paper we present an implementation of personal wiki as a collaboration service on the semantic peer network. The basic idea is that in each one's desk top a personal wiki is maintained having wikitext along with the associated metadata in SIOC. On the other hand, the personal wiki can simultaneously share information collaboratively with other ones through the underlying semantic peer network.

The remainder of this paper is organized as follows. We first describe the architecture of the P2P-based social collaboration platform. We then describe the distributed knowledge based of the platform: a two-level ring architecture based on distributed hash technology for managing distributed RDF store. We describe the design of social collaboration service based on the distributed RDF store and then the implementation of the platform. Finally, we investigate the future work, and conclusions are made.

SEMANTIC P2P ARCHITECTURE FOR SOCIAL COLLABORATION

The technical architecture (Figure 1) of the platform is built on a structured peer-to-peer network, where each node has the same function provides common collaborative services for user to perform social collaborations. Through the collaboration service interfaces user can access the services used to see in social software, including weblog, wiki, annotation, bookmark sharing, etc. In addition, personal reading tools, like MS PowerPoint, Adobe Acrobat Reader, plugged with annotation service can be used to do annotation taking and sharing as well. For example, User A takes notes on selected objects in PowerPoint slides and would like to share the annotation with other peer users. The annotation result is then stored as local RDF file and simultaneously published to the distributed RDF store by using the interface of the knowledge management layer. On the other hand, suppose user B wants to know what others have noted on the pieces of slides she is interested. User B may right click on the interested objects and select the request for notes from other peer nodes. The request is then carried out by querying the distributed RDF store and the result is presented to User B.

In the peer-to-peer network, a distributed RDF store is maintained based the distributed hash technology (DHT) used in Chord (Stoica et al., 2003). The structure of Chord network is ring. In this paper, we employ the DHT for distinguishing

Figure 1. Technical architecture of semantic peer network

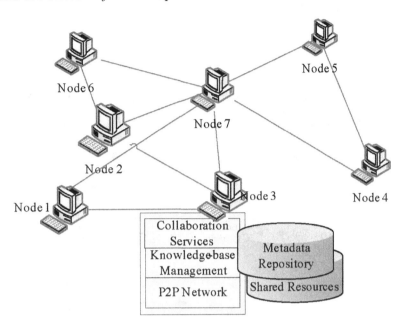

the ontology of RDF triples and then the same technology is applied again to for distributing triples belong to the same ontology. The double application of DHT thus forms two-level ring structure. The nodes in each group are responsible for storing and routing triples of a specific ontology. Each node in a group connects to its inter-successors in other groups.

To support the above technical architecture, we design three-layered system architecture in each peer node, as shown in Figure 2. At bottom is the P2P Network Layer which is implemented using JXTA (Oaks, Traversat, & Gong, 2002), which allows any device to join the network as a peer node to exchange messages with each other and collaborate independently of the underlying network topology. JXTA protocols are employed to communicate with other peers. In the middle is the RDF Management Layer which performs two tasks, the DHT topology management and distributed RDF store management. The former is carried out by the two-level Routing Module

along with Intra- and Inter-finger tables. The latter is divided into management functions on distributed store and local RDF store.

Service Layer provides two kinds of interface for user. One is for the system administration of the underlying distributed knowledge-base and the other is for user to access the social collaboration service on the semantic peer network. The system administration services consist of boot and configure the peer itself and access both the local and distributed RDF stores. The boot function is used for joining to the network. It starts up the two-level ring module and load RDF files to the memory and publishes the RDF triples to the network. Local RDF Management provides add, modify and delete function for users. Distributed RDF Management is use to managing the triples of other peer. Users can views the triples store in their distributed RDF store in Distributed RDF Management. The System Configure service is used to set up the system property.

Figure 2. Layered system architecture of peer node

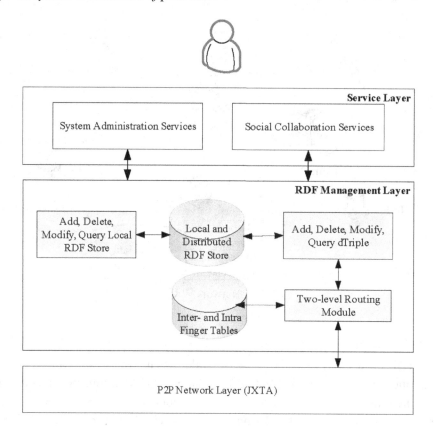

DISTRIBUTED RDF STORE ON THE TWO LEVEL RING NETWORK

We extend the RDF ring mechanism in Chord-based RDF Peers (Stoica et al., 2003; Cai & Frank, 2004) by distinguishing the ontology schema in RDF triples. The basic idea is that the RDF triples of a schema are distributed to the same ring. We therefore design a two-level ring on the P2P network, where the first level is based on the ontology schema and the next one is on the RDF itself. Each peer node is assigned two identifiers, GROUP_ID and PEER_ID, to recognize to which schema group the peer node belongs and identify itself within the belonging group. According to the concept of M-Ring (Lin., Ho, Chan, & Chung, 2008), we treat the nodes of the same group as a virtual region or a "node" of the two-level ring. Thus, in each peer node we

maintain two finger tables: the inter-finger table recording the connections to other groups, and the intra-finger table storing connection information of other peer nodes in the same group. According to the Chord mechanism, the entries in the intra-finger store the next 1, 2, 4...and 2^n-1 nodes in a group, where n is the bit-length of PEER_ID. The entries in the intra-finger table record the next 1, 2, 4...and 2^m-1 nodes on the group ring, where m is the bit-length of GROUP_ID.

Consider the case that if in two linking groups a specific node in each group is designated as the linking point, then either node being off would result in problem of missing link between both sides, a kind of hot spot problem. In this paper, we employ the concept of virtual region in M-Ring to avoid the above hot spot problem (Lin, Ho, Chan, & Chung, 2008). Instead of single linking point, each node in a group connects to

Figure 3. Example of two-level ring

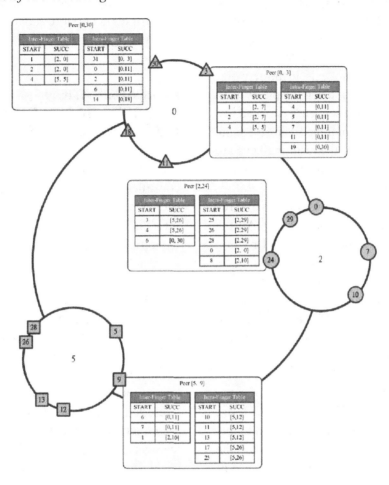

its nearest successor in the succeeding group, or the inter-successor. For example, in Figure 3, the first entry of Nodes (0, 30) and (0, 3) point to their inter-successors (2, 0) and (2, 7), respectively.

When a peer sends a message to other peer, it checks whether the target peer is in the same group. If not, the sending peer starts inter-group routine. Sender gets the nearest predecessor from the inter-finger table and forwards the message to the linking point in the predecessor group until the target group is reached. The intra-group routing is employed in order to pass the message to the target node. In Figure 3, for example, Node (0, 30) wants send a message to Node (2, 24). Node (0, 30) starts the inter-group routing, and find the nearest node, Node (2, 0). Then Node (0,

30) sends its message to Node (2, 0), and consign the sending process to it. Here, Node (2, 0) and the target node, Node (2, 24) belong to the same group. Node (2, 0) begins the intra-group routing. Because Node (2, 24) is in Node (2, 0)'s intra-finger table, the message is passed to Node (2, 24) by way of Node (2, 0).

Having developed the structured P2P network based on the two-level ring architecture, we then construct the distributed RDF store based on the structured P2P network. The basic idea is to deposit the triples in RDF files produced by peer nodes using the distributed hash computations. When an RDF file is newly created, modified, or deleted by a peer node, it is first of all parsed into the sequence of triples. The new, modified, and obsolete triples are computed to obtain their key which is formed by the GROUP_ID and PEER_ID as described previously. Then these keys are routed properly to be stored in the destination peer nodes. Noted that the destination peer may not exist in the network; we choose the successor to store the triple. When a query in the form of triple sequence is issued in a peer node, each triple is computed using the same distributed hash function, the resulting keys are used to route to the peers having the information about the triples in question.

Each of the above operations consists of three steps, hashing, routing, and processing. All the operations have the same hashing and routing steps, while each one has its own processing. In the following, we first of all describe the hashing and routing steps, and then describe each of the processing. Each peer node can nominate an RDF file in its persistent storage for storing the distributed triples.

- **Hashing:** Given the S, P, and O of a triple, where P is predicate and has the form, namespace:label, the namespace and label are hashed to be a pair of keys, which is the GROUP_ID and PEER_ID, for the use of routing step.
- **Routing:** The triple is routed to the target group using the GROUP_ID and then routed to the target peer node in the group using PEER_ID. The former inter-group routing is achieved by consulting the inter-finger tables, while the latter intra-group routing is accomplished through the help of intra-finger table.
- **Processing:** If the destination peer receive message, it process the add, delete or query function according to the message type. For the add operation, the triple is inserted into the distributed RDF store directly. For delete and modify operations, the process is similar except the insert operation is replaced by delete and modify operations, respectively. As for the query operation,

the query pattern is used to match with the triples in the store and then return the result to original peer.

COLLABORATION SERVICES BASED ON THE DISTRIBUTED RDF STORE

As mentioned previously services interfaces are constructed upon the distributed RDF store, as shown in Figure 2, for system administration and for social collaborations, respectively. The former is described in further details in the implementation section. In this section we describe the design of social collaboration service.

When viewing lecture videos and reading lecture notes, hereafter termed lecture content, the information flow is in general from producer to consumer, i.e., from instructor to learner only. In other words, learner can only read the content but is not able to express her opinions to the interested parts of the video or lecture notes. We argue that the collaboration among instructors and learners can be greatly improved by taking into of the structure of lecture content in the sharing of annotations. For example, learner may want to insert question about specific part of a lecture video and look for help from others. Learner may want to pause reading a learning content and take down her notes and share them with others of the lecture. Also instructor can have a look at the annotations about the lecture video to see the response from learners. In this paper we develop collaboration services for lecture content annotation taking and sharing to facilitate collaborations among instructors and learners. In this paper, the lecture contents are in various formats, including video, slide presentations in for example, Microsoft PowerPoint, and Adobe PDF.

The annotation plug-ins to existing tools, for example MS PowerPoint, provides user interface for the management of annotations on lecture content. The service programs deal with two kinds of objects. One is the physical files containing the knowledge sources, and the other is associated metadata, i.e., the annotations in RDF formats. Thus in a peer node the annotations are in RDF formats and stored as local files. We employ the SIOC (Semantically Interlinked Online Communities) ontology which is used describes the entities found in Web 2.0 services, including blogs, wikis, forums, social web site, etc., as the basis schema to develop the management functions of the knowledge produced by the annotation tools (Bojars, Breslin, Peristeras, Tummarello, & Decker, 2008). Each service program then extends the vocabulary according to their specific requirements.

In this paper, we intend to develop a personal wiki system built on the distributed RDF store described in the preceding section. In the following, we summarize the design goals of the personal wiki system.

- **Personal Blog/Wiki Content and Annotation Management:** This service is used to manage her articles and the associated annotations at her own site.
- **Peer-To-Peer Topology:** The service is working over peer-to-peer network for the purpose of autonomy of content created by user.
- **Conceptual Search and Semantic Navigation:** User can discover what they are interested in by specifying the concept in mid instead of keyword.
- **Ontology-Based Annotation Management:** Enhance the management of annotation created by user.
- **Common Annotation Representation:** This is to make the annotated result be reused in various ways.

According to the design goals, we therefore construct three components, Content Manager, Annotation Manager, and Query Manager, in the personal wiki system, which are described as below.

- Content Manager is responsible for dealing with all information that is input from user, and it converts the resulting information into physical files and associated metadata in RDF format. When Content Manager detects newly created objects, it analyzes the in the objects and creates the metadata RDF file. Content Manager gathers information specific to the new objects, including creator, title and tags *etc.* from the lecture objects. With SIOC ontology, Content Manager stores the information as RDF files in the local repository. For example, the RDF file associated with a newly created article is shown as below.

- Annotation Manager is used to manage all the annotations user inserts into lecture objects. User inserts annotations to selected part of the lecture object, for example, some object in a PowerPoint slide, and indicate whether shares it with others. When inserting an annotation of the question type, it then sends off the question to the peer-to-peer network to ask for help. The annotations are stored as RDF file using SIOC schema. For example, a comment inserted to the article having the above metadata is represented as RDF as below.

- Query Manager is used to process the query issued by node user. When user wants to look for something interested to her, she enters the conditions through the user interface and the conditions are converted into SPARQL statement as input to the Query Manager. It then converts the SPARQL statement into triple patterns recognized by the query function of the search function in the underlying distributed RDF management system. It waits for the result returning from the query function and presents rendered result in the user interface for user to consume.

IMPLEMENTATION

In this section we first describe the implementation of the distributed RDF store and then the personal wiki system as a collaboration service as shown in Figure 2.

The Distributed RDF Store

The distributed RDF store is made up of the RDF Management Layer and the peer-to-peer network layer of the system architecture as shown in Figure 2. The later is implemented using an XML-based peer-to-peer network JXTA (Oaks, Traversat, & Gong, 2002). In this paper, we employ the JXTA peer ID as the basis to construct the peer ID in the two-level ring. We mainly use EndpointService protocol for the exchange of messages between peer nodes (Chang, 2009).

The RDF Management Layer performs two tasks, the DHT topology management and distributed RDF store management. The former is carried out by the Two-level Routing Module along with Inter- and Intra-finger tables. The latter is divided into management functions on dTriple and local RDF store. The former, dTriple, consists of the triples of distributed RDF store dispatched to a node according to of the DHT topology. The local RDF store consists of the RDF generated by the node. Both management functions support their user interfaces in the administration service of the service layer in Figure 2.

As shown in Figure 2, in the top layer of the system architecture provides of two kinds of services. One is for system administration and the other is for user's social collaborations. The former provides of a number of user interfaces: Boot, System Configuration, Local RDF Management, and Distributed RDF Management.

- The Boot interface is used for joining to the network. It starts up the two-level ring module and load RDF files in the memory and publishes the RDF triples to the network. The System Configuration interface is used to set the system properties, including setting the port, peer name and the directory of local RDF file.

- The Local RDF Management interface provides add, modify and delete function for users to manage the RDF files stored locally. The Distributed RDF Management interface is used to view the triples being stored in the peer node according to the two-level DHT algorithm. User can query in SPARQL the distributed RDF store as shown in Figure 4.

We have conducted experiments using these interfaces to measure the performance of the distributed RDF store. We set up an 8-peer node network to carry out the experiment. Each peer node is a personal computer commonly found in lab. We summarize the measuring results in Table 1.

To test the joining time, we divide the 8 nodes into two groups, DC and FOAF, and each node uses the same RDF file. At first, we create a boot peer, PEER 1. Then PEER2 to 7 join the network in turn by PEER 1. The result is of each peer join the network is shown in the row of Join with average 4.7 sec. For the test of publishing time, we employ an SIOC and a FOAF RDF file, each with 63 and 12 triples, respectively. The result is shown in the row Publish. We perform two query tests.

```
PREFIX dc:<http://purl.org/dc/ele-
ments/1.1/>
SELECT ?r using the following SPARQL
statements, and the result is shown in
the rows of Query1 and Query2, respec-
tively.
WHERE { ?r dc:title ?z.}
PREFIX dc:<http://purl.org/dc/ele-
ments/1.1/>
SELECT ?r ?c
WHERE { ?r dc:title ?z.?r dc:creator ?c}
```

Figure 4. Querying the distributed RDF store using SPARQL

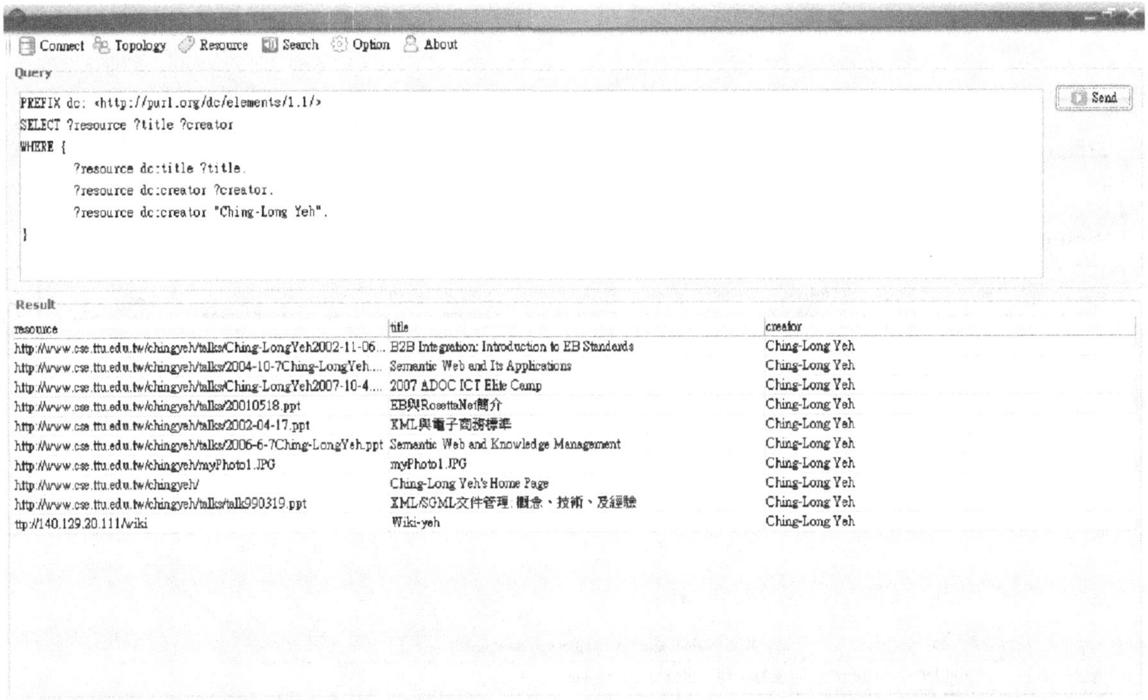

Table 1. Performance measurement for accessing an 8-node distributed RDF store (ms)

	Join	Publish	Qry1	Qry2
1	2266	625	516	1056
2	2251	6047	1801	1104
3	6938	1390	172	667
4	5094	5766	1280	1196
5	5390	3531	944	1045
6	4735	3953	797	856
7	5657	1406	612	944
8	5328	4719	1409	956
Avg	4707	3430	941	978

The Personal Wiki System

The performance measurement to some extent shows a preliminary success in building the prototype for the development of social collaboration services for e-learning. The social collaboration services are implemented by using the application program interfaces when dealing with distributed RDF store.

We add an annotation plug-in to TiddlyWiki (a reusable non-linear personal web notebook, http://www.tiddlywiki.com/) a personal wiki tool using HTML with Javascript technology, as the tool for managing personal notebooks. Behind the user interface, the associated annotations in RDF is created and stored as local file. The RDF file is then published to the distributed RDF store for other to share. As described previously, the annotations associated with wikitext are represented in RDF with vocabulary mainly reusing the SIOC ontology (Bojars, Breslin, Peristeras, Tummarello, & Decker, 2008; Bojars, Breslin, Finn, & Decker, 2008). For example, in Figure 5 is an example of blog post with the attached annotations.

When Content Manager detects a newly created text, it then gathers the metadata about the new resource, author information, created time,

Figure 5. A blog post with the associated annotations

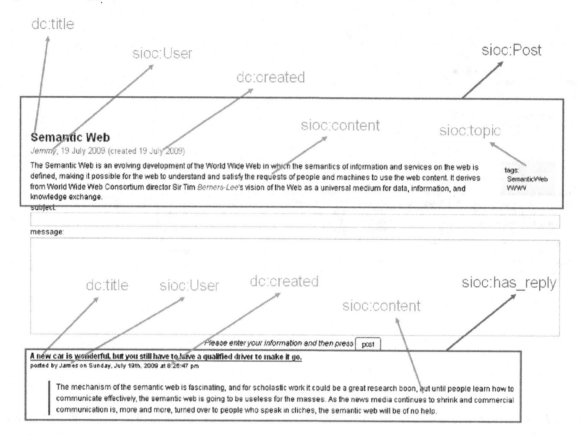

the source address etc., and save the information in a local RDF file using the vocabulary just described. Whenever any text is modified, the Content Manager repeats the action that it has done before and save the new RDF file.

We use the Comment Plugin tool in Tiddly-Wiki to support the annotation function for user to annotate posts in found the wiki system. When a user published some comments, the Annotation Manager reads the original HTML article about comments and turned these comments into RDF format as shown in Figure 6 and inserts it in the original RDF file of the post. See Box 1.

The comments published by readers except the original creator, should not connect with the original articles immediately. Actually, Annotation Manager preserves these new comments and related RDF files in the local repository for a while,

and checks whether the original article creator is online. The new annotations will then write in repository which located in the peer that belongs to the original article creator.

On the top right of the wiki page is a search window for user to enter condition of interested objects. The search result is displayed in the left-hand side of the wiki page as shown in Figure 6. User can choose any object returned and show it as shown in Figure 6. The comments in the bottom of the article can be used to share personal opinions with anyone. If the post is a Wiki article, which means it has no Journal tag, user can modify this article directly with the edit button. When annotation or modify operation is toggled, the new contents can join in the repository that belongs to the original creator through peer-to-peer network.

Figure 6. Search result and comments

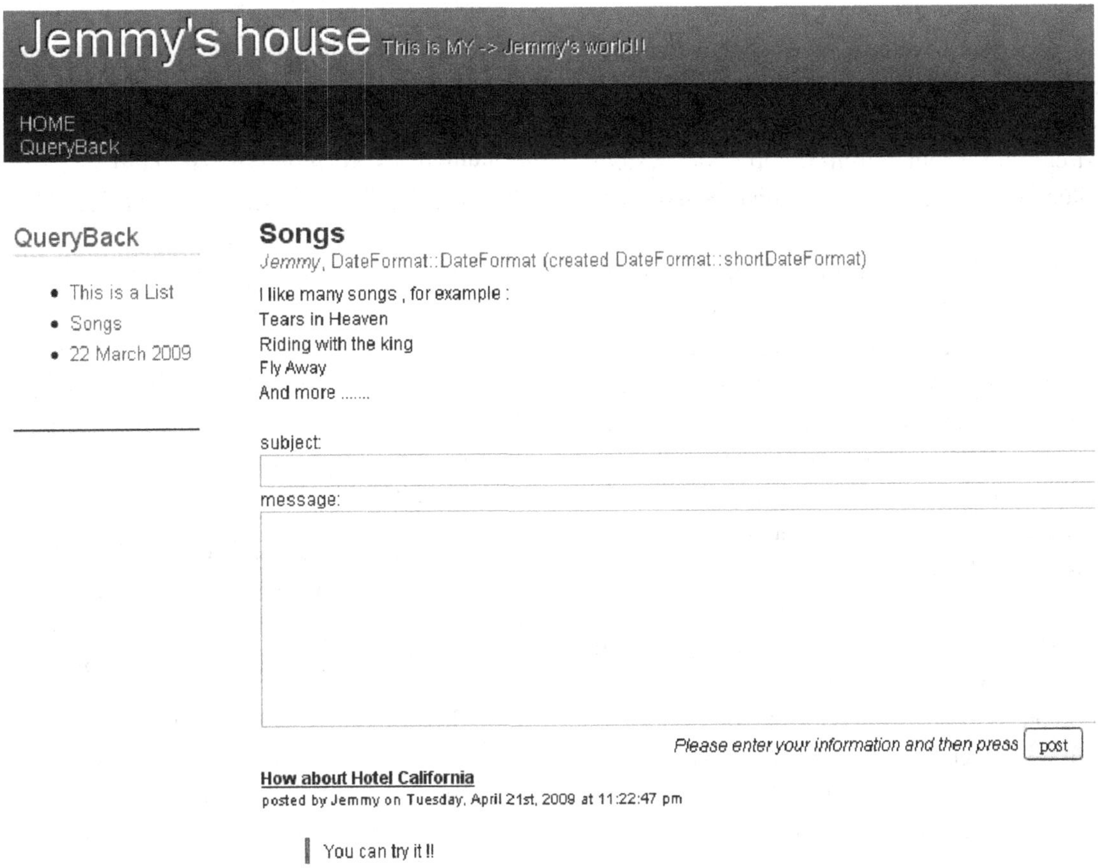

Box 1.

```
<sioc:has_reply>
 <sioc:Post rdf:about=http://IamGod?p=7#comment-1
    dc:title="I like Cats">
  <sioc:description>posted by maligbi on Friday,
    March 13th, 2009 at 12:18:39 pm </sioc:description>
  <sioc:content>It's my favorite animals</sioc:content>
  <rdfs:seeAlso rdf:resource="http://IamGod?p=7"/>
 </sioc:Post>
</sioc:has_reply>
```

CONCLUSION

In this paper, we develop 3-layer system architecture on peer-to-peer network for building social collaboration services to facilitate collaborations in e-Learning. From bottom to top in the system architecture are peer network, distributed RDF management, and service layers. We combine the Semantic Web and Web 2.0 technologies to develop the distributed ontology-based RDF store of the system architecture. At the service layer, social collaboration services invoke the application program interfaces provided by the underlying distributed RDF management layer to make the annotation sharable to other user on the peer-to-peer network. We have implemented and tested the distributed RDF store and get preliminary success in operating the system. We design and implement a personal wiki system by extending TiddlyWiki to show a typical social collaboration service on the distributed RDF store.

We will extend tools used to be accompanied with lecture contents in various formats to make the consumption-only way of using lecture content to become collaborative among instructors and learners, including Microsoft PowerPoint, video player and Microsoft Live Messenger, to make the personal tools have annotation sharable. In the future, we will investigate how to make use of the in e-Learning and study the acceptability of using personal social collaborations tool in teaching.

ACKNOWLEDGMENT

Research for this paper was financed by National Science Council, Taiwan, under contract NSC 98-2221-E-036-042.

REFERENCES

Anderson, P. (2007). *What is Web 2.0? Ideas, technologies and implications for education.* Retrieved from http://www.jisc.ac.uk/media/documents/techwatch/tsw0701b.pdf

Beckett, D. (Ed.). (2004). *RDF/XML syntax specification (Revised).* Retrieved from http://www.w3.org/TR/REC-rdf-syntax/

Bojars, U., Breslin, J., Finn, A., & Decker, S. (2008). Using the Semantic Web for linking and reusing data across Web 2.0 communities. *Journal of Web Semantics*, *6*(1), 21–28. doi:10.1016/j.websem.2007.11.010

Bojars, U., Breslin, J., Peristeras, V., Tummarello, G., & Decker, S. (2008). Interlinking the social web with semantics. *IEEE Intelligent Systems*, *23*(3), 29–40. doi:10.1109/MIS.2008.50

Cai, M., & Frank, M. (2004). RDFPeers: A scalable distributed RDF repository based on a structured peer-to-peer network. In *Proceedings of the 13th International Conference on World Wide Web* (pp. 650-657). New York, NY: ACM Press.

Chang, C. (2009). *Design and implementation of an ontology-based distributed RDF store based on Chord network.* Unpublished doctoral dissertation, Tatung University, Taipei, Taiwan.

d'Aquin, M., Motta, E., Dzbor, M., Gridinoc, L., Heath, T., & Sabou, M. (2008). Collaborative semantic authoring. *IEEE Intelligent Systems*, *23*(3), 80–83. doi:10.1109/MIS.2008.43

Franklin, T., & van Harmelen, M. (2007). *Web 2.0 for content for learning and teaching in higher education.* Retrieved from http://www.jisc.ac.uk/publications/reports/2007/web2andpolicyreport.aspx

Gillet, D., Helou, S., Yu, C., & Salzmann, C. (2008). Turning Web 2.0 social software into versatile collaborative learning solutions. In *Proceedings of the First International Conference on Advances in Computer-Human Interaction* (pp. 170-176). Washington, DC: IEEE Computer Society.

Gillet, D., Ngoc, A., & Rekik, Y. (2005). Collaborative web-based experimentation in flexible engineering education. *IEEE Transactions on Education*, *48*(4), 696–704. doi:10.1109/TE.2005.852592

Handschuh, S., & Staab, S. (2003). CREAM: CREAting metadata for the Semantic Web. *Computer Networks*, *42*(5), 579–598. doi:10.1016/S1389-1286(03)00226-3

Handschuh, S., Staab, S., & Ciravegna, F. (2002). S-CREAM - semi-automatic CREAtion of metadata. In A. Gomez-Perez & V. R. Benjamins (Eds.), *Proceedings of 13th International Conference Knowledge Engineering and Knowledge Management Ontologies and the Semantic Web* (LNCS 2473, pp. 358-372).

Kahan, J., & Koivunen, M. Prud'Hommeaux, E., & Swick, R. (2001). Annotea: An open RDF infrastructure for shared web annotations. In *Proceedings of the International Conference on the World Wide Web*, Hong Kong.

Krötzsch, M., & Vrandecic, D. (2009). Semantic Wikipedia. *Social Semantic Web*, 393-421.

Krötzsch, M., Vrandecic, D., & Völkel, M. (2006). Semantic MediaWiki. In I. Cruz, S. Decker, D. Allemang, C. Preist, D. Schwabe, P. Mika et al. (Eds.), *Proceedings of the 5th International Semantic Web Conference* (LNCS 4273, pp. 935-942).

Larusson, J. A., & Alterman, R. (2009). Wikis to support the "collaborative" part of collaborative learning. *International Journal of Computer-Supported Collaborative Learning*, *4*(4), 371–402. doi:10.1007/s11412-009-9076-6

Lin, T., Ho, T., Chan, Y., & Chung, Y. (2008). M-ring: A distributed, self-organized, load-balanced communication method on super peer network. In *Proceedings of the 9th International Symposium on Parallel Architectures, Algorithms, and Networks*, Sydney, Australia (pp. 59-64). Washington, DC: IEEE Computer Society.

O'Reily, T. (2005). *What is Web 2.0: Design patterns and business models for the next generation of software.* Retrieved from http://oreilly.com/web2/archive/what-is-web-20.html

Oaks, S., Traversat, B., & Gong, L. (2002). *JXTA in a nutshell.* Sebastopol, CA: O'Reilly Media.

Owen, M., Grant, L., Sayers, S., & Facer, K. (2006). *Social software and learning.* Retrieved from http://www.futurelab.org.uk/resources/documents/opening_education/Social_Software_report.pdf

Souzis, A. (2005). Building a semantic wiki. *IEEE Intelligent Systems*, *20*(5), 87–91. doi:10.1109/MIS.2005.83

Stahl, G., & Hesse, F. (2009). Paradigms of shared knowledge. *International Journal of Computer-Supported Collaborative Learning*, *4*(4), 365–369. doi:10.1007/s11412-009-9075-7

Stoica, I., Morris, R., Liben-Nowell, D., Karger, D., Kaashoek, M., & Dabek, F. (2003). Chord: A scalable peer-to-peer lookup protocol for internet applications. *IEEE Transactions on Networking*, *11*(1), 17–32. doi:10.1109/TNET.2002.808407

Yang, S. (2006). Context aware ubiquitous learning environments for peer-to-peer collaborative learning. *Journal of Educational Technology & Society*, 9(1), 188–201.

Yang, S., Chen, I., & Shao, N. (2004). Ontological enabled annotations and knowledge management for collaborative learning in virtual learning community. *Journal of Educational Technology & Society*, 7(4), 70–81.

This work was previously published in the International Journal of Handheld Computing Research (IJHCR), Volume 2, Issue 2, edited by Wen-Chen Hu, pp. 20-33, copyright 2011 by IGI Publishing (an imprint of IGI Global).

Chapter 11
A Petri–Net Based Context Representation in Smart Car Environment

Jie Sun
Ningbo University of Technology, China

Yongping Zhang
Ningbo University of Technology, China

Jianbo Fan
Ningbo University of Technology, China

ABSTRACT

Driving is a complex process influenced by a wide range of factors, especially complex interactions between the driver, the vehicle, and the environment. This paper represents the complex situations in smart car domain. Unlike existing context-aware systems which isolate one context situation from another, such as road congestion and car deceleration, this paper proposes a context model which considers the driver, vehicle and environment as a whole. The paper tries to discover the inherent relationship between the situations in the smart car environment, and proposes a context model to support the representation of situations and their correlation. The detailed example scenarios are given to illustrate our idea.

1. INTRODUCTION

The smart car is becoming a promising application domain of ubiquitous computing, which aims at assisting the driver with easier driving, less workload and less chance of getting injured (Moite, 1992). For this purpose, a smart car must collect and analyze the relevant information about the driving task, i.e., context. Driving is a complex process influenced by a wide range of factors: traffic, vehicle speed, distraction, fatigue, errors, and even capabilities and experience of the driver. How to represent the complexity is the basis for a car to be smart.

Lots of smart cars have been developed in the past decade. However, the knowledge representation in the smart car area is not paid enough attention. The most commonly used approach

DOI: 10.4018/978-1-4666-2785-7.ch011

is getting low level context from sensors, which is applicable in direct service development, but is not sufficient for representation of complex knowledge in the changeable and complex driving environment. New representation approach is required to describe the intricate and complex interactions between the driver, the car and the environment.

The remainder of the paper is organized as follows. A general introduction of a smart car is given in Section 2. Section 3 proposes a novel representation approach for definition and classification of information in a smart car environment. We give the model of the smart car to illustrate how our approach works in Section 4. The performance evaluation is shown in Section 5. Section 6 introduces the related work of context representation approaches in smart car. The conclusions are given in Section 7.

2. SMART CAR

A smart car is a comprehensive integration of many different sensors, control modules, actuators, and so on (Wang, 2006). A smart car can monitor the driving environment, assess the possible risks, and take appropriate actions to avoid or reduce the risk. A general architecture of a smart car is shown in Figure 1.

The information, i.e., context, needing to be collected for a smart car includes:

1. **Traffic:** A variety of scanning technologies are used to recognize the distance between the car and other road users. Active environments sensing in- and out-car will be a general capability in near future (Tang, Wang, & Miao, 2006). Lidar-, radar- or vision-based approaches can be used to provide the positioning information. The radar and lidar sensors provide information about the relative position and relative velocity of an object. Multiple cameras are able to eliminate blind spots, recognize obstacles,

Figure 1. The general architecture of a smart car

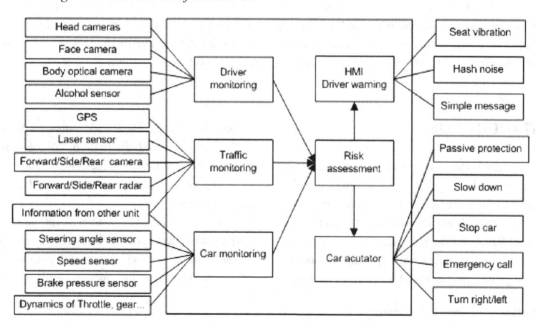

and record the surroundings. Besides the sensing technology described above, the car can get traffic information from Internet or nearby cars.

2. **Driver:** Drivers represent the highest safety risk. Almost 95% of the accidents are due to human factors and in almost three-quarters of the cases human behaviour is solely to blame (Rau, 1998). Smart cars present promising potentials to assist drivers in improving their situational awareness and reducing errors. With cameras monitoring the driver's gaze and activity, smart cars attempt to keep the driver's attention on the road ahead. Physiological sensors can detect whether the driver is in good condition.

3. **Car:** The dynamics of a car can be read from the engine, the throttle and the brake. These data will be transferred by controller area networks (CAN) in order to analyze whether the car functions normally.

Assessment module determines the risk of driving task according to the situation of traffic, driver and car. Different levels of risk will lead to different responses, including notifying the driver through HMI (Human Machine Interface) and taking emergency actions by car actuators. HMI warns the driver of the potential risks in non-emergent situations. For an example, a fatigue driver would be awakened by an acoustic alarm or vibrating seat. Visual indications should be applied in a cautious way, since complex graph or long

text sentence will seriously impair the driver's attention and possibly cause harm. The actuators will execute specified control on the car without the driver's commands. The smart car will adopt active measures such as stopping the car in case that the driver is unable to act properly, or applying passive protection to reduce possible harm in abrupt accidents, for example, popping up airbags.

3. CONTEXT REPRESENTATION

Context is any information that can be used to characterize the situation of an entity (Dey, Salber, & Abowd, 2001). As a kind of knowledge about the driving environment, context data can be separated into three layers according to the degree of abstraction and semantics: sensor layer, context atom layer and context situation layer, as shown in Figure 2.

3.1. Context Atoms

Sensor layer is the source of context data. Context atoms are retrieved from sensors, serving as an abstraction between the physical world and semantic world. For each type of context atom, a descriptive name must be assigned for applications to use the context. The name comes from the ontology library and is domain-specific. In smart car, we use three classes of ontologies to provide agreed understanding of contexts, as shown in Figure 3:

Figure 2. The three-layer of context data

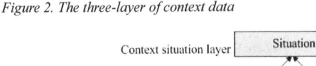

Figure 3. Ontology for context atoms

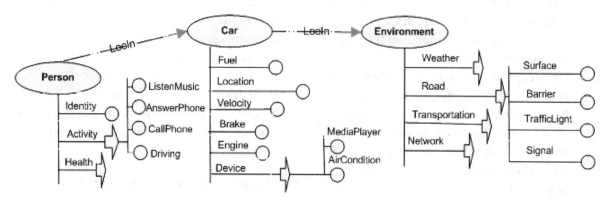

- **Ontology for Environment Contexts:** Environmental contexts are related to physical environments. The ontology of physical environments in the smart car includes the description of weather status, the road surface status, the traffic information, the road signs, the signal lamp and the network status.

- **Ontology for Car Contexts:** The ontology of the smart car includes three parts: the power system, the security system and the comfort system. The power system concerns the engine status, the accelerograph, the power (gasoline) etc.. The security system includes factors impacting the safety of the car and the driver, such as the status of the air bag, the safe belt, the ABS, the reverse-aids, the navigation system, the electronic lock, etc.. The comfort system is about entertainment, the air condition, and windows.

- **Ontology for Driver Contexts:** The driver contexts detect the user's physiological status, including heart beating, blood pressure, density of carbon dioxide, diameter of pupils etc. From the information we want to deduce the healthy status and mental status of the driver to determine whether he is sick or is able to continue driving. We implement the ontology using protégé, as shown in Figure 4.

3.2. Context Situation

Context atoms layer provides the elementary conceptual data pieces and use ontology to provide the definition of attributes and classes. However, it is unable to represent complex knowledge, such as the potential danger of two-car collision. Thus a situation layer is built on the top of atom layer. The main purpose of situation layer is to fuse individual context atoms into meaningful context situations. A context situation represents the current state of a specific entity.

We choose Petri net (Murata, 1989) to represent context situation for its excellent performance in modeling concurrent computation and distributed system. Each place of the Petri net corresponds to a context situation, and each transition of the Petri net is extended to represent the correlation between context situations. We can formalize the definition of context situations into: A context situation model is a quintuple:

$$SN = (S, T, F, W, M_0) \tag{1}$$

where S is the finite set of places to define context situations and

$$S = \{s_1, s_2; ..., s_m\}$$

Figure 4. Smart car ontology implemented with protégé

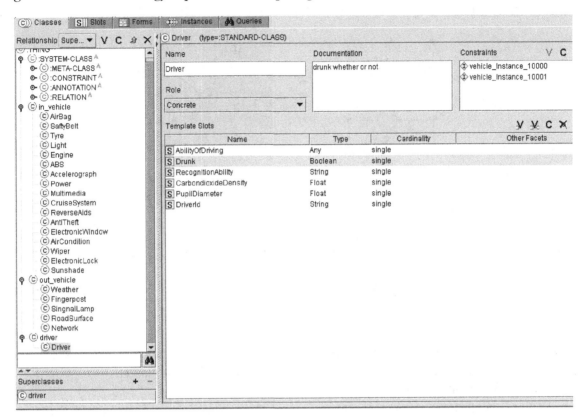

T is the finite set of transitions between context situations and

$$T = \{t_1, t_2, ..., t_n\}$$

F is a function that defines the correlation by directed arcs form transitions to places, and

$$F \subseteq (S \times T) \cup (T \times S)$$

W is the weight of each transition and

$$W \rightarrow \{1, 2, 3, ...\}$$

M_0 is the initial marking of the system, i.e., the initial assignment of tokens, and

$$M_0 : S \rightarrow \{0, 1, 2, 3, ...\}$$

A transition indicates a situation change and can represent the correlation between context situations. We can summarize the correlations between situations into three classes:

1. **Concurrency:** Context situation S_1 and S_2 are independent of each other, as shown in Figure 5a. They can take place in the same time and be responded at any sequence without influence. S_3 contain two tokens, which is indicated in black dot in the figure. The number of black dot represents the number of tokens. We mention all the transition weight is 1, so the transition T_1 and T_2 can be enabled and fired at the same time. Consequently, the model has the same probability to reach context situation S_1 and S_2, and the reach order does not matter.

Figure 5. The different correlations of two context situations represented using Petri net: (a) concurrency situations; (b) conflict situations; (c) - (e): causal situations

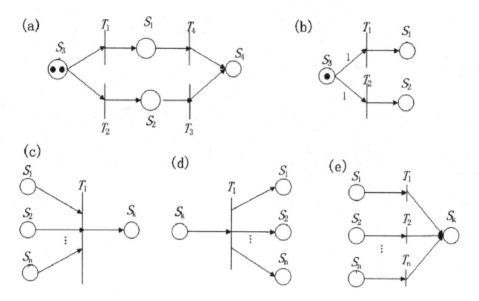

2. **Conflict:** If context situation S_1 and S_2 take place in the same time and they will compete for the user's attention or the resource, we call the two contexts are concurrent contexts. Figure 5b shows that S_3 contain only one token, so the transition T_1 and T_2 must compete for the token to be enabled and fired. Consequently, there is only one context situation for the model to reach. If transition T_1 is enabled and fires, the token of S_3 is consumed and thus transition T_2 is dead. The reverse is also true. So we use conflict correlation to represent the either-or condition.

3. **Causality:** Context situation S_2 is derived from S_1, we call this relationship is causality. If context S_2 derives from S_1, then there is a path from context S_1 to S_2. We call context S_1 the parent of context S_2. We call the nodes that have not any input edges as leaves, which represent context atoms coming from sensors. So the reasoning process will generate a graph, where each of the intermediate nodes represents an intermediate sub-target of reasoning and a sub-graph represents a local reasoning process for the corresponding intermediate sub-target.

As shown in Figures 5c through 5e, the causality is represented as:

c: IF S1 and S2,…, and Sn THEN Sk (CF=ui)

where

$$p_k = \min(p_1 * u_i, p_2 * u_i, …, p_n * u_i)$$

d: IF Sk THEN S1 and S2,…, and Sn (CF=ui)

where

$$p_1 = p_2 = … = p_n = p_k * u_i$$

e: IF S1 or S2,…, or Sn THEN Sk (CF=ui)

where

$$p_k = \max(p_1 * u_i, p_2 * u_i, …, p_n * u_i)$$

in which p_i is the confidence of context situation S_i and CF is the confidence of the inference rule.

4. AN APPLICATION SCENARIO: THE SMART CAR

Petri net is a powerful tool for the description and analysis of concurrence and synchronization in parallel systems. Using the approach described in Section 3, we can represent the smart car environment as shown in Figure 6. Part of the definitions of the situations and transitions are listed in Table 1.

The first step in building the context model is to specify the situations demanding the smart car to recognize and react to. For a specified application environment, the situations are definite and can be defined beforehand. The variables that cannot be pointed out beforehand are changes of the situations.

The model describes the scenarios from the driver enters the smart car to the driver arrives at the destination, including:

Scenario 1: The Smart Car Sets Out

The initial situation is that the driver enters the smart car. From the initial situation, we have five concurrent output arcs which represent five sequential situations independent of each other. So we assign the token of the initial situation to be 5.

The five concurrent situations are: 1) the weather detection inside the car, including temperature, humidity and the quality of the air; 2) the new device detection, such as the personal digital assistant (PDA), iPhone and MP4 player; 3) Network detection, such as Wi-Fi and Bluetooth. 4) Speech detection, to determine whether the user in the car is speaking. If the user is speaking, the detection will determine whether the speech is vocal command. If it is, the voice command will be executed; 5) drive the car.

The situations 1 to 4 are circular, ready to detect the new changes in the smart car.

Scenario 2: Driving Environment Inside the Car

During the driving, there are four concurrent arcs which represent four scenarios independent of each other: 1) fuel supply detection, to estimate whether the residual fuel is enough to drive to the destination. There are two statuses: "normal" means the residual fuel is more than a threshold amount and "insufficient" means the residual fuel is less than a threshold amount. The smart car can only in one of the two statuses at any given instance, so the two statuses are conflict. We model the statuses to be of conflict correlation; 2) engine detection, to determine whether the engine functions normally. There are two statuses: "normal" means the engine functions well and "fault" means the engine is in trouble. The two statuses are exclusive options, so we model the statuses to be of conflict correlation. If the engine is in status of "fault," the smart car will execute operation of "StopCar"; 3) driver detection, to estimate whether the driver is competent for the driving. The driver situation includes "normal" and four abnormal statuses: "Drowsy," "Abstracted," "Drunk," and "Sick." In the first two situations, the smart car will awake the driver by harsh noise. In the latter two scenarios, the smart car has to slow down and park on the roadside; and 4) the driving behavior (detail is described in next scenario).

Scenario 3: The Driving Behavior

Driving behavior represents the process of driving and adapting speed according to the traffic and road situation. We take information that will impact the driving task into account, such as traffic light, the distance between cars, and the lane.

Driving scenario is specified into: 1) the car drives at a constant speed; 2) when the car arrives at an intersection or T-junction, it recognizes the

Figure 6. Petri net model of the smart car, where each node is a context situation, and each directed edge indicates the transition between different situations. Symbol ⊥ in the node represents the conflict relationship.

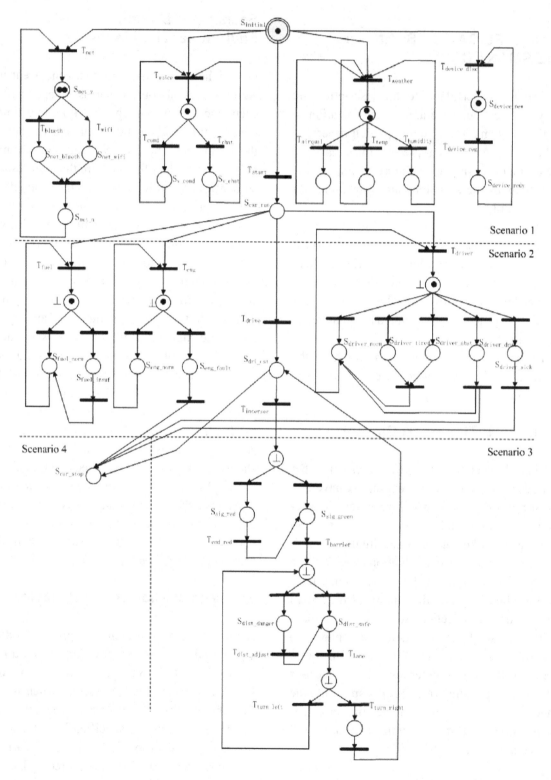

Table 1. Part of the description of situations and transitions in the Petri net model of smart car

Identity	Type	Detail
$S_{initial}$	situation	The initial situation of the smart car
S_{car_run}		The smart car is ready to operate
S_{dri_cst}		The car drives at a constant speed
S_{sig_red}		The traffic light is red
S_{sig_green}		The traffic light is green
S_{dist_danger}		The distance between cars is dangerous
S_{dist_safe}		The distance between cars is safe
S_{turn_left}		The smart car is ready to turn left
S_{turn_right}		The smart car is ready to turn right
S_{car_stop}		The smart car is stopped
T_{start}	transition	The smart car begins to operate
T_{drive}		The smart car begins to drive
$T_{intersec}$		The smart car enters a intersection
$T_{barrier}$		There is a barrier in front of the car
T_{lane}		the car needs to change the lane
T_{dist_adjust}		The car adjusts its speed

traffic light. If the light is red, the car must evaluate the distance from the car ahead and from the stop line. If the distance is less than 1 meter the smart car will slow down. If the distance is less than 0.5m, the car will stop. Thus the smart car will always maintain safe distance from the car ahead; 3) If the traffic light is green or the traffic light turns green from red, the smart car must determine the direction and the lane. If the car needs to turn left, there will be two lane-changing. The first lane change occurs when the car turns left and leaves the left lane, and the second occurs when the car drives into the right lane. Afterwards, the smart car continues to drive at a constant speed.

Scenario 4: The Smart Car Arrives at the Destination

After the smart car arrives at the destination, it will turn down the engine and check all the digital devices in the car. The anti-theft system will work.

5. ANALYSIS AND PERFORMANCE EVALUATION

The main phase of the analysis of the model is the study of its reachability graph and transient analysis.

The reachability graph of a Petri net is a graph representation of its possible firing sequences, including a node for each reachable system state and an arc for each possible transition from one situation to another. It suffers from the state space explosion problem. In our experiment, the modules are independent of each other, so we can construct the reachability graph separately without lose of legality. The graphs are shown in Figure 7.

From Figure 7, the liveness of the Petri net model can be analyzed. Smart car is a special application domain with the definition of conflict correlation, so some transition cannot be fired in particular condition. According to the liveness definition, each of the transitions is L1-liveness for it can be fired at least once in some firing sequence, so the Petri net model of the smart car is L1-liveness (Murata, 1989).

A transient analysis shows the behavior of the model from a starting point in time to a predefined time. We use the time consumption for the processing sequence of context sensing, processing, situation clustering, and service invoking to assess efficiency of context recognition. The response time equation is denoted as:

$$T = T_s + T_{wap} + T_{recg} + T_{parse} + T_{net} \qquad (2)$$

where T_s is the delay time for conveying the data from the sensor to the context infrastructure, T_{wap} is the delay time for context atom management, including mapping sensor data into semantic atoms and publishing them to those subscribing the context, T_{recg} is the delay time to match the current context atoms into a situation pattern, T_{parse} is the delay time to parse the situation profile to find out the appropriate service of the situation,

Figure 7. The reachability graph of the smart car model

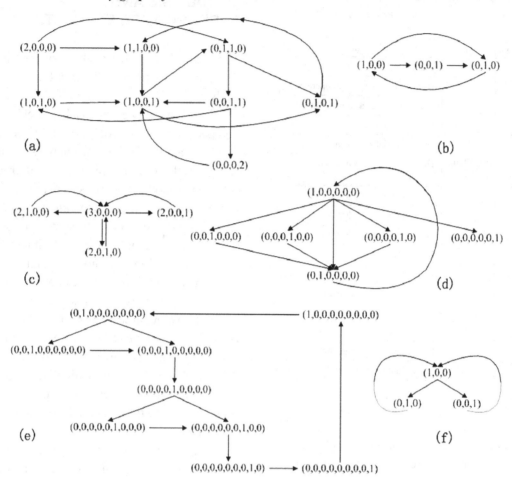

and T_{net} is the delay time for conveying the service message to a certain actuator.

Table 2 shows the response time performance of two people in the smart car environment. The delay appears close in each session, so the averages are listed for evaluation. The data rate of the sensor network using ZigBee is 250 kbit/s in the 2.4 GHz band, with a 14-byte format including

Table 2. Response time in the smart car (unit: second)

Person	T_s	T_{wap}	T_{recg}	T_{parse}	T_{net}	Total
One	0.1	0.2	0.8	0.08	0.2	1.4
Two	0.1	0.4	0.8	0.09	0.2	1.6

the head and the data. The data rate of CAN bus is 1Mbps, with an 8- byte format including the message identity in the first byte and the priority and length in the second byte. So the communication delay from the sensor network to the context infrastructure and the delay from the context infrastructure to the actuator are less than 1ms and can be neglected.

Context representation and recognition is a computationally intensive task. However, it is still feasible for non-time-critical applications, so the efficiency (with delay of nearly 1.5 second) is acceptable. For time-critical applications such as collision avoidance systems and navigating systems, we need to improve the sensing technology and the analysis method of situation recognition.

6. RELATED WORK

The early cognition of context comes from (Schilit & Theimer, 1994) and (Dey et al., 2001), which mostly defines location, time and identity. Strang and Popien (2004) give a survey of the current approaches to modeling context.

Ontologies are powerful tool to specify concepts and interrelationships, providing a uniform way for specifying the model's concepts, relations, and properties. Hence ontology-based context model is commonly applied by different context-aware projects.

CoOL (Context Ontology Language) (Strang, Popien, & Frank, 2003) includes two subsets: CoOL Core and CoOL Integration. CoOL Core uses aspect, scale and context information to define concepts. Each aspect aggregates one or more scales, and each scale aggregates one or more context information. CoOL Integration is a collection of schema and protocol extensions.

CONON (Context Ontology) (Wang, Zhang, Gu, & Pung, 2004) defines 14 core classes to model Person, Location, Activity and Computational Entities. It provides an upper context ontology that captures general concepts about basic context, and also provides extensibility for adding domain-specific ontology in a hierarchical manner.

SOUPA (Standard Ontology for Ubiquitous and Pervasive Applications) (Chen, Perich, Finin, & Joshi, 2004) consists of two distinctive but related set of ontologies: SOUPA Core and SOUPA Extension. The SOUPA Core ontologies define generic vocabularies that are universal for building pervasive computing applications. The SOUPA Extension ontologies define additional vocabularies for supporting specific types of applications and provide examples for defining new ontology extensions.

As for interrelationship among contexts, reference (Henricksen, Indulska, Rakotonirainy, 2002) argues that context information is highly interrelated when discussing the characteristics of context information.

Recently more and more research attention is paid on smart car. Intelligent transportation systems and Advanced Driver Assistant System have made rapid progresses over the past two decades in both transportation infrastructures and vehicles themselves (Wang, Zeng, & Yang, 2006). The Europe Union has sponsored many projects of the smart car. Many projects use state vectors to model and represent the information collected from the driving environment.

The AIDE (Adaptive Integrated Driver-vehicle InterfacE) (Broström, Engström, Agnvall, & Markkula, 2006) project uses DVE state vector obtained from the Driver-vehicle-environment monitor to model context. Christopher (Graff & Weck, 2006) presents a modular state-vector-based modeling. The exhaust system model includes an input file and output file. The input file describes the engine exhaust conditions over time to model different engine/vehicle combinations under different drive cycles. The system converts the information from the input data file into the state vector. After the process ends, the program creates an output file. The output data can be used to validate and tune the model to the actual components in production or under development.

ADAS (Advanced Driver Assistance Systems) project (Küçükay & Bergholz, 2004) defines the environment, vehicle, and driver states as driver's gazing region, time to lane crossing, and time to collision. These states are estimated by enhanced detection and tracking methods from in- and out-of-vehicle vision systems.

Reference Kim, Choi, Won, & Oh (2009) defines three major states from in- and out-of-vehicle vision systems: 1) Traffic environment state, estimated from the position and relative velocity of the front vehicle; 2) Vehicle state, estimated from the lane position and lateral offset; and 3) Driver state, including Driver's gazing region estimated from the driver's gazing direction.

7. CONCLUSION AND FUTURE WORK

This paper presents a context representation approach in smart car, consisting of context atoms that define the basic concept and attributes of each entity and context situations that represent complex knowledge. Context situations try to discover the inherent relationship between different elements in the smart car environment,

This paper attempts to represent complex scenarios in smart car using Petri-net. We describe the theory and then apply the theory in the smart car domain. After building the context model, we analyze it and make a performance evaluation.

Our future work includes applying more sophisticated sensing technologies to detect the physiological and psychological status of the driver to enhance the smart car prototype. Knowledge-based inference approaches will be developed and asserted into the risk assessment module for more reliable decision-making. More considerations and efforts will be made for driver intention prediction.

ACKNOWLEDGMENT

This research is supported by Ningbo Natural Science Foundation (2010A610108), NSF of Zhejiang Province, China (No.Y1080123), Foundation of MHRSS of China (No. 2009-416), and Scientific Research Foundation for the Returned Overseas Chinese Scholars, State Education Ministry(SRF for ROCS, SEM, No. 2009-1590).

REFERENCES

Broström, R., Engström, J., Agnvall, A., & Markkula, G. (2006). Towards the next generation intelligent driver information system (IDIS): The VOLVO car interaction manager concept. In *Proceedings of the ITS World Congress* (p. 47).

Chen, H., Perich, F., Finin, T., & Joshi, A. (2004). SOUPA: Standard ontology for ubiquitous and pervasive applications. In *Proceedings of the First Annual International Conference on Mobile and Ubiquitous Systems: Networking and Services* (pp. 258-267).

Dey, A. K., Salber, D., & Abowd, G. D. (2001). A conceptual framework and a toolkit for supporting the rapid prototyping of context-aware applications. *Human-Computer Interaction Journal*, *16*(2), 97–166. doi:10.1207/S15327051HCI16234_02

Graff, C., & de Weck, O. (2006). A modular state-vector based modeling architecture for diesel exhaust system design, analysis and optimization. In *Proceedings of 11th AIAA/ISSMO Multidisciplinary Analysis and Optimization Conference* (pp. 6-8).

Henricksen, K., Indulska, J., & Rakotonirainy, A. (2002). Modeling context information in pervasive computing system. In *Proceedings of the First International Conference on Pervasive Computing* (pp. 167-180).

Kim, S. Y., Choi, H. C., Won, W. J., & Oh, S. Y. (2009). Driving environment assessment using fusion of in- and our-of-vehicle vision systems. *International Journal of Automotive Technology*, *10*(1), 103–113. doi:10.1007/s12239-009-0013-5

Küçükay, F., & Bergholz, J. (2004). Driver assistant systems. In *Proceedings of the International Conference on Automotive Technologies*.

Moite, S. (1992). How smart can a car be. In *Proceedings of the Intelligent Vehicles Symposium* (pp. 277-279).

Murata, T. (1989). Petri nets: Properties, analysis and applications. *Proceedings of the IEEE, 77*(4), 541–580. doi:10.1109/5.24143

Rau, P. S. (1998). A heavy vehicle drowsy driver detection and warning system: scientific issues and technical challenges. In *Proceeding of the 16th International Technical Conference on the Enhanced Safety of Vehicles.*

Schilit, B. N., & Theimer, M. (1994). Disseminating active map information to mobile hosts. *IEEE Network, 8*(5), 22–32. doi:10.1109/65.313011

Strang, T., & Popien, C. L. (2004). A context modeling survey. In *Proceedings of the Workshop on Advanced Context Modelling, Reasoning and Management as part of UbiComp* (pp. 33-40).

Strang, T., Popien, C. L., & Frank, K. (2003). CoOL: A context ontology language to enable contextual interoperability. In *Proceedings of the 4th IFIP International Conference on Distributed Applications and Interoperable Systems* (pp. 236-247).

Tang, S. M., Wang, F. Y., & Miao, Q. H. (2006). ITSC 05: Current issues and research trends. *IEEE Intelligent Systems, 21*(2), 96–102. doi:10.1109/MIS.2006.31

Wang, F. Y. (2006). Driving into the future with ITS. *IEEE Intelligent Systems, 21*(3), 94–95. doi:10.1109/MIS.2006.45

Wang, F. Y., Zeng, D., & Yang, L. Q. (2006). Smart cars on smart roads, an IEEE intelligent transportation systems society update. *IEEE Pervasive Computing/IEEE Computer Society and IEEE Communications Society, 5*(4), 68–69. doi:10.1109/MPRV.2006.84

Wang, X. H., Zhang, D. Q., Gu, T., & Pung, H. K. (2004). Ontology based context modeling and reasoning using OWL. In *Proceedings of the Second IEEE Annual Conference on Pervasive Computing and Communications Workshops* (pp. 18-22).

This work was previously published in the International Journal of Handheld Computing Research (IJHCR), Volume 2, Issue 2, edited by Wen-Chen Hu, pp. 34-46, copyright 2011 by IGI Publishing (an imprint of IGI Global).

Section 4
Mobile Human Computer Interaction (HCI)

Chapter 12
Tool–Supported User–Centred Prototyping of Mobile Applications

Karin Leichtenstern
Augsburg University, Germany

Elisabeth André
Augsburg University, Germany

Matthias Rehm
University of Aalborg, Denmark

ABSTRACT

There is evidence that user-centred development increases the user-friendliness of resulting products and thus the distinguishing features compared to products of competitors. However, the user-centred development requires comprehensive software and usability engineering skills to keep the process both cost-effective and time-effective. This paper covers that problem and provides insights in so-called user-centred prototyping (UCP) tools which support the production of prototypes as well as their evaluation with end-users. In particular, UCP tool called MoPeDT (Pervasive Interface Development Toolkit for Mobile Phones) is introduced. It provides assistance to interface developers of applications where mobile phones are used as interaction devices to a user's everyday pervasive environment. Based on found tool features for UCP tools, a feature study is described between related tools and MoPeDT as well as a comparative user study between this tool and a traditional approach. A further focus of the paper is the tool-supported execution of empiric evaluations.

DOI: 10.4018/978-1-4666-2785-7.ch012

INTRODUCTION

Recent years have brought the tendency to develop mobile applications in human-centred iterations (ISO norm 13407) in order to increase the user-friendliness (Nielsen, 1994; Rogers, Sharp, & Preece, 2002) of the final product. Usually, this process includes a phase to (1) understand and specify the context of use, (2) specify the user and organisational requirements (3) produce design solutions, and (4) evaluate these design solutions against the requirements. With the continuous involvement of end-users in empiric user evaluations, these phases are iteratively executed until the intended application fulfils all requirements of the end-users (ISO norm 13407). All phases of this human-centred process can be very time-consuming and error-prone and thus they can increase the development costs and time. In particular, the third phase requires experienced programming skills since different kinds of prototypes need to be efficiently and effectively designed and implemented. Also, the fourth phase involves expert knowledge in usability engineering due to the fact that user evaluations need to be properly conducted and analysed in order to efficiently and effectively verify whether the requirements are met.

These are causes why Myers (1995) argues for the development and application of tools or toolkits in order to save time and money. Myers points out two main requirements for tools. Firstly, tools need to improve the result of a development process: the quality of the resulting product. Secondly, tools should also enhance the process itself: the ease of use and efficiency to run through the process.

In this paper we present the approach of all-in-one tools which support both the third and fourth phase of the human-centred design process. We call these tools user-centred prototyping (UCP) tools since they support the design and implementation of prototypes (third phase) as well as the prototypes' evaluation and analysis (fourth phase) with end-users. As an example we introduce our UCP tool called MoPeDT (Pervasive Interface Development Toolkit for Mobile Phones) that supports the UCP of mobile applications where mobile phones are used as interaction device in a user's everyday environment. This concept follows Alan Kay's term *Third Paradigm Computing* and the concept of *Pervasive* or *Ubiquitous Computing* (Weiser, 1991) where users can either (1) directly interact with real world objects of the user's everyday life or (2) mediated via an interaction device.

In the following, we illustrate typical steps of the process that we call user-centred prototyping – the third and fourth phase of the human-centred design process. These steps base on our practical experience (Rukzio, Leichtenstern, Callaghan, Schmidt, Holleis, & Chin, 2006). Then we describe existing tool-support for the UCP and typical tool features. The main part of the paper addresses MoPeDT and its validation via a feature and user study. Finally, we cover ideas to tool-supported execute empiric evaluations.

Design Specification of Prototypes

When producing a design solution, the application's appearance and behaviour is typically first designed and later on implemented. For mobile applications, the design is often done by defining a model for the different screens and the application flow which typically resembles a state chart. In this model, each state represents a screen of the mobile application and from each of these screen states user interactions can call other screen states. Consequently, user interactions (e.g. the execution of a keyboard command) are represented by transitions in the model. With respect to the level of detail, the design specification can be modelled with low or high fidelity. Low-fidelity models are often specified with pen and paper since thereby the models can quickly and easily be changed. High-fidelity prototypes are often designed with the support of graphical tools (Rogers, Sharp, & Preece, 2002). Their look and feel is often quite similar to the final product. At early stages of the

UCP, the design specification characteristically is less detailed than in later iterations. Figure 1 illustrates a high-fidelity design specification of a mobile application.

Implementation of Prototypes

After having specified the mobile application in the design phase, the prototype can be implemented. In terms of interface design, a prototype represents a partial simulation of a product with respect to its final appearance and behaviour (Houde & Hill, 1997). Prototypes can be classified by their level of detail, range of functions and reusability. Similar to the design specification, the prototypes can be implemented with a low or high level of detail (Nielsen, 1994; Rogers, Sharp, & Preece, 2002). Low-fidelity prototypes normally are prototypes which are implemented with paper (paper prototypes) while high-fidelity prototypes typically are running applications for the intended interaction device (e.g. mobile phones). High-fidelity prototypes for mobile phones can be implemented for different platforms (e.g. J2ME, Android or Windows Mobile) with the support of an integrated development environment (e.g. Eclipse, Netbeans or Visual Studio). Both of these kinds of prototypes are usually restricted by their range of functions. Their supported functions can be limited vertically or horizontally (Nielsen, 1994; Rogers, Sharp, & Preece, 2002). Horizontal prototypes provide an overview of all functionalities but do not provide these functionalities in depth whereas vertical prototypes provide functionality in depth but not for all functionalities. A further prototype classification aims at the prototype's reusability in the following iterations (Davis, 1992). Throwaway prototypes are not reused. For instance, low-fidelity prototypes (e.g. paper prototypes) are often thrown away after a single iteration. In contrast to throwaway prototypes, evolutionary prototypes are reused, modified and retested with experts and end-users in several

iterations until they meet all requirements of the end-users (Davis, 1992).

Evaluation of Prototypes

After having produced a design solution, this solution is typically evaluated with end-users in empiric evaluations (Nielsen, 1994). Afterwards, the results of the evaluation are analysed whether the user's requirements are fulfilled or not. In empiric evaluations, the end-users apply the implemented prototypes to either execute predefined tasks or freely use it. During this usage, observation techniques (Rogers, Sharp, & Preece, 2002) are applied in order to record different objective data of the user evaluation, such as user behaviour via cameras and microphones or user interactions via logging mechanisms. In addition to the observation techniques, inquiry techniques (e.g. interviews or questionnaires) are also often applied to gather subjective data (Rogers, Sharp, & Preece, 2002). The subjective data can often help to interpret the gathered objective data, such as why the users preferred a specific functionality of the prototype.

Analysis of Data

The last step is the analysis of the captured objective and subjective data (Rogers, Sharp, & Preece, 2002) in order to find usability problems and whether the user requirements are met. The gathered data either provide qualitative or quantitative content. Quantitative content (e.g. time or error measurements) can easily be analysed by different analysis techniques, such as statistical analyses. In contrast to quantitative content, qualitative content (e.g. audio and video files) frequently needs to be quantified by using so-called annotation schemas. During the process called annotation, these schemas provide information about the characteristics which need to be identified and labelled for the qualitative data, such as specific

Figure 1. High-fidelity design specification of the appearance and behaviour of a mobile application

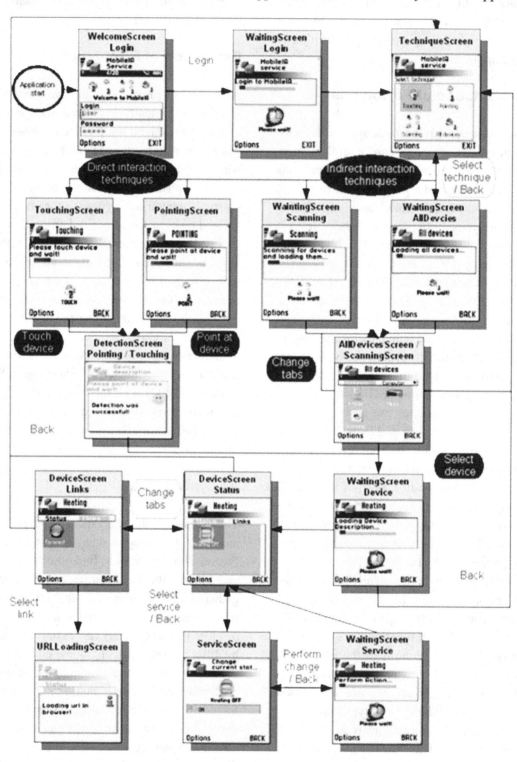

user behaviour (e.g. body movements). The results of the annotation reveal where, when, how often and how long the intended characteristics were recognised in the qualitative data (e.g. a video file). Now, these quantified data can be analysed statistically in order to find significant differences between prototypes, such as an increased user involvement when using a prototype compared to the others. After the analysis and interpretation of the gathered data, a new iteration is started with a modification of the design specification if the requirements of the users are still not completely fulfilled otherwise the iterations terminate.

TOOL SUPPORT FOR USER-CENTRED PROTOTYPING

In the remaining paper we call the design and implementation of prototypes as well as their evaluation and analysis as the four main phases that need to be conducted when executing UCP iterations. In our research we aim at all-in-one tools which support all of these steps. Using these UCP tools, however, the implementation phase is not required anymore since evolutionary prototypes with high level of detail are automatically generated as results of the design specification. The generated prototypes provide the designed functionalities and directly run on the respective interaction devices, such as mobile phones. For instance, MoPeDT generates applications that run on mobile phones which support Java2MircoEdition (J2ME). More details about MoPeDT's generated prototypes are given in the next section. Applying functional high-fidelity prototypes enables rather realistic empiric evaluations and gives the user a clearer picture of the prototype (Holmquist, 2005) than prototypes with a low fidelity and a strongly limited range of functions. A decisive advantage of all-in-one user-centred-prototyping (UCP) tools is the strong link between the design, evaluation and analysis component. For instance, the evaluation component supports the conduction

of evaluations with prototypes which were automatically generated during the tool-based design whereas the analysis component assists interface designers with the interpretation of synchronously captured qualitative and quantitative data of the evaluation. Consequently, problems of compatibility between the different components can be prevented that would typically happen when using separate tools for the different phases. A further benefit is that interface developers do not need to learn different tools for the different phases since they can use a single software to run all phases which can reduce training periods. A last important aspect of UCP tools is the support to execute remote evaluations. A remote evaluation normally means a spatial and / or temporal separation of the subjects and evaluators (Andreasen, Nielsen, Schroder, & Stage, 2007) during the execution of an empiric evaluation in a laboratory or in a field setting (in-situ, i.e. at home).

Tools for User-Centred Prototyping

The only known tools which support all phases of the UCP are d.tools (Hartmann et al. 2006) and SUEDE (Klemmer, Sinha, Chen, Landay, Aboobaker, & Wang, 2000). Klemmer et al. (2000) developed SUEDE, that assists in the iterative development of speech interfaces whereas Hartmann and colleagues implemented d.tools that supports the design, evaluation and analysis of physical computing applications. SUEDE is used to design dialogue examples, evaluate the examples in a Wizard of Oz setting and later on to analyse the evaluation, such as the user's used dialogue path during the test. SUEDE focuses on the execution of local evaluations in a laboratory setting similar to d.tools. D.tools can be primarily applied to develop, test and analyse new information appliances, such as new media players or cameras and their buttons and sliders. Using d.tools, the interaction devices (e.g. a media player), however, need to be connected to the designer's desktop PC. Thus, the generated prototypes do not directly run on the intended device that, however, is a precondition for a tool-supported assistance to execute remote evaluations.

Apart from the mentioned UCP tools, there are further tools which only support one or two phases of the UCP. In the remaining paper we concentrate on tools which address the development of mobile applications. Topiary (Li, Hong, & Landay, 2004), MScape (Hull, Clayton, & Melamed, 2004), TE-RESA (Chesta, Paternò, & Santoro, 2004), OIDE (McGee-Lennon, Ramsay, McGookin, & Gray, 2009), iStuff Mobile (Ballagas, Memon, Reiners, & Borchers, 2007), OmniSCOPE (de Sá & Carriço, 2009) and MakeIT (Holleis & Schmidt, 2008) assist the design of mobile applications. Topiary and MScape support developers of location-based applications for PDAs. The main difference between MScape and Topiary is the fact that Topiary only provides a mock-up of the application whereas MScape provides functional prototypes which do not require a Wizard to manually call GPS events of the application. MScape's prototypes directly run on PDAs. TERESA provides assistance for the tool-based design of functional nomadic interfaces (e.g. websites for mobile phones) and OIDE assists the generation of multimodal input that, for instance, can be used for mobile applications. The result of a design specification via OIDE, however, still requires comprehensive programming skills to generate a functional prototype which is similar to OmniSCOPE, MakeIT and iStuff Mobile. iStuff Mobile supports the design of mobile applications but only with a focus on the sensor-based input. OmniSCOPE and MakeIT similarly support the design specification of a prototype's appearance but MakeIT additionally assists in the specification of the behaviour and the execution of analytic evaluations (Holleis, Otto, Hussmann, & Schmidt, 2007).

For the evaluation and analysis phases, we only found few tools. These tools primarily focus on a support for the evaluation in-situ (field studies): MyExperience (Froehlich, Chen, Consolvo, Harrison, & Landay, 2007), ContextPhone (Raento, Oulasvirta, Petit, & Toivonen, 2005) and Momento (Carter, Mankoff, & Heer, 2007). Via these tools, different user interactions as well as active and passive user contexts can be logged for later on analyses. For instance, GPS locations of the users (passive user context) can be logged as well as text notes or images (active user context).

Tool Features for the User-Centred Prototyping

All mentioned tools provide ideas for tool features of a UCP tool.

For the design, a graphical user interface is usually provided that enables the modelling of the appearance and behaviour. Characteristically, a state-chart is visualised (e.g. see d.tools) that can be edited by the interface designers in order to specify the screens and user interactions. Since most of the mentioned tools (e.g. d.tools and SUEDE) rather focus on local stand-alone applications for specific devices with static content, the content of their prototypes is known during the development time. This aspect simplifies the design specification because the multimedia content can directly be assigned to a screen. Content, however, is often not known at the specification time but instead just at runtime. For instance, MScape dynamically displays content dependent on the user's outdoor location. To specify such a dynamic appearance and behaviour, interface designers make use of a scripting language during the design.

In the context of the *Third Paradigm,* a tool is additionally expected to support the specification of user interactions which base on novel input channels. Classically, users can interact with an application via a keyboard or mouse. In the context of the *Third Paradigm* users rather apply more natural input channels to interact with an application (e.g. via a camera, microphone, GPS receiver and NFC reader). In order to enable interactions via these input channels, the design tools need to support the specification of novel user interactions. For instance, MScape and OIDE provide assistance for the specification and application of pervasive or ubiquitous techniques. MScape enables the design specification of GPS-based locations which can be applied as user input in a mobile application. For their specification they make use of maps which can be loaded and edited in order to input so-called location-based points of interest, such as places with interesting sightseeing in a town.

Usually, the synchronous recording of all user interactions is supported in the evaluation phase (e.g. MyExperience or Momento). The only analysis of the logged user interactions often does not provide a covering insight to user behaviour and preferences. Instead, the analysis of the recorded user interactions in combination with the audio-visual content can provide valuable data for interpretations. D.tools is an example that supports the synchronised recording of user interactions and audio-visual content but does not consider user context. MyExperience or Momento are examples which enable the logging of active and passive user context, such as passively the user's GPS locations or actively the user's captured images and text notes. Momento additionally enables a remote communication between the evaluator and the subjects during the execution of a user evaluation whereas MyExperience enables the recording of the prototype's appearance since screenshots of the subject's mobile phone can be captured.

The main objective of the analysis phase is to find problems of a prototype. Additionally, the analysis also helps answer questions about user behaviour or preferences in different situations. A software tool is expected to appropriately display all captured data to easily and quickly enable the analysis. A common feature is to synchronously display the audio-visual content as well as the other captured data in a time-line based GUI. By this means, the audio-visual content is pre-annotated. The developers do not need to perform this by hand anymore which is a time-consuming and annoying task. Now, the developer can immediately scroll through the pre-annotated video or jump

to intended data. Additionally, the developer can add, edit or remove annotations. D.tools applies the interactive time-line based visualisation of the logged data as well as the audio-visual content.

MOPEDT: A USER-CENTRED PROTOTYPING TOOL FOR MOBILE PHONES

Based on the knowledge of the related tools and their features, we investigated further aspects of the tool-supported UCP, such as additional desirable tool features as well as aspects on user acceptance and problems when using UCP tools. The focus of our research was not to find generic solutions for the mobile application development but instead to develop a UCP tool as a test bed to investigate and improve the tool-supported UCP process.

MoPeDT's application domain bases on Alan Kay's term *Third Paradigm Computing* and the concept of *Pervasive* or *Ubiquitous Computing* (Weiser, 1991) where users can either (1) directly interact with real world objects of the user's everyday life or (2) mediated via an interaction device. MoPeDT (Leichtenstern & André, 2010) focuses on the second aspect where mobile phones are used as interaction devices to a pervasive environment (e.g. a store) and their physical objects (e.g. products in the store). The generated applications of MoPeDT can be used to select a physical object and call its services. Services offer different kinds of contents about the physical object (e.g. a description about the ingredients of a product) but also provide opportunities to generate and provide multimedia content to other users (e.g. user reviews about a product). In contrast to d.tools and SUEDE, MoPeDT generates evolutionary prototypes with a high fidelity that directly run on the intended interaction. These prototypes can support different mobile user interactions (Leichtenstern & André, 2009) to select and use services of physical objects (e.g. via the mobile phone's keyboard, NFC reader, microphone).

The generated prototypes run on mobile phones which support Java2Microedition (J2ME) and the corresponding hardware for the intended mobile user interactions (e.g. NFC reader). Since mobile devices of the different operators (e.g. Nokia, Samsung or Sony Ericson) often have different operating systems and implementations of J2ME, we decided to focus on S40 and S60 devices from Nokia. We successfully tested our automatically generated prototypes on Nokia 6131 NFC and Nokia N95 phones. In empiric evaluations, the generated prototypes do not require a Wizard as with SUEDE or a connected desktop PC as with d.tools but instead the subjects and the evaluation can be remotely.

The following requirements for a UCP tool should be fulfilled for the design: (1) Static and dynamic specifications of the appearance and behaviour should be enabled; (2) Also, different mobile user interaction should be supported; (3) High-fidelity prototypes should automatically be generated as results of the design specification; (4) The generated applications should enable remote access to load and display remote content; (5) Approved interface guidelines should automatically be considered to increase the user-friendliness of the resulting prototypes. For the evaluation, the following requirements should be fulfilled: Recordings of (1) user interactions, (2) audio-visual content, (3) live comments and the prototype's appearance, (4) active and passive user context and (5) environmental contexts should be supported. Finally, the tool should support field and laboratory studies (6) locally and (7) remotely. The analysis component should (1) synchronously display all recorded data, (2) automatically annotate the audio-visual content, (3) enable a modification of the annotations and enable the (4) export of the data for statistical analyses. To enable a UCP tool that support these features, an appropriate architecture and several software modules are required.

Architecture and Software Modules

Our UCP tool bases on a plug&play client-server architecture that contains physical objects, mobile clients, a server, a database as well as sensors, actuators and evaluators as components (Figure 2).

Physical objects are real objects in a pervasive environment (e.g. objects of art or products in a store). In order to address physical objects (e.g. via the mobile phone's keyboard or NFC reader), users can apply their mobile phones with an application that bases on the software module of the mobile client. After having selected a physical object, the mobile client communicates with the server to load services and contents which are stored in the database. Sensors, a further component, can also be plugged in to the server in order to provide knowledge about the user's environmental context (e.g. lighting condition or loudness). This knowledge can help to interpret user behaviour in the analysis phase. Another component called *evaluator* is applied whenever tool-supported evaluations are executed. Several evaluators can connect to the server and register their interest in other connected components: mobile clients and sensors. Now, the evaluators can synchronously log all contexts of the selected mobile clients (user interactions as well as active and passive user context) and sensors (environmental context) for a later on analysis. The last plug&play component of the architecture is the actuator that can be used as an additional output channel for multimedia content, such as to display video content on a public display (Leichtenstern & André, 2009).

For all components we provide software modules in J2SE or J2ME (Leichtenstern & André, 2009). One interesting software module is the mobile client. This module handles the whole mobile client-server communication and contains the implementation of different mobile user interactions (e.g. based on the keyboard,

Figure 2. The client-server architecture of MoPeDT

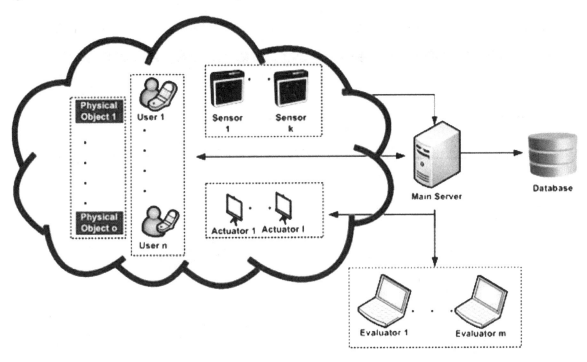

microphone, NFC reader and GPS receiver) for the mobile phone. For the specification of the intended mobile user interactions and user context as well as the static and dynamic application's appearance and behaviour, XML files are used. The software module of the mobile client can interpret these XML specifications and correctly display the application at the runtime. For the specification of the appearance and behaviour, the mobile client provides screen templates with different layouts (e.g. ItemScreen, MediaScreen or InfoScreen). These templates base on mobile phone guidelines from Nokia's Design and User Experience Library (http://library.forum.nokia.com). Due to these templates, the generated prototypes fulfill approved interface guidelines. For instance, reversibility is considered as well as a consistent layout, soft key usage and navigation style.

User-Centred Prototyping with MoPeDT

Based on our architecture and the software modules, the UCP tool was built.

Similar to the related tools, the design component of MoPeDT (Figure 3) supports the GUI-based specification of the appearance and behaviour as well as the specification of different novel user interactions (e.g. based on the keyboard, NFC and microphone). Dynamic appearance and behaviour are specified by making use of a scripting language (Leichtenstern & André, 2009). At the runtime, the dynamic specified screens display local content or remote content from the database. This database content can also be specified via the GUI of MoPeDT. The automatically generated prototypes of the design specification are executable JAR files which directly run on the end-device.

Figure 3. The design component of MoPeDT. The left part of the component displays the state chart view of the application that contains the screen states and transitions while the right part of the design component provides a preview of the selected screen and options to modify the design specification.

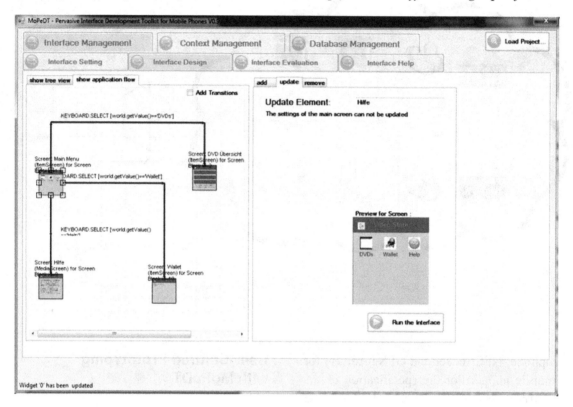

Besides the generation of prototypes, MoPeDT also provides a GUI to capture user interactions and audio-visual content in the evaluation phase (Figure 4). In addition to most other evaluation and analysis tools, MoPeDT assists the synchronous logging of live comments and task descriptions as well as the capturing of the prototype's appearance at the evaluation time. To capture the prototype's appearance, a cloned screen of the subject's mobile phone is displayed on the desktop PC of the evaluator. Now, whenever a new screen is loaded for the subject's device, the cloned screen is updated and captured. A last difference to the former mentioned tools is the possible recording of the already mentioned environmental context via the sensor component of the architecture. To execute an evaluation with MoPeDT, all required components of the evaluation, however, require synchronous clocks.

In the final step of the UCP, the developer can analyse captured user evaluations. MoPeDT's analysis component (Figure 5) also provides a time-line based visualisation of the recorded data in order to navigate through and interact with them (e.g. to find usability problems or user preferences). For the analysis component we extended ANVIL (Kipp, 2001), which supports the display of audio-visual content as well as the visualisation and modification of annotations at various freely definable time-line based tracks. Since the determination of significant results is often an important analysis task, a further feature of MoPeDT's analysis component is to support statistical analyses. MoPeDT supports the export of the annotated data in different formats of statistic tools (e.g. SPSS) in order to investigate typical behaviours in particular contexts.

Figure 4. The evaluation component of MoPeDT. The upper part displays the two selected cameras while the lower part shows the selected mobile client and sensors as well as incoming messages from the server. The right part displays the cloned screen view of the selected mobile client.

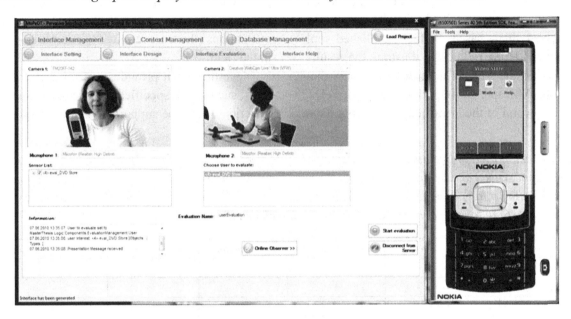

Figure 5. The analysis component of MoPeDT. The upper part displays the captured videos and screen-shots (right). The lower part provides a time-line based visualisation of the audio track as well as the labeling (annotation) of tasks, user interactions or contexts in the corresponding track.

Feature Study with MoPeDT

We downloaded and tested all former mentioned tools which were available for download and reviewed them based on the established tool features. Tables 1, 2, and 3 illustrate the results of this feature study. The three tables show that d.tools, MScape, MyExperience and MoPeDT fulfill several of the identified features but have

limitations. D.tools, for example, does not give support for the development of high-fidelity prototypes which directly run on the intended interaction device and thus does not enable remote evaluations which are required to execute studies in the field.

Nevertheless, it provides several useful features, such as the specification of novel user interactions. MScape supports dynamic specifica-

Table 1. Supported tool features for the design of prototypes: (1) static; and (2) dynamic specifications of the appearance and behaviour; (3) specifications of novel user interactions; (4) auto-generation of high-fidelity prototypes as result of the design specification; (5) remote access for the generated prototypes; and (6) automatic compliance of approved interface guidelines

Tool	Static	Dynamic	Novel Interactions	Auto-Generation	Remote Access	Interface Guidelines
d.tools	+	-	+	-	-	-
SUEDE	+	-	+	-	-	-
TERESA	+	-	-	+	-	-
OIDE	-	-	+	-	-	-
MakeIT	+	-	+	-	-	-
MScape	+	+	+	+	-	-
Topiary	+	-	+	-	-	-
OmniSCOPE	+	-	-	-	-	-
iStuff Mobile	-	-	+	-	-	-
MoPeDT	+	+	+	+	+	+

Table 2. Supported tool features for the evaluation of prototypes: recording of (1) user interactions and (2) audio-visual content; (3) recordings of live comments, tasks and appearance of the prototype; (4) remote communication of evaluator and subjects; recording of (5) active and passive user contexts as well as (6) environmental context; execution of (7) local and (8) remote evaluations

Tool	User Interactions	Audio-Visual Content	Live Comments, Tasks and Appearance	Remote Communication	User Context	Environmental Contex	Local Evaluations	Remote Evaluations
d.tools	+	+	-	-	-	-	+	-
SUEDE	+	-	-	-	-	-	+	-
My-Experience	+	+	+	-	+	-	+	+
Context-Phone	+	-	-	-	+	-	+	+
Momento	+	+	-	+	+	-	+	+
MoPeDT	+	+	+	-	+	+	+	+

Table 3. Supported tool features for the analysis of prototypes: (1) synchronous display of all recorded data; (2) automatic annotation of audio-visual content; (3) modification of annotations; (4) multi-display of several recorded user evaluations; (5) export to execute statistical analyses

Tool	Synchronous Display	Automatic Annotation	Modifications	Multi-Display	Statistical Analyses
d.tools	+	+	+	+	-
SUEDE	+	+	-	-	-
MyExperience	-	-	-	-	-
MoPeDT	+	+	+	-	+

tions of the prototype's appearance and behaviour as well as the automatic generation of high-fidelity prototypes but does not consider interface guidelines or a remote access. Additionally, it neither provides assistance for the evaluation nor for the analysis. MyExperience covers several useful features for the execution of remote evaluations, such as the logging of user interactions but does neither provide a synchronised display of the recorded data in the analysis phase nor support for the complete design phase.

User Study with MoPeDT

We not only wanted to evaluate MoPeDT based on a feature study but also via a user study with potential end-users of MoPeDT: interface developers (Leichtenstern & André, 2010). With such a user study, we wanted to answer the question whether interface developers can quicker (Efficiency) develop user-friendlier products (Effectiveness) with a higher satisfaction (Satisfaction) compared to traditional approaches (e.g. via an IDE). Additionally, we searched for benefits and problems of UCP tools.

Twenty subjects participated in our user study and used MoPeDT and the traditional approach (with-in subjects approach) for the design, evaluation and analysis of a pervasive shopping assistant which helps users to receive information about products in a shopping store (e.g. about the ingredients of products). To prevent any positioning effect, ten subjects started with the traditional

approach and afterwards used MoPeDT whereas the other ten students used MoPeDT first. The subjects of our user study were students of our three-month course *Usability Engineering* with the age between 22 and 29 (M = 24.15, SD = 1.90). They rated themselves as medium skilled in object-oriented programming and usability engineering.

To keep comparability, the subjects received a detailed description about the intended prototype. For example, the screen contents were pre-defined as well as the user interactions which support must be given: keyboard-based and NFC-based. The subjects were also instructed to implement a logging mechanism when using the traditional approach to enable a recording of the user interactions in the evaluation. For the evaluation, the subjects were instructed to audio-visually capture three end-users and their user interactions while they were executing pre-defined tasks. The logging of user and environmental context were not considered. After the evaluation, the subjects had to analyse their captured data to find usability problems of the two prototypes. For instance, they validated the logged user interactions to find problems of efficiency.

Before we ran the one-month user study, we conducted tutorials within our course and taught all subjects how to use MoPeDT for the UCP and how to implement and evaluate mobile phone prototypes using the IDEs Eclipse and Netbeans. During this period and the study, the subjects were not informed that we developed MoPeDT. Also, we comprehensively taught all subjects about usability in general, the human-centred design, mobile phone usability and Nokia's mobile phone guidelines. We reminded the subjects to apply these guidelines in our user study.

During the study, we used a post-task questionnaire to acquire subjective data whereas protocol recordings and a guideline review were utilised to collect objective data. In the questionnaire, we asked the subjects to rate statements for both levels in terms of efficiency, effectiveness, satisfaction, learnability, transparency, and user control. By using our protocol, the subjects documented their required time for the UCP's phases. Also, emerged problems had to be noted. Finally, we conducted a guideline review and investigated the resulted prototypes. An independent usability expert who was not involved in the development or evaluation of MoPeDT used the generated prototypes and investigated their violation against the mentioned Nokia guidelines.

After the conduction of the user study, we analysed the data in order to shed light in our main objectives whether MoPeDT can improve the usability for the interface developer and which benefits and problems emerge when using a UCP.

When analysing the protocols with regard to the interface developer's efficiency, on average, the required UCP time in minutes with traditional approaches (M=816.60, SD=318.81) was significantly higher compared to MoPeDT (M=266.65, SD=208.14), t(19)=9.2, p<0.001. During the design, the programming of the GUI and network communication required much more time whereas the annotation task impaired the interface developer's efficiency in the analysis phase. The qualitative and quantitative feedback of the questionnaire substantiates the results (Figure 6). Most subjects found the tool usage *quick and easy* and see a benefit in *the very quick prototyping and evaluation of applications*.

Regarding effectiveness, the qualitative and quantitative feedback (Figure 6) reveals no clear differences between the two levels. While most of the subjects highlighted the generated prototypes of MoPeDT as *beautiful* which *follow design guidelines*, they also claimed the limitation caused by the screen templates. Despite this moderate subjective data, prototypes from MoPeDT had, on average, less violations against the 22 guidelines from Nokia (M = 0.85, SD = 0.93) than prototypes from the traditional approaches (M =

Figure 6. User ratings for effectiveness and efficiency. The subjects rated the provided statements based on a scale from 1 (strongly disagree) to 5 (strongly agree). In terms of design, efficiency and time gain were significantly better rated for MoPeDT compared to the traditional approach (* p < 0.05, ** p < 0.01, *** p < 0.001).

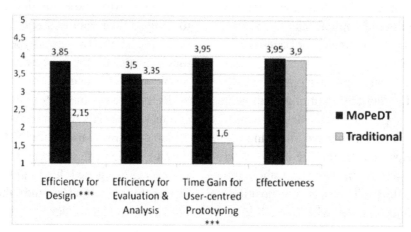

4.35, SD = 2.52), t (19) = 5.48, p < 0.001. Typical problems were the inconsistent usage of the soft keys (17 of 20 subjects).

Despite the positive results in terms of effectiveness and efficiency, the interface developer's satisfaction turned out to be a problem (Figure 7). The weak satisfaction was mainly caused by a lack of an appropriate user control and transparency of MoPeDT. The interface developers *want to see what is going on in the background* and they want to have an increased level of freedom. Nevertheless, the subjects considered benefits for the learnability when using MoPeDT since they realised less required skills compared to traditional approaches.

Finally, we asked for the preferred levels: MoPeDT or traditional approach. Figure 8 shows the distribution. Most subjects either liked Mo-PeDT or both levels. Finally, we asked for the preferred components of MoPeDT (Figure 9). Most subjects liked all components and thus the all-in-one tool solution. A subject mentioned that *only the combination of all components meaningfully supports the iterative prototyping* whereas another subject mentioned that the prototyping can be improved by *the close interleaving of the three components* in order to prevent *the induction in several programs*.

Based on our results we conclude that MoPeDT improves the interface developer's efficiency by reducing the required time for the UCP as well as

Figure 7. User ratings for satisfaction, user control and learnability. The subjects rated the provided statements based on a scale from 1 (strongly disagree) to 5 (strongly agree). In terms of design, satisfaction and user control were significantly better rated for the traditional approach compared to MoPeDT while learnability was better rated for MoPeDT (* p < 0.05, ** p < 0.01, *** p < 0.001).

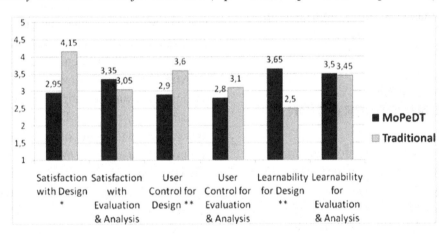

Figure 8. Preferred approach (MoPeDT or the traditional) for the design (left) as well as evaluation and analysis (right)

Figure 9. Preferred components of MoPeDT

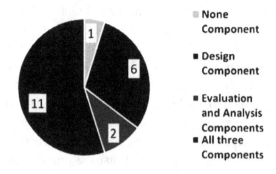

effectiveness by reducing the prototypes' non-compliance of interface guidelines. The results additionally indicate that interface developers accept UCP tools for all steps of the UCP and that they give a priority to an all-in-one tool for the UCP instead of separate tools for the different phases. The results of our user study, however, also suggest problems when applying UCP tools. The interface developer's satisfaction is strongly linked to an appropriate user control as well as transparency that need to be indispensability considered.

EXECUTING EMPIRIC EVALUATIONS: OBSERVATION

One interesting aspect of the tool-supported UCP is the assistance to execute empiric evaluations. Most tools only support the execution of real world simulations in laboratories or real world evaluations in the field. A new idea is to also support a hybrid simulation with partial simulations in a virtual world. In the following we illustrate the tool-supported real world evaluation and simulation as well as the hybrid simulation.

Real World Evaluations and Simulations

The execution of empiric evaluations in the field (Häkkilä & Mäntyjärvi, 2006) is the most reasonable method. There is evidence that these studies

provide realistic and valuable data because they are performed with real contextual constraints. Despite promising benefits, their conduction, however, might also lead to uncontrolled contextual settings rendering the outcome useless. Also, field tests are often time-consuming and expensive. Consequently, the idea is to simulate the real world in a more controlled environment and execute laboratory studies instead. Laboratory studies can be used during minor iterations of the UCP if appropriate methods are applied. Field tests, however, cannot be completely substituted and should be used at least at the end of the development process. There is a complementing effect between field and laboratory studies and thus both settings should be supported by UCP tools.

In the following we present an example of a laboratory study that we executed with MoPeDT. We simulated a DVD store (Figure 10) to investigate user trust in two different mobile user interactions: Keyboard-based and NFC-based (Figure 10). Figure 11 shows screens of the prototype that can be used to select and buy DVDs. If this evaluation should be executed in the field, the prototype needs to be generated in the same way.

After the generation, we asked 20 subjects to participate in our evaluation and perform different tasks (e.g. buy a DVD). During the execution of these different tasks the subjects were introduced to apply *the method of thinking aloud* (Nielsen, 1994) to express their thoughts in the different situations. We used MoPeDT to audio-visually capture the users and their user interactions. As a second method we applied a post-task questionnaire with questions in terms of user trust. This evaluation can be conducted in the same way if it is executed in the field but the experimenter needs to be locally present. A remote evaluation can also be executed but in this case the audio-visual recordings need to be dropped.

Later on, we reflected the captured user comments and interactions as well as the questionnaire to find differences of user trust. The procedure to analyse the captured data would have been the same for a field study. The results of the ques-

Figure 10. The user while interacting with the mobile phone's keyboard (left) and NFC reader (right)

Figure 11. Some screens of the mobile prototype used in the user evaluation

tionnaire revealed no significant differences. The analysis of the video showed that most subjects mentioned the NFC-based interaction as easier and quicker to use but not more trustworthy compared to the keyboard-based interaction.

Hybrid Simulations

Besides the traditional evaluations, MoPeDT can also be used to run hybrid simulations. Morla and Davies (2004) give a first impression of a hybrid simulation. Hybrid simulation means an integration and combination of the real and the virtual world. Morla and Davies (2004) used a virtual world to test real sensors which were attached to a wearable medical monitoring system. A similar approach called "dual reality" is introduced by Lifton and Paradiso (2009). They used SecondLife as a virtual visualisation tool of streams which are generated from real world sensors (e.g. the

temperature in the real building or sound and lighting conditions). Driving simulators (http://www.carrsq.qut.edu.au/simulator) also aims at the idea of the hybrid simulation. Users interact with a real steering wheel and dashboard while they are driving through a virtual presented test route. In our work, we do not use the virtual world as a visualisation platform for the performance of real devices as Davies or Lifton. We apply the virtual world as a platform for evaluations similarly as used for driving simulators but we use real mobile phones as interaction devices and a virtual simulation of a pervasive environment. The user still interacts with real mobile phones similar as in a real simulation but the pervasive environment and the physical objects are only virtually represented.

MoPeDT can be used to generate prototypes for mobile phones which are evaluated via the hybrid simulation but some adaptations are required

(Figure 12). There is a need to shift the pervasive environment from the real to the virtual world using a platform that supports virtual simulations. The simulation contains the virtual representation of the physical objects but their services are still stored in the database and can be accessed with real mobile phones. Another difference to the former setting is the need for a representation of the user in the virtual world. With this avatar the user can interact via the PC's keyboard within the virtualised pervasive environment.

To simulate the pervasive environment, we make use of an open source version of SecondLife: Open Simulator (http://opensimulator.org). It allows setting up a virtual world that behaves exactly like SecondLife and can be accessed with the same viewers. Hence, in the remainder of this paper we will use SecondLife and Open Simulator as synonyms. SecondLife represents a multiplayer platform that is not primarily concerned with gaming but aims at establishing a general virtual meeting place (e.g. buying or selling virtual or real goods). Central feature of SecondLife is the use of avatars which represent the real user in the virtual environment.

We propose to employ SecondLife to simulate a real environment which has been augmented for context dependent interactions. Apart from setting up the simulation server, three steps are necessary for simulating a pervasive environment in a hybrid simulation. The environment itself has to be modelled, it has to be equipped with physical objects and sensors, and it has to allow for communicating with the outside world such as the real mobile device. Standard modelling tools or in-world modelling tools can be used to model the virtual world. Figure 13 shows a snapshot from a pervasive environment. The challenge of a hybrid simulation is to realise the complex interplay between sensors, physical objects, and the mobile device, which can be seen as the inherent characteristic of a pervasive environment. The general idea is to use the real mobile phone for the interaction with the virtual world. This is not always possible. For NFC-based interactions, objects are equipped with RFID tags to allow NFC with the mobile phone (Figure 13). Creating a virtual RFID tag is no challenge but of course this tag cannot be read out by the real mobile device. Thus, it is necessary to create a virtual representation of the

Figure 12. Modifications of the architecture to execute a hybrid simulation

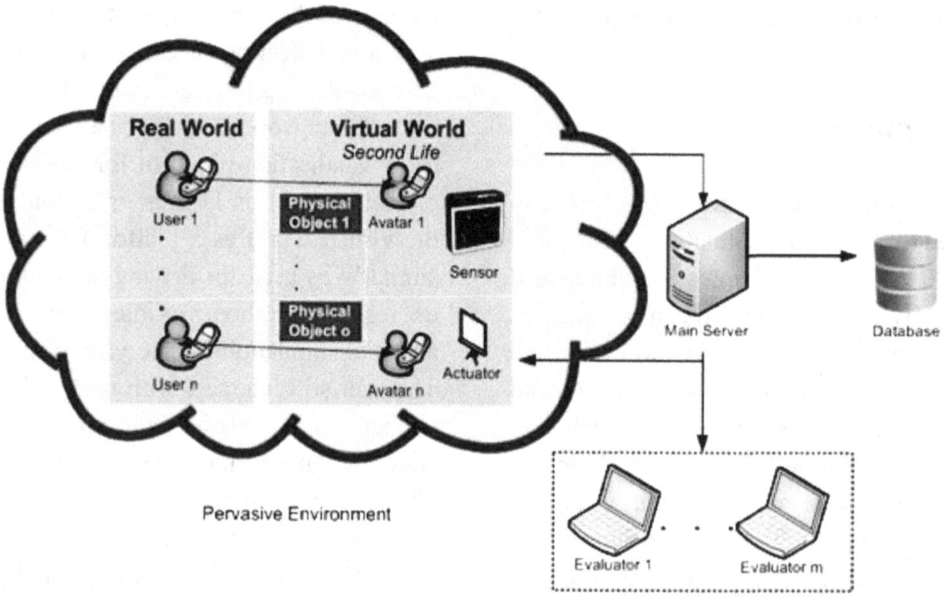

mobile device for some of the contextual input. In the current version, a virtual mobile device is used for registering the contextual input that is provided by the simulated environment. Then, the virtual phone communicates with MoPeDT's *main server* (Figure 12) in order to transmit events which lead to an adaptation of the real phone.

As a proof-of-concept study of the hybrid simulation, we replicated a former evaluation that was conducted in a laboratory (Rukzio, Leichtenstern, Callaghan, Schmidt, Holleis, & Chin, 2006) this time we made use of a hybrid simulation (Leichtenstern, André, & Rehm, 2010). In the original and in the replicated evaluation we compared three different mobile user interactions in different contextual situations, such as different locations of users and physical objects. The results of the hybrid simulation provided similar insights in user preferences and behaviour as the laboratory setting: users tended to switch the mobile user interactions dependent on the respective context with location as the most important context. Based on this proof-of-concept study, we see a first indicator that hybrid simulations might be a useful evaluation setting.

Overall, in some points the hybrid simulation benefits compared to a laboratory setting. (1) There is no need to physically rebuild the environment in a laboratory which can save money and time;

(2) Relying on the hybrid simulation, even initial ideas can easily and efficiently be mediated and investigated because the real mobile application can be tried out and demonstrated in the corresponding simulated environment; (3) Another benefit is the ease of changing the environment. Different models of physical objects can rapidly be generated, modified and deleted. Using SecondLife as virtual world adds further advantages; (4) Due to its widespread use, it is known to a great number of users who do not have to be introduced to the specifics of using the virtual environment; (5) Because the application realises a multi-player platform over the internet, it can be accessed anywhere anytime; (6) This can also reduce the organisational effort of subject recruiting since the subjects do not need to be physical present in a laboratory. Of course some restrictions apply like the necessity of compatible mobile devices; and (7) Finally, in contrast to a virtual simulation alone approach, the hybrid simulation can be performed more realistic.

Despite these promising benefits, there are also problems. Of course, an offset inevitably emerge between a real world and a hybrid simulation: (1) The user requires less motivation and less physical effort to move and explore the virtual setting; (2) Also, user interactions might be different to its real usage (e.g. NFC-based); (3) Having interactions

Figure 13. Snapshot of the virtualized environment and the user while interacting via NFC with the virtualized physical object

in the virtual simulation can also lead to usability problems; and (4) Finally, the virtual world needs to be modelled as realistic as possible to reduce side effects.

We consider the need that UCP should also support hybrid simulations. Supporting evaluations in the field, laboratory or via a hybrid simulation can meet several objectives. At the beginning of the UCP the tool can help to execute hybrid simulations to virtually simulate first ideas of applications. Later on, real world simulations can be performed to increase the realism for the users. At the end of the process, the tool can be used to execute evaluations in the field.

CONCLUSION

In this paper we covered the idea to support interface developers with an all-in-one software solution to more efficiently and effectively execute the third and fourth phases of the human-centred design process (produce design solutions and evaluate them against user requirements) which we call user-centred prototyping (UCP) process: the design, evaluation and analysis of prototypes with the involvement of end-users. The paper provides a review of existing and new tool features for UCP tools as well as insights to their typical benefits and problems.

Based on our proof-of-concept tool called MoPeDT, we extended ideas of other UCP tools. For instance, we introduced ideas of an architecture to enable the generation of network-based mobile applications. By this means remote content of a database can be loaded and displayed but this also enables the conduction of remote user evaluations. Additionally, we described the usage of screen templates to consider the compliance of approved interface guidelines. Finally, we described concepts how UCP tools can be extended to support field and laboratory studies as well as hybrid simulations. Our new ideas for

UCP tools, however, also revealed some problems. For instance, if a tool bases on screen templates the interface developers felt a strong limitation of their user control which in turn impaired their satisfaction with the tool.

ACKNOWLEDGMENT

This research is partly sponsored by OC-Trust (FOR 1085) of the German research foundation (DFG). A special thank to Sebastian Thomas and Dennis Erdmann for their work on MoPeDT and to Julia Karcher and Michael Goj for their support during the conduction of some of the introduced studies.

REFERENCES

Andreasen, M. S., Nielsen, H. V., Schroder, S. O., & Stage, J. (2007). What happened to remote usability testing?: An empirical study of three methods. In *Proceedings of the SIGCHI Conference on Human Factors in Computing Systems* (pp. 1405-1414). New York, NY: ACM Press.

Ballagas, R., Memon, F., Reiners, R., & Borchers, J. (2007). iStuff mobile: Rapidly prototyping new mobile phone interfaces for ubiquitous computing. In *Proceedings of the SIGCHI Conference on Human Factors in Computing Systems* (pp. 1107-1116). New York, NY: ACM Press.

Carter, S., Mankoff, J., & Heer, J. (2007). Momento: Support for situated ubicomp experimentation. In *Proceedings of the SIGCHI Conference on Human Factors in Computing Systems* (pp. 125-134). New York, NY: ACM Press.

Chesta, C., Paternò, F., & Santoro, C. (2004). Methods and tools for designing and developing usable multi-platform interactive applications. *PsychNology Journal, 2*(1), 123–139.

Davis, A. M. (1992). Operational prototyping: A new development approach. *IEEE Software, 9,* 70–78. doi:10.1109/52.156899

de Sá, M., & Carriço, L. (2009). Mobile support for personalized therapies - OmniSCOPE: Richer artefacts and data collection. In *Proceedings of the 3rd International Conference on Pervasive Computing Technologies for Healthcare* (pp. 1-8). New York, NY: ACM Press.

Froehlich, J., Chen, M. Y., Consolvo, S., Harrison, B., & Landay, J. A. (2007). MyExperience: A system for in situ tracing and capturing of user feedback on mobile phones. In *Proceedings of the 5th International Conference on Mobile Systems, Applications and Services* (pp. 57-70). New York, NY: ACM Press.

Häkkilä, J., & Mäntyjärvi, J. (2006). Developing design guidelines for context-aware mobile applications. In *Proceedings of the 3rd International Conference on Mobile Technology, Applications and Systems* (p. 24). New York, NY: ACM Press.

Hartmann, B., Klemmer, S. R., Bernstein, M., Abdulla, L., Burr, B., Robinson-Mosher, A., et al. (2006). Reflective physical prototyping through integrated design, test, and analysis. In *Proceedings of the 19th Annual ACM Symposium on User Interface Software and Technology* (pp. 299-308). New York, NY: ACM Press.

Holleis, P., Otto, F., Hussmann, H., & Schmidt, A. (2007). Keystroke-level model for advanced mobile phone interaction. In *Proceedings of the Conference on Human Factors in Computing Systems* (pp. 1505-1514). New York, NY: ACM Press.

Holleis, P., & Schmidt, A. (2008). MakeIt: Integrate user interaction times in the design process. In *Proceedings of the 6th International Conference on Pervasive Computing* (pp. 56-74).

Holmquist, L. E. (2005). Prototyping: Generating ideas or cargo cult designs? *Interaction, 12*(2), 48–54. doi:10.1145/1052438.1052465

Houde, S., & Hill, C. (1997). What do prototypes prototype? In Helander, M., Landauer, T., & Prabhu, P. (Eds.), *Handbook of human-computer interaction.* Cambridge, MA: Elsevier Science.

Hull, R., Clayton, B., & Melamed, T. (2004). Rapid authoring of mediascapes. In *Proceedings of the 6th International Conference of Ubiquitous Computing* (pp. 125-142).

Kipp, M. (2001). Anvil - a generic annotation tool for multimodal dialogue. In *Proceedings of the 7th European Conference on Speech Communication and Technology* (pp. 1367-1370). New York, NY: ACM Press.

Klemmer, S. R., Sinha, A. K., Chen, J., Landay, J. A., Aboobaker, N., & Wang, A. (2000). Suede: A Wizard of Oz prototyping tool for speech user interfaces. In *Proceedings of the 13th Annual ACM Symposium on User Interface Software and Technology* (pp. 1-10). New York, NY: ACM Press.

Leichtenstern, K., & André, E. (2009). Studying multi-user settings for pervasive games. In *Proceedings of the 11th International Conference on Human-Computer Interaction with Mobile Devices and Services* (pp. 190-199). New York, NY: ACM Press.

Leichtenstern, K., & André, E. (2009). The assisted user-centred generation and evaluation of pervasive. In *Proceedings of the Third European Conference on Ambient Intelligence* (pp. 245-255).

Leichtenstern, K., & André, E. (2010). MoPeDT - features and evaluation of a user-centred prototyping tool. In *Proceedings of the 2nd ACM SIGCHI Symposium on Engineering Interactive Computing Systems* (pp. 93-102). New York, NY: ACM Press.

Leichtenstern, K., André, E., & Rehm, M. (2010). Using the hybrid simulation for early user evaluations. In *Proceedings of the 6th Nordic Conference on Human-Computer Interaction* (pp. 315-324). New York, NY: ACM Press.

Li, Y., Hong, J. I., & Landay, J. A. (2004). Topiary: A tool for prototyping location-enhanced applications. In *Proceedings of the 17th Annual ACM Symposium on User Interface Software and Technology* (pp. 217-226). New York, NY: ACM Press.

Lifton, J., & Paradiso, J. A. (2009). Dual reality: Merging the real and virtual. In *Proceedings of the First International ICST Conference on Facets of Virtual Environments* (pp. 12-18).

McGee-Lennon, M. R., Ramsay, A., McGookin, D., & Gray, P. (2009). User evaluation of OIDE: A rapid prototyping platform for multimodal interaction. In *Proceedings of the 1st ACM SIGCHI Symposium on Engineering Interactive Computing Systems* (pp. 237-242). New York, NY: ACM Press.

Morla, R., & Davies, N. (2004). Evaluating a location-based application: A hybrid test and simulation environment. *IEEE Pervasive Computing/IEEE Computer Society and IEEE Communications Society*, *3*(3), 48–56. doi:10.1109/MPRV.2004.1321028

Myers, B. A. (1995). User interface software tools. *ACM Transactions on Computer-Human Interaction*, *2*(1), 64–103. doi:10.1145/200968.200971

Nielsen, J. (1994). *Usability engineering*. San Francisco, CA: Morgan Kaufmann.

Raento, M., Oulasvirta, A., Petit, R., & Toivonen, H. (2005). ContextPhone: A prototyping platform for context-aware mobile applications. *IEEE Pervasive Computing / IEEE Computer Society and IEEE Communications Society*, *4*(2), 51–59. doi:10.1109/MPRV.2005.29

Rogers, Y., Sharp, H., & Preece, J. (2002). *Interaction design: Beyond human-computer interaction*. New York, NY: John Wiley & Sons.

Rukzio, E., Leichtenstern, K., Callaghan, V., Schmidt, A., Holleis, P., & Chin, J. (2006). An experimental comparison of physical mobile interaction techniques: Touching, pointing and scanning. In *Proceedings of the Eighth International Conference on Ubiquitous Computing* (pp. 87-104).

Weiser, M. (1991). The computer for the 21st century. *Scientific American*.

This work was previously published in the International Journal of Handheld Computing Research (IJHCR), Volume 2, Issue 3, edited by Wen-Chen Hu, pp. 1-21, copyright 2011 by IGI Publishing (an imprint of IGI Global).

Chapter 13
Sampling and Reconstructing User Experience

Panos Markopoulos
Eindhoven University of Technology, The Netherlands

Vassilis-Javed Khan
NHTV Breda University of Applied Sciences, The Netherlands

ABSTRACT

The Experience Sampling and Reconstruction Method (ESRM) is a research method suitable for user studies conducted in situ that is needed for the design and evaluation of ambient intelligence technologies. ESRM is a diary method supported by a distributed application, Reconexp, which runs on a mobile device and a website, enabling surveying user attitudes, experiences, and requirements in field studies. ESRM combines aspects of the Experience Sampling Method and the Day Reconstruction Method aiming to reduce data loss, improve data quality, and reduce burden put upon participants. The authors present a case study of using this method in the context of a study of communication needs of working parents with young children. Requirements for future developments of the tool and the method are discussed.

INTRODUCTION

The current trend towards pervasive and context sensitive applications where information and computational technology are embedded in our social and physical environments presents substantial methodological challenges for researchers, designers, or technologists, wishing to design, analyze, or evaluate, corresponding user experiences. Available research methods have been shaped in past decades to support the design and evaluation of the cognitive ergonomics of task-oriented interaction, usually contained within a contained time span. Extending such methods to study user experiences as these occur *in situ*, unfolding over days or weeks, capturing social interactions between several people and diverse environmental and technical contingencies, requires a substantial scaling up the data sampling in terms of frequency, duration, and the richness of records made.

The objectives of system evaluation have also changed significantly. Transcending usability, evaluations of applications and services that are

DOI: 10.4018/978-1-4666-2785-7.ch013

mobile and often context sensitive, typically examine higher level aspects of user experiences and user needs relating to persuasion, fun, engagement, trust, etc. Contextualized methods of data collection need to support the reporting of attitudes, opinions, or appraisals, close to the moment that a particular experience occurs and in the context where events and activities unfold. Such surveying of user attitudes can occur repeatedly over time, allowing the study of behaviors and experiences over medium or long periods of time, even enabling researchers to examine patterns of use over time.

A well established method that addresses these requirements to a large extent is the "diary method" whereby informants are asked to keep a journal or a log, where they record events, activities and experiences regularly over a specified period of time (Rieman, 1993). In traditional diary studies informants record data, usually in writing, but often combining or even replacing written records with other recording media, see for example Carter and Mankoff (2005).

In diary studies, the initiative for capturing information is left up to the study participants who have to remember and take the initiative to report in their diaries. This may be detrimental to the quality of the data collected for several reasons. Participants may forget to enter information in diaries, or entries may be made at moments that they have the time and appetite to do so, rather than the ones of interest to the researcher. This can lead to loss of data and systematic response biases.

For these reasons, the Experience Sampling Method (ESM) is gaining ground in human-computer interaction studies for understanding human behavior to design better products and services and for studying use in the field. The ESM is a quasi-naturalistic method that involves signaling questions at informants repeatedly throughout the sampling period. In its original form (Cziekszentimihalyi & Larson, 1987) Experience Sampling method required participants to carry a pager or any another notification device through which they would be reminded to fill in a set of questions in a paper diary. With developments in handheld computing, this method has become computerized and a variety of tools have been developed for handheld computing devices to support it (Barret & Barret, 2001). Participants are typically requested to carry a dedicated handheld device for the whole study period through which a predefined question-asking protocol is applied.

The ESM method is gaining in popularity in the field of human-computer interaction. Consolvo and Walker (2003) have used the ESM for evaluating an Intel Research system called Personal Server. Hudson et al. (2002) have used the ESM to explore attitudes about availability of managers at IBM Research. Froehlich et al. (2006) used it to investigate the relationship between explicit place ratings and implicit aspects of travel such as visit frequency. The list is longer and growing rapidly as this field turns its attention towards the design and evaluation of mobile and context sensitive applications.

CHALLENGES AND PITFALLS OF EXPERIENCE SAMPLING

Although very useful in prompting the reporting of subjective experiences over time and in context, ESM also has shortcomings such as interrupting the subject at inappropriate moments, the onus of repeatedly answering the same or similar questions, the difficulty of entering self-report data in inconvenient social and physical contexts. The main consequence of these is loss of data: participants ignoring the alert and refraining from providing the requested self-report. Loss of data seems to be a major problem. Froehlich et al. (2006) report completion rate of 80.5% similar to Consolvo and Walker (2003) who report an 80% completion rate (on average 56 out of 70) with as low as 28.5% (20 out of 70). Even worse, these numbers are silent regarding the significance of the data lost. It is reasonable to assume that data loss occurs when people are busy or engaged in social or professional activities. Depending on the goals of the investigator, these might be precisely the situations that researchers are interested in studying.

Another problem with the traditional form of experience sampling when applied to the study of specific user experiences has to do with the density and coverage of the sampling. Contrary to studies of emotions and moods, as was the original application of the method, usage of applications and services may be sporadic and tied to specific contexts. A homogeneous sampling is then inappropriate and researchers need to sample more frequently when specific activities take place and less frequently otherwise, or to adapt the frequency to specific contexts. This requires the use of more sophisticated sampling schemes that can be pre-programmed or even context sensitive. Thus, Intille et al. (2003) have developed software that enables researchers to acquire feedback from participants only in particular situations that are detected by sensors connected to a handheld computing device. Froehlich et al. (2007) developed MyExperience, a system for capturing both objective and subjective in situ data on mobile computing activities. A common characteristic of these works is that they aim to optimize the choice of when to prompt informants with a question. This choice can be based on previous answers of a participant or on inferences made regarding their activity based on context sensing, see for example, Kapoor and Horvitz (2008). Results on this direction are positive but the fundamental limitation of ESM noted above resulting in loss of data remains.

Regardless of the possibility to vary and adapt the timing and frequency of sampling ESM is inherently expensive; it puts a high burden on participants who may have to interrupt own activities and spend effort in situ reporting on their experiences. Further, despite the technical developments described above, it can only provide limited information about uncommon or brief events, which are rarely sampled (Kahneman et al., 2004).

These problems lead to loss of data, inaccurate reporting and nuisance to participants. Current research in this field is concerned with developing methodological innovations and corresponding tools to remedy these shortcomings. On the other hand, the unique advantage of ESM is its ability to capture daily life as it is directly perceived from one moment to the next (Froehlich et al., 2007), providing a rich set of data to researchers.

THE DAY RECONSTRUCTION METHOD

An alternative to ESM, proposed by Kahneman et al. (2004) is the Day Reconstruction Method (DRM), which was designed to assess how people experience their various activities and settings of their lives. Participants in this case are asked to record a detailed diary of activities and events during one day. These do not relate directly to the focus of inquiry of the researcher, which is not disclosed to them at this point, but are meant as a memory aid, a kind of scaffolding, to allow informants to recall and reconstruct the experiences and feelings of the last day during a follow up interview the day after. This is an in-depth semi-structured interview, during which the researcher probes regarding experiences and feelings that the investigation aims to explore. Kahneman produced strong evidence regarding the efficacy of this method; however by its nature, DRM is more appropriate for short studies. Its efficacy for providing rich and contextualized accounts of user emotions in the last 24 hours is achieved by means of an elaborate interview which is not meant to be carried out repeatedly in a study and is practically difficult to repeat over longer sampling periods. Field studies in the domain of ambient intelligence typically exceed two weeks in duration, reaching sometimes even half a year. For such cases, and for studying patterns in the data over time, DRM can help understand only a small fraction of the activities and experiences of informants, missing out a lot of information regarding the context in which it takes place.

A combination of ESM and DRM has the potential to compensate for their complementary weaknesses. Such a combination is the Experience Sampling and Reconstruction Method (ESRM) introduced below.

Experience Sampling and Reconstruction Method

ESRM is a combination of Experience Sampling and the Day Reconstruction methods described in the previous section. Following this hybrid method participants follow the same procedure as with ESM through which a partially complete record of their experiences is created (given the data loss issues discussed above). At regular intervals (e.g., daily) participants are required to reconstruct, elaborate and even reflect on the reported experiences using the partially complete ESM log as a scaffold. The experience can be reconstructed in a manner similar to DRM by completing gaps in the data collection of the day and partly by elaborating and reflecting on this recent data. Because this part of the self-reporting does not have to be done in situ and can be postponed slightly, it can be done using a desktop device that supports more efficient visualization and data entry than handheld devices. Crucially, and thank to system support, this stage is still lightweight enough that it can be repeated daily. The queries which are missed during the sampling day can then be recovered in this way.

The motivation for the reconstruction is two-fold. The first motive is to retrieve data which are lost, e.g., because participants are prompted to report at moments inconvenient for them, e.g., when in a meeting, or when driving, etc. The second motive is to provide a way for participants to annotate the samples taken during the day. By its nature, the ESM requires that responses so-licited from informants should be brief to reduce disruption and encourage frequent reporting. This though can come at the expense of capturing rich and detailed information about the experiences studied. The reconstruction step allows for more comprehensive reporting and more reflective comments to be provided by informants than by the traditional ESM protocol alone.

In contrast to the classical DRM the purpose of the sampling is divulged to participants when they construct the log of experiences. Further, the sampling and reconstruction activities may take place for long durations as with ESM. Contrary to ESM, informants are allowed to report data post-hoc and even elaborate and modify their earlier responses. This, it is hoped, should address potential loss of data of the ESM and recollection problems that may occur with DRM.

ESRM PROCEDURE

The procedure is described below from the perspective of the participant; the required initialization of the application by the researcher is self-evident so it is omitted.

The exact procedure for an informant is as follows (Figure 1):

- Personalization of experience sampling protocol on desktop device.
- Combined ESM and DRM data collection.
- Debriefing interview.

These steps are discussed in detail below.

Personalization Step

The purpose of having a personalization step is to subsequently minimize the time and effort needed by participants to respond to the mobile device. Reducing the effort required is expected to help prevent data loss but also encourage accurate reporting.

Figure 1. The steps of the ESRM method. First, participants insert information to personalize the experience sampling step. Then the experience sampling is executed using a hand-held device and for each experience sampling day participants are asked to access the web application to review their answer and fill out the data which were lost during the sampling day.

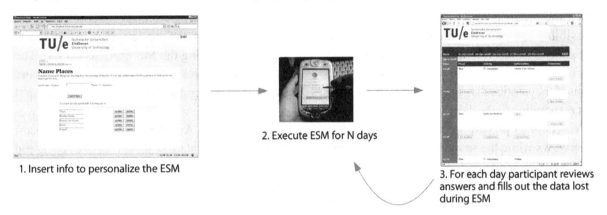

1. Insert info to personalize the ESM

2. Execute ESM for N days

3. For each day participant reviews answers and fills out the data lost during ESM

Personalization can mean a few things: adjusting the timing of the sampling procedure or personalizing defaults and choice items offered during the experience sampling part. The personalization step also gives insight into the differences of self-report obtained out of context prior to the sampling period and the data obtained through experience sampling. This comparison can be interesting in itself as it relates participants' expectations with what actually happens and is experienced. Further, the information obtained is used to populate list-boxes offering choices to participants at sampling time rather than requiring text entry. Personalization of the timing for the sampling events can help, e.g., to limit sampling to waking hours, to avoid times where it is known a priori that the participant is inhibited from answering questions, or even to enable dense sampling at times of most interest for the investigator.

Sampling and Reconstruction Step

During the day the device prompts participants to enter information as in standard experience sampling. Prompting can be programmatically controlled to occur in regular intervals, at random moments or when some conditions regarding the context and the informant activity have been specified.

The informant can respond by selecting between choices of items describing his activity, context or emotions, or even by free text entry to answer more open questions.

The information entered on the handheld device is stored on the online database and is available for retrieval and review directly.

The reconstruction step should happen as close as possible to the collection of data through experience sampling, e.g., within 24 hours. It requires the visualization of the experience sampling logs, the ability to edit them and provide extra information. The interaction requirements for the tool support are different than those applicable for experience sampling: whereas mobility and speed of entry of some brief information is the priority during experience sampling, it is now required to have a good visualization, and efficient ways of editing and inputting text, e.g., using a desktop computer. Of course, one could also allow revision and editing of answers using a small handheld device also for the reconstruction, but this could be at the expense of obtaining richer and more extensive descriptions from informants.

Appropriate visualization of earlier answers can help informants reconstruct their experiences and provide richer descriptions about them. Also important, such visualization can help researchers track the progress of the study, opening up the possibility to modify the sampling protocol while the study unfolds. Researchers can, for example, provide additional incentives or further instructions if they notice that a particular participant is not responding to the daily queries. It also allows researchers to prepare questions for debriefing interviews while the sampling is still unfolding.

Debriefing Interview Step

During the debriefing interview participants are asked to reflect upon their opinions to the queries posed during the sampling period, to solicit in depth explanations and reflection. This step becomes even more useful if the logs of answers are reviewed before approaching each participant. For example, researchers might spot in the log a pattern in the way a participant had answered to a particular question. Based on such an observation the researcher has a unique opportunity in discussing the pattern in detail with the participant. Moreover, the researcher conducting the interview can go through the logs together with each participant and let participants give further explanations of the underlying reasons behind the answers.

Implementation: The RECONEXP Tool

The Reconexp ("reconstructing experience") tool was developed to support the ESRM method. It is a distributed application partly running on a Smartphone (from now on mentioned as "device") and partly on a website. This section describes how Reconexp embodies the characteristics of the hybrid method and how we used it for the purposes of our research.

Currently the application has been deployed on QTek 9090 and HTC Touch P3450 smartphones.

QTek runs Windows Mobile 2003 Second Edition while HTC runs Windows Mobile 6 Professional. The Reconexp application has been programmed using Microsoft's .NET Compact Framework in C#, and OpenNetCF libraries for controlling the WiFi adapter of the device. Participant's data are managed using the MS SQL Server CE. The replication features of Microsoft's SQL Server are used for merging data collected with the central database. For the website part, Windows XP was used as platform, Apache as web server, Microsoft SQL Server as database server, PHP as back-end scripting language and the jQuery framework for implementing user interface features.

CASE STUDY: SURVEYING COMMUNICATION NEEDS OF BUSY PARENTS

ESRM was applied in an investigation of intra-family communication needs and the way pervasive computing would be able to support them to have awareness of each other through the day. This study was aimed at investigating the potential benefits but also the potential disadvantages and detriments to user acceptance of technologies that rely on context sensing to provide frequent updates of the activities and whereabouts of a person to her family see for example Markopoulos et al. (2005). The target group for this investigation was that of "busy parents" who are considered here as pairs of individuals who are married or cohabiting, both working at least part-time and have at least one young, dependent child.

There are several reasons why a sustained and in situ survey of user attitudes was necessary to understand the issues surrounding the awareness needs between couples. First, communication needs vary dynamically; one might wish to reveal her location only when a certain event occurs, e.g., departing from workplace, but not the rest of the day. The same holds for the recipient's interest in such information; it only becomes relevant at particular times and in relation to specific activities. This issue was highlighted during an interview study with 20 busy parents (Khan et al., 2006), a field study of a simple system for helping parents be aware of their children's whereabouts while at school (Khan et al., 2007a) and an online survey with 69 participants (Khan et al., 2007b). These three studies produced results that were sometimes conflicting, only low granularity of information relating to how participants assess experiences for a whole week divorced from its specific time and space context.

Initially it was attempted to answer this question using ESM. However, during pilots with this method involving two participants and two members of the research team for a period of one week, the inherent shortcomings of ESM as discussed above emerged clearly. Pilot study participants perceived the protocol as tedious, which in turn led to repetitive and uninformative answers. There was substantial data loss because of inappropriate timing. Finally, participants missed seeing that their input was actually used and acknowledged by the system. In light of these problems, it was decided to apply the ESRM method using Reconexp as discussed above.

Participants

Participants fitting the profile of a busy parent were recruited through social relations, through a local community group (scouts) and through a participant database managed by the university. All participants were Dutch citizens, married or cohabiting and were thanked with a gift voucher at the end of the study.

A complication that arose while conducting the study was that some participants had problems synchronizing data. Factors such as firewalls, antivirus applications or having a proxy prevented 8 out the 20 participants (40%) to synchronize their data and have the opportunity of filling out the unanswered questions posed in the device at the website. Thus the results reported concern 12

out of the 20 participants initially recruited (seven men and five women).

The participants' average age was 38 years (max: 44, min: 28, sd = 5.72); they had on average approximately two children (mean: 1.91, max: 4, min: 1, sd = 0.79) whose average age was 5.47 years (max: 8.5, min: 0.7, sd = 2.57). Participants were married on average approximately 10 years (mean: 10.86, max: 20, min: 2, sd = 5.22) and on average worked close to 30 hours per week (mean: 28.18, max: 40, min: 20, sd = 6.63). Finally, the participant's spouse worked on average also roughly the same hours per week (mean: 30.91, max: 50, min: 20, sd = 8.92).

Procedure

Personalization

In this section the use of Reconexp for this case study is described. After participants accepted to take part in the study, they were directed to the website where they were initially asked to choose what information they would be interested in sharing with their partner (Figure 2). Rather than a free text entry, participants could select from an extensive list of 41 different types of information to share. This list was compiled from related literature to represent the diversity of information types that is shared through research prototypes described in related literature as well as those mentioned during the interview study mentioned above (Khan et al., 2006). Statements that participants did not wish to share during the personalization step were removed from the options offered during the experience sampling part of the study.

After completing the first part participants were asked to provide information about their context by using the website (Figure 3). The term "context" refers to places and activities (Figure 4) participants visit and perform during a usual working day of theirs. In the final part of this boot-strapping phase on the website participants

Figure 2. First part of personalization step: participants choose information statements they would be generally (in any context) willing to share with their partner, do not mind sharing, or do not want to share

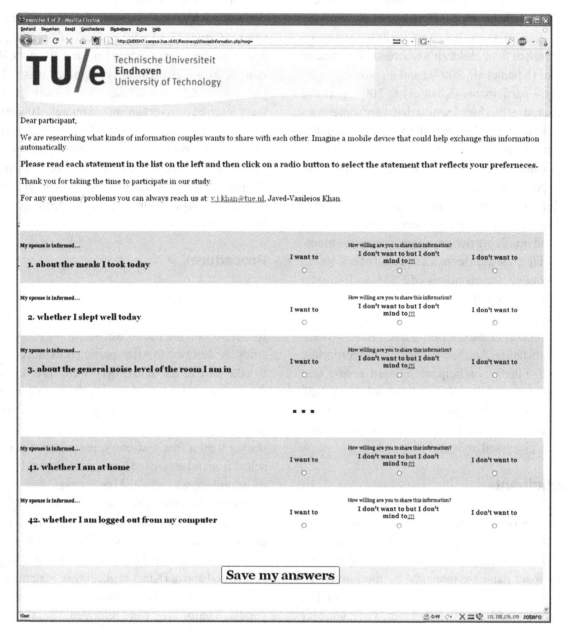

were asked to imagine what information they would like to communicate while being in a specific place doing a certain activity (Figure 5). This list of information statements was carried from the first part of the personalization step including only those statements that they want or don't mind to share. Participants could add information state-

ments to the list in case they would think of other information that they wanted to share with their spouse in a specific context (Figure 5). Added statements would be presented in the list for all available contexts. This final step was left optional for participants since it would need to be repeated for every place and activity. If for

Figure 3. Second part of personalization step: naming places visited on a typical working day

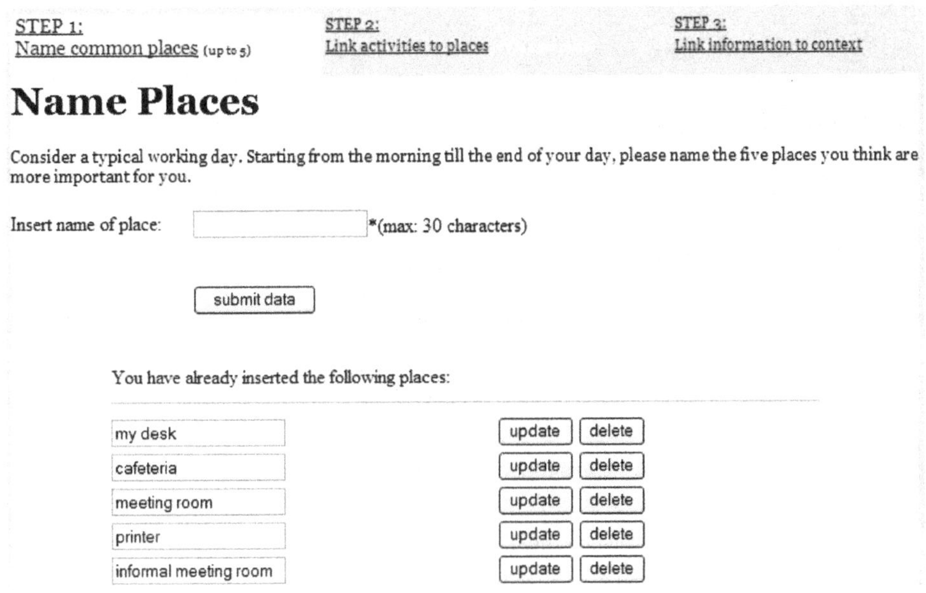

Figure 4. Third part of personalization step: naming activities performed on a typical day

Figure 5. Fourth part of personalization step: linking information to context, participants can insert additional information statements to the existing list

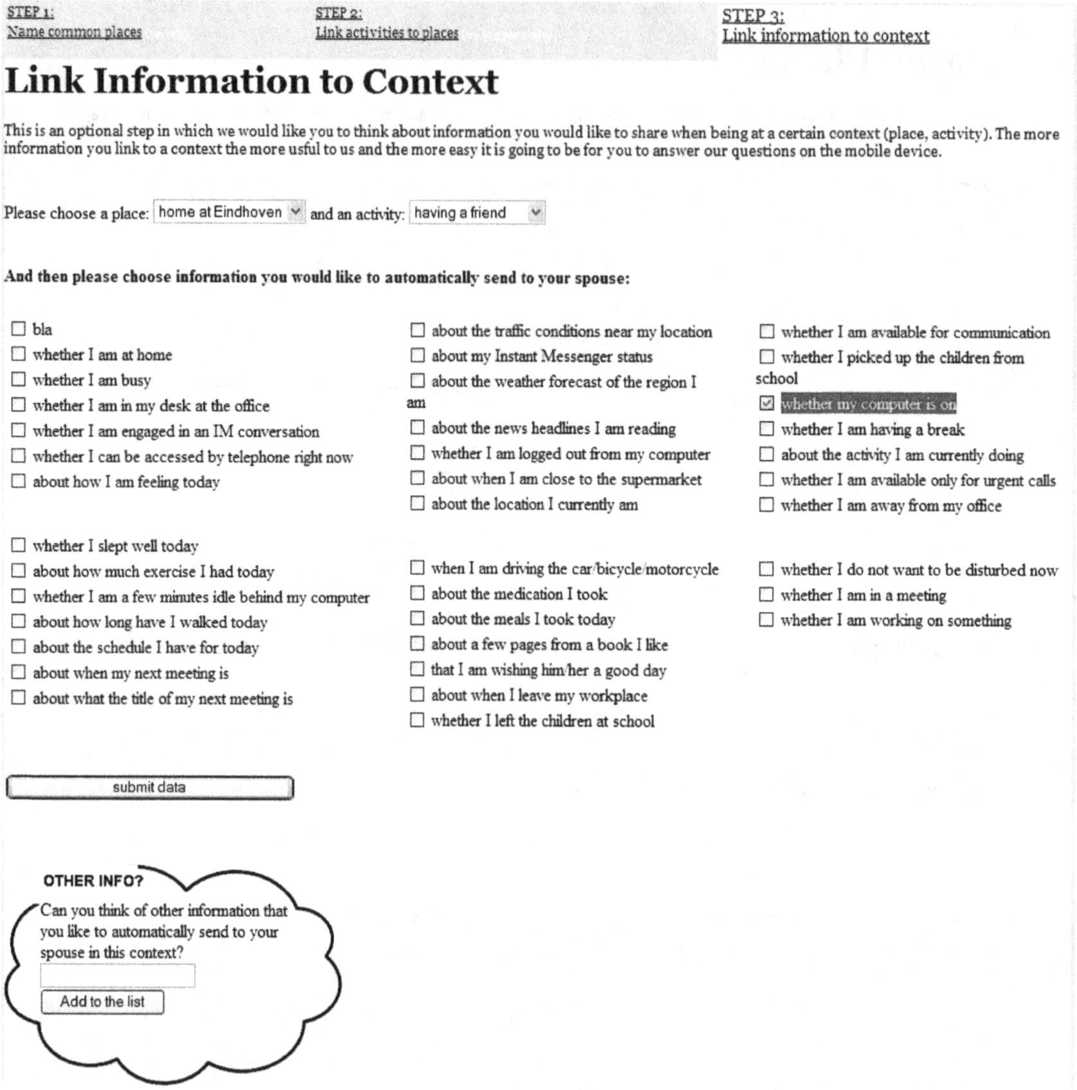

STEP 1:   STEP 2:   STEP 3:
Name common places   Link activities to places   Link information to context

Link Information to Context

This is an optional step in which we would like you to think about information you would like to share when being at a certain context (place, activity). The more information you link to a context the more usful to us and the more easy it is going to be for you to answer our questions on the mobile device.

Please choose a place: [home at Eindhoven ▾] and an activity: [having a friend ▾]

And then please choose information you would like to automatically send to your spouse:

☐ bla
☐ whether I am at home
☐ whether I am busy
☐ whether I am in my desk at the office
☐ whether I am engaged in an IM conversation
☐ whether I can be accessed by telephone right now
☐ about how I am feeling today

☐ whether I slept well today
☐ about how much exercise I had today
☐ whether I am a few minutes idle behind my computer
☐ about how long have I walked today
☐ about the schedule I have for today
☐ about when my next meeting is
☐ about what the title of my next meeting is

☐ about the traffic conditions near my location
☐ about my Instant Messenger status
☐ about the weather forecast of the region I am
☐ about the news headlines I am reading
☐ whether I am logged out from my computer
☐ about when I am close to the supermarket
☐ about the location I currently am

☐ when I am driving the car/bicycle/motorcycle
☐ about the medication I took
☐ about the meals I took today
☐ about a few pages from a book I like
☐ that I am wishing him/her a good day
☐ about when I leave my workplace
☐ whether I left the children at school

☐ whether I am available for communication
☐ whether I picked up the children from school
☑ whether my computer is on
☐ whether I am having a break
☐ about the activity I am currently doing
☐ whether I am available only for urgent calls
☐ whether I am away from my office

☐ whether I do not want to be disturbed now
☐ whether I am in a meeting
☐ whether I am working on something

[submit data]

OTHER INFO?
Can you think of other information that you like to automatically send to your spouse in this context?
[   ]
[Add to the list]

example a participant would have named five places and for each place named five activities then this participant would need to repeat 25 times the last step. That is why participants were allowed to choose the most important contexts and link information statements to those only. In any case, those links would be created during the ESM part.

After a participant inserted the initial information on the website a device was synchronized with the provided data and given to this participant for one week. Participants were requested to keep the device in close proximity constantly.

An audio notification alerted the participant when it was time to record information. Then the application (Figure 6) gave the participant five minutes to respond to three questions: about the place where the participant is at the moment, about the participant's current activity and about the information the participant would like to exchange automatically with his/her partner.

Figure 6. The three questions posed at the device (top part of the screens). The drop-down lists in the first two screens and the checkbox list are populated by the personalization step.

Activities and information statements were adapted according to the previous answer. For example, if a participant would answer that he was at his office the next question would present him with the activities he had previously indicated to be doing being at the office. For every question presented on the device, the participant had the option to answer "Other." If a participant would choose "Other" then for the next question no possible answers would be short-listed; rather this participant would be presented with a list-box containing all the activities known to the system (regardless of place). For the last question if the participant did not check any item from the list, this would be recorded as "Nothing," meaning that the participant did not want to send any information to her/his partner at all.

Furthermore, participants were requested to place the device in its cradle at the end of every day in order to synchronize the data. After synchronizing the data participants were instructed to log onto the website to review the data (Figure 7) correcting omissions of the survey data obtained during the day.

There were two kinds of omissions that participants could correct at this stage. Answers as "Other" (if for example the participant had indicated to be in a place not listed in the drop-down menu offered during the experience sampling) and unanswered questions. When a participant repairs an omission, for example names a place which had previously not been identified, then this new place is appended to the list of places she has made during the personalization phase. This new place will also be present on the mobile device when the participant synchronizes the data. Participants were asked to annotate their answers while reviewing them in order to capture the reasoning behind them.

At the end of the week a semi-structured interview was conducted. In this interview participants were asked whether it had been difficult for them to remember unanswered queries, what information they thought they usually wanted to share and tried to address other issues they might have had with the whole study.

Sampling Protocol

The sampling protocol used combines time based and event based sampling. If an hour has elapsed since the last answer that the participant has given a query is issued. If only 30 to 60 minutes have elapsed the system issues a question only if the location detected by the system has changed. Location changes are detected by comparing the list of Wi-Fi access points to those detected by the system during the most recent answer given by the informant[1].

Reconstruction Step

Participants were requested to login to a website and review the log of the sampling each day. Participants would then view a log of the sampling obtained during the day (Figure 7). While reviewing this log, participants could fill out missing queries.

Figure 7. The day reconstruction interface

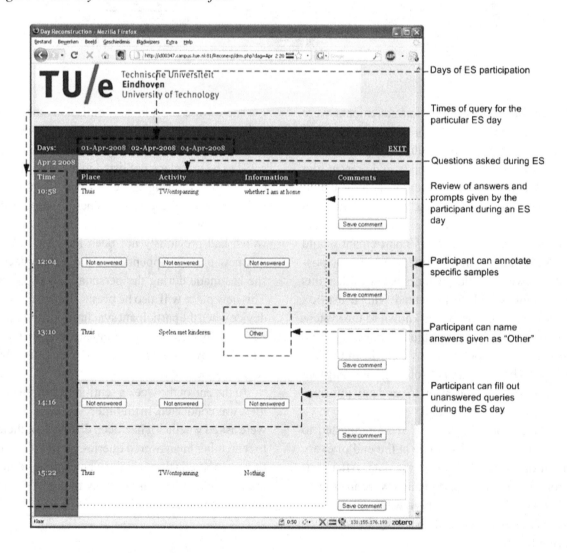

RESULTS

The twelve participant actions which were logged and their occurred frequency are presented in Table 1.

A first inspection of the data suggests that the website was used heavily. The mean number of actions performed: 55.64 (excluding actions such as logging onto the system).

As had been expected, the website was used mostly (64%) during the evening for the 'reconstruction' component of the method.

Furthermore, it turns out that participants logged in 2.75 times in the period of one week to review the data they had inserted during the day. This means that participants would check their log on average approximately once per two sampling days.

In Table 2 the results of the data logging are summarized. It is evident that several questions were not answered during the Experience Sampling component of the method the mean percentage of non response to the second question ("What

Table 1. Type of participant actions which were logged and their occurrence for the reported study

Logged Participant Action	Occurrence
Login to the system	33
Name a new Location which was not answered	11
Name a Location which was not answered using existing value	155
Name "Other" Location using existing value	10
Name "Other" Location	8
Name a new Activity which was not answered	14
Name Activity which was not answered using existing value	155
Name "Other" Activity using existing value	26
Name "Other" Activity	14
Name Information which was not answered using existing value	155
Name "Other" Information	1
Name "Other" Information using existing value	4
Total participant actions	**586**

are you active in now?") was 49.41%. However, a substantial proportion of those (55.38%) were recovered by the use of the website, i.e. via the reconstruction component of the method.

It was about one out of two times that participants could not answer when prompted by the device and it was a just over a half of those that they recovered with the help of the website. When considering answers given for all questions using the device and the website then the overall response improvement of the website to the Experience Sampling Method is: 27.36%. Data pertaining to the second question posed on the device are analyzed since there was an equal number of times participants were asked the three questions; the recovery of the data for the other two questions differs very slightly.

Places which were named on the website but were never selected during the experience sampling part on the handheld device includes among other: "home," "doctor," "train," "fitness," "meeting room," "car." Activities which were named on the website but were never selected during the experience sampling part on the handheld device include among others: "housekeeping," "eating," "cooking," "shopping," "check internet," "put coat on," "fitness training," "driving," "reading news." Information statements which were named on the website and then never reported using the device were: "dinner time," "about when he expects to be at home," "changes in working hours," "if I can bring something from shop," "how late I will be home," "when dinner is ready," "whether I need to use the car," "whether the children are going to play elsewhere," "whether the children are ok."

The above information illustrates that with the reconstruction phase we were able to obtain reports regarding places and activities where the ESM alone would result in omitted responses. On the one hand this shows that data loss was reduced; on the other it gives us insight into the places and activities in which ESM is less applicable. It is interesting to note that beyond

Table 2. Results of logged data

Description		Max	Min	SD
Mean number of actions performed (logins not counted in this number)	46	155	1	64
Mean logins (in 5 days)	2.75	7	1	1.92
Mean times participants were questioned (at least 5 days)	62.58	124	14	27.67
Mean times the 2nd question (about activity) was not answered	30.92	65	1	16.8
Percentage of mean number of activities (2nd question) not answered:	49.41%			
Mean of percentage of activities recovered (through the website)	55.38%	97%	0%	40%
Overall improvement (data recovery) of the website to the method	**27.36%**			
Total comments (number of participants who commented: 8 out of 12)	33	4.13	15	4.51

some expected places and activities such as "car," "meeting room," "eating," "fitness training" for which we would not have expected participants to be able to respond to the device, we observe places such as "train" and "home" and activities such as "check internet" and "reading news" for which participants were not able to respond as well. Thus, the proposed hybrid method helps in extending the scope of ESM as a survey method and its coverage over different contexts.

Furthermore, the content analysis illustrates how reconstruction of places and activities can be extremely precise. In terms of activities for example, the reported activity "put coat on" is extremely detailed and it could not be expected that participants could recall such a fine grained activity in a diary study. Both log data and stated opinions seem to corroborate the opposite.

In terms of recalling events we found that the hybrid method does not pose considerable difficulties. During the debriefing interview we asked participants how difficult it was for them to remember the location, activity, and information they would like to exchange when they were asked in the past on the device but could not answer. All but one said that it was easy for them to remember and accurately answer a question they could not answer at the device. They offered two reasons for that. The first one was that when trying to remember and fill out the unanswered questions they concerned situations that were not too remote

in the past. In all cases except for one they provided missing information one or two days later.

The second reason they mentioned was that the overview on the website was providing them with a frame of reference, (which was the motivation for doing so based on the Day Reconstruction Method). For example, when a participant could not answer a question posed at 13:30 but did answer several questions before and after that, these answers would help recall whereabouts, activities and what information this participant would have liked to exchange in such a context.

RELATED WORK

In this section we will review current support for computerized Experience Sampling with the aim to identify directions for the development of related tool support in the future.

Notable examples of such tools include: PsycLab Mobile, a tool which supports audio recording on a pre-defined schedule (Mehl et al., 2001), CAES (Intille et al., 2003), a pioneer in combining sensor input to trigger queries based on events recorded by sensors, Purdue Momentary Assessment Tool (PMAT), which was developed Military Family Research Institute at Purdue University (Weiss et al., 2004) and supports both time based and event based studies and it also provides a desktop application to configure parameters of the ESM study and finally ESP, an open-source tool developed at Boston College (Barrett and Barrett, 2005). The latest version (ESP 4.0) includes a suite of software packages, a Palm OS based application that interacts with participants and a PC application for designing experiments, configuring ESP settings, and collecting data which runs on Windows and Linux. Although these tools have been pioneers in trying to support researchers in conducting ESM they nowadays seem outdated and one of them (CAES) has even discontinued and has joined forces with a more recent tool which will be reviewed in the following paragraphs (MyExperience). Here, more recent studies which have followed and improved upon the early tools are reviewed.

We will begin this review with two projects which have developed tools to integrated aspects of both easily creating Ubicomp prototypes as well as evaluating them (Carter et al., 2007, de Sá et al., 2008).

The framework of de Sá et al. (2008) supports both prototyping and evaluation. It provides a log of events taking place on the mobile device and a researcher has also the ability to execute an ESM. It runs in Windows Mobile, Palm OS and Symbian OS devices. The logging engine stores a variety of events. Events range from each tap on the screen, each button press or even each character that was typed by the user. It supports audio and video capture. To analyze the collected data, a log player is provided. The log player resembles a "movie player" which re-enacts every action that took place while the user was interacting with the prototype. ESM can be event-triggered. The framework is publicly available.

A similar open source tool, supporting both mobile prototype creation and remote evaluation of those prototypes is Momento (Carter et al., 2007). It was created to support remote testing of Ubicomp applications. Momento can also gather log data, experience sampling data, diary data, and other qualitative data. One of the requirements the researchers found while conducting interviews

with ubicomp developers to elicit requirements for their system is the need for integrated tools to allow participants to annotate and review qualitative data. Momento can run on participants' existing networked mobile devices. Researchers can use a desktop application to configure experimental details, to monitor incoming information from participants, send information to participants and review data or export it for further analysis. Momento uses SMS and MMS and HTTP (if available) to share information between participants and researchers. It supports audio, photo and video capture and situated annotation of captured media. The mobile client is configured using a text file. Momento can also support the review and annotation of data by the participants after they have been collected. However it does not support the recovery of data lost during ESM and the developers have not researched the potential benefits of such a feature. A disadvantage is that it needs a desktop installation for both client and server and therefore requires support for troubleshooting and version track management.

MyExperience (Froehlich et al., 2007) is an open source software that supports passive logging of device usage, user context, and environmental sensor readings and active context-triggered user experience sampling to collect in situ, subjective user feedback. Queries can be targeted to moments of interest by triggering sensor readings. Using XML researchers can define survey questions and configure sensors, triggers, and actions. Embedded scripts are used to provide flexibility and expressiveness in specifying the conditions to trigger surveys. MyExperience supports sophisticated survey logic including multiple branching, parameterized questions, and persistent states. It supports audio, photo and video capture. Although it synchronizes collected data opportunistically, it does not provide a web interface for participants to review their answers and fill out the missing queries of the ESM part. While it is designed to run on participant phones and although there extensive work has been done to simplify the installation process it still remains difficult installing it in a Windows Mobile device. Even though it has an incredible amount of features it still requires a long installation process. Moreover, in the case a researcher wants to have real time access it requires to have and manage a MS SQL database server with replication features. Another shortcoming is that it requires removal of the security lock from the phone which might allow programs such as dialers to run without the users' consent.

Another study has revealed that showing participants of experience sampling studies their own collected information helped in increasing compliance rate. In a 25-day field study, Hsieh et al. (2008) found that users who saw visualizations of their own data were more likely to respond to sampling requests compared to users who did not see visualizations. The compliance rate of those receiving feedback was 23% higher than the rate of those who did not receive feedback. This result suggests that showing participants visualizations of their own answers has a positive effect to the study itself. The researchers do not report the use of any tool for conducting the ESM therefore we assume that it was a custom made application.

In their quest to measure and evaluate emotional responses to user interactions with mobile device applications Isomursu et al. (2007) administered an ESM study in which participants answered questions by selecting an appropriate emoticon on the mobile phone's screen. It was a custom made ESM tool which allowed both system and user initiated experience recording. It supported event based triggering of queries. Participants could not insert text or any multimedia input. Further, the tool logged user interactions on the mobile device. The tool was running at the participants' phones. Additional comments could not be provided to clarify the answers. One important finding of their evaluation regarding the method was that participants would accidentally press a button and give an unwanted response in situations like driving, or when having the phone in their coat, or in cases that they would receive a phone call. Such a finding suggests that ESM tools must support defining inconvenient moments for participants. Having such an option would be beneficial to researchers because they would not collect data which were accidentally inserted and they would not disturb participants with queries during inconvenient moments.

Isomursu et al. (2007) have also created the Experience Clip method. In this method, pairs of participants are recruited. Both of them are provided with mobile phones. One carries the application which is under evaluation and the other is instructed to take short video clips of the usage of the first participant. The participant taking the video clips was a friend of the other participant. Time stamps of the videos were used to match interaction events with the expressions of emotions captured by videos. As a conclusion the researchers state that having the users decide which usage situations to record did not seem to spur versatile and innovative usage. This approach had the disadvantage that it was not clear which captured situations represent real usage situations and which not. They suggest that a combination of the Experience Clips with other non-intrusive methods would perhaps yield better results. This method brings a different perspective into the computerized version of ESM. It advocates the involvement of people in the surrounding of the subject whose experience is sampled.

DISCUSSION

The arguments presented above and the case study show that using Reconexp to support ESRM allows for reduction of data loss and also, some streamlining of the effort required by participants.

One could raise the question whether the DRM possibly induces "postponing behavior" where participants would opt to postpone answering, exactly because the option is available to them to fill out information easier on the website later. This would mean that the ESM component of the combined method, underperforms, and perhaps accuracy of data is lost as a whole. While there has been no evidence found of such behavior, it is important from a methodological perspective to eliminate that such a behavior undermines the quality of the data obtained. In our future investigations we aim to compare ESM on its own against its adapted version with DRM.

One should note that completion rates with the ESM part of Reconexp (so prior to reconstruction) were considerably lower than those typically reported in the literature. For example, Froehlich et al. (2006) report completion rate of 80.5% and Consolvo and Walker (2003) 80% whereas ours was approximately 52%. This is probably due to the difference in participant groups. The previous studies cite University students as participants. In our case we recruited people who had young children, were working and had a truly hectic schedule. These participants have much less time to respond to an experience sampling study and this is what our results portray. However, further applications of the method are needed to obtain a better understanding of how this set up influences compliance to the amount and quality of the data obtained.

While promising, one needs to keep in mind that Reconexp was a first attempt to support ESRM. There are several usability issues that still need to be tackled and as previously mentioned, there were some issues with synchronizing the data. We hope that ongoing improvements of the functionality and ease of use of the tools should greatly enhance the performance of the method.

CONCLUSION, FUTURE WORK, AND DIRECTIONS FOR IMPROVEMENT

ESM tools can support behavioral as well as requirements elicitation process for the design and iterative evaluation of mobile applications. Although there are an increasing number of tools available, most of the studies we reviewed still rely on custom-made software to execute ESM studies, which suggests that existing tools do not yet meet the requirements of researchers and practitioners interested to apply the method. Based on the case study findings and the review of existing experience sampling tools a list of directions for future developments can be derived.

From our experience in installing the available tools, we find them still quite challenging to install. It would be even more challenging for a researcher without extensive computer literacy to deploy any of them. All of the reviewed tools require extensive installation procedures and in most cases management of database or web servers. In tools that combine mobile phones and desktop computers, installation procedures have to be followed in both devices. This presents a major obstacle to researchers who lack the relevant technical skills. Quick and easy installations on mobile devices and even no installation procedures for desktops would be another important requirement for such tools. Therefore, special effort should be spent in minimizing requirements for installation and maintenance.

Another important conclusion for future ESM tools is providing an interface for participants to give feedback during the ESM period. Although some of the aforementioned studies have used synchronization of collected data (Froehlich et al., 2007; Hsieh et al., 2008) none of them gave the opportunity to participants to fill out queries they could not answer during the day. This property that was supported by ESRM introduced in this paper, will not only increase compliance rate but it will also recover lost data, as it was presented. Environmental sensor data would also be useful in inferring with greater accuracy the participant's context and in that way make the presentation of queries even more sophisticated at a more convenient time.

The event-triggering of queries, for example when a participant is in a particular location, is supported by some of the tools reviewed however it is either a built-in function, in the case of Reconexp for example, or in the best case (MyExperience) it is programmable by using XML. Although XML is in many ways easier than programming in C or VB it still presents a high threshold for researcher who are not programmers. An appropriate interface for an easily programmable sampling protocol seems an important requirement for the wider uptake of such tools.

The tools reviewed all present the prompts and questions to participants without taking into account whether they are in fact able to observe the screen at that time; some context awareness specifically with regards to the attention of the participant would help optimize the sampling procedure and yield better results.

Support for multimodal participant input has been already included in several tools. Text, audio, photos, and video can provide richer data to the researcher (Carter et al., 2007). On the other hand, participants can choose the most efficient and convenient modality for addressing the query. In the case of CAES, MyExperience, and Reconexp among others, user context factors are automatically captured. The location and possibly the activity of a participant can serve as examples. Automatic capture of participants' context would provide different perspectives for researchers to look at the gathered data and obviously provide more in-depth results.

Another requirement is support for optional, user initiated input. In the case of Reconexp, participants could not initiate the queries at will. That would be useful in cases where participants would recognize the importance, in terms of research, of the context they currently are and initiate the research queries. In that way salient information will be gathered.

An important shortcoming of Reconexp was the difficulty participants had in synchronizing data. Automatic synchronization of data captured on the device to a remote server would both secure the data as well as provide the grounds for feed-

ing the data back to participants as the case with Reconexp was. MyExperience already supports such a feature.

Automatic and configurable information visualization tools of the collected data would be a crucial feature for helping researchers disambiguate the data and quickly provide useful results. Alternative visualizations, e.g. in the form of graphs, can enable researchers to view the data in new fresh ways, and provide insights during analysis. This analysis tool should be able to support visualization of events that occur both frequently and infrequently (Barrett & Barrett, 2001).

Participants might become less motivated during the course of the research study. Programmable by researchers email or SMS notifications to the participants can help to keep participants highly motivated. In addition, support for notifications for researchers when certain events occur would also be of added value.

In most cases, ESM tools which use a mobile device force participants to carry an extra to their own mobile device. It would be more convenient if such tools would run on participants' phones. Such studies could involve larger sample of participants and the reliability of results would be enhanced. However, a researcher would thoroughly need to have tested the tool so that it would not hamper the operation of the participant's device. Moreover, agreements with the mobile service provider must be made in advance so that participants are not burdened with the cost of the service.

Another important feature beyond the ability of participants to review the collected data would be to annotate the data and also to fill-in gaps. The potential benefits of this feature which was implemented in Reconexp have been extensively discussed above.

Mobile devices have limited processing and memory in comparison to desktops. The data collection tool on the mobile device should not noticeably impact the performance of the participant's mobile phone (Carter et al., 2007). If that happens it might affect the results of the study since participants will experience a lag in the presented queries.

Finally, Carter et al. (2007) identify some other important user requirements: In case where a mobile device is lost, the tool on the mobile device should offer mechanisms to protect the security and privacy of the data. Further, to ensure ease of use, participants should be able to increase the color contrast, the font size, etc.

This paper has presented a distributed platform (Reconexp – reconstructing experience) developed to support the combination of two research methods for collecting subjective data in field studies regarding experiences and feelings of informants: Experience Sampling and Day Reconstruction. This novel method (ESRM) is part of a line of research to support Experience Sampling tools with the use of mobile devices and context sensing technology.

Compared to related systems, Reconexp is the only one complementing the use of a handheld device for reporting brief notes in situ, with surveying recollection of informants using a website. Related is the work of Froehlich et al. (2007) who have developed a tool which synchronizes data captured during an experience sampling study. Compared to their system Reconexp innovates by allowing participants to review and fill out the gaps created by ESM as well as annotate the captured data.

There is a lot of scope to develop Reconexp further. Our first experiences with Reconexp and ESRM confirm the value of complementing the mobile experience sampling tools with data collection on a website, leading to a reduction of data loss and the improvement of the quality of the data collected. Follow up studies are needed to consolidate these methodological results and to effect relevant improvements on the tools.

REFERENCES

Barrett, L. F., & Barrett, D. J. (2001). An introduction to computerized experience sampling in psychology. *Social Science Computer Review*, *19*(2), 175–185. doi:10.1177/089443930101900 204doi:10.1177/089443930101900204

Barrett, L. F., & Barrett, D. J. (2005). *ESP: The experience sampling program*. Retrieved from http://www.experience-sampling.org

Carter, S., & Mankoff, J. (2005). When participants do the capturing: The role of media in diary studies. In *Proceedings of the SIGCHI Conference on Human Factors in Computing Systems* (pp. 899-908).

Carter, S., Mankoff, J., & Heer, J. (2007). Momento: Support for situated ubicomp experimentation. In *Proceedings of the SIGCHI Conference on Human Factors in Computing Systems* (pp. 125-134).

Consolvo, S., & Walker, M. (2003). Using the experience sampling method to evaluate ubicomp applications. *IEEE Pervasive Computing/IEEE Computer Society and IEEE Communications Society*, *2*(2), 24–31. doi:10.1109/MPRV.2003.1203750doi:10.1109/MPRV.2003.1203750

Cziekszentimihalyi, M., & Larson, R. (1987). Validity and reliability of the experience-sampling method. *The Journal of Nervous and Mental Disease*, *56*, 5–18.

de Sá, M., Carriço, L., Duarte, L., & Reis, T. (2008). A framework for mobile evaluation. In *Proceedings of the SIGCHI Conference on Human Factors in Computing Systems* (pp. 2673-2678).

Froehlich, J., Chen, M., Smith, I., & Potter, F. (2006). Voting with your feet: An investigative study of the relationship between place visit behavior and preference. In *Proceedings of the Conference on Ubiquitous Computing* (pp. 333-350).

Froehlich, J., Chen, M. Y., Consolvo, S., Harrison, B., & Landay, J. A. (2007). MyExperience: A system for in situ tracing and capturing of user feedback on mobile phones. In *Proceedings of the 5th International Conference on Mobile Systems, Applications and Services* (pp. 57-70).

Hsieh, G., Li, I., Dey, A., Forlizzi, J., & Hudson, S. E. (2008). Using visualizations to increase compliance in experience sampling. In *Proceedings of the Conference on Ubiquitous Computing* (pp. 164-167).

Hudson, J. M., Christensen, J., Kellogg, W. A., & Erickson, T. (2002). I'd be overwhelmed, but it's just one more thing to do: Availability and interruption in research management. In *Proceedings of the SIGCHI Conference on Human Factors in Computing Systems* (pp. 97-104).

Intille, S. S., Rondoni, J., Kukla, C., Ancona, I., & Bao, L. (2003). A context-aware experience sampling tool. In *Proceedings of the SIGCHI Conference on Human Factors in Computing Systems* (pp. 972-973).

Isomursu, M., Tähti, M., Väinämö, S., & Kuutti, K. (2007). Experimental evaluation of five methods for collecting emotions in field settings with mobile applications. *International Journal of Human-Computer Studies*, *65*(4), 404–418. doi:10.1016/j.ijhcs.2006.11.007doi:10.1016/j.ijhcs.2006.11.007

Kahneman, D., Krueger, A. B., Schkade, D. A., Schwarz, N., & Stone, A. A. (2004). A survey method for characterizing daily life experience: The day reconstruction method. *Science*, *306*, 1776. doi:10.1126/science.1103572doi:10.1126/science.1103572

Kapoor, A., & Horvitz, E. (2008). Experience sampling for building predictive user models: A comparative study. In *Proceedings of the SIGCHI Conference on Human Factors in Computing Systems* (pp. 657-666).

Khan, V. J., Markopoulos, P., de Ruyter, B., & IJsselsteijn, W. (2007). Expected information needs of parents for pervasive awareness systems. In B. Schiele, A. K. Dey, H. Gellersen, B. de Ruyter, M. Tscheligi, R. Wichert et al. (Eds.), *Proceedings of the European Conference on Ambient Intelligence* (LNCS 4794, pp. 332-339).

Khan, V. J., Markopoulos, P., & Eggen, B. (2007). On the role of awareness systems for supporting parent involvement in young children's schooling. *International Federation for Information Processing*, *241*, 91–101.

Khan, V. J., Markopoulos, P., Mota, S., IJsselsteijn, W., & de Ruyter, B. (2006). Intra-family communication needs: How can awareness systems provide support? In *Proceedings of the 2nd International Conference on Intelligent Environments* (Vol. 2, pp. 89-94).

Markopoulos, P. (2005). Designing ubiquitous computer human interaction: The case of the connected family. In H. Isomaki, A. Pirhonen, C. Roast, & P. Saariluoma (Eds.), *Future interaction design* (pp. 125–150). New York, NY: Springer. doi:10.1007/1-84628-089-3_8doi:10.1007/1-84628-089-3_8

Mehl, M. R., Pennebaker, J. W., Crow, M. D., Dabbs, J., & Price, J. H. (2001). The electronically activated recorder (EAR): A device for sampling naturalistic daily activities and conversations. *Behavior Research Methods, Instruments, & Computers*, *33*, 517–523. doi:10.3758/BF03195410doi:10.3758/BF03195410

Rieman, J. (1993). The diary study: A workplace-oriented research tool to guide laboratory efforts. In *Proceedings of the SIGCHI Conference on Human Factors in Computing Systems* (pp. 321-326).

Weiss, H. M., Beal, D. J., Lucy, S. L., & Mac-Dermid, S. M. (2004). *Constructing EMA studies with PMAT: The Purdue momentary assessment tool user's manual.* Retrieved from.http://www.ruf.rice.edu/~dbeal/pmatusermanual.pdf

ENDNOTES

[1]   If in a previously stored place A the system has detected N1 access points and if N2 is the number of access points of N1 which were not found in the surrounding access points at the query moment, the following formula was used to approximate the probability of the user being at A:

$$p = \log \frac{(N_1 - N_2) * 10}{N_1},$$
$$for \frac{(N_1 - N_2) * 10}{N_1} > 1 \wedge N_1 > 0$$

It appears that this formula can easily flag when an informant moves to a different building, floor or part of town based on the current density of wi-fi points. This approach does not help track fine grain movements, e.g., away from and to a desk, for which other approaches may be more appropriate.

This work was previously published in the International Journal of Handheld Computing Research (IJHCR), Volume 2, Issue 3, edited by Wen-Chen Hu, pp. 53-72, copyright 2011 by IGI Publishing (an imprint of IGI Global).

Section 5
Mobile Health

Chapter 14
Mobile E–Health Information System

Flora S. Tsai
Singapore University of Technology and Design, Singapore

ABSTRACT

A mobile e-Health information system (MEHIS) aims to speed up the operations of health care in medical centers and hospitals. However, the proper implementation of MEHIS involves integrating many subsystems for MEHIS to be properly executed. A typical MEHIS can consist of many components and subsystems, such as appointments and scheduling; admission, discharge, and transfer (ADT); prescription order entry; dietary planning; and smart card sign-on. This paper describes the development of a MEHIS with open-source Eclipse, using currently available health care standards. The author discusses the issues of building a mobile e-Health information system which can help achieve the goal of ubiquitous and mobile applications for the personalization of e-Health.

INTRODUCTION

The explosive proliferation of information sources such as blogs (Chen et al., 2007), social networks (Tsai et al., 2009), and medical records, have been spurred by the growth of the information technology. An electronic health record (EHR) system allows a patient's medical reports and results to be ubiquitously available to clinicians anytime. The EHR is a collection of electronic health care data about patients and general populations, and is the main technology used to integrate health care information in both paper format and in electronic medical records (EMR) for improving the quality of health care (Gunter & Terry, 2005). The

EHR is the key in development of a truly digital hospital, where daily record-keeping and operations are performed exclusively with computers. Although EHRs have been a health care priority in many countries, progress in implementing EHRs has lagged behind the latest information technologies available, and only a few health care establishments actually use them (Weiss, 2002). The issues that impede the widespread implementation of EHRs include data sharing among departments, security, privacy, and confidentiality (Weiss, 2002). In addition seamless integration of e-Health services allow the creation, exchange and manipulation of medical data (Amoretti & Zanichelli, 2009).

DOI: 10.4018/978-1-4666-2785-7.ch014

A typical EHR system (EHRS) consists of several subsystems, which may be independently designed and developed (Tsai, 2010). One problem with an EHRS is that the subsystems are usually integrated in an ad-hoc fashion, without proper consideration of the overall system. For example, subsystems that are designed using conventional non-object-oriented design may be difficult to fully integrate with those subsystems designed using object-oriented design. An EHRS that integrates the various heterogeneous subsystems together can never achieve the quality and reliability of a system that is designed from the ground up. Contrary to popular belief, such a system may not necessarily be more expensive to implement than trying to integrate an existing system.

The vast growth of wireless and mobile devices has created a large demand for mobile information content that is personalized for the user (Tsai et al., 2010) as well as a mobile Web of semantic data and services (Yee et al., 2009), context-aware mobile learning (Chia et al., 2011), geographic search (Tsai, 2011), anomalous behavior detection (Thing et al., 2011), and other topics in handheld computing research (Hu et al., 2010). Thus, the mobile e-Health information system (MEHIS) can be developed based on the design of the EHRS services. MEHIS can consist of many different components and subsystems, such as appointments and scheduling; admission, discharge, and transfer (ADT); prescription order entry; dietary planning; routine clinical notes; lab and radiology orders; picture archiving, and smart card sign-on.

In this paper, we first use Eclipse and UML (Unified Modeling Language) to design and implement an EHRS based on existing EHR models and frameworks. Once the design of all the subsystems have been completed, the implementation and integration of the EHRS are relatively straightforward and likely to be less costly than patching together multiple legacy subsystems. Using Eclipse as the design environment can also speed up the implementation phase, as the tight integration with the

programming language Java using the Eclipse JDT (Java Development Tools) can ease in the transition from design to implementation (Tsai, 2006). These tools include perspectives, project definitions, editors, views, wizards, refactoring tools, a Java compiler, a scrapbook for expression evaluation, and search tools (D'Anjou et al., 2005) that can work together with design and testing tools to create a highly integrated development environment suitable for software development across heterogeneous platforms.

After the EHRS is designed, the MEHIS can be implemented using the similar platforms, by implementing smaller services (Kwee & Tsai, 2009). This paper describes the development of a MEHIS with open-source Eclipse, using currently available health care standards. We discuss the issues of building an adaptive and personalized e-Health information system and the use of ubiquitous and mobile applications for the personalization of e-Health.

This paper is organized as follows. First, we describe the different e-Health architectures, which include CEN ENV, openEHR, HL7, and Eclipse Open Healthcare Framework (OHF). Then, the design of the mobile e-Health Information System is presented. Next, we describe the implementation of MEHIS with different components. Finally, the last section summarizes the entire paper.

RELATED WORK

Due to the popularity of mobile applications, many healthcare applications now are available to run on mobile devices. A handheld system that provides medical staff with information based on their location was used to retrieve medical information relevant to the user's current activity, such as a patient's medical record made available when the physician is near her bed (Rodriguez et al., 2004). The Intelligent Control Assistant for Diabetes (INCA) created a mobile intelligent

Personal Assistant for continuous self-monitoring glucose and subcutaneous insulin infusion integrated into a telemedicine diabetes management service (Hernando et al., 2006). A mobile outpatient service system (MOSS) was initiated in a local hospital in Taiwan, and focuses on illness treatment, illness prevention and patient relation management for outpatient service users (Jen et al., 2007). A Mobile Intelligent Medical System (MIMS) supports mobile nursing applications and clinical decision support, and includes RFID-based mobile applications for monitoring physiological instruments (Trappey et al., 2009). 3G Doctor is a service that allows video consultation with General Medical Council registered Doctors using a 3G Video Mobile, which can also create, store and manage personal health records (http://www.3gdoctor.com). However, many of the mobile e-Health systems do not focus on the entire e-Health process which can consist of many components and subsystems.

EHR ARCHITECTURE

The EHR architecture is based on the generic component model which form the agreed vocabularies enabling inter-operability as well as the Reference Information Model (RIM) (Blobel, 2004). This architecture facilitates an unambiguous presentation of medical concepts by preserving the original context (Blobel, 2002), and must be sufficiently designed along with its structure and. behavior.

One approach for EHRS design uses a single comprehensive model of all the structures, functions, and terminology, such as the CEN "Electronic Healthcare Record Communication" standard. Another approach uses many specialized models reflecting organizational, functional, operational, contextual, and policy requirements (Blobel, 2002), such as openEHR and Health Level Seven (HL7) standards.

CEN ENV 13606 Architecture

CEN, the European Standards body, has targeted the Electronic Healthcare Record (EHCR) as a primary area for European standards establishment. The ENV 12265 Electronic Health care Record Architecture (1995) was a foundation standard defining the basic principles of electronic health care records. In 1999, a four-part successor standard for Electronic Health Care Record Communication, ENV 13606, was published (European Committee for Standardization, 1999).

Part 1, the Extended Architecture, defined the Reference Architecture to allow the construction, usage, sharing and maintenance of EHCR content, where the subject of the record is an individual person. Part 2, the Domain Termlist, defined a set of domain-specific terminologies to support EHCR interoperability in different systems (European Committee for Standardization, 1999). Part 3, the Distribution Rules, specified the means by which security policies and attributes can be defined as well as a set of data objects that represent the rules for defining access privileges to EHCRs (European Committee for Standardization, 1999). Part 4 defined a list of messages to enable the communication of EHCRs in response to a particular request (European Committee for Standardization, 1999).

OpenEHR Architecture

OpenEHR is a set of open specifications for EHRs which enables semantic interoperability of health information. OpenEHR enables effectively support for healthcare, medical research and related areas by allowing the representation of the semantics of the healthcare sector. A knowledge-oriented computing framework that includes ontologies, terminology and a semantically enabled health computing platform is part of openEHR. OpenEHR also supports the construction of maintainable and adaptable health computing systems and patient-centric EHRs.

Clinical knowledge concepts are captured as archetypes, which support the recording required for common clinical activities. Data built according to these archetypes are stored in an EHR in larger composition structures which result from clinical events. The archetypes contain a maximum data set about each clinical concept, including attendant data required such as: protocol, method of measurement, related events, and context required for the clinical data to be interpreted.

Aggregations of archetypes are combined in openEHR templates to capture the dataset corresponding to a particular clinical task. Templates can be used to build generic forms to represent the approximate layout of the EHR, which can be used by vendors to contribute to their user interface development. Both archetypes and templates can be linked to terminologies that will support appropriate term selection by healthcare providers (http://www.openehr.com).

HL7 Architecture

"Level Seven" is the highest level of the International Organization for Standardization (ISO) communications model for Open Systems Interconnection (OSI). The application level defines the information to be exchanged, the timing, and the communication of errors to the application. The seventh level supports security and availability checks, participant identification, and data exchange (HL7, 2005).

The Reference Information Model (RIM), the cornerstone of the HL7 development process, is a complete visual model of the clinical data (domains). The RIM identifies the life cycle of events that a message can carry, and is a shared model between all the domains. The RIM also explicitly represents the connections that exist between the data carried in the fields of HL7 messages, and consists of templates, vocabulary, and XML standards (HL7, 2005).

HL7 templates are data structures based on the HL7 RIM which express the data content needed in a specific clinical or administrative context. They are prescribed patterns by which multiple observation result (OBX) segments may be combined to describe selected observations. Some simple observations, such as blood pressure measurements, involve a set of expected observations. Other more complex diagnostic procedures may involve many different pieces of information. Templates will couple the multiple OBX segments with separately encapsulated rules. Based on the preferences and needs of the user, the template can define the collection of OBX segments needed and the set of validation rules only once. Templates can also be "plug and play" at a given user's site (HL7, 2005).

While data can be exchanged between systems, its usefulness is compromised unless there is common vocabulary of the information being transferred. HL7 provides an organization and repository for maintaining a coded vocabulary that, when used in conjunction with HL7 and related standards, will enable the exchange of clinical data and information so that sending and receiving systems have a shared, well defined, and unambiguous knowledge of the meaning of the data transferred. The purpose of the exchange of clinical data includes, but is not limited to: provision of clinical care, support of clinical and administrative research, execution of automated transaction oriented decision logic (medical logic modules), support of outcomes research, support of clinical trials, and to support data reporting to government and other authorized third parties (HL7, 2005).

HL7 also uses XML standards for all of its platform- and vendor-independent health care specifications. In January 2005, the HL7 approved Version 2 of the Clinical Document Architecture (CDA), which defines an XML architecture for exchange of clinical documents. The encoding is based on XML Document Type Definitions (DTDs) included in the specification and its semantics are defined using the HL7 RIM and HL7 registered coded vocabularies (HL7, 2005).

Eclipse Open Healthcare Framework (OHF)

The Eclipse Open Healthcare Framework (OHF) is an Eclipse project that can improve the levels of interoperability between healthcare applications and systems. This project used extensible frameworks, tools and health informatics standards as plug-ins. The Eclipse platform is designed to serve as an open tools platform so that its components could be used to build almost any client application. The set of plug-ins needed to build a rich client application is known as the Rich Client Platform (RCP). Using the OHF components, the Eclipse RCP application with user interface and workflow logic can be created. The National Health Information Infrastructure (NHII) was connected to the repository database, interoperability stack. They can be accessed by the OHF plug-ins using Patient Identifier Cross-referencing and Cross-enterprise Document Sharing (XDS).

However, this method of implementation could pose certain problems, such the need to interconnect. With the prevalence of Internet connectivity and mobile communication devices, the market demands a client/server solution if multiple users are required. With the traditional method of Eclipse RCP application development, the client plug-ins would cause big problems in the future. Thus, it is necessary to adopt a better framework from the beginning; otherwise, large amounts of time, money and effort could be invested in projects and applications which may not succeed. With this issue in mind, a team of IBM software engineers developed a solution they called the OHF Bridge.

OHF Bridge

The OHF Bridge is a sub-project of the OHF, and uses Axis and Server-Side Eclipse allow OHF components to be accessed via Web services. The OHF bridge is an "Open Services Gateway initiative (OSGi) on Server," embedding OHF Plug-ins and exposing some of their functions as Web services. Any application using Simple Object Access Protocol (SOAP) can invoke the available functions with Web services by the OHF Bridge.

The OHF Bridge runs the "OSGi on server" servlet inside a Web container, which may be any container, such as the Apache Tomcat Web container. The OSGi contains the runtime plug-ins along with the OHF plug-ins and the Axis Engine. This Axis Engine has a web service that accepts SOAP calls and translates the calls to the relevant API that the OHF plug in exposes.

Thus all functions are being transported from the client side to the server side. This would create even higher interoperability between applications when they attempt to transfer clinical data or search patient records on a database server.

MEHIS DESIGN

In this section, we focus on the design of a subset of the application layer of MEHIS. A typical MEHIS can consist of many subsystems, such as appointments and scheduling; admission, discharge, and transfer (ADT); prescription order entry; dietary planning; routine clinical notes; lab and radiology orders; picture archiving, and smart card sign-on. Figure 1 shows the high level diagram for some of the various subsystems of a typical MEHIS.

Eclipse for UML Design

Eclipse users can choose from a variety of Unified Modeling Language (UML) plug-ins for object-oriented analysis and design. If Eclipse is to be used throughout the entire software engineering process, the benefits are manyfold. Advantages in using Eclipse for design include automatic code generation, tight integration with the implementation language, refactoring, configuration management, debugging, and testing support. Although the Eclipse Integrated Development Environment (IDE) is open source, many of the

Figure 1. EHR subsystems

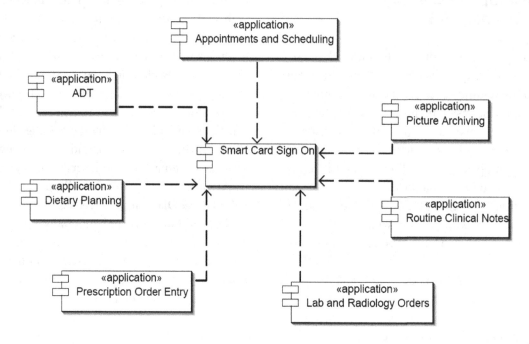

UML plug-ins are not. Therefore, users who cannot afford to purchase a commercial UML tool are left with fewer choices. Many vendors who offer a free version of their UML plug-in for Eclipse only provide limited options, such as a subset of the full functionality, inability to generate source code from design models, or inability to import or export models to different formats. In this paper, we use Omondo's EclipseUML Studio, which uses a standard UML2 metamodel. The EclipseUML2 metamodel is the same as IBM Rational, and represents over 95% of the UML tools market (Omondo, 2005).

MEHIS SUBSYSTEMS

This section presents the description and design of some of the MEHIS subsystems using the EclipseUML Studio. The design of MEHIS conforms to the HL7 RIM Version 2.08. The class diagram of the various roles in the MEHIS is shown in Figure 2. This diagram shows the various roles such as

LicensedEntity, Patient, Access, and Employee which are used in the subsystems described in the following sections.

Appointments and Scheduling

The appointments and scheduling subsystem is used to manage and record the appointment scheduling of patients and doctors. It will be able to record information about the patients, set the schedule of the doctor(s) on duty at the respective specialist clinics for the day, verify the availability of a doctor for a selected time slot, and allow patients to create, change and cancel appointment accordingly.

Prescription Order Entry

The prescription order entry subsystem will be used to manage prescription ordering activities in a hospital. The subsystem can add, change, and delete prescriptions given to the patients. The issued prescriptions will be stored in the database

Figure 2. MEHIS class diagram of HL7 RIM roles

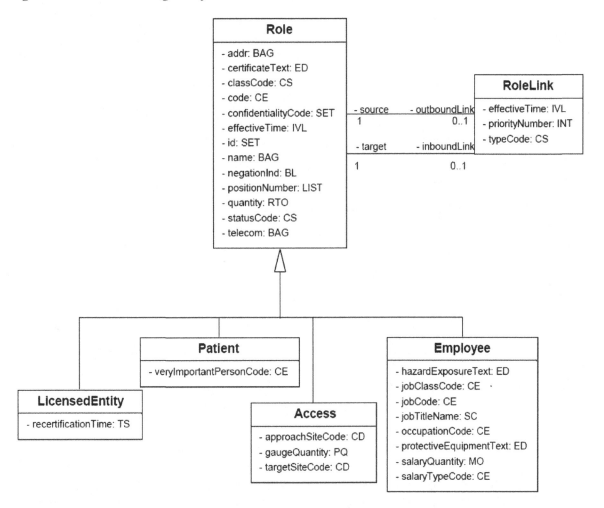

for future reference. When a patient consults a doctor, the doctor will be able to search the database for the patient's record. In the record, the doctor is able to view the patient's particulars and the patient's medical history. From there, the doctor will be able to add in new prescriptions based on his diagnosis. The nurses and the pharmacists will be able to verify, and confirm the prescriptions intended by the doctor.

The design process primarily constituted the development of various UML diagrams. In the first step, various use case diagrams and use case scenarios were created.

Use Cases

Order Creation

The use case shown in Figure 3 provides an understanding of the different steps involved in giving a new prescription to a patient. This involves the doctor checking the patient's information, test results, prescription history and then finally giving out the prescription.

Figure 3. Order creation

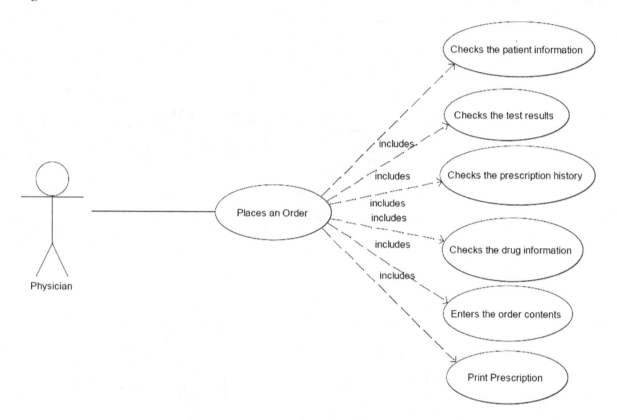

Order Modification/Cancellation

The use case shown in Figure 4 provides an understanding of the different steps involved in modifying or canceling an existing patient prescription. When the doctor sees a need to modify or cancel a prescription, he or she acquires the existing prescription contents and thereafter modify/cancel it.

Order Management

The use case shown in Figure 5 provides an understanding of the different steps involved in managing a prescription by the pharmacist. After the doctor enters the prescription, the pharmacist has the option to issue an enquiry, request for a prescription change or simply dispense the medicines prescribed.

Class Diagrams

During the second step the class diagram for the system was designed with the attributes derived from the Health Level 7 Reference Information Model (RIM). HL7 RIM Act class structure was found to be the class intended to contain the information that gets communicated to support clinical shared care. Figure 6 shows the class diagram for the CPOE System with the Drug Records Class having a two way multiple cardinality association with the Drug Class.

The attributes of the above two classes were obtained from Substance Administration and Supply Class, sub classes of the Act Class. Also the Entity Class and its subclass Material were referred to obtain the attributes for the Drugs class.

Figure 4. Order modification/cancellation

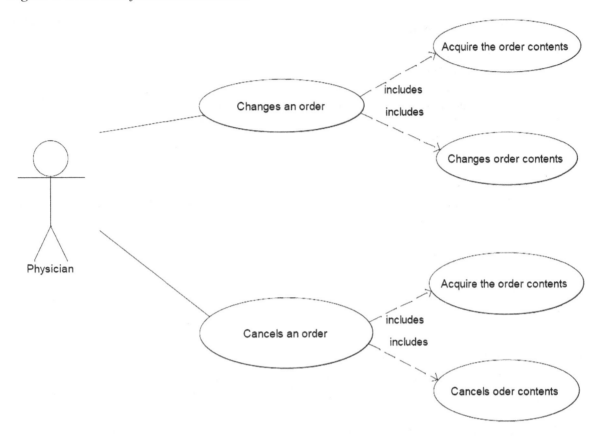

Figure 5. Order management

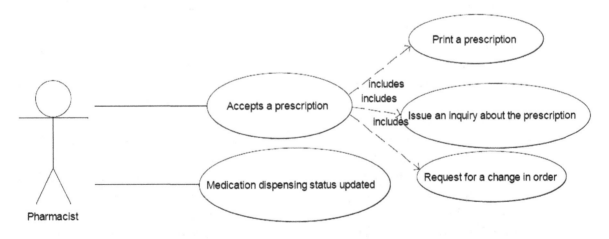

Figure 6. Drug record and drug class diagram

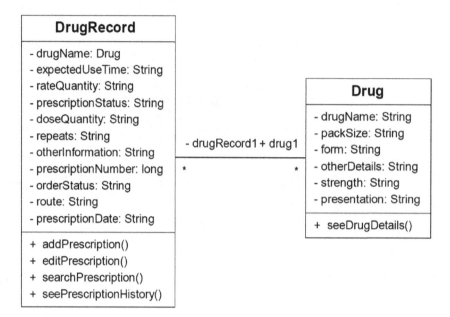

DrugRecord

- drugName: Drug
- expectedUseTime: String
- rateQuantity: String
- prescriptionStatus: String
- doseQuantity: String
- repeats: String
- otherInformation: String
- prescriptionNumber: long
- orderStatus: String
- route: String
- prescriptionDate: String

+ addPrescription()
+ editPrescription()
+ searchPrescription()
+ seePrescriptionHistory()

- drugRecord1 + drug1

*   *

Drug

- drugName: String
- packSize: String
- form: String
- otherDetails: String
- strength: String
- presentation: String

+ seeDrugDetails()

Class Name: Substance Administration

Description: The act of introducing or otherwise applying a substance to the subject. Attributes used were:

- **routeCode:** The method of introducing the therapeutic material into or onto the subject. Common routes are Per os (PO), sublingual (SL), rectal (PR), per inhalationem (IH), ophtalmic (OP), nasal (NS), otic (OT), vaginal (VG), intra-dermal (ID), subcutaneous (SC), intra-venous (IV), and intra-cardial (IC).

- **doseQuantity:** The amount of the therapeutic agent or other substance given at one administration event. The dose may be specified either as a physical quantity of active ingredient (e.g., 200 mg) or as the count of administration-units (e.g., tablets, capsules, "eaches"). Which approach is chosen depends upon the player of the 'consumable' participation (which identifies the drug being administered). If the consumable has a non-countable dosage form (e.g., measured in milligram or litre) then the dose must be expressed in those units. If the consumable has a countable dosage form (tablets, capsules, "eaches"), then the dose must be expressed as a dimensionless count (i.e., with no other unit of measure specified).

- **rateQuantity:** Identifies the speed with which the substance is introduced into the subject. Expressed as a physical (extensive) quantity over elapsed time (e.g., examples are 100 mL/h, 1 g/d, 40 mmol/h, etc.). This is appropriate for continuously divisible dose forms (e.g., liquids, gases). If specified as an interval, the rate should be in some specified range. This attribute is specified as an extensive physical quantity over elapsed time, i.e., a quantity that could be used for the doseQuantity divided by a duration quantity.

Class Name: Supply

Description: An act that involves provision of a material by one entity to another. The attribute used was:

○ **expectedUseTime:** Identifies the period time over which the supplied product is expected to be used, or the length of time the supply is expected to last. In some situations, this attribute may be used instead of Supply. quantity to identify the amount supplied by how long it is expected to last, rather than the physical quantity issued e.g., 90 day supply of medication (based on an ordered dosage), 10 hours of jet fuel, etc. NOTE: When possible, it is always better to specify Supply.quantity, as this tends to be more precise. Supply.expectedUseTime will always be an estimate that can be influenced by external factors

Class Name: Entity and its subclassMaterial.

Description: Name and quantity attribute was selected from the Entity Class. The attribute was:

○ **formCode:** Represents the state (solid, liquid, gas) and nature of the material was selected from the Material Class. For therapeutic substances it is basically the dose form, such as tablet, ointment, gel, etc.

In the last step, sequence, collaboration, and activity diagrams were designed for the CPOE system.

Sequence and Collaboration Diagrams

Add Prescription

The sequence diagram for adding a prescription is shown in Figure 7. The object initiating a sequence of message is a healthcare professional, who through the MEHIS GUI sends an add prescription request to the drug record class.

Figure 7. Add prescription sequence diagram

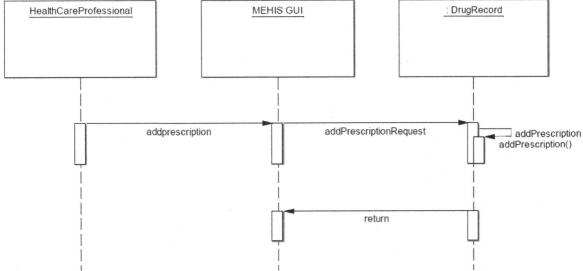

The drug record class in return generates the add prescription window.

Search Prescription

The collaboration diagram for searching prescription is shown in Figure 8. The healthcare professional sends a search prescription request to the drug record class by entering the prescription number. Once the drug record class verifies the existence of the prescription number in its drug record database, it returns the prescription details.

Edit Prescription

The sequence diagram for editing a prescription is shown in Figure 9. The healthcare professional can edit the prescription and send an edit prescription request for the drug record class to save the edited prescription details and then return back to the system GUI.

View Drug Details

The collaboration diagram for viewing drug details is shown in Figure 10. The healthcare professional can view the drug details by sending a request for seeing the drug record details to the drug class. If the drug details for that particular drug exist, the details are returned to the system GUI.

ADMISSIONS, DISCHARGE, AND TRANSFER

The ADT subsystem manages the admission, discharge, and transfer of patients in a hospital. The subsystem can create, display, change, and delete patients' admission, discharge, and transfer records. The activity diagram is shown in Figure 11.

An ADT System is used to maintain a log of patients admitted to the hospital. It helps the hospital in carrying out ward management and tasks such as generating health record analysis reports, patient registers and ADT Reports. It also helps in tracking patients as the health care information systems generates daily admission, discharge and transfer reports listing patients by name, bed number or location, and health record number.

In the first step of the design, the business requirements of the system were captured using the use case diagrams and use case scenarios. The requirements were obtained from chapter 3 – Patient Administration of HL7 Version 2.5 and from the Japanese Association of Healthcare Information Systems Industry (JAHIS, 2005).

Functionalities included from the Patient administration document are listed as follows:

Code: Description
3.3.1 ADT/ACK: Admit/visit notification (Event A01)
3.3.2 ADT/ACK: Transfer a Patient (Event A02)
3.3.3 ADT/ACK: Discharge/end visit (Event A03)
3.3.4 ADT/ACK: Register a Patient (Event A04)
3.3.11 ADT/ACK: Cancel admit/visit notification (Event A11)
3.3.12 ADT/ACK: Cancel transfer (Event A12)
3.3.13 ADT/ACK: Cancel discharge/end visit (Event A13)
3.3.15 ADT/ACK: Pending transfer (Event A15)
3.3.16 ADT/ACK: Pending discharge (Event A16)
3.3.25 ADT/ACK: Cancel pending discharge (Event A25)
3.3.26 ADT/ACK: Cancel pending transfer (Event A26)

The different use case diagrams are:

- **Patient Admission:** This use case shown in Figure 12 provides an understanding of the different steps involved in admitting a patient. This basically involves the doctor checking with the ward manager on the room to be assigned as per the patient preference and then placing a new admission appointment. In a similar way, the doctor

Figure 8. Search prescription collaboration diagram

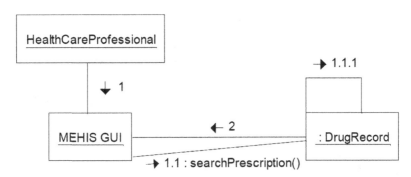

Figure 9. Edit prescription sequence diagram

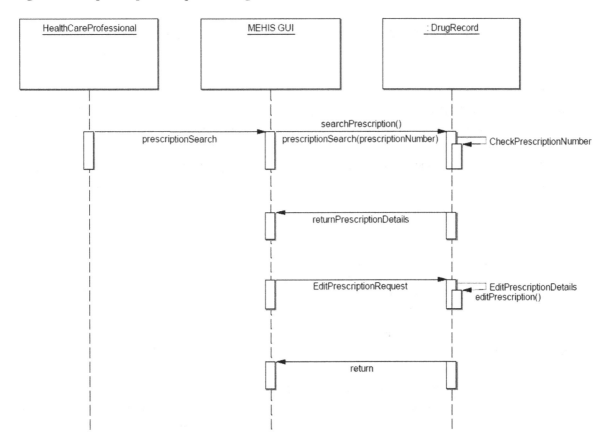

can change or cancel the admission appointment for the patient.

- **Patient Transfer:** This use case shown in Figure 13 provides an understanding of the different steps involved in transferring a patient. If there is a need to transfer a pa-

tient then the doctor checks with the ward manager on the new room to be assigned as per patient preference and then place a transfer order. In a similar way, the doctor can change or cancel the transfer order for the patient.

Figure 10. Drug details collaboration diagram

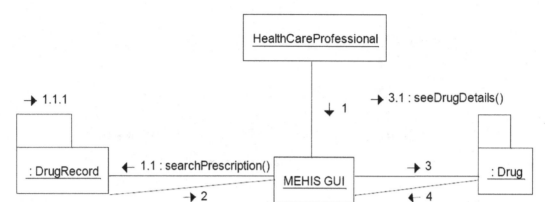

- **Patient Discharge Order:** This use case shown in Figure 14 provides an understanding of the different steps involved in discharging a patient. The doctor can discharge a patient by placing an order for discharge and entering the necessary discharge order information. The doctor also has the option to either cancel or change the discharge order and related information.

Class Diagram

In the second step the class diagram for the system was designed with the attributes derived from the Health Level 7 Reference Information Model (RIM). HL7 RIM Act class structure was found to be the class intended to contain the information that gets communicated to support clinical shared care. Figure 15 shows the class diagram for the ADT System with the Patient Class having a one way multiple cardinality association with the ADT Class.

The Patient encounter class which is a generalization of the Act class is illustrative of the hospital-based ADT ancestry of this model. The attributes found in this class are the administrative details typically recorded on a patient's admission to the hospital, and could not readily be generalized to any kind of health care encounter.

The attributes used were:

1. **Length of Stay Quantity:** Identifies the total quantity of time when the subject is expected to be or was resident at a facility as part of an encounter. The actual number of days cannot be simply calculated from the admission and discharge dates because of possible leaves of absence.

2. **Discharge Disposition Code:** The code depicts the actual disposition of the patient at the time of discharge (e.g., discharged to home, expired, against medical advice, etc).

3. **Special Courtesies Code:** The code identifies those special courtesies extended to the patient, such as extended courtesies, professional courtesy, or VIP courtesies.

DIETARY PLANNING

The dietary planning subsystem helps dietitians to perform dietary planning. The subsystem will be used to plan daily menu for the patients, which can be viewed and printed by other users in the hospital. When a new patient is admitted to the hospital, a new dietary record is added to the patient record database for future reference. Information like the patient particular and medical record is keyed and store in patient database and medical record database respectively. The system then calculates the dietary intake requirement and

Figure 11. Activity diagram for ADT subsystem

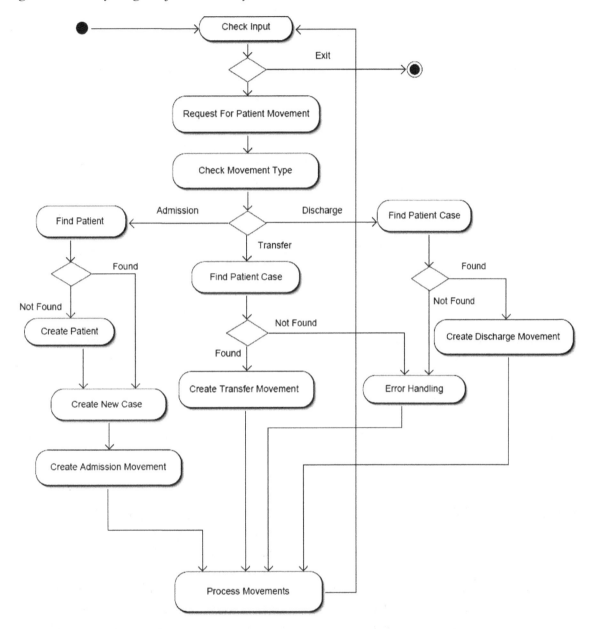

gives dietitians a suggested meal with restriction of food taken into consideration. The dietitian will verify the menu before the menu is printed.

Based on our MEHIS design, the system is first divided into several independent subsystems, then the subsystems are designed concurrently from the ground up, with the integration of the subsystems occurring during the implementa-

tion of MEHIS. If enough time is spent on the proper design of the MEHIS subsystems, the implementation of MEHIS is straightforward. The system is implemented using Java with the JDT implementation of Eclipse, which allows for editing, compiling, testing, and debugging in an integrated environment.

Figure 12. Patient admission order

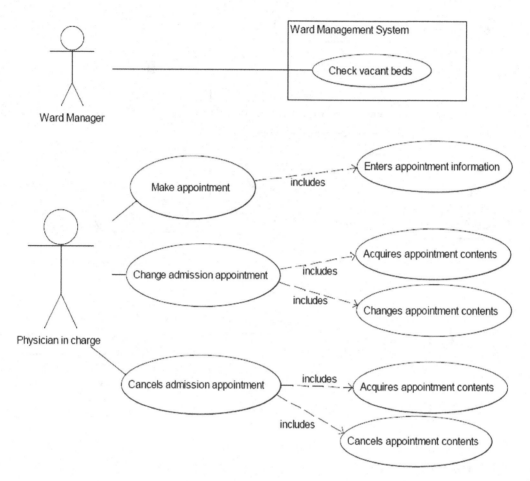

MEHIS IMPLEMENTATION

In the implementation of MEHIS, Eclipse was selected as the development environment for implementing the MEHIS system. It fulfills the needs of a large product development team, from product manager to content developer to product tester.

The biggest advantage of using Eclipse was found to be its integration with external tools such as the UML and Visual Editor. Eclipse can also be configured with many action, wizard, view and editor extensions for any number of languages, tasks, roles, and phases within the development life cycle.

The JDT project (D'Anjou et al., 2005) provides the tool plug-ins that implements a Java IDE supporting the development of any Java application, including Eclipse plug-ins. It adds a Java project nature and Java perspective to the Eclipse Workbench as well as a number of views, editors, wizards, builders, and code merging tools. It allows Eclipse to be a development environment for itself.

The JDT project is broken down into four components. Each component operating like a project unto its own, with its own set of committers, bug categories and mailing lists.

Figure 13. Transfer order

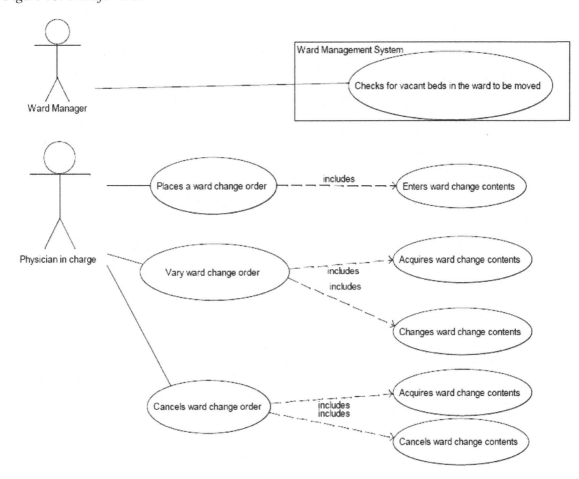

- **APT:** Java 5.0 annotation processing infrastructure.
- **Core:** Java IDE headless infrastructure.
- **Debug:** Debug support for Java.
- **UI:** Java IDE User Interface.

For the mobile development, the tools utilized were Mobile Tools for Java (MTJ) which replaces its predecessor EclipseME plugin and Java ME platform SDK 3.0. The implementation was carried out in three parts. Part 1 consisting of the patient record form and the starting windows for the MEHIS system. In part 2, the computerized prescription order entry system was implemented, followed by admission, discharge and transfer system implemented in part 3.

Graphical User Interface (GUI)

The Eclipse Visual Editor was used to design the system GUI. It is a vendor-neutral, open development platform that supplies frameworks for creating GUI builders, and exemplary, extensible tool implementations for Swing/JFC and SWT/RCP. The Swing component library was used to implement the GUI.

The implementation was carried out in three parts. Part 1 consists of the patient record form and the starting windows for the EPR system. In Part 2, the computerized prescription order entry system was implemented, followed by admission, discharge and transfer system implemented in part 3. It should be noted that different healthcare

Figure 14. Patient discharge order

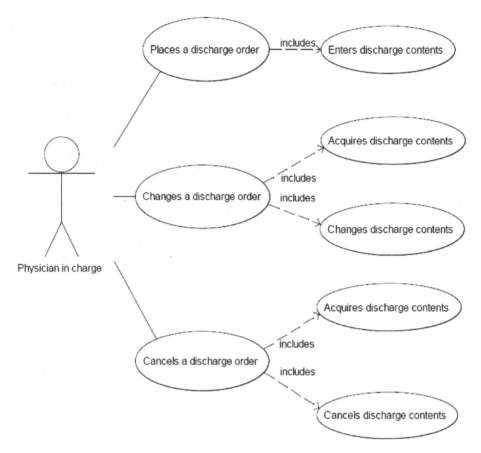

professionals are not given different authorization levels in this system and therefore the system will behave the same for all users. Although this is currently a limitation for the system, it could be easily removed given more time and resources.

Multiple Level Authorization

A multiple level authorization system is used to maintain confidentiality of patient details. It does this by differentiating different roles each healthcare personnel performs and allocates a level of authorization to the personnel.

There are different levels of authorization to be allocated. The different levels of authorization to be allocated are based on the role and responsibility the personnel hold in the healthcare organization.

Table 1 depicts some of the roles and responsibilities found in a typical healthcare scenario.

The procedure is as follows:

1. End user accesses the EHR system. The user enters his username and password.
2. Class loginDialog checks with the database what category of position the user holds.
3. The loginDialog class returns the level of authorization and the appropriate forms are opened to the user.

Three levels of authorization are created:

1. Personnel Officer (PO)
2. Health Care Provider (HC)
3. Administrator (AD)

Figure 15. ADT class diagram

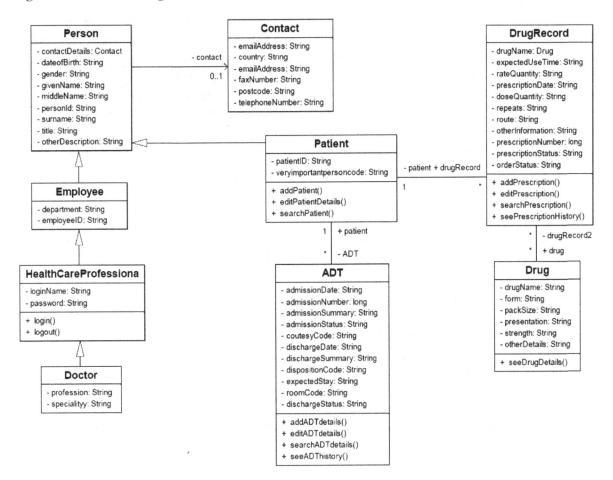

Table 1. Table of roles and responsibilities

End User	Locations	Activities
Health Care Professional Doctor Nurse Administrator	Hospital clinic or ward, including intensive care units; GP surgery or health care centre; Patient's home; In their car or at roadside	Review patient records, including: Outpatient clinic consultations, medical history taking, clinical examinations, requests and results of laboratory test, radiology, biosignal, angiography records and diagrams, operation notes, anesthetic records, nursing observations, dietary history, education by specialist nurses, outpatient clinical notes, discharge reports. Revise data sets and templates for clinical care and research. Clinical care audit of patients.
Patient	Home, Office, in transit	Review and edit personal record
Administrator	Hospital office or GP surgery office	Register new patients, edit registration details. Arrange admissions, clinic appointments
Personnel Officer	Hospital office or GP surgery office	Register new clinical staff, edit staff details.

Smart Card Sign-on

A few assumptions were made to successfully develop the sign-on interface. Firstly, it is presumed that the mobile device has some form of card scanner or reader and the information would be readily available in the mobile terminal's screen upon the user scanning the card. To simulate this process, the user still has to enter the data such as user name and password via the mobile's native keypad.

The system is primarily designed for three groups of people consisting of doctors, patients and staff in the medical field. After the successful sign-on process, different levels of services are accessible by the users. Doctors naturally being the highest in the hierarchy can access all services followed by staff and lastly patients. Most services accessed by patients are only for their viewing pleasure and mainly contain their medical history. Therefore, to emulate these three groups, three users are present in the implementation. They are: 1) Doctor, 2) Staff, and 3) Patient.

When the Midlet runs, the login screen will be displayed as shown in Figure 16.

After pressing the Sign-in button, the user waits for the status of the login. The user must wait until the progress bar is 100% before getting the result. The successful login attempt is shown in Figure 17. A marquee runs at the top screen displaying the appropriate message.

This system starts with the user logging into the MEHIS system. The user enters his user name and password. When the login button is pressed, the system generates an SQL statement to query the database. A successful login will return a value, namely POSITION. Variable POSITION has been set to PUBLIC and can be of three values.

Case 1: Personnel Officer

If the user is a personnel officer, the database query will return POSITION as PO and

Figure 16. Smart card sign-on

Figure 17. Successful login

generate the Staff Search Window. Personnel Officers are allowed to add new hospital staff and amend hospital staff details but they are not allowed to add, edit or view patient particulars. They belong to the human resource department of the hospital. They handle all issues pertaining to hospital staff. Thus, access to patient particulars is denied.

○ **Viewing Staff Details:** In this window, personnel officers are able to view all staff details. He or she can also sort by the various attributes such as gender, date of birth, NRIC, Surname, Country etc. in ascending or descending order. He or she can search the particular attribute, the surname.

○ **Add New Staff:** The Staff Details Screen allows the personnel officer to edit various staff related details such as Mobile number, address, department etc. Clicking on the save button after changing certain details updates the database with the new particulars.

○ **Delete Staff Window:** The personnel officer can delete staff from the Staff Search Window. A warning message is shown when attempting to delete a staff.

Case 2: Administrator

○ **Viewing Patient Details:** If the user is an administrator, the database query will return POSITION as AD and generate the Patient Search Window. As administrators to the hospital, they are allowed to add and delete patients. They are also allowed to amend patient details. Patient Search Window is for Administrators as well as Healthcare Professionals.

○ **Administrator:** They can also carry admission, discharge and transfer matters for the patient. They are also al-

lowed to schedule new appointments, edit existing appointments as well as delete appointments. However, as they are not qualified physicians, they are not allowed to prescribe medication. As such, the Computerized Prescription Order Entry System has been disabled. In the Patient Details Window, Administrator's view, the Computerized Prescription Order Entry System has been disabled whereas the Admission Discharge and Transfer and the Appointment Scheduling System are enabled.

Case 3: Healthcare Professional

If the user is a healthcare Professional, he/she is brought to the healthcare professional view. Here he/she can access two systems. The user can choose to browse the Patient Search Window or he/she can choose to enter the Appointment Search Window.

○ **View Patient Details:** A healthcare professional is allowed to add, edit and delete patients. Also, the healthcare professional is allowed all aspects of the system pertaining to the patient's care. This includes the ADT, CPOE, appointments as well as adding, editing and deleting patients from the database. In the Patient Details Window for a healthcare professional, all systems are enabled for the healthcare professional's access.

The user may press the Menu command to access the services he/she is entitled to. In this case, Doctor has the highest access level and therefore has all options shown in his menu. For the case of the patient, this might not be the case. The commands are all linked up in such a way user can navigate to the previous screen and are self-explanatory as seen in Figure 18.

Figure 18. Main menu

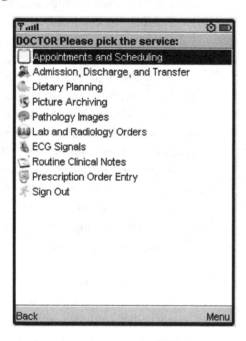

Appointment Scheduling Subsystem

The Appointment Scheduling System allows users to add appointments for patients and edit existing appointments. The appointment Scheduling system works from the patient's perspective as well as the healthcare professional's perspective. The user can choose to view the appointment based on the patient or based on the healthcare professional.

Patient's Perspective

1. **View Appointment Details:** Figure 19 shows the Appointment Scheduling System based on the patient selected. The user can view or make changes to patient's appointment and also add new appointment for the patient.

2. **Add New Appointment:** Figure 20 shows the Add New Appointment Window. Here users can add new appointment details such as appointment date, appointment time as well as appointment duration and other details pertaining to the appointment to be made.

3. **View and Edit:** This shows an appointment record window that allows users to view or make changes to existing appointments. Clicking save on the form updates the current details of the appointment to the database.

Healthcare Professional's Perspective

Healthcare professionals may also choose to access the appointment scheduling system from the healthcare professional's window. Doing so will bring the user to the appointment scheduling system healthcare professional's window.

Figure 19. Appointment scheduling system

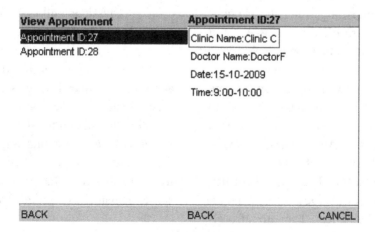

Figure 20. Add new appointment

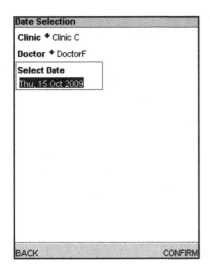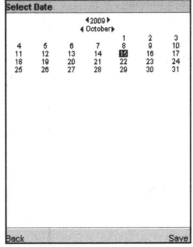

1. **View Appointments:** Figure 21 shows the Search Appointments Window for healthcare professionals. Double clicking on the appointment allows the healthcare professional to edit the details of the appointment.
2. **Edit Appointment Details:** This shows how a user can edit the appointment details. Clicking on the save button updates the database of the updated appointment detail.

Prescription Order Entry System

CPOE is a system required for direct entry of patient care orders by a physician into a computer system and is used in both inpatient and outpatient settings.

CPOE is usually integrated with a collection of other clinical information system components such as a clinical data repositories, clinical decision support system, laboratory system, or pharmacy systems. The degree of integration varies from

Figure 21. Doctor view appointments

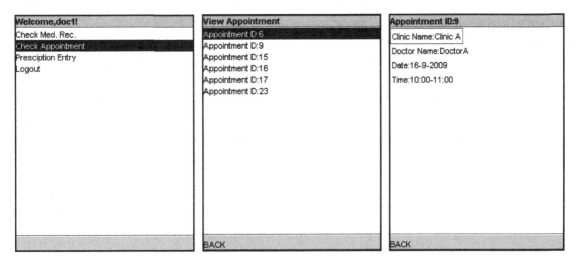

loosely coupled arrangement in which orders are passed to other systems using standard message structures such as HL7.

Order entry is subdivided into order creation and order management. Order creation includes the process of creating, modifying, or discontinuing orders while order management includes processes related to order routing, maintaining a profile of a patient's order and reporting.

The basics functions of a CPOE system are:

- Order creation, modification and discontinuation.
- Order dictionary management.
- Management of a patient's order profile.
- Order routing to various departments and programs that carry them out.
- Reporting and summarization.

CPOE offers several advantages over traditional paper-based systems, including process improvements, improved resource utilization, and clinical decision support. Firstly, CPOE improves the process of order writing by generating legible order that require less clarification. The order workflow is streamlined by eliminating valueless steps such as an order transcription and secondly, CPOE can improve resource utilization by modifying provider ordering.

Studies have shown reductions in charges after the implementation of CPOE systems. These savings results from a variety of features, including faster administration of medication, more effective monitoring, more appropriate test and medication choices, and reduced redundancy.

Figure 22 shows a patient's prescription details window that allow users to view or make changes to patient's existing prescription details.

Figure 22. Prescription details window

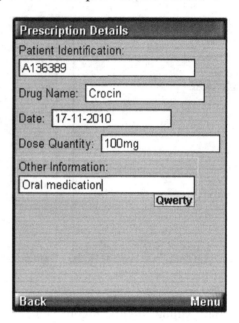

Admission, Discharge and Transfer System

The admission, discharge and transfer system allows users to create ADT records for the patient, check patient's existing ADT history and view/edit ADT records. It should be noted that for every admission record created, there are multiple transfers possible for a patient.

Figure 23 shows the patient ADT history. The user can either view or make changes to patient's specific ADT details and also add new ADT form.

- **Add New ADT Form:** Figure 24 shows the window created to add new ADT record for a patient.
- **View Existing ADT Record for a Patient:** Figure 25 shows an ADT record window that allow users to view or make changes to patient's ADT details.

Figure 26 shows a new transfer record window created when a patient is being transferred to another ward or room.

Figure 23. ADT main window

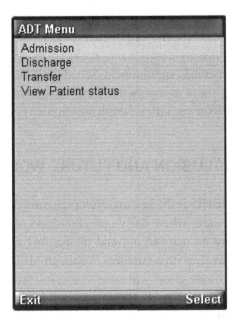

Figure 25. ADT record details window

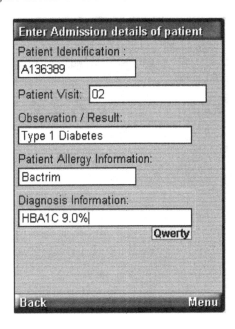

Figure 24. New ADT record window

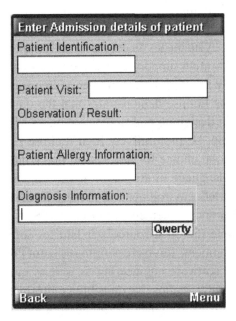

Figure 26. New transfer record window

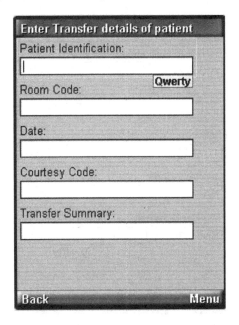

Figure 27 shows the patient transfer history for a specific admission number.

Figure 28 shows the transfer record details for the patient. The user has the option to edit and save the transfer details.

MEHIS EVALUATION

For evaluating the effectiveness of MEHIS, local health organizations were contacted to collect more information on the subsystems to have an

Figure 27. Patient transfer history window

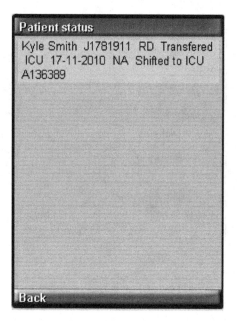

Figure 28. Transfer record details

effective and an efficient system design. A pharmacist from the National Skin Centre in Singapore, was interviewed to get an understanding of the procedure involved in prescribing a medicine and to map the business requirements for the system.

For the Appointment Scheduling System, a visit to a Raffles Medical Group Clinic, Singapore Toa Payoh Branch, allowed a better understanding on the necessary components required by a well-implemented Appointment Scheduling System. All of these components were integrated into MEHIS for a seamless health information system.

CONCLUSION AND FUTURE WORK

The MEHIS is the key to developing truly digital health care, where daily operations and record-keeping are carried out and maintained almost exclusively with computers. Although MEHIS is a health care priority in many countries, progress in implementing MEHIS has lagged behind the latest information technologies available. Furthermore, the diverse heterogeneous environments of many MEHIS make maintaining the systems very difficult.

The development and implementation of MEHIS is by no means an easy task, and requires the collaborative support of HCEs and governments to truly succeed. Using an open source software such as Eclipse demonstrates that the time and cost in designing and implementing an MEHIS need not be prohibitively high. Eclipse is seen as a good choice for any system that requires a multi-language, multi-platform, and multi-vendor supported environment. By using Eclipse for design and implementation of MEHIS, we can promote the open source standard for software development, thus increasing the usage of Eclipse in the software engineering as well as medical community. Eclipse's extensible architecture easily lends itself to the incorporation of individual components which can be updated easily to add new functionality.

Future work can involve extending the architecture and design of the MEHIS for increased security, privacy, and mobility. In addition, possible integration to the open source project, Eclipse Open Healthcare Framework (OHF) Project is possible. However, the success of MEHIS will depend on standardization of a common health information infrastructure, initially at the national level, so that the goal of a truly global MEHIS can be realized.

REFERENCES

Amoretti, M., & Zanichelli, F. (2009). The multi-knowledge service-oriented architecture: Enabling collaborative research for e-health. In *Proceedings of the 42nd Hawaii International Conference on System Sciences* (pp. 1-8).

Blobel, B. (2002). Architecture of secure portable and interoperable electronic health records. In. *Proceedings of the International Conference on Computational Science-Part, II*, 982–994.

Blobel, B. (2004). Authorisation and access control for electronic health record systems. *International Journal of Medical Informatics*, 251–257. doi:10.1016/j.ijmedinf.2003.11.018

Chen, Y., Tsai, F. S., & Chan, K. L. (2007). Blog search and mining in the business domain. In *Proceedings of the International Workshop on Domain Driven Data Mining* (pp. 55-60).

Chia, Y., Tsai, F. S., Ang, W. T., & Kanagasabai, R. (2011). Context-aware mobile learning with a semantic service-oriented infrastructure. In *Proceedings of the 7th International Symposium on Web and Mobile Information Services.*

D'Anjou, J., Fairbrother, S., Kehn, D., Kellerman, J., & McCarthy, P. (2005). *The Java developer's guide to Eclipse* (2nd ed.). Reading, MA: Addison-Wesley.

European Committee for Standardization. (1999). *Health informatics: Electronic healthcare record communication - Part 1: Extended architecture (Tech. Rep. No. CEN TC251/WG1)*. Brussels, Belgium: European Committee for Standardization.

Gunter, T. D., & Terry, N. P. (2005). The emergence of national electronic health record architectures in the United States and Australia: Models, costs, and questions. *Journal of Medical Internet Research*, 7(1), 3. doi:10.2196/jmir.7.1.e3

HL7. (2005). *Health level seven (HL7)*. Retrieved from http://www.hl7.org/

Hernando, M. E., Gómez, E. J., García-Olaya, A., Torralba, V., & Pozo, F. (2006). A mobile telemedicine workspace for diabetes management. In Micheli-Tzanakou, E. (Ed.), *Topics in biomedical engineering* (pp. 587–599). New York, NY: Springer.

Hu, W.-C., Zuo, Y., Chen, L., & Yang, H.-J. (2010). Contemporary issues in handheld computing research. *International Journal of Handheld Computing Research*, 1(1), 1–23. doi:10.4018/jhcr.2010090901

Jen, W.-Y., Chao, C.-C., Hung, M. C., Li, Y.-C., & Chi, Y. P. (2007). Mobile information and communication in the hospital outpatient service. *International Journal of Medical Informatics*, 76(8), 565–574. doi:10.1016/j.ijmedinf.2006.04.008

Kwee, A. T., & Tsai, F. S. (2009). Mobile novelty mining. *International Journal of Advanced Pervasive and Ubiquitous Computing*, 1(4), 43–68. doi:10.4018/japuc.2009100104

Rodriguez, M. D., Favela, J., Martinez, E. A., & Munoz, M. A. (2004). Location-aware access to hospital information and services. *IEEE Transactions on Information Technology in Biomedicine*, 8(4), 448–455. doi:10.1109/TITB.2004.837887

Thing, V., Subramaniam, P., Tsai, F. S., & Chua, T.-W. (2011). Mobile phone anomalous behaviour detection for real-time information theft tracking. In *Proceedings of the Second International Conference on Technical and Legal Aspects of the e-Society.*

Trappey, C. V., Trappey, A. J., & Liu, C. S. (2009). Develop patient monitoring and support system using mobile communication and intelligent reasoning. In *Proceedings of the IEEE International Conference on Systems, Man and Cybernetics* (pp. 1195-1200).

Tsai, F. S. (2006). *Object-oriented software engineering*. Singapore: McGraw-Hill.

Tsai, F. S. (2010). Security issues in e-healthcare. *Journal of Medical and Biological Engineering, 30*(4), 209–214. doi:10.5405/jmbe.30.4.04

Tsai, F. S. (2011). Web-based geographic search engine for location-aware search in Singapore. *Expert Systems with Applications, 38*(1), 1011–1016. doi:10.1016/j.eswa.2010.07.129

Tsai, F. S., Etoh, M., Xie, X., Lee, W.-C., & Yang, Q. (2010). Introduction to mobile information retrieval. *IEEE Intelligent Systems, 25*(1), 11–15. doi:10.1109/MIS.2010.22

Tsai, F. S., Han, W., Xu, J., & Chua, H. C. (2009). Design and development of a mobile peer-to-peer social networking application. *Expert Systems with Applications, 36*(8), 11077–11087. doi:10.1016/j.eswa.2009.02.093

Weiss, G. (2002). Welcome to the (almost) digital hospital. *IEEE Spectrum*, 44–49. doi:10.1109/6.988704

Yee, K. Y., Tiong, A. W., Tsai, F. S., & Kanagasabai, R. (2009). OntoMobiLe: A generic ontology-centric service-oriented architecture for mobile learning. In *Proceedings of the IEEE Tenth International Conference on Mobile Data Management: Systems, Services and Middleware, Workshop on Mobile Media Retrieval* (pp. 631-636).

This work was previously published in the International Journal of Handheld Computing Research (IJHCR), Volume 2, Issue 4, edited by Wen-Chen Hu, pp. 1-28, copyright 2011 by IGI Publishing (an imprint of IGI Global).

Chapter 15
Integration of Health Records by Using Relaxed ACID Properties between Hospitals, Physicians and Mobile Units like Ambulances and Doctors

Lars Frank
Copenhagen Business School, Denmark

Louise Pape-Haugaard
Aalborg University, Denmark

ABSTRACT

This paper describes an architecture for integrating both stationary health units like hospitals and group physicians with health records of mobile health units like ambulances and doctors at emergency call service. This paper focuses on how it is possible to have high availability in all the integrated health units and at the same time keep the consistency between the health records in the different locations at an acceptable level. In central databases the consistency of data is normally implemented by using the Atomicity, Consistency, Isolation, and Durability (ACID) properties of a Data Base Management System (DBMS) (Gray & Reuter, 1993). This is not possible if mobile databases are involved and the availability of data also has to be optimized. Therefore, this paper describes using relaxed ACID properties across different locations. The objective of designing relaxed ACID properties across different database locations is to make it possible for all the involved locations to operate in disconnected mode and at the same time give the users a view of the data that may be inconsistent across different locations but anyway better than the data in a centralized database with low availability for the users.

DOI: 10.4018/978-1-4666-2785-7.ch015

1. INTRODUCTION

Electronic Health Records (EHRs) have high requirements to storage and reusability of information. The stored information has to be available for health care professionals at any given time, and furthermore, the health care professionals also need to be allowed to store new information gathered in a patient-near setting. (Koch, 2006; Digital Health, 2008) To meet the requirements for storage and reusability EHRs are based on large databases. These databases are often created, maintained and used through a DBMS. When using DBMS to manage databases an important aspect involves transactions which are any logical operation on data and per definition database transactions must be atomic, consistent, isolated, and durable in order for the transaction to be reliable and coherent.

The ACID properties of a database are delivered by a DBMS to make database recovery easier and make it possible in a multi user environment to give concurrent transactions a consistent chronological view of the data in the database. The ACID properties are consequently important for users that need a consistent view of the data in a database. However, the implementation of ACID properties may influence performance and thereby slow down the availability of a system in order to guarantee that all users have a consistent view of data even in case of failures. In several situations, the availability and the response time will be unacceptable if the ACID properties of a DBMS are used without reflection. Especially, in the case of distributed and/or mobile databases where a failure in connections of a system should not prevent the system to operate in a meaningful way in so-called disconnected mode.

Information systems that operate in different locations can be integrated by using more or less common data and/or by exchanging information between the systems involved. In both situations, the union of the databases of the different systems may be implemented as a database with so-called relaxed ACID properties where temporary inconsistencies may occur in a controlled manner. However, when implementing relaxed ACID properties it is important that from a user's point of view it must still seem as if traditional ACID properties were implemented, which therefore will keep EHRS trustworthy for decision making.

In the following part of the introduction, we will give an overview of how relaxed ACID properties may be implemented and used to integrate EHRs from different types of health institutions.

1.1. Relaxed ACID Properties

The Atomicity property of a DBMS guarantees that either all the updates of a transaction are committed/executed or no updates are committed/executed. This property makes it possible to re-execute a transaction that has failed after execution of some of its updates. This property is especially important in replicated databases where inconsistency will occur if only a subset of data is replicated. The Atomicity property of a DBMS is implemented by using a DBMS log file with all the database changes made by the transactions. The global Atomicity property of databases with relaxed ACID properties is implemented by using compensatable, pivot and retriable subtransactions in that order as explained in Section 2.1. By applying these subtransactions it is allowed to commit/execute only part of the transaction and still consider the transaction to be atomic.

As explained in Section 2.2 the global Consistency property is not defined in databases with relaxed ACID properties because normally such databases are inconsistent and this inconsistency may be managed in the same way as the relaxed Isolation property.

The Isolation property of a DBMS guarantees that the updates of a transaction cannot be seen by other concurrent transactions until the transaction is committed/executed. That is the inconsistencies cause by a transaction that has not executed all its updates cannot be seen by other transactions. The Isolation property of a DBMS may be implemented by locking all records used by a transaction. That is the locked records cannot be used by other transactions before the locks are released when the transaction is committed. The global Isolation property of databases with relaxed ACID properties is implemented by using countermeasures against the inconsistencies/anomalies that may occur. This is explained in more details in Section 2.3.

The Durability property of a DBMS guarantees that the updates of a transaction cannot be lost if the transaction is committed. The Durability property of a DBMS is implemented by using a DBMS log file with all the database changes made by the transactions. By restoring the updates of the committed transactions it is possible to recover a database even in case it is destroyed. The global Durability property of databases with relaxed ACID properties is implemented by using the local Durability property of the local databases involved.

1.2. Health Units and Need for Information

A utilized tool for health care professionals is health records to document the clinical process, i.e., documentation of diagnostics, treatment, medication, progress, para-clinical recordings

etc. (Kalra, 2009) Furthermore due to needs for consecutively continuity of care the health care professional also use health records as a communicative tool to enhance patient treatment. It is a commonly accepted hypothesis that having relevant patient information at health care professionals' disposal will optimize patient treatment (Dobrey et al., 2009; Black et al., 2011). This accepted hypothesis has been key in developing EHRs and mobile handheld devices to capture patient information on site, i.e., in patient-near settings. EHRs and mobile handheld devices are used in several settings in health care, and in this paper we define a health unit as a generalization of hospitals, nursing centers, group physicians, emergency call service centers and mobile as ambulances, handheld devices, emergence service doctors, and hospital doctors where information for the EHRs is generated and/or the information is re-used. This generalization is illustrated in the subtype diagram of Figure 1.

1.3. Objective

The objective in this paper is to describe an architecture for how it is possible to implement the replication of EHRs with relaxed ACID properties.

By describing this architecture we enable the possibility for stationary health units and mobile handheld devices to operate in disconnected mode

Figure 1. Subtype diagram illustrating relations between different types of health units

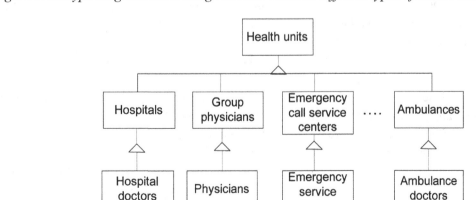

to central EHR databases and at the same time ensuring trustworthy information. These benefits are achieved through automatic replication of EHRs to the locations where they are needed by making it possible for all types of locations to operate in disconnected mode and the relaxed ACID properties make the replicated EHRs trustworthy for decision making.

In order to optimize EHR replication it may be practical that the EHRs of a person are stored in a primary location for that person. From there other stationary health units may subscribe the health records of the person. In contrast to stationary health units, mobile health units may be within the jurisdiction of a stationary health unit and in this case it should only subscribe and deliver health records to its superior stationary health unit from where the EHRs may be redistributed to the primary location of the person.

The paper is organized as follows: First, we will describe the transaction model used in this paper. In Section 3, we will describe the most important asynchronous replication methods and in Section 4 we will describe how to implement these by using e.g., SOA services. Next, we will describe the most important integration architecture for distributed EHRs where replication methods with relaxed ACID properties are used to optimize both availability and consistency of the data. The special needs of mobile health records are taken into account. Finally, we have conclusions and future work.

Related Research. There are several mobile transaction models used, however we have decided to use the countermeasure transaction model (Frank & Zahle, 1998) which is described in Section 2. This model owes many of its properties to Garcia-Molina and Salem (1987), Mehrotra, Rastogi, Silberschatz, and Korth (2002), Weikum and Schek (1992), and Zhang, Nodine, Bhargava, and Bukhres (1994). The replication designs used in this paper are described in more details by Frank (2005, 2010).

Frank and Munck (2008) have given an overview and an evaluation of the integration architectures we are mixing and using in this paper. Frank and Andersen (2010) have given an evaluation of different methods to store incompatible heterogeneous health records in a common database.

2. THE TRANSACTION MODEL

A multi-database is a union of local autonomous databases. Global transactions (Gray & Reuter, 1993) access data located in more than one local database. In recent years, many transaction models have been designed to integrate local databases without using a distributed DBMS. The countermeasure transaction model (Frank & Zahle, 1998) has, among other things, selected and integrated properties from these transaction models to reduce the problems, as reliability and consistency, caused by the missing ACID properties in a distributed database that is not managed by a distributed DBMS. In the countermeasure transaction model, a global transaction involves a root transaction (client transaction) and several single site subtransactions (server transactions). Subtransactions may be nested transactions, i.e., a subtransaction may be a parent transaction for other subtransactions. All communication with the user is managed from the root transaction, and all data is accessed through subtransactions. The following subsections will give a broad outline of how relaxed ACID properties have to be implemented.

2.1. The Atomicity Property

An updating transaction has the atomicity property and is called atomic if either all or none of its updates are executed. In the countermeasure transaction model, the global transaction is partitioned into the following types of subtransactions executed in different locations:

The pivot subtransactions manage the atomicity of the global transactions. A global transaction is said to be committed when the pivot subtransaction is committed locally. Therefore, a global transaction can only have one pivot subtransaction. If the pivot subtransaction aborts, all updates of other subtransactions must be compensated/removed in order to obtain the atomicity property.

The compensatable subtransactions ensure the possibility for compensation. Compensatable subtransactions must always be executed before the pivot subtransaction is executed to make it possible to compensate them if the pivot subtransaction cannot be committed. A compensatable subtransaction may be compensated by executing a so-called compensating subtransaction that removes the database changes made by the corresponding compensatable subtransaction. In integrated EHR systems, we will recommend to use compensatable subtransactions to make local EHR updates that later must be controlled for conflicts with primary copy EHRs before they can be committed globally.

The retriable subtransactions are designed in such a way that the execution is guaranteed to commit locally (sooner or later) if the pivot subtransaction has been committed. Therefore, it is possible (without violating the atomicity property) to execute retriable subtransactions after the global transaction is committed by the pivot subtransaction. Retriable subtransactions may for example be used to update replicated data or execute compensating subtransactions.

The global atomicity property is implemented by executing the compensatable, pivot and retriable subtransactions of a global transaction in that order. For example, if the global transaction fails before the pivot has been committed, it is possible to remove the updates of the global transaction by compensation. If the global transaction fails after the pivot has been committed, the remaining retriable subtransactions will be (re)executed automatically until all the updates of the global transaction have been committed.

2.2. The Consistency Property

A database is consistent if its data complies with the consistency rules of the database. If the database is consistent both when a transaction starts and when it has been completed and committed, the execution has the consistency property. Transaction consistency rules may be implemented as a control program that rejects the commitment of transactions, which do not comply with the consistency rules.

The above definition of the consistency property is not useful in distributed databases with relaxed ACID properties because such a database is almost always inconsistent. However, a distributed database with relaxed ACID properties should have asymptotic consistency, i.e., the database should converge towards a consistent state when all active transactions have been committed/compensated. Therefore, the following property is essential in distributed databases with relaxed ACID properties:

If the database is asymptotically consistent when a transaction starts and also when it has been committed, the execution has the relaxed consistency property.

2.3. The Isolation Property

The isolation property is normally implemented by using long duration locks, which are locks that are held until the global transaction has been committed (Frank & Zahle, 1998). In the countermeasure transaction model, long duration locks cannot instigate isolated global execution as retriable subtransactions may be executed after the global transaction has been committed in the pivot location. Therefore, short duration locks are used, i.e., locks that are released immediately after a subtransaction has been committed/aborted locally. To ensure high availability in locked data, short duration locks should also be used in compensatable subtransactions, just as locks should be released before interaction with a user. This is not a problem in the countermeasure transaction model as the traditional isolation property in retriable subtransactions is lost anyway. If only short duration locks are used, it is impossible to block data. (Data is blocked if it is locked by a subtransaction that loses the connection to the "coordinator" (the pivot subtransaction) managing the global commit/abort decision). When transactions are executed without isolation, the so-called isolation anomalies may occur. In the countermeasure transaction model, relaxed isolation can be implemented by using countermeasures against the isolation anomalies. If there is no isolation and the atomicity property is implemented, the following classical isolation anomalies may occur (Berenson et al., 1995; Breitbart et al., 1992).

- **Lost Update Anomaly:** A situation where a first transaction reads a record for update without using locks. Subsequently, the record is updated by another transaction. Later, the update is overwritten by the first transaction. In extended transaction models, the lost update anomaly may be prevented, if the first transaction reads and updates the record in the same subtransaction using local ACID properties. Unfortunately, the read and the update are sometimes executed in different subtransactions belonging to the same parent transaction. In such a situation, a second transaction may update the record between the read and the update of the first transaction.

- **Dirty Read Anomaly:** A situation where a first transaction updates a record without committing the update. Subsequently, a second transaction reads the record. Later, the first update is aborted (or committed), i.e., the second transaction may have read a non-existing version of the record. In extended transaction models, this may happen when the first transaction updates a record by using a compensatable subtransaction and later aborts the update by using

a compensating subtransaction. If a second transaction reads the record before it has been compensated, the data read will be "dirty".

- **Non-Repeatable Read Anomaly, or Fuzzy Read:** A situation where a first transaction reads a record without using locks. Later, the record is updated and committed by a second transaction before the first transaction has been committed. In other words, it is not possible to rely on the data that have been read. In extended transaction models, this may happen when the first transaction reads a record that later is updated by a second transaction, which commits the update locally before the first transaction commits globally.

- **Phantom Anomaly:** A situation where a first transaction reads some records by using a search condition. Subsequently, a second transaction updates the database in such a way that the result of the search condition is changed. In other words, the first transaction cannot repeat the search without changing the result. Using a data warehouse may often solve the problems of this anomaly (Frank, 2003).

The countermeasure transaction model (Frank & Zahle, 1998) describes general countermeasures that reduce the problems of the anomalies. In Section 5, we will describe in details how some of these may be used to prevent or reduce the problems of isolation anomalies in the recommended integration architectures for EHRs.

2.4. The Durability Property

Updates of transactions are said to be durable if they are stored in a stable manner and secured by a log recovery system. In case a global transaction has the atomicity property (or relaxed atomicity), the global durability property (or relaxed durability property) will automatically be implemented, as it is ensured by the log-system of the local DBMS systems (Breitbart et al., 1992).

3. MOST IMPORTANT ASYNCHRONOUS REPLICATION METHODS FOR DISCONNECTED SYSTEMS

The n-safe and quorum safe replication designs are not suited for systems where it should be possible to operate in disconnected mode. Therefore, we will only describe the asynchronous replication methods in the following. For all these replication methods it is possible to implement them by using compensatable and retriable services from other locations.

3.1. The Basic 1-Safe Design

In Figure 2, the basic 1-safe design (Gray & Reuter, 1993) is illustrated. In the basic 1-safe design the primary transaction manager goes through the standard commit logic and declares completion when the commit record has been written to the

local log. In the basic 1-safe design, the log records are asynchronously spooled to the locations of the secondary copies. In case of a primary site failure in the basic 1-safe design, production may continue by selecting one of the secondary copies but this is not recommended in distributed ERP systems as it may result in lost transactions.

3.2. The 0-Safe Design with Local Commit

In Figure 3 the 0-safe design is illustrated. The 0-safe design with local commit is defined as n table copies in different locations where each transaction first will go to the nearest database location, where it is executed and committed locally.

If the transaction is an update transaction, the transaction propagates asynchronously to the other database locations, where the transaction is re-executed without user dialog and committed locally at each location. This means that all the table copies normally are inconsistent and not up to date under normal operation. The inconsistency must be managed by using countermeasures against the isolation anomalies. For example, to

Figure 2. The basic 1-safe database design

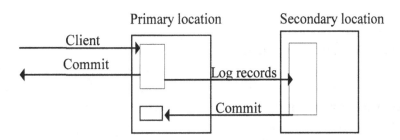

Figure 3. The 0-safe database design

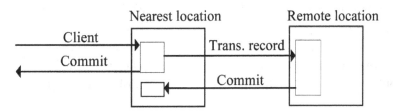

prevent lost updates in the 0-safe design, all update transactions must be designed to be commutative (Frank & Zahle, 1998).

3.3. The 1-Safe Design with Commutative Updates

The 1-safe design can transfer updates to the secondary copies in the same way as the 0-safe design. In such a design, lost transactions cannot occur because the transaction transfer of the 0-safe design makes the updates commutative. The properties of this mixed design will come from either the basic 1-safe or the 0-safe design. The 1-safe design with commutative updates does not have the high update performance and capacity of the 0-safe design. On the other hand, in this design the isolation property may be implemented automatically as long as the primary copy does not fail. Normally, this makes it much cheaper to implement countermeasures against the isolation anomalies, because it is only necessary to secure that "the lost transactions" are not lost in case of a primary copy failure.

3.4. The 0-Safe Design with Primary Copy Commit

The "0-safe designs with primary copy commit" is a replication methods where a global transaction first must update the local copy closest to the user by using a compensatable subtransaction. Later, the local update may be committed globally by using a primary copy location. If this is not possible, the first update must be compensated. The primary copy may have a version number that also are replicated. This may be used to control that an updating user operates on the latest version of the primary copy before an update is committed globally in the location of the primary copy. If an update cannot be committed globally in the primary copy location, the latest version of the primary copy should be send back to the updat-

ing user in order to repeat the update by using the latest version of the primary copy.

4. IMPLEMENTATION OF INTERNET REPLICATION SERVICES WITH RELAXED ACID PROPERTIES

In order to implement flexibility (fault tolerance in case a back up unit is used) it is not acceptable that the different health units communicate directly with each other. All communication between local internet systems must be executed by applications offered as e.g., SOA services. Therefore, each health unit should offer the following types of services as these are the basis for implementing the Atomicity property:

- Read only services that are used when a health unit wants to read data managed by another health unit.
- Compensatable update services that are used when a health unit wants to make compensatable updates in tables managed by another health unit.
- Retriable update services that are used when a health unit wants to make retriable updates in tables managed by another health unit.

However, these three services may also be used to build the replication services described in Section 3. This is summed up and described in the following:

- All types of 1-safe designs can be implemented if the primary copy location uses the retriable services of the secondary copy locations that subscribe to the patients EHR.
- A retriable service call to a primary copy location can implement the 0-safe design with local commit if all the secondary lo-

cations of a patient subscribe to the EHR from the primary copy location of the patient (it is assumed that any citizen/patient at any time has one and only one "home hospital" or "hospital of residence" that should be responsible as a primary copy location).

• A compensatable update service call to a primary copy location from where retriable service to all the secondary copy locations may be used to implement the o-safe design with primary copy commit.

5. DESCRIPTION OF THE MOST IMPORTANT INTEGRATION ARCHITECTURES FOR DISTRIBUTED EHRS

Frank and Munck (2008) have a description and evaluation of seven different integration architectures for distributed health records. Among them, two are best suited for our purposes. Both store EHR information locally where they are needed and in both architectures it is possible for a health unit to subscribe to EHRs from another health unit. In both architectures, it is assumed that EHRs from other locations should be replicated to databases, where the patient is under current or planned treatment. A given patient's EHR is only replicated to locations subscribing to this exact patient's EHRs. The main difference between these two architectures is that in the "decentralized

redistribution" it is possible for any health unit to subscribe to EHRs from any other health unit. In the other "centralized redistribution" it is only a central database location that may subscribe to EHRs from any other health unit. Another difference is that it is assumed that the "centralized redistribution" architecture may have some ERH services stored and executed in the central location and thus save software costs by not having all types of EHR services executed in all locations that need the service. It is assumed that the "decentralized redistribution" architecture does not use such a possibility and therefore has full local autonomy in selecting (and paying for) software.

In the following the two integration architectures are described and evaluated in more details.

5.1. Decentralized Redistribution Approach

In this solution, any location that needs EHRs from a patient should subscribe to the patient's EHRs stored in the home location of the patient/ hospital of residence. At the same time the hospital of residence should subscribe to all the subscribing locations for new or changed EHRs for the patient. This is illustrated in Figure 4. We will recommend that diagnostics, surgeries, treatment in general, prescriptions and other decisions that may cause severe conflicts with other stored EHRs are replicated using the 0-safe design with deferred primary copy commit in the hospital of residence, and secondary copies in all the loca-

Figure 4. Decentralized redistribution

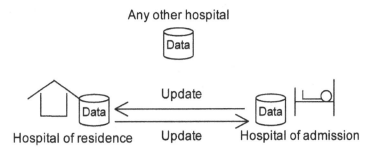

tions subscribing to this patient's EHRs. All other types of the EHR information can be replicated using the 0-safe design with local commit as these health records only need local validation.

5.2. Centralized Redistribution Approach

In this solution, a central database subscribe to all locations for new or changed EHRs. At the same time, may any location that needs EHRs from a patient subscribe to the patient's EHRs stored in the central location. This is illustrated in the Figure 5. Also in this solution we will recommend that diagnoses, surgeries, treatment in general, prescriptions and other decisions that may cause severe conflict with other stored EHRs are replicated using the 0-safe design with deferred primary copy commit in the hospital of residence and secondary copies in all the locations subscribing to this patient's EHRs. All other types of the EHR information can be replicated using the 0-safe design with local commit as these health records only need local validation. If SOA is used, the central location can be used as a real time back up database for the subscribing hospitals without any change.

5.3. Evaluation and Recommendation of the Solutions

Both solutions have the same read performance as the patient's EHR may be read locally. Both solutions also have the same high tolerance against isolation anomalies. However, the solution with centralized redistribution has less autonomy than the solution with decentralized redistribution because it is more likely that hospitals with centralized redistribution also are forced to use centrally developed EHR functionality for economical reasons. Therefore, the centralized redistribution solution also has lesser software costs than the decentralized redistribution solution where all the EHR functionality is executed locally.

As both of the described integration architectures have exactly the same problems with isolation anomalies we can describe a common solution for dealing with them.

EHRs are very special compared with other types of application records because they are normally not updated. Instead new patient relevant information is added. Therefore the lost update anomaly cannot occur. That is the cheap basic 1-safe design can be used instead of the more complex 1-safe design with commutative updates.

Figure 5. Centralized redistribution

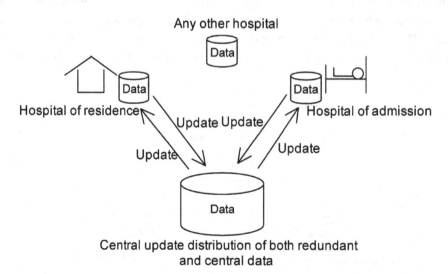

The dirty read anomaly may occur when the 0-safe design with primary copy commit is used as a local update may be rejected in the primary copy location and later compensated by a retriable subtransaction. However, if the dirty data is marked as "dirty" and demarked by a retriable committing or compensating subtransactions, then it is possible to warn the users as long as they are using dirty data.

The non-repeatable read anomaly may occur when a secondary copy is read after the primary copy has been updated and before the secondary copy has been updated. However, this situation is not normal as EHRs normally are not updated. Anyway, the user may be warned by using the AI (Artificial Intelligence) countermeasure as described below.

The phantom anomaly may occur when retriable updates from another location are delayed. This may be prevented by only reading EHRs from a home location of the patient. However, this is not realistic and therefore, we will recommend using simple decision support tools or AI to control for conflicts each time EHRs from another location are stored. In this way, it is possible to warn against a possible conflict even though it may be too late. Especially disconnected mobile users may experience this situation.

6. DISCUSSION OF IMPLICATIONS AND POTENTIALS WITH THE PROPOSED ARCHITECTURAL SOLUTION

In this section we wish to discuss the implications and potentials with the proposed architectural solution.

We do recommend a mix of these two architectures, as argued in previous section. However, if we consider the implications following an implementation of our solution to the problem of how information can be reliable in disconnected modes, we need to acknowledge that in practice, it may be a political must to select the solution with decentralized redistribution when hospital locations with different leadership are to be integrated even though this solution has higher costs.

In contrast to the political agenda, regional hospitals or chains of hospitals may agree on more corporations and select the cheaper solution with centralized redistribution. This contrast occurs because different hospitals typically have autonomy in selecting EHR systems (Frank & Andersen, 2010), and thereby also autonomous economics. The implications mentioned before will influence the strategies for implementing the proposed architecture, and awareness of cost influence ought to be included and also compared. Anyway, if we consider the potentials for implementing our solution – a solution ensuring trustworthy data despite disconnected modes, we can see that clinicians can benefit from trustworthy ubiquitous information regardless of this information is accessed from stationary health units or mobile health units. Even though mobile health units are special as they normally relate to a superior health unit from where they can subscribe to the needed EHRs. That is the mobile units may use a special case of centralized redistribution, where they only can subscribe to EHRs stored in the superior health unit. On the other hand, if an ambulance often is driving patients from one hospital to another it may be convenient that the ambulance may subscribe and deliver EHRs to any hospital that it operate for and this can be viewed as a special case of the decentralized redistribution solution.

In general there has to our knowledge not been made any central or national decisions about how to integrate mobile health units. However, we see a potential in integrating mobile health units in the same way as the stationary health units. That is it is up to both the stationary and the mobile health units to implement the benefits of using relaxed ACID properties. Therefore, it should at least in theory be possible for all users to be mobile as they may be viewed as any other location that are integrated with other locations by using relaxed ACID properties.

7. CONCLUSION AND FUTURE RESEARCH

In this paper, we have described an architecture for integrating the health records from stationary health units like hospitals and group physicians with mobile health units like ambulances and emergency service doctors. The proposed architecture enables the possibility for stationary health units and mobile handheld devices to operate in disconnected mode to both central and distributed EHR databases and at the same time ensuring trustworthy information in order for the clinicians to take and make reliable decisions regarding the patient's treatment. We have also described how to deal with problems as isolation anomalies across different health units. However, there are more types of anomalies than the classical types described in this paper (Frank, 2010). Anyway, the theory of these anomalies is not consolidated, and therefore, we have not analyzed how to handle these. Furthermore, we have described how the existing theory of relaxed ACID properties may be used to improve the integration of distributed EHRs. This is a solution proposal which acknowledge the need for solutions and supporting architectures which can include already implemented systems.

ACKNOWLEDGMENT

This research is partly funded by the 'Danish Foundation of Advanced Technology Research' as part of the 'Third Generation Enterprise Resource Planning Systems' (3gERP) project, a collaborative project between Department of Informatics at Copenhagen Business School, Department of Computer Science at Copenhagen University, and Microsoft Business Systems. Partly funded by Digital Health, Denmark.

REFERENCES

Berenson, H., Bernstein, P., Gray, J., Melton, J., O'Neil, E., & O'Neil, P. (1995). A critique of ANSI SQL isolation levels. In *Proceedings of the ACM SIGMOD International Conference on Management of Data*, San Jose, CA (pp. 1-10).

Black, A. D., Car, J., Pagliari, C., Anandan, C., Cresswell, K., & Bokun, T. (2011). The impact of ehealth on the quality and safety of health care: A systematic overview. *PLoS Medicine*, 8(1). doi:10.1371/journal.pmed.1000387

Breitbart, Y., Garcia-Molina, H., & Silberschatz, A. (1992). Overview of multidatabase transaction management. In. *Proceedings of the Conference of the Centre for Advanced Studies on Collaborative Research*, 2, 23–56.

DigitalHealth. (2008). *Projectdescription in Danish of shared medication record*. Retrieved from http://www.sdsd.dk/det_goer_vi/Faelles_Medicinkort.aspx

Dobrev, A., Jones, T., Stroetmann, K., Vatter, Y., & Peng, K. (2009). *The socio-economic impact of interoperable electronic health record (EHR) and ePrescribing systems in Europe and beyond*. Retrieved from http://www.ehealthnews.eu/publications/latest/1839-the-socio-economic-impact-of-interoperable-ehr-and-eprescribing-systems-in-europe-and-beyond

Frank, L. (2003, July 1-2). Data cleaning for datawarehouses built on top of distributed databases with approximated acid properties. In *Proceedings of the International Conference on Computer Science and its Applications*, San Diego, CA (pp. 160-164).

Frank, L. (2005). Replication methods and their properties. In Rivero, L. C., Doorn, J. H., & Ferraggine, V. E. (Eds.), *Encyclopedia of database technologies and applications*. Hershey, PA: IGI Global. doi:10.4018/978-1-59140-560-3.ch092

Frank, L. (2010). *Design of distributed integrated heterogeneous or mobile databases*. Saarbrücken, Germany: Lambert Academic Publishing.

Frank, L., & Andersen, S. K. (2010). Evaluation of different database designs for integration of heterogeneous distributed electronic health records. In *Proceedings of the IEEE/ICME International Conference on Complex Medical Engineering* (pp. 204-209).

Frank, L., & Mukherjee, A. *(2002, June). Distributed electronic patient encounter with high performance and availability. In* Proceedings of the 15th IEEE Symposium on Computer-Based Medical Systems *(pp. 373-376)*.

Frank, L., & Munck, S. (2008). An overview of architectures for integrating distributed electronic health records. In *Proceeding of the 7th International Conference on Applications and Principles of Information Science* (pp. 297-300).

Frank, L., & Zahle, T. U. (1998). Semantic acid properties in multidatabases using remote procedure calls and update propagations. *Software, Practice & Experience, 28*(1), 77–98. doi:10.1002/(SICI)1097-024X(199801)28:1<77::AID-SPE148>3.0.CO;2-R

Garcia-Molina, H., & Salem, K. (1987). Sagas. In *Proceedings of the SIGMOD Conference on Management of Data* (pp. 249-259).

Gray, J., & Reuter, A. (1993). *Transaction processing: Concepts and techniques*. San Francisco, CA: Morgan Kaufmann.

Kalra, D., Lewalle, P., Rector, A., Rodrigues, J. M., Stroetmann, K., Surjan, G., et al. (2009). *Semantic interoperability for better health and safer healthcare*. Retrieved from http://ec.europa.eu/information.../health/.../2009semantic-health-report.pdf

Koch, S. (2006). Home Telehealth –Current state and future trends. *International Journal of Medical Informatics, 8*, 565–576. doi:10.1016/j.ijmedinf.2005.09.002

Mehrotra, S., Rastogi, R., Silberschatz, A., & Korth, H. F. (2002). A transaction model for multidatabase systems. In *Proceedings of the 12th International Conference on Distributed Computing Systems* (pp. 56-63).

Weikum, G., & Schek, H. J. (1992). Concepts and applications of multilevel transactions and open nested transactions. In *Proceedings of the Database Transaction Models for Advanced Applications* (pp. 515-553).

Zhang, A., Nodine, M., Bhargava, B., & Bukhres, O. *(1994). Ensuring relaxed atomicity for flexible transactions in multidatabase systems. In* Proceedings of the ACM SIGMOD International Conference on Management of Data *(p. 78)*.

This work was previously published in the International Journal of Handheld Computing Research (IJHCR), Volume 2, Issue 4, edited by Wen-Chen Hu, pp. 29-41, copyright 2011 by IGI Publishing (an imprint of IGI Global).

Section 6
Pervasive Computing

Chapter 16
DSOA:
A Service Oriented Architecture for Ubiquitous Applications

Fabricio Nogueira Buzeto
Universidade de Brasília (UnB), Brazil

Carlos Botelho de Paula Filho
Universidade de Brasília (UnB), Brazil

Carla Denise Castanho
Universidade de Brasília (UnB), Brazil

Ricardo Pezzuol Jacobi
Universidade de Brasília (UnB), Brazil

ABSTRACT

Ubiquitous environments are composed by a wide variety of devices, each one with different characteristics like communication protocol, programming and hardware platforms. These devices range from powerful equipment, like PCs, to limited ones, like cell phones, sensors, and actuators. The services provided by a ubiquitous environment rely on the interaction among devices. In order to support the development of applications in this context, the heterogeneity of communication protocols must be abstracted and the functionalities dynamically provided by devices should be easily available to application developers. This paper proposes a Device Service Oriented Architecture (DSOA) as an abstraction layer to help organize devices and its resources in a ubiquitous environment, while hiding details about communication protocols from developers. Based on DSOA, a lightweight middleware (uOS) and a high level protocol (uP) were developed. A use case is presented to illustrate the application of these concepts.

DOI: 10.4018/978-1-4666-2785-7.ch016

1. INTRODUCTION

The presence of computer devices endowed with processing power and communication capabilities grows every day. Taking advantage of these capabilities in order to assist the user in its tasks demanding as minimum attention as possible is the basis of ubiquitous computing (Weiser & Brow, 1995). A smart space (Weiser, 1993) is an environment where the resources made available by the devices are organized and coordinated in order to provide intelligent services to the users.

Such intelligence is implemented in the smart space through applications. In order to abstract and simplify the complexity for building these applications, it is common place the adoption of middlewares that facilitate the implementation of smart spaces. We can highlight some important requirements (Abowd, Atkeson, & Essa, 1998) that must be addressed by middlewares (Bernstein, 1996) for ubicomp:

- Limited devices are part of the smart space, so its limitations like CPU, memory, bandwidth and battery life must be considered. Such issues can be seen in middlewares like the MundoCore (Aitenbichler, Kangasharju, & Mühlhäuser, 2007).
- Interaction details on how applications and resources cooperate must be addressed. Not only synchronous but asynchronous communication must be taken into account. An example of a middleware in the literature that focus on these issues is MoCA (Sacramento et al., 2005). Communication can occur not only in small messages, but also can be streamlined as in the MediaBroker project (Modahl, Bagrak, Wolenetz, Hutto, & Ramachandran, 2004).
- Platform heterogeneity among devices. The hardware and software must be considered in order to allow the integration of as many devices as possible to the smart space. This type of concern is observed in projects like

the MundoCore (Aitenbichler et al., 2007) and WSAMI (Issarny, Sacchetti, Chibout, Dalouche, & Musolesi, 2005) which provide solutions to multiple programming platforms.

Many middleware solutions, like the WSAMI project (Issarny et al., 2005) and the Home-SOA (The OSGi Alliance, 2009) utilize a SOA (MacKenzie, 2006) based solution. The use of SOA assists in the abstraction of the smart space functionalities in order to simplify the development of smart applications. One problem with a pure SOA approach lies in the fact that it does not address some specific problems of ubicomp environments. For example, SOA does not define the way services are accessed, which is an important aspect in the ubicomp context.

This work proposes an extension of the SOA architecture, denominated DSOA (Device Service Oriented Architecture), to model a smart space taking into account the requirements outlined above. Founded on DSOA, we also developed a lightweight multi-platform communication interface, called uP, and the middle-ware uOS to support the development of smart applications. To illustrate some characteristics of the proposed model a use case, named Hydra Application, is also shown.

This paper is organized as follows. Some related works are presented in Section 2 while Section 3 describes the DSOA architecture. The protocol uP and the middleware uOS are presented in Sections 4 and 5 respectively. Section 6 brings into focus the Hydra Application. Some results and final considerations are addressed in Sections 7 and 8.

2. RELATED WORK

A ubicomp environment is permeated by applications that may run in a variety of devices, including static and mobile ones. In order to easy the development of this kind of software many initiatives can be found in the literature (Garlan, 2002; Cerqueira, 2001; Helal et al., 2005; Brumitt, Meyers, Krumm, Kern, & Shafer, 2000). Among these, we highlight four of them which achieve relevant contribution towards the three requisites presented on section 1.

2.1. MundoCore

The MundoCore (Aitenbichler et al., 2007) middleware focuses on the development of applications for limited devices. It proposes the creation of an environment where applications exchange data through service objects. Such objects are available through a distributed environment and are accessed using application labels. The MundoCore middleware is available in three versions (Java, C++ and Phyton) and service objects can be serialized using either XML or a binary format.

2.2. MoCA

The MoCA (Mobile Collaboration Architecture) (Sacramento et al., 2005) project focuses on context aware applications. The middleware captures QoS, localization, connectivity and battery information from devices and make them available to applications. Such information is transmitted through asynchronous event-base communication. Events are notified through XML based messages using various types of channels like SMS (Short Message Service), TCP (Transmission Control Protocol), UDP (User Datagram Protocol), JMS (Java Message Service) and WAP (Wireless Application Protocol).

2.3. MediaBroker

The MediaBroker (Modahl et al., 2004) is a D-Stampede (Adhikari, Paul, & Ramachandran, 2002) based middleware with focus on efficient stream data transport. It provides a structure that allows applications do communicate through data channels. Such channels are associated with types which allow consumers (data "sinks") to find and choose which producers (data "sources") that best suit their needs. Along with the typed data channels the middleware provides ways to specify conversion algorithms between types in order to allow the creation of virtual data conversion channels.

2.4. WSAMI

The WSAMI (Issarny et al., 2005) is a SOAP based middleware with focus on PCs and Palm devices in a distributed environment. The middleware provides a P2P algorithm to easy the service discovery in the smart space. The use of a web-service architecture allows an interoperable access to services available in the devices.

2.5. Comparative Analysis

Among these initiatives we can highlight some points of interest, like how the environment capabilities are accessed (environment vision), platform support and communication. With respect to the architecture, all the above projects are distributed middlewares, reinforcing the fact that the volatile characteristics of a ubicomp environment influence the way devices are organized.

Environment vision. The way each architecture provides access to the capabilities available in the environment defines how applications interact and cooperate. The use of services is a clear tendency among most of the projects, although we can observe some particularities (Table 1).

Table 1. Environment vision

Project	Vision
Media Broker	Streams
WSAMI	Synchronous Services
MundoCore	Objects
MoCA	Events

- The MediaBroker provides access to services through typed data streams. This approach makes access to continuum data easy and is adequate to large amount of data or stream applications like audio and video.
- A synchronous access is provided by WSAMI middleware through the web-service architecture. This approach is very similar to the functional programming model and applies to data queries and command requests.
- MundoCore makes the environment capabilities available through labeled service objects which allow applications to share state information through these objects.
- Event notifications are used by the MoCA project which allows applications to be aware of environment changes as they occur.

Platform. An important characteristic of a middleware is the software and hardware platform it provides support for. With respect to this criterion, four main characteristics were observed and are summarized in Table 2.

- **Hardware Platform:** Considering computing power the middlewares that were analyzed target three different kinds of hardware platforms:
  - **High Platforms:** Devices with higher performance and higher energy consumption, like PCs and laptops.
  - **Intermediate Platforms:** Devices like handhelds and smart-phones which are in the middle of our scale sharing characteristics with both classes.
  - **Limited Platforms:** Devices with limitations in processing power, battery, memory and connectivity like most of mobile devices (soft phones, watches and jewelry), sensors and actuators.
- **Development Platform:** Refers to which development platform was supported by each project.

Communication. Communication plays an important role in a ubiquitous environment. The way each initiative establishes how to exchange data among applications can limit which platforms and capabilities will be available in the smart space. Table 3 summarizes the communication formats chosen by each project. As can be observed, most of them have elected a communication format based on XML, for it allows the interoperation among different platforms, which is an essential aspect of ubicomp environments.

3. THE DSOA ARCHITECTURE

As stated in Section 1, a smart space may be composed by a large and dynamic variety of devices. Applications must be aware of such devices, the capabilities they make available in

Table 2. Middlewares platform support

Project	High Platforms	Intermediate Platforms	Limited Platforms	Development Platforms
Media Broker	Yes	No	No	C++
WSAMI	Yes	Yes	No	Java, C++
MundoCore	Yes	Yes	Yes	Java, C++,Phyton
MoCA	Yes	No	No	Java

Table 3. Communication format

Project	Communication Format
Media Broker	Binary
WSAMI	SOAP/HTTP
Mundo	XML, Binary
MoCA	XML

the environment as well as the user behavior and activity in order to decide the best actions to take. A single smart space can have many applications in place which will have access to the same group of available devices. The goal of an architecture in this scenario is to provide a common way to organize applications and the capabilities provided by the devices in order to coordinate the inherent dynamics of the environment.

The SOA concepts (MacKenzie, 2006) are well suited for this scenario where distributed applications must exchange common interests. However, SOA concerns to a more generic distributed environment and does not specify particular issues of ubicomp environments. The DSOA (Buzeto, Filho, Castanho, & Jacobi, 2010) (Device Service Oriented Architecture) presents an extension of the SOA architecture focusing on two main points:

- **Modeling the Smart Space:** SOA suggests that all capabilities are available to the applications and must be accessed through services. In a ubicomp environment, however, some entities may present a set of interrelated capabilities, like those tied to physical interface, for instance. In order to better model these features, DSOA proposes an extension of SOA concepts to provide a more cohesive way to access them (Stevens, Myers, & Constantine, 1979).
- **Interactions:** SOA does not specify how interactions between applications should occur. On the other hand, ubicomp environments have some specific interac-

tion details that must be handled. DSOA gives a set of strategies to deal with these situations.

3.1. Basic Concepts

The Smart Space. The smart space is a neither empty nor unitary set of devices provided with computing power and interconnected by a communication network in a collaborative way.

It defines the environmental scope addressed by DSOA from which we can highlight three points:

1. A smart space cannot be defined as a single device. One of the main purposes of ubicomp is the coordination of available capabilities in the environment, thus without a set of devices there is no point in coordinating them.
2. All devices must have a way to communicate with each other. This can be done with any type of technology or a set of them. Being capable of exchanging information is the basis for devices to interact and coordinate actions.
3. Devices collaborate in order to provide services to the user in a transparent way. By transparent we mean services that require the least attention from the user, according to the concept of peripheral attention (Weiser & Brow, 1995).

The smart space is where ubicomp happens and the definition presented above narrows the scope to the main focus of this work, which is the construction of an intelligent environment where computational power is used in favor of the user as a new helping layer.

Devices. A device is a computing equipment with communication capabilities, which must host applications or make resources available in the smart space.

Devices are the basis of a smart space. Their ability to process information and exchange data is what makes the intelligence possible. It is the device's responsibility to host applications which implement the required intelligence of the smart space and to provide its available capabilities through resources.

Resources. A resource is a group of functionalities logically related. These functionalities must be accessible in the environment through pre-defined interfaces.

The concept of a resource allows applications to be aware of a set of functionalities (services) in a more cohesive way. Some functionalities present dependencies or correlation with each other. Consider, for example, an air conditioner as a resource. It can provide functionality for setting the desired temperature and another for setting the fan speed. These are two different functionalities, but both are associated to the same resource.

Resources can be either physical (like screen, keyboard, speaker, etc.) or logical (like user positioning, converters, etc.). Logical resources can provide new functionalities by accessing other resources. In SOA this is known as "composition". Imagine a camera resource that provides video stream functionality in raw format. A video conversion resource can provide a compression functionality of the same stream in a different format, like H.264/AVC.

A resource must be available to the smart space applications through a known public interface. In the DSOA a resource is uniquely identified by a name (or identifier) and the set of services it provides. This interface allows applications to verify if available resources match the expected set of capabilities desired.

Services. A service is the implementation of a functionality made available in the smart space through a resource with a known public interface. A service is responsible for providing the execution of a resource's functionality. A service is only relevant if it is capable of producing an exect that can be verified by other entities in the smart

space. This effect can be an exchange of information between a resource and an application, or a change in the state of the environment (perceived by another application or user).

The interface of a service is defined by the resource the service is part of. A name (or identifier) is responsible for uniquely identifying the service in a resource. The parameters that specify the information required for the execution of a service are also part of its interface. This interface establishes an agreement in how a service provides its capability to the applications in the smart space.

Application. An application consists of the implementation of a set of rules and behaviors related to the resources and users in the smart space.

An application establishes a set of rules and behaviors which produce actions based on the users and resources present in the smart space. Such actions must be taken in accordance to the ubicomp principles like aiding the user in a way that requires as minimum attention as possible. Applications are hosted in the devices and take advantage of the resources and services available.

The Environment. Figure 1 exemplifies the organization of a DSOA environment according to the above definitions. The smart space is composed by devices responsible for providing computing power to the environment. These devices host applications and resources that are used to bring intelligence to the environment. Resources, on the other hand, have their functionalities represented as services.

As an example of a DSOA environment consider the following. A smart space is defined as a meeting room. The available devices are a Dell Laptop, a Mac-Book, a Sony Ericson w580i cell phone and a Sony LCD 52" TV connected to a PC. The Dell Laptop can be made available to the smart space its resources like webcam, keyboard, mouse, microphone, and storage. Besides that, the laptop can host an application which allows attendees to interact with its available resources.

Figure 1. A DSOA smart space environment and its entities

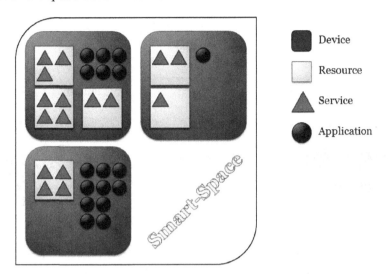

The webcam resource of the Dell Laptop can provide three different services like snapshot, video stream and motion capture.

An application1 addresses the services in the smart space through the re-source's interface. Suppose an application needs to access all available cameras in the environment. It starts by querying the smart space for available resources with the corresponding identifier. Once the application knows all the available cameras, it verifies which ones are compatible with the expected interface. From such list, the application can choose the one that fits best and access its choice.

Roles. DSOA defines three roles for devices during communication (Figure 2):

- A device acts as a consumer if it has a resource or application accessing services in the smart space.
- A device plays the role of a provider when it enables the access to a service through its defined interface. In addition, it is the provider responsibility to supply the information about its resources and services.
- The entity that keeps track of information about the resources and services available in the smart space is the register. This can

be done by monitoring devices and their available resources, or by providing means for each device to do that spontaneously.

A single device can act in any combination of these roles at any time, but any device must fit in at least one of these roles in order to be considered part of the smart space.

3.2. Communication Strategies

DSOA defines two groups of strategies to deal with the interaction details stated in Section 1.

Data Transport. In the communication between two devices DSOA defines two different strategies regarding the way a provider and a consumer can exchange data. It is worth highlighting that DSOA (as SOA) does not impose any restrictions on the kind of communication among devices.

- **Discrete Messages:** Carry information with known size. They are characterized by mutual knowledge of the beginning and the end of data in the communication. This strategy is used for small scale and discrete information, like requests for actions or queries.

Figure 2. Interaction according to the devices' roles in the DSOA

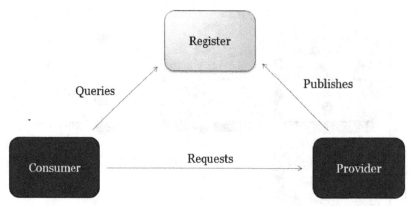

- **Continuum Data:** Characterized by an information flow with no previous knowledge of its ending. The common usage of this type of communication is in data streams transmissions, like audio/video, or large scale data transfers.

The data transport strategies are independent of the method used to implement communication among devices (message queue, shared memory, etc.) or the transmission technology (Bluetooth, ethernet, etc.). Considering the nature of the exchanged information, DSOA defines two types of data channels (Figure 3):

- **Control Channel:** There is only one control channel where discrete messages are exchanged. This channel always exists between two devices since it is through it that service requests (control messages) are communicated.
- **Data Channels:** Responsible for transporting continuum data. There can be as many data channels as needed, since each of them enclosures a different data stream that applications and resources must deal with.

Device Interaction. Applications running on one device usually call services provided by resources on other devices. In this context, the applications behave like a distributed system. Considering the synchronization of the distributed processes, DSOA proposes the use of both synchronous and asynchronous calls.

- A synchronous interaction (Figure 4) occurs when a consumer starts its service re-

Figure 3. Communication channels between devices A and B

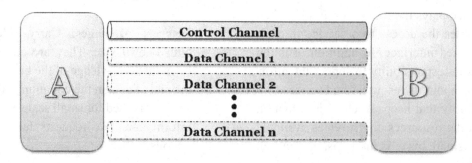

Figure 4. A synchronous interaction

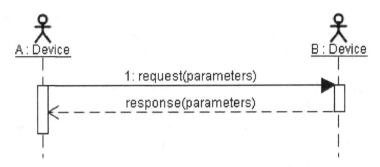

quest and a provider responds in a request-response manner. Common uses of this type of interaction strategy can be found in command requests (like lowering the air conditioner temperature) or in a simple information query (like asking for the current room temperature).

- Some interactions don't start immediately after the client request. This type of interaction is defined as asynchronous (Figure 5) and has the following steps. (1) A client asks a provider to be informed about the occurrence of an event. (2) The provider receives the request and waits for the event to happen. (3) In an arbitrary moment (which may never happen) the provider notifies the client of the occurrence

of the event. This type of interaction is commonly used to identify changes on the environment like user actions or temperature changes.

4. UBIQUITOUS PROTOCOLS (UP)

UP (ubiquitous protocols) consists of a set of protocols created for interfacing communications in a DSOA compliant smart space. It provides a set of messages and a way to format data in order to standardize how interaction in the smart space occurs. Along with that a set of protocols is defined to provide means to execute and discover services in accordance with the DSOA strategies.

Figure 5. An asynchronous interaction

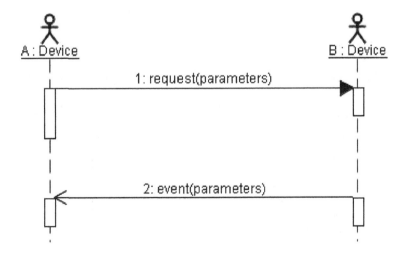

4.1. JSON

The uP messages are in JSON (JavaScript Object Notation) format. JSON consists of a structured data format, and was chosen mainly because of its characteristics of portability in a variety of computing and programming platforms. The use of UTF8 (Yergeau, 2003) encoding and a message format specification tackle the platform heterogeneity issue mentioned in Section 1.

JSON has a structured format of representing data which facilitates message handling and future expansion of any predefined message. Since these characteristics are also found in XML (eXtensible Markup Language) (Bray et al., 2006) format, which is commonly used, we have performed a set of tests2 to evaluate the relation between the message format and the consumption of two types of resources. Results have shown a clear difference between both formats regarding message size (Figure 6a) and processing time (Figure 6b). The size of the message generated in each format is directly related to the amount of memory needed to handle it and the bandwidth allocated for its

Figure 6. Number of fields vs. (a)message size and (b) time to deal represented in JSON and XML with a message in JSON and XML

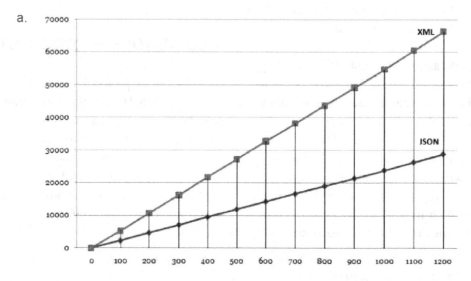

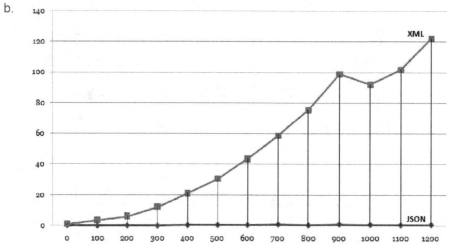

communication. The time needed to process each message is related to the processing power and the amount of energy demanded. Since limitations in CPU, memory, battery and bandwidth are common in the devices that are part of a ubicomp environment, protocols which focus on such platforms must minimize their consumption. Experiments have shown that JSON is a suitable choice for this task.

4.2. Representations

The uP establishes a representation for the basic concepts defined by DSOA. Using the JSON format, each one of these concepts is defined as a data map with the following information:

- **Device:** Represents the device.
  - ◦ **"name":** A name responsible for identifying the device in the smart space.
  - ◦ **"networks":** A list of network interfaces available in the device. It's composed by a pair of network address and network type.
- **Driver:** Represents the interface of a resource.
  - ◦ **"name":** A name responsible for identifying a resource available in the smart space.

- ◦ **"services":** A list of synchronous services available by the resource.
- ◦ **"events":** A list of asynchronous services available by the resource.
- **Service:** Represents the interface of a service.
  - ◦ **"name":** A name responsible for identifying a service available in the resource.
  - ◦ **"parameters":** A list of parameters needed for the execution of the service. Such parameters can be either mandatory or optional.

Box 1 shows an example of an uP driver and its services.

4.3. Protocols

The uP is composed by a set of protocols with the purpose of enabling the interaction and discovery of resources in the smart space similar to what is proposed by the SLP (Service Location Protocol) (Perkins, 1998). These protocols are categorized in two groups:

1. The base protocols focus on establishing the interaction between applications and resources through services.

Box 1. uP driver and services

```
{   ''name'':''br.unb.ubiquitos.webcam.ns60'',
  ''services'':[ {
    ''name'':''snapshot'',
    ''parameters'': {
      ''width'':''MANDATORY'',
      ''height'':''MANDATORY'',
      ''encoding'':''OPTIONAL''
     }
  } ]
}
```

2. The discovery of services and resources available in the environment as well as some other information about the smart space are made available through the complementary protocols.

4.4. Base Protocols

There are two base protocols in the uP, the SCP (Service Call Protocol), responsible for establishing the communication with synchronous services, and the EVP (Event Protocol), responsible for the access to asynchronous services.

The SCP is composed by two messages:

1. The Service Call message carries the information needed to initiate a service request. It is composed by the driver name, the service to be invoked and the parameters needed for its execution.
2. The Service Response message is responsible for transmitting the information about the service execution and returning data for the client who sent the Service Call.

The EVP starts with a SCP call for the service "registerListener" which is responsible for registering the client for receiving notifications for a specific event. When such notifications are no longer needed a SCP call for the service "unregisterListener" will remove the client from the notification list. Every event notification is informed using the Notify message, which carries the data regarding the event conditions.

Along with Service Call, Service Response and Notify messages, uP has defined a fourth message type called Encapsulated Message. This type of message is responsible for exchanging coded information and is useful for encryption and compression.

4.5. Complementary Protocols

The main purpose of the complementary protocols relies on sharing information about resources available in the smart space and other device information. These protocols are accessed through SCP calls to specific drivers. Such drivers are:

- **DeviceDriver:** Responsible for implementing protocols that share information about the device itself. Available protocols are:
  - **"ListDrivers":** This protocol returns a list of all drivers available in the device. It can receive as an optional parameter a driver name which will restrict the returned list only to the instances of that driver on that device.
  - **"Handshake":** Through this protocol two devices exchange information about themselves. It receives a parameter with the representation of the caller device and returns a representation of the device which received the call.
  - **"Goodbye":** This protocol informs that the caller device information is no longer valid and must be removed from the database. It's commonly used when a device is leaving the smart space.
  - **"Authenticate":** This protocol is responsible for establishing a security context among two devices according to the chosen security scheme. One implementation of this protocol can be found at (Ribeiro, Gondim, Jacobi, & Castanho, 2009).
- **RegisterDriver:** Responsible for sharing information about the smart space and the device neighborhood. Protocols available are:
  - **"ListDrivers":** This protocol returns a list of drivers available in devices

in the smart space. It can receive as an optional parameter a driver name which will restrict the returned list to only the instances of that driver in those devices.

○ **"Publish":** Allows a device to publish a driver to be available in the smart space.

○ **"UnPublish":** Removes a driver's availability from the smart space.

5. UBIQUITOUS OS (UOS)

In order to assist the construction and coordination of applications and resources in a DSOA smart space, a middleware named uOS (ubiquitous OS) was developed. Its current version is available in JSE and JME platforms.

The uOS utilizes a layered architecture (Figure 7) for abstracting the lower level details of the environment. The lower layer of the middleware is the network layer, responsible for managing the network plugins. Each plugin encapsulates the implementation details for a specific communication technology. It is also responsibility of the plugin to provide mechanisms for device discovering in the smart space. This mechanism is

Figure 7. Overview of responsibilities according to the uOS middleware

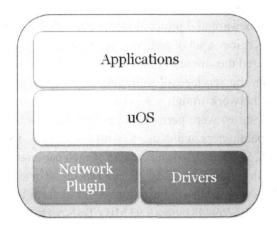

called radar and helps to monitor continuously the set of devices present in the smart space. The current implementation of the uOS provides plugins for Bluetooth, TCP, UDP and RTP technologies.

The connectivity layer is responsible for translating the data from and to the network layer using uP protocol message formats.

The adaptability layer manages the lifecycle of applications and drivers in the device. uOS handles all service calls and events sent to the drivers. It also provides an interface for applications and resource drivers to access the smart space functionalities like resource discovery, service call and event management.

UOS provides full implementations of the Device Driver and Register Driver specified in the uP protocol, making available the services for sharing information among devices. The Device Driver provides an implementation (Ribeiro et al., 2009) of a simple authentication protocol for limited devices, which establishes a security context for communication.

6. USE CASE: THE HYDRA APPLICATION

In order to analyze the use of the DSOA architecture and the uOS /uP operation a case of study is proposed. The Hydra application enables a computer to redirect its common I/O peripherals, like keyboard and mouse, to resources available in the smart space. Taking advantage of the resources available the user can decouple its computer into better suited resources in the smart space. Consider the following scenario.

Suppose the smart space is a meeting room, where we have available a Dell Laptop, a MacBook, a Sony Ericson w580i cell phone and a Sony LCD 52" TV connected to a PC. These devices were modeled as follows:

• The Dell Laptop and the MacBook make available in the smart space their keyboard, mouse, webcam and video output (display) as resources.

- The Sony Ericson w580i cell phone provides keyboard and mouse resources to the smart space.
- The PC provides a video output (display) resource as the Sony LCD 52" TV.

In this room, there is a meeting being held with the presence of Estevão (with the MacBook), Lucas (with the Dell laptop) and Ricardo (at the moment with the w580i cell phone). Lucas has installed in his laptop the Hydra application. He starts the meeting and wants to show a piece of his source code to the others. Using Hydra, Lucas searches for available screens in the smart space, and finds three available, the Mac Book, the Dell Laptop (his own) and the LCD TV. Lucas chooses to redirect his display to the TV which is more suitable for visualization. During the meeting, Ricardo wants to highlight some part of the source code. Following the same process Lucas redirects its mouse to Ricardo's cell phone. Later Estevão also wants to make a change in the source code. In the same way, Lucas redirects its keyboard to Estevão's MacBook. At the end of the meeting, Lucas' laptop became a composition of the resources it accessed in the smart space. The display is the Sony LCD 52" TV, the processing element is the Dell Laptop, the mouse is controled by the Sony Ericson w580i cell phone and the keyboard is Mac Book's one. All these resources are integrated by the Hydra application running on the Dell Laptop.

7. RESULTS

From the Hydra application use case we can highlight some key features of DSOA architecture. The way resources were defined and their services made available in the meeting room are strongly related to what the devices were capable of sharing between each other. This model of resources is not influenced by the applications built on top of the smart space. Other applications could use the same resources for different purposes. This is related to the fact that the DSOA has a more natural way[3] of representing the smart space among applications in contrast to other propositions like the MPACC (Roman, Manuel, Campbell, & Roy, 2001).

In Hydra all drivers and applications can be implemented using the API provided by the middleware. The middleware is responsible for handling the device and resource discovery process along with delegating service calls to the respective drivers.

Beyond the features brought by the DSOA, the performance of the developed uP/uOS solution was analyzed and compared to other available middlewares. The comparison is focused on two main goals that must be considered when the solution includes limited devices, the middleware footprint and the communication latency.

7.1. Latency

The use of the uP/uOS solution handles the heterogeneity of platforms (both hardware and software) with the use of a structured format (JSON) and UTF8 encoding. JSON tests showed that JSON messages are smaller than XML messages aiding with memory and bandwidth restrictions of devices. Another common limitation of devices lies on its computational capabilities, and this was the target of some tests on the uP/uOS solution.

In order to validate the overhead added by the middleware and protocol in the communication of services a set of tests was built. Such set considered the amount of time the solution demands from the service call to its properly delivery to the network trough a service call. Twenty five test suites were performed varying the number of parameters in messages (with values of 1, 16, 128, 512 and 1024) and the size of each parameter (with values of 16, 128, 512, 1024 and 2048 bytes). This resulted in a range of message size from 195 bytes e 7269448 bytes (6.93 MB). Tests performed in a Dell Vostro 1500, Intel(R) Core(TM)2 Duo CPU T7500 @ 2.20GHz 2x2 GB DD2 333MHz, Windows Vista SP2.

The results of such tests are summarized on Table 4 and Figure 8. Common ubicomp applications have its parameters varying from 1 to 16 and its size from 16 bytes to 128 bytes, which leads to an average overhead of 1,935203 ms. Considering this average data and the most common transmission technologies we can see in Table 5 that in most of the cases the uP/uOS solution achieves a less than 1% overhead in the service communication.

For comparison purposes we've chosen the MundoCore, WSAMI and Home-SOA solutions for their focus on limited devices and the use of SOA compatible architectures. MundoCore is a multi platform solution available in Java, Phyton and C++ versions and with XML and Binary communications available, for comparison purposes the Java+XML data were chosen. According to results presented (Schmitt, 2008) Mundo-Core achieves and average overhead of 2.67 ms. WSAMI is a SOAP based middleware which runs on a Java platform, since no specific tests on this solution were found we considered the results of other SOAP solutions available (Head, 2005) leading to an overhead of 4.70 ms. In (Bottaro & Gérodolle, 2008) is presented a SOA solution built upon the OSGi middleware which achieves an average of 5,23 ms. These results can be better observed in Figure 8.

Table 4. Load tests results using the uP/uOS solution

NP x BP*	16	128	512	1024	2048
1	1,537 ms	1,876 ms	2,239 ms	2,624 ms	1,843 ms
16	1,957 ms	2,369 ms	3,380 ms	3,313 ms	5,840 ms
128	2,530 ms	4,299 ms	9,665 ms	17,050 ms	41,135 ms
512	3,962 ms	7,580 ms	32,148 ms	63,395 ms	349,411 ms
1024	3,755 ms	11,951 ms	61,145 ms	135,668 ms	561,951 ms

NP: Number of parameters.
BP: Bytes per parmeter.

Figure 8. Time overhead of the uP/uOS solution

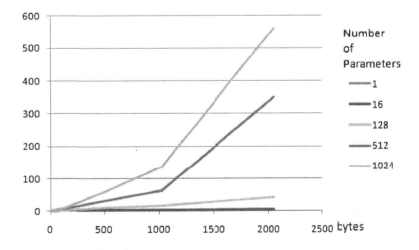

Table 5. Overhead of the uP/uOS solution versus transmission technologies

Transmission	Speed	Transfer Time	% Overhead
ZigBee	250 Kbit/s	254464 ms	0,0008%
Bluetooth 1.2	1 Mbit/s	20976 ms	0,0092%
Bluetooth 2.0	3 Mbit/s	6992 ms	0,0276%
802.11 (Wi-Fi)*	54 Mbit/s	399,6 ms	0,4829%
Ethernet*	100 Mbit/s	206,2 ms	0,8943%
GigE*	1000 Mbit/s	20,62 ms	8,9435%

* Using TCP

Figure 9. Time overhead on presented solutions

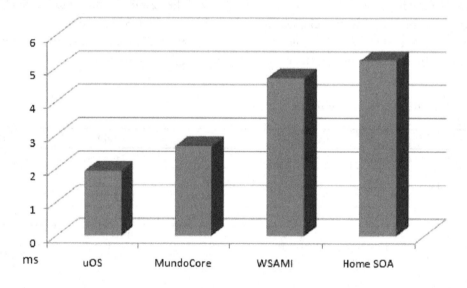

Figure 10. Solution footprint

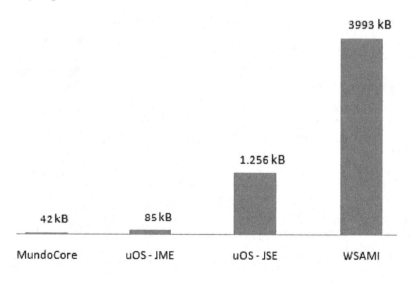

7.2. Solution Footprint

The utilization of middlewares aims in the ease in the development effort needed to build applications. Along with its benefits the use of middlewares carries some downsides, one of them reside in the memory overhead brought by its footprint which can limit which applications can be available in some limited devices.

The uOS middleware is available in two distinct versions. A JSE version aims more capable devices and it utilizes many development facilities provided by the platform (like annotations, generics and reflections) which leads to a footprint for this version of 1.256 KB. The JME version aims for more limited devices which leads to a footprint of 85 KB from which 61,7 KB are only responsible for the uP message handling.

Observing other solutions we've found that the MundoCore have a footprint of 42 KB and the WSAMI of 3.9 MB. Figures 9 and 10 show a comparative of this solutions footprint and uOS.

8. CONCLUSION

DSOA, uOS and uP are part of a larger project called UbiquitOS which aims to the development of ubiquitous applications. This project ranges from the definition of smart space organizations to application development involving artificial intelligence, user profile analysis, and alternative user interfaces, among others.

We presented the DSOA architecture and how it assists modeling a ubiquitous environment using a more natural approach. Besides, we described the DSOA strategies to deal with common interactions issues in a smart space. The use of uP and uOS for building applications in this scenario was presented and compared to other solutions. The results have shown a consistency of the implemented solution to the established requisites of a ubiquitous environment.

REFERENCES

Abowd, G., Atkeson, C., & Essa, I. (1998). *Ubiquitous smart spaces.* Paper presented at the DARPA Workshop.

Adhikari, S., Paul, A., & Ramachandran, U. (2002). *D-stampede: Distributed programming system for ubiquitous computing* (Tech. Rep. No. GIT-CC-01-04). Atlanta, GA: Georgia Institute of Technology.

Aitenbichler, E., Kangasharju, J., & Mühlhäuser, M. (2007). MundoCore: A Light-weight Infrastructure for Pervasive Computing. *Pervasive and Mobile Computing, 3*(4), 332–361. doi:10.1016/j.pmcj.2007.04.002

Bernstein, P. A. (1996). Middleware: A model for distributed system services. *Communications of the ACM, 39*(2), 86–98. doi:10.1145/230798.230809

Bottaro, A., & Gérodolle, A. (2008). Home Soa: Facing protocol heterogeneity in pervasive applications. In *Proceedings of the 5th International Conference on Pervasive Services* (pp. 73-80). New York, NY: ACM Press.

Bray, T., Paoli, J., Sperberg-McQueen, C. M., Maler, E., Yergeau, F., & Cowan, J. (Eds.). (2006). *Extensible markup language (xml) 1.1 (Second edition).* Retrieved from http://www.w3.org/TR/xml11/

Brumitt, B., Meyers, B., Krumm, J., Kern, A., & Shafer, S. (2000). Easyliving: Technologies for intelligent environments. In P. Thomas & H.-W. Gellerson (Eds.), *Proceedings of the Second International Symposium on Handheld and Ubiquitous Computing,* Bristol, UK (LNCS 1927, pp. 12-29).

Buzeto, F., Filho, C., Castanho, C, & Jacobi, R. (2010). DSOA: A service oriented architecture for ubiquitous applications. In P. Bellavista, R. Chang, H. Chao, S. Lin, & P. Sloot (Eds.), *Proceedings of the 5th International Conference of Advances in Grid and Pervasive Computing* (LNCS 6104, pp. 183-192).

Cerqueira, R. (2001). *Gaia: A development infrastructure for active spaces*. Paper presented at the Joint Workshop of UBICOMP and Application Models and Programming Tools for Ubiquitous Computing, Atlanta GA.

Garlan, D. (2002). Project aura: Toward distraction-free pervasive computing. *IEEE Pervasive Computing/IEEE Computer Society and IEEE Communications Society*, *1*(2), 22–31. doi:10.1109/MPRV.2002.1012334

Head, M. R., Govindaraju, M., Slominski, A., Liu, P., Abu-Ghazaleh, N., van Engelen, R., et al. (2005). A benchmark suite for soap-based communication in grid web services. In *Proceedings of the ACM/IEEE Conference on Supercomputing* (p. 19). Washington, DC: IEEE Computer Society.

Helal, S., Mann, W., El-Zabadani, H., King, J., Kaddoura, Y., & Jansen, E. (2005). The gator tech smart house: A programmable pervasive space. *IEEE Computer Magazine*, *38*(3), 50.

Issarny, V., Sacchetti, D., Chibout, R., Dalouche, S., & Musolesi, M. (2005). *WSAMI: A middleware infrastructure for ambient intelligence based on web services*. Retrieved from https://www-roc.inria.fr/arles/index.php/software/64-wsami-a-middleware-infrastructure-for-ambient-intelligence-based-on-web-services.html

MacKenzie, C. M., Laskey, K., McCabe, F., Brown, P. F., Metz, R., & Hamilton, B. A. (2006). *Reference model for service oriented architecture 1.0*. Retrieved from http://www.oasis-open.org/committees/download.php/19679/soa-rm-cs.pdf

Modahl, M., Bagrak, I., Wolenetz, M., Hutto, P., & Ramachandran, U. (2004). Mediabroker: An architecture for pervasive computing. In *Proceedings of the 2nd Annual Conference on Pervasive Computing and Communication* (pp. 253-262). Washington, DC: IEEE Computer Society.

Perkins, C. (1998). *SLP white paper topic*. Retrieved from http://playground.sun.com/srvloc/slp_white_paper.html

Ribeiro, B., Gondim, J., Jacobi, R., & Castanho, C. (2009). *Autenticação mútua entre dispositivos no middleware uos*. SBSEG – Simpósio Brasileiro em Segurança da Informação e de Sistemas Computacionais.

Roman, M., & Campbell, R. H. (2001). *A model for ubiquitous applications*. Retrieved from http://gaia.cs.uiuc.edu/papers/ubicomp01-c.pdf

Sacramento, V., Endler, M., Rubinsztejn, H. K., Lima, L. S., Gonçalves, K., Nascimento, F. N. et al. (2004). MOCA: A middleware for developing collaborative applications for mobile users. *IEEE Distributed Systems Online*, *5*(10).

Schmitt, J., Kropff, M., Reinhardt, A., Hollick, M., Schafer, C., Remetter, F., et al. (2008). *An extensible framework for context-aware communication management using heterogeneous sensor networks* (Tech. Rep. No. TR-KOM-2008-08). Darmstadt, Germany: Technische Universität Darmstadt.

Stevens, W., Myers, G., & Constantine, L. (1979). *Structured design: Fundamentals of a discipline of computer program and systems design* (pp. 205–232). Upper Saddle River, NJ: Prentice Hall.

The OSGi Alliance. (2009). *OSGi service platform core specification*. Retrieved from http://www.osgi.org/download/r4v41/r4.core.pdf

Weiser, M. (1993). The world is not a desktop. *Interactions (New York, N.Y.)*, *1*(1), 7–8. doi:10.1145/174800.174801

Weiser, M., & Brow, J. S. (1995). *Designing calm technology*. Retrieved from http://nano.xerox.com/weiser/calmtech/calmtech.htm

Yergeau, F. (2003). *Utf-8, a transformation format of ISO 10646*. Retrieved from http://www.ietf.org/rfc/rfc2279.txt

ENDNOTES

[1] Regarding interaction, resources act as applications since they are allowed to access other resources.

[2] Tests were performed in a Gateway P-172X FX 2.40 GHz Intel Core 2 Duo T8300 2x2 GB DDR II SDRAM 667 MHz 3MB L2 Cache, Windows Vista.

[3] This refers to the fact that DSOA resource representations are closer to the real resource in the smart space.

This work was previously published in the International Journal of Handheld Computing Research (IJHCR), Volume 2, Issue 2, edited by Wen-Chen Hu, pp. 47-64, copyright 2011 by IGI Publishing (an imprint of IGI Global).

Chapter 17
A Generic Context Interpreter for Pervasive Context-Aware Systems

Been-Chian Chien
National University of Tainan, Taiwan

Shiang-Yi He
National University of Tainan, Taiwan

ABSTRACT

Developing pervasive context-aware systems to construct smart space applications has attracted much attention from researchers in recent decades. Although many different kinds of context-aware computing paradigms were built of late years, it is still a challenge for researchers to extend an existing system to different application domains and interoperate with other service systems due to heterogeneity among systems This paper proposes a generic context interpreter to overcome the dependency between context and hardware devices. The proposed generic context interpreter contains two modules: the context interpreter generator and the generic interpreter. The context interpreter generator imports sensor data from sensor devices as an XML schema and produces interpretation scripts instead of interpretation widgets. The generic interpreter generates the semantic context for context-aware applications. A context editor is also designed by employing schema matching algorithms for supporting context mapping between devices and context model.

INTRODUCTION

With the rapid growth of wireless sensors and mobile devices, the research of pervasive computing is becoming important and popular in recent years. The vision of ubiquitous computing (Weister, 1993) or pervasive computing (Weiser, 1991) is to integrate hardware, network systems, and information technologies to provide appropriate service for our lives in a vanishing way. A context-aware system is a pervasive computing environment in which users' preference services can be detected by making use of context including location, time, date, nearby devices and other

DOI: 10.4018/978-1-4666-2785-7.ch017

environmental activities to adapt users' operations and behavior (Chen & Kotz, 2000). All kinds of context-aware architectures and frameworks have been designed and employed for a wide spectrum of applications (Baldauf, Dustdar, & Rosenberg, 2007). Since most of the systems focus on their specific application domains; the current context-aware systems are heterogeneous in all aspects, such as hardware, mobile resources, operating systems, application software, and platforms. The serious heterogeneous characteristics of context-aware computing are especially important and become significant drawbacks while developing or integrating context-aware applications in pervasive computing environments.

The concept of context independence was revealed in CADBA architecture (Chien, Tsai, & Hsueh, 2009; Chien et al., 2010). Two types of context independence, the physical context independence and the logical context independence, are classified in the article. The physical context independence is to prevent misinterpreting raw data from sensors with various specification standards; whereas the logical context independence is to allow context to be understood and applied by applications. As a result of context independence, cross-domain service applications will be able to be integrated into a unified context-aware system regardless of the heterogeneity in pervasive environment.

An OSGi-based service platform (Gu, Pung, & Zhang, 2004) based on Java VM is one of the practical solutions for accomplishing logical context independence. In this paper, a generic context interpreter is proposed to overcome the physical context dependence problem between context and sensor devices in the framework of context-aware computing. The context generic interpreter is composed of two modules: the context interpreter generator and the generic interpreter. First, the context interpreter generator imports sensor data from sensor devices as an XML schema and produces interpretation scripts instead of interpretation widgets. Then, the generic interpreter generates

the semantic context for context-aware applications. An interface tool, the context editor, is also designed by employing automatic XML schema matching schemes for supporting smart context mapping between devices and context model.

The remainder of this paper is organized as follows. First we introduce the foundation and summary of context-aware architectures. The architecture of proposed generic context interpreter is presented next. The detailed design of system components is described. Then, the performance evaluation of context mapping for proposed generic context interpreter is demonstrated and discussed. Finally we conclude the work and express future work.

PRELIMINARIES

The term context-aware first appeared in 1994 mentioned by Schilit and Theimer (1994). Since then, various context-aware computing architectures were proposed: The Context Toolkit (Dey & Abowd, 1999; Dey, Abowd, & Salber, 2001) provided context interpretation using widgets and a set of object-oriented APIs to offer the creation of service components. The Hydrogen (Hofer et al., 2002) is a framework based on layered architecture in which contains the adaptor layer, the management layer and the application layer. The Gaia project (Roman et al., 2002) is a middle-ware based architecture; the system consists of Gaia kernel and application framework to support the development and execution of mobile applications. Another middle-ware system, SOCAM (Gu, Pung, & Zhang, 2004) uses a central server called context interpreter to obtain context data for building and prototyping of context-aware services. The CORTEX system (Biegel & Cahill, 2004) is also a middle-ware structure based on sentient object model which supports context-aware services in an ad-hoc mobile environment.

The above context-aware system architectures generally follow the framework presented in

(Ailisto et al., 2002; Baldauf, Dustdar, & Rosenberg, 2007). The five-layer model in (Ailisto et al., 2002) consists of the physical layer, the data layer, the semantic layer, the inference layer and the application layers. These layers in this model focus on the descriptions of functions in a context-aware system. The abstract five-layer proposed in (Baldauf, Dustdar, & Rosenberg, 2007) described a conceptual framework of a context-aware system containing the sensors layer, the raw data layer, the preprocessing layer, the storage/management layer, and the application layers.

A context-aware system was utilized on its individual specific application domain of employment whatever the framework or architecture it used. Thus, the problem of heterogeneity issues was generally characterized not only on the physical devices but also on the logical context interoperability. To bridging the communication between heterogeneous devices, some researches on adapting and integrating systems sprout recently. A OSGi-based infrastructure (Gu, Pung, & Zhang, 2004) was proposed to adapt to changing

context. Bartelt et al. (2005) proposed a system that integrates devices dynamically and enable interoperability between them. Nakazawa et al. (2006) presented a framework for heterogeneity handling. Schmohl and Baumgarten (2008) further derived a generalized context-aware architecture for pervasive computing environments to resolve the heterogeneity issues in both context-awareness and interoperability domains.

This work is based on another generalized context-aware architecture CADBA (Chien, Tsai, & Hsueh, 2009; Chien et al., 2010). The architecture of CADBA is also structured by a five-layer framework consisting of the device layer, the interpretation layer, the context layer, the storage layer, and the application layer, as shown in Figure 1.

The main modules and components for each layer are described as follows.

1. **Device Layer:** This layer contains the physical equipments and devices operated and used in the context-aware systems including

Figure 1. The CADBA architecture for five-layer framework

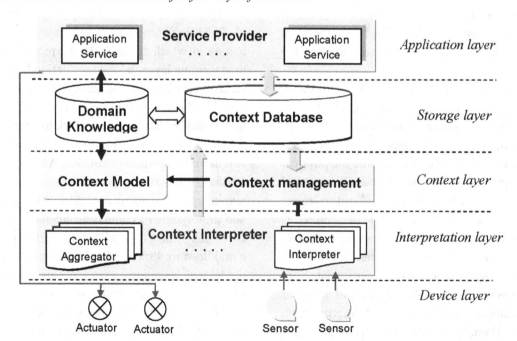

sensors, identifiers, mobile devices, and actuators, etc. The possible hardware and the drivers used in the domain that could be mobile devices like PDAs, smart cell phones, detecting sensors like thermometers, RFIDs, or actuators like alarms, temperature regulators.

2. **Interpretation Layer:** A semantic mapping layer between the device layer and the context layer is needed. The main components in this layer contain context interpreters and context aggregators.

    ◦   **Context Interpreter:** The raw signals from sensors or mobile devices cannot work as their original format. They have to be transformed to context information in context-aware system. The context interpreter is used to interpret the structures of raw data from sensors and represent the information as low-level context called sensor context.

    ◦   **Context Aggregator:** The context aggregator then gathers the related low-level sensor context data to form a higher-level context. As soon as the context model is provided, context can be generated by context interpreters and context aggregators.

3. **Context Layer:** Context processing is the core of a context-aware system. A context model and efficient context management are the main two components.

    ◦   **Context Model:** The context model is essential for a context-aware system to progress the context processing. The CADBA uses ontology to represent the context model with OWL.

    ◦   **Context Management:** Context-aware computing implies the interconnection of people, devices, and environments. It makes possibility of sharing the users' context among

each other in heterogeneous systems. The issues and challenges of context management include context configuration, device fault, context conflict, context inconsistency and security (Hegering, Küpper, Linnhoff-Poien, & Reiser, 2003).

4. **Storage Layer:** The storage layer stores not only the context data of the current status but also the historical context data in the context-aware system.

    ◦   **Domain Knowledge:** It contains the entire resources of the context-aware computing including places, persons, devices, and objects.

    ◦   **Places:** All possible locations where activities will occur in the environment of applications.

    ◦   **Persons:** The possible people including the application users and other persons who interact in the system.

    ◦   **Devices:** The description for all existing devices that support the device manager to record the status of the hardware.

    ◦   **Context Database:** The context database is to provide the integration of current context and the storage of historical context data. The main function of the context database is to provide an efficient context access mechanism to store and retrieve context data using context queries.

5. **Application Layer:** In this layer, application service can be built and executed by the service provider through querying the current status of context and the related historical context data.

    ◦   **Service Provider:** The functions of the service provider consist of context querying, service construction, and service coordination. To satisfy the user's need and complete the application, the service provider starts context process by context querying. Then, the returning context data from the context database are collected to determine and prepare the necessary application services. Finally, the service coordinator is used to arrange and perform the application services. The logical context independence can be accomplished by applying context queries to create application services.

THE ARCHITECTURE OF GENERIC CONTEXT INTERPRETER

Two types of context independence are revealed in the previous section: the physical context independence and the logical context independence. The main goal in this paper is to achieve the physical context independence using the proposed generic context interpreter. The physical context independence is defined as the immunity of context from the changing or updating of physical sensor devices. A traditional context interpreter is usually dedicated to interpreting the context designated for a specific device in an application. Although such a tightly coupled structure results in good performance of fast reaction for a system, its drawback is that the application does not work and the service cannot be extended to other devices without completely re-packing function program of the context interpreter after changing or equipping a new sensor device. The physical context independence tries to maintain normal operations of the context-aware system regardless of changing, reconstructing, or replacing sensor devices anytime.

Before constructing and performing context interpreter, the context model is essential in a context-aware system and the representation of context must be determined. For designing a generic context interpreter, ontology with OWL representation is considered as the context model for the proposed architecture (Melnik, Garcia-Mo-

lina, & Rahm, 2002). The techniques of ontology fusion can be applied to integrate heterogeneous context models and the representation, OWL, is XML based scripts. It is formatted, unified, and standardized. While importing data from sensor devices, the XML format is also easy to convey the information extracted by data acquisition interface to the context interpreter. The context used in the CADBA architecture is divided into three levels by means of XML-based context model:

1. **Sensor Context ($Context_{sensor}$):** This type of context is the essential raw information triggered by sensors and interpreted by context interpreters directly. The definition is as follows:

$$Context_{sensor} = Interpret\{sensor_i | sensor_i \in D_j \text{ and } D_j \subseteq Dev\}$$

where $sensor_i$ is a kind of sensors in the type of devices D_j, and Dev is the set of all types of devices in the context aware system.

2. **Event Context ($Context_{event}$):** This type of context is aggregated by different sensor context. We define the behavior of aggregating various actions of sensors as an event context.

$$Context_{event} = Aggregate \left\{ \bigcup_{j=1}^{m} Context_{sensor_j} \right\}$$

3. **Scenario Context ($Context_{scenario}$):** The scenario context is another high-lever context containing not only sensor context but also at least one event context. The scenario context is defined as follows:

$$Context_{scenario} = Aggregate \left\{ \bigcup_{i=1}^{n} Context_{event_i} \right\} \cup \left\{ \bigcup_{i=1}^{m} Context_{sensor_i} \right\}$$

where $1 \leq |Context_{event}| \leq n$ and $1 \leq |Context_{sensor}| \leq m$, or $2 \leq |Context_{event}| \leq n$ and $0 \leq |Context_{sensor}| \leq m$.

The architecture of the proposed generic context interpreter is shown in Figure 2. The main components are the context interpreter generator and the generic interpreter. The context interpreter generator further contains three functions: the mapping operators, the context editor and the schema matching algorithm. In this model, context interpreters generated by the interpreter generator are interpretation scripts instead of interpretation functions or programs. The interpretation script draws linking relationships and transforming methods between sensor data and the semantics in the context model. The generic interpreter then translates sensor raw data into context by the corresponding interpretation scripts while the context interpretation process proceeds. We depict each component block as follows and the detailed design of function blocks.

- **Sensor Data Model:** Sensor models can be provided by some techniques of connectivity standards, for example, UPnP and SOAP, which enable data transfer in

XML-based procedure call. Each type of sensor delivers its sensor data by the predefined XML schema according to the hardware specification. Such a schema will be sent to the context editor for linking relationship rules with the context model.

- **Context Model:** The context model is built for different application environment. For mapping context schema into sensor data schema, ontology with XML-based representation is used here.

- **Context Interpreter Generator:** Interpreter generator produces interpretation scripts which contain the mapping relationships between context schema and sensor data schema. The mapping interface tool is the context editor. The context editor supports a graphical user interface to assist users to link the relationship between context schema and sensor data schema. The mapping job can be finished manually or automatically by applying schema matching schemes.

- **Interpretation Script:** The interpretation script uses XSL (eXtensible Stylesheet Language) to describe the schema mapping. The content of the script is the rules

Figure 2. The architecture of generic context interpreter

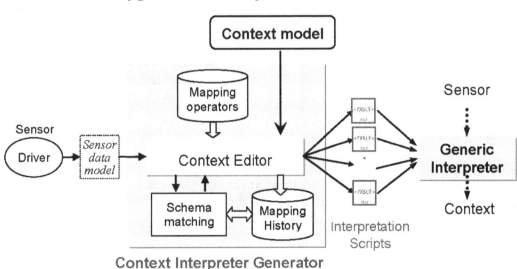

of transforming source data schema (sensor data) into target data schema (context).

- **Generic Interpreter:** Since interpretation scripts are stated in XSL, the XSLT (XSL Transformation) Processor can be used to be the generic interpreter directly. The XSLT-Processor will read the sensor raw data with XML tag and interpret the context according to the corresponding interpretation script.

The interpretation flow consists two phrases: interpretation scripts generation and context interpretation. In the interpretation scripts generation phrase, mobile sensors prepared circumstances data schema in XML format and delivered them to the context interpreter generator. After sensor data schema was received by the context interpreter generator, the context editor is used to build the relationship between sensor data and sensor context according to the definitions in context model. An XSL script will be generated for interpreting the sensor context if the confirmation is done. Then, in the context interpretation phrase, the XSLT-Processor is used to translate original XML sensor raw data into XML sensor context defined in the context model. After completing the definitions of sensor context, event context and scenario context can be aggregated from sensor context by the context editor as the same way.

FUNCTIONAL DESIGN OF CONTEXT INTERPRETER GENERATOR MAPPING OPERATORS

The function of mapping operators is to transfer the sensor raw data into semantic context. Four types of mapping operators are defined in the system. The operators are designed by XSL descriptions. These operators thus can be easily reused and modified to a new one.

Operator 1: *Concat*

The concatenation operator is used to merge two separated data into one. For instance: date-time or firstname-lastname. The rule of concatenation operator in XSL is shown in Box 1.

Operator 2: *Value-Map*

This operator maps the raw data with a item value into the semantic context described in context model. It states that the item value is equal to the context in the system. For example, if "RFID-tag:0X8001" represents the context "Person: John". The definition of the operator in XSL description is shown in Box 2.

Operator 3: *Value-Semantic*

The purpose of this operator is to process numerical values from senor raw data. For example, if temperature is lower than 16°C, the context "cold" will be interpreted and used in the system instead of the original numerical value. The XSL definition of the above context can be described as in Box 3.

Box 1. Concat

```
<AllName>
    <xsl:value-of select="concat(concat(string(LastName), ' '),
string(FirstName))"/>
    </AllName>
```

Box 2. Value-map

```
<xsl:template name="Value_map:Person">
            <xsl:param name="input" select="/.."/>
            <xsl:choose>
                    <xsl:when test="$input='0X8001'">
                            <xsl:value-of select="'John'"/>
                    </xsl:when>
                    <xsl:when test="$input='0X8002'">
                            <xsl:value-of select="'Peter'"/>
                    </xsl:when>
                    <xsl:when test="$input='0X8003'">
                            <xsl:value-of select="'George'"/>
                    </xsl:when>
                    <xsl:otherwise>
                            <xsl:value-of select="'none'"/>
                    </xsl:otherwise>
            </xsl:choose>
</xsl:template>
```

Operator 4: *Conversion-Function*

This operation allows users to define formula for some specific requests. A practical example is the transform between degree centigrade and Fahrenheit scale. The transform formula is usually used to convert 20°C into 68°F. See Box 4.

SCHEMA MATCHING

The main function of schema matching algorithm is to assist users to retrieve related context from the historical mapping repository. Since the number of context is usually large in a context-aware application, the management of context is an important work. One of the issues is to find similar or even the same context mapping to develop context interpretation scripts. The advantage of using XML to represent context here is that XML schema matching method can be applied to search the sensor schema with high similarity. Making use of reusing the similar schema of

context mapping, the new context mapping will be built quickly. A good XML schema matching algorithm can help users to construct intelligent context interpreter generator.

The XML schema and ontology matching problem is one of the research issues in data manipulation. There are many researches and discussion on such a topic. Schema matching was done manually by domain experts in the past. However, specifying schema matching manually is tedious and time-consuming even supported by graphical tools. While dealing with the integration of a large number of schemas, e.g. web data sources, E-business, and XML documents, it is impossible for such a manual approach to be performed well due to its slow processing speed. Hence, an automatic schema matching scheme is required for handling the applications with complex, large, and rapidly increasing number of schema data.

Many schema matching method have been proposed. A taxonomy of schema matching approaches was given by Rahm and Bernatein (2001). They classified the matchers into schema-

Box 3. Value-semantic

```
<xsl:template name="Value_Semanitc:Temperature">
      <xsl:param name="input_value" select="/.."/>
      <xsl:param name="lessThan" select="/.."/>
      <xsl:param name="greaterThan" select="/.."/>
      <xsl:param name="High_Semantic" select="/.."/>
      <xsl:param name="Medium_Semantic" select="/.."/>
      <xsl:param name="Low_Semantic" select="/.."/>
      <xsl:choose>
          <xsl:when test="string(($input_value &lt; $lessThan)) != 'false'">
                  <xsl:value-of select="$Low_Semantic"/>
          </xsl:when>
          <xsl:otherwise>
          <xsl:choose>
                  <xsl:when test="string(($input_value &gt; $greater-
Than)) != 'false'">
                  <xsl:value-of select="$High_Semantic"/>
                  </xsl:when>
                  <xsl:otherwise>
                    <xsl:value-of select="$Medium_Semantic"/>
                  </xsl:otherwise>
          </xsl:choose>
          </xsl:otherwise>
      </xsl:choose>
  </xsl:template>
```

Box 4. Conversion-function

```
<xsl:template name="Conversion:CelsiusToFahrenheit">
      <xsl:param name="CelsiusValue" select="/.."/>
      <xsl:value-of select="((($CelsiusValue * 9) div 5) + 32)"/>
  </xsl:template>
```

only based and instanced based, and the approaches of combining matchers was divided into hybrid matchers and composite matchers. A hybrid matcher is integrated by individual matchers who respectively handle names, types and structures. A composite matcher combines multiple match results produced by different matching algorithms.

Some famous schema matching methods, like Similarity Flooding (Melnik, Garcia-Molina, &

Rahm, 2002) and Cupid (Madhavan, Bernstein, & Rahm, 2001), are hybrid matchers. The methods COMA (Do & Rahm, 2002) and COMA++ (Aumuller, Do, Massmann, & Rahm, 2005) are composite matchers. Two schema matching algorithm were applied in this work. The first is Cupid proposed by Madhavan et al. (2001). The other is COMA++ by Aumuller et al. (2005). We will discuss the effectiveness and efficiency of the two methods after applying to the proposed context interpreter generator.

CONTEXT EDITOR

The context editor is designed for defining mapping operators, managing context mapping between sensor data schema and context model, and generating context interpretation scripts. A graphical user interface is provided to help users to finish the sensor binding efficiently and easily. The context mapping operation in a context editor is demonstrated as Figure 3. First, the editor reads the XML sensor data or others context from the

context model, as shown in Figure 3a. Second, the corresponding context in the context-aware system is selected from the context model, as Figure 3b shows. The next step shown in Figure 3c is to arrange the interpreting operations. Some predefined transformation operators and user defined operators can be added into the system for transferring data values to meaningful context. At last, the finished context mapping, as Figure 3d, can be stored in XSL scripts to interpret the context in the context-aware system. Owing to the context interpreter is done by interpretation scripts, the context interpreter only builds a new context interpretation script instead of full-function program while mobile devices are updated or changed.

Figure 3. The context editor

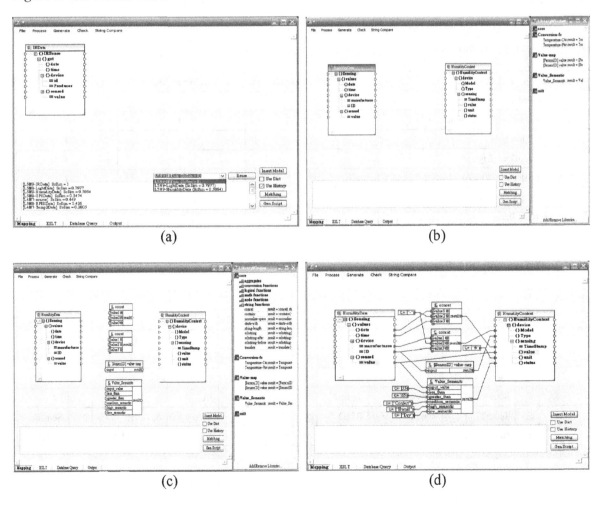

(a) (b)

(c) (d)

The goal of the physical context independence can be enforced.

SYSTEM CONSTRUCTION AND EVALUATION

The generic context interpreter proposed in this paper used Mapforce API to develop the context mapping in the context editor. The context editor provides both manual mapping and automatic mapping tools. The initial blank system needs build context mapping manually. Once more mapping datasets were accumulated in the mapping history, Users will be able to refer to the existing schema mapping cases and a similar sensor schema mapping was selected to modify as a new context mapping.

The test schema sets include seven different sensor schemas listed in Table 1. The depths of schema structures are four levels. The number of leaves is between the range four and six. The number of nodes is in the range of seven to ten. We first ranked the similarity degree of each schema by experts as shown in Table 2. Then the schema matching algorithms, Cupid and COMA++, are tested on each schema. The matching results of similarity are evaluated and ranked, as shown in Table 3. To evaluate the performance of ranking, we refer to R_{norm} (Gudivada & Raghavan, 1995) values as the criterion of effectiveness. The estimation of running time is also shown in Table 4.

The experimental results show that COMA++ is generally superior to Cupid in both effectiveness and efficiency. The R_{norm} values of COMA++ are better than Cupid for 5 schemas except S_5:RFID

Table 1. The test schema sets

	Schema	Number of Leaves	Number of Nodes	Tree Depth
S_1:	GPSData (GPS)	5	9	4
S_2:	HumidityData (Humid)	5	9	4
S_3:	IRData (IR)	5	9	4
S_4:	LightData (Light)	5	9	4
S_5:	RFIDData (RFID)	5	8	4
S_6:	SensorData (Sensor)	6	10	4
S_7:	Temp2Data (Temp2)	4	7	4

Table 2. The ranking results of experts for the test sets

Experts Ranking	Schema													
	S_1	GPS	S_2	Humid	S_3	IR	S_4	Light	S_5	RFID	S_6	Sensor	S_7	Temp2
2	S_2	Humid	S_1	GPS	S_4	Light	S_2	Humid	S_1	GPS	S_1	GPS	S_4	Light
3	S_4	Light	S_4	Light	S_2	Humid	S_3	IR	S_2	Humid	S_2	Humid	S_1	GPS
4	S_3	IR	S_3	IR	S_1	GPS	S_1	GPS	S_4	Light	S_7	Temp2	S_2	Humid
5	S_6	Sensor	S_6	Sensor	S_7	Temp2	S_7	Temp2	S_3	IR	S_4	Light	S_3	IR
6	S_7	Temp2	S_7	Temp2	S_5	RFID	S_5	RFID	S_6	Sensor	S_3	IR	S_6	Sensor
7	S_5	RFID	S_5	RFID	S_6	Sensor	S_6	Sensor	S_7	Temp2	S_5	RFID	S_5	RFID

Table 3. The evaluation results of the of schema matching algorithms

Schema Ranking	Cupid							COMA++						
	S_1	S_2	S_3	S_4	S_5	S_6	S_7	S_1	S_2	S_3	S_4	S_5	S_6	S_7
2	S_6	S_4	S_1	S_2	S_1	S_1	S_1	S_2	S_1	S_1	S_2	S_1	S_1	S_4
3	S_3	S_1	S_4	S_1	S_3	S_7	S_6	S_4	S_4	S_2	S_1	S_7	S_2	S_1
4	S_7	S_6	S_2	S_6	S_2	S_2	S_2	S_3	S_3	S_4	S_7	S_2	S_4	S_3
5	S_2	S_3	S_6	S_3	S_4	S_4	S_4	S_6	S_6	S_7	S_3	S_4	S_3	S_2
6	S_4	S_7	S_7	S_7	S_6	S_3	S_3	S_7	S_7	S_6	S_6	S_3	S_7	S_6
7	S_5	S_5	S_5	S_5	S_7	S_5	S_5	S_5	S_5	S_5	S_5	S_6	S_5	S_5
R_{norm}	0.667	0.905	0.801	0.801	0.905	0.952	0.762	1.000	1.000	0.857	0.857	0.81	0.905	0.952
Average	0.828							0.912						

Table 4. The estimated running time for Cupid and COMA++

Methods	Schema													
	S_1	GPS	S_2	Humid	S_3	IR	S_4	Light	S_5	RFID	S_6	Sensor	S_7	Temp2
Cupid		27.892		27.402		36.010		25.954		32.339		33.072		21.029
COMA++		2.844		2.865		1.998		2.632		2.559		2.545		2.538

and S_6:Sensor. The reason is that the type of value (xs:decimal) in S_7:Temp matched the type of value(xs:string) in S_5 and S_6. This mistake causes the higher rank of S_7:Temp. It shows that COMA++ is relatively weak in the matching of types on leaves.

CONCLUSION

The main contribution of this paper is to propose a generic context interpreter to accomplish physical context independence. This work is based on the context-aware architecture, CADBA. Ontology based context model is used in this architecture. We design a generic context interpreter including context interpreter generator and a generic interpreter. A context interpretation script is proposed to replace function-based context interpreter. As we know, this is the originality of context provider or interpretation in context-aware computing. We also design a context editing tool for support the context mapping operation and devices maintenance. By introducing mapping operators and schema matching schemes, the generic context interpreter performs a more intelligent operating interface for users. The heterogeneity in pervasive context-aware computing will gain a graceful solution.

The problem of heterogeneity is a bottleneck while developing and extending context-aware systems in pervasive computing environment. The enforcement of context independence resolves dependency of devices and improves interoperability of applications. This work is intended as a starting point of future research on context generation. The problems of context management for context-aware computing will be paid more attention in the future.

ACKNOWLEDGMENT

This research was supported in part by the National Science Council of Taiwan, R.O.C. under contract NSC 98-2221-E-024-012.

REFERENCES

Ailisto, H., Alahuhta, P., Hataja, V., Kylloenen, V., & Lindholm, M. (2002, September). *Structuring context aware applications: Five-layer model and example case*. Paper presented at the Workshop on Concepts and Models for Ubiquitous Computing, Goteborg, Sweden.

Aumuller, D., Do, H., Massmann, S., & Rahm, E. (2005). Schema and ontology matching with COMA++. In *Proceedings of the ACM SIGMOD International Conference on Management of Data* (pp. 906-908).

Baldauf, M., Dustdar, S., & Rosenberg, F. (2007). A survey on context-aware systems. *International Journal of Ad Hoc and Ubiquitous Computing*, *2*(4), 263–277. doi:10.1504/IJA-HUC.2007.014070

Bartelt, C., Fischer, T., Niebuhr, D., Rausch, A., Seidl, F., & Trapp, M. (2005). Dynamic integration of heterogeneous mobile devices. In *Proceedings of the Workshop on Design and Evolution of Autonomic Application Software* (pp. 1-7).

Biegel, G., & Cahill, V. (2004). A framework for developing mobile, context-aware applications. In *Proceedings of the 2nd IEEE Conference on Pervasive Computing and Communication* (pp. 361-365).

Chen, G., & Kotz, D. (2000). *A survey of context-aware mobile computing research* (Tech. Rep. No. TR2000-381). Hanover, NH: Dartmouth College.

Chien, B. C., He, S. Y., Tsai, H. C., & Hsueh, Y. K. (2010). An extendible context-aware service system for mobile computing. *Journal of Mobile Multimedia*, *6*(1), 49–62.

Chien, B. C., Tsai, H. C., & Hsueh, Y. K. (2009). CADBA: A context-aware architecture based on context database for mobile computing. In *Proceedings of the International Workshop on Pervasive Media and the Sixth International Conference on Ubiquitous Intelligence and Computing* (pp. 367-372).

Dey, A. K., & Abowd, G. D. (1999). The context toolkit: Aiding the development of context-aware applications. In *Proceedings of the SIGCHI Conference on Human Factors in Computing Systems* (pp. 434-441).

Dey, A. K., Abowd, G. D., & Salber, D. (2001). A conceptual framework and a toolkit for supporting the rapid prototyping of context-aware applications. *Human-Computer Interaction*, *16*, 97–166. doi:10.1207/S15327051HCI16234_02

Do, H., & Rahm, E. (2002). COMA - a system for flexible combination of schema matching approaches. In *Proceedings of the 28th Conference on Very Large Data Bases* (pp. 610-621).

Gu, T., Pung, H. K., & Zhang, D. Q. (2004). Toward an OSGi-based infrastructure for context-aware applications. *IEEE Pervasive Computing/IEEE Computer Society and IEEE Communications Society*, *3*(4), 66–74. doi:10.1109/MPRV.2004.19

Gudivada, V. N., & Raghavan, V. V. (1995). Design and evaluation of algorithms for image retrieval by spatial similarity. *ACM Transactions on Information Systems*, *13*(2), 115–144. doi:10.1145/201040.201041

Hegering, H. G., Küpper, A., Linnhoff-Poien, C., & Reiser, H. (2003). Management challenges of context-aware services in ubiquitous environments. In M. Brunner & A. Keller (Eds.), *Proceedings of the 14th IFIP/IEEE International Workshop on Distributed Systems: Operations and Management* (LNCS 2867, pp. 321-339).

Hofer, T., Schwinger, W., Pichler, M., Leonhartsberger, G., Altmann, J., & Retschitzegger, W. (2002). Context-awareness on mobile devices - the hydrogen approach. In *Proceedings of the 36th Annual Hawaii International Conference on System Sciences* (pp. 292-302).

Lim, W., Nguyen, T. V., Yu, M., & Choi, D. (2007). Converting sensor data into ontology using DOM model in home network. In *Proceedings of the International Symposium on Information Technology Convergence* (pp. 101-105).

Madhavan, J., Bernstein, P. A., & Rahm, E. (2001). Generic schema matching with Cupid. In *Proceedings of the 27th International Conference of Very Large Data Bases* (pp. 49-58).

Melnik, S., Garcia-Molina, H., & Rahm, E. (2002). Similarity flooding: A versatile graph matching algorithm and its application to schema matching. In *Proceedings of the 18th International Conference on Data Engineering* (pp. 117-128).

Nakazawa, J., Tokuda, H., Edwards, W. K., & Ramachandran, U. (2006). A bridging framework for universal interoperability in pervasive systems. In *Proceedings of the 26th IEEE International Conference on Distributed Computing Systems* (pp. 3-3).

Rahm, E., & Bernstein, P. A. (2001). A survey of approaches to automatic schema matching. *International Journal on Very Large Data Bases*, *10*(4), 334–350. doi:10.1007/s007780100057

Roman, M., Hess, C., Cerqueira, R., Ranganathan, A., Campbell, R. H., & Nahrstedt, K. (2002). Gaia: A middle-ware infrastructure for active spaces. *IEEE Pervasive Computing/IEEE Computer Society and IEEE Communications Society*, *1*(4), 74–83. doi:10.1109/MPRV.2002.1158281

Schilit, A., & Theimer, M. (1994). Disseminating active map information to mobile hosts. *IEEE Network*, *8*(5), 22–32. doi:10.1109/65.313011

Schmohl, R., & Baumgarten, U. (2008). A generalized context-aware architecture in heterogeneous mobile computing environments. In *Proceedings of the Fourth International Conference on Wireless and Mobile Communications* (pp. 118-124).

Weiser, M. (1991). The computer for the twenty-first century. *Scientific American*, *265*(3), 94–100. doi:10.1038/scientificamerican0991-94

Weister, M. (1993). Hot topics: Ubiquitous computing. *IEEE Computer*, *26*(10), 71–72.

This work was previously published in the International Journal of Handheld Computing Research (IJHCR), Volume 2, Issue 2, edited by Wen-Chen Hu, pp. 65-77, copyright 2011 by IGI Publishing (an imprint of IGI Global).

Section 7
Mobile Green Computing, Location–Based Services (LBS), and Mobile Networks

Chapter 18

Reducing Power and Energy Overhead in Instruction Prefetching for Embedded Processor Systems

Ji Gu
University of New South Wales, Australia

Hui Guo
University of New South Wales, Australia

ABSTRACT

Instruction prefetching is an effective way to improve performance of the pipelined processors. However, existing instruction prefetching schemes increase performance with a significant energy sacrifice, making them unsuitable for embedded and ubiquitous systems where high performance and low energy consumption are all demanded. This paper proposes reducing energy overhead in instruction prefetching by using a simple hardware/software design and an efficient prefetching operation scheme. Two approaches are investigated: Decoded Loop Instruction Cache-based Prefetching (DLICP) that is most effective for loop intensive applications, and the enhanced DLICP with the popular existing Next Line Prefetching (NLP) for applications of a moderate number of loops. The experimental results show that both DLICP and the enhanced DLICP deliver improved performance at a much reduced energy overhead.

INTRODUCTION

On-chip cache has been widely used in modern microprocessor systems to bridge the speed gap between the processor and main memory. Cache exploits the spatial and temporal locality of memory reference to avoid the long latency of memory access from the processor. The high cache hit ratio plays a vital role in the overall system performance. This is especially essential for the instruction cache (I-cache) because of frequent instruction fetch operations. An instruction cache miss will cause the processor stall, hence slowing down the system.

DOI: 10.4018/978-1-4666-2785-7.ch018

Plenty of techniques have been proposed to reduce I-cache misses for high system performance. Among them is the instruction prefetching (Smith, 1978; Hennessy & Patterson, 2003) -fetching instructions from memory into the cache before they are used so that cache misses can be avoided or reduced. However, existing instruction prefetching schemes mainly focus on improving cache performance, often suffering significant energy losses due to a large amount of wasteful over-prefetching operations and/or the complicated prefetching hardware components.

Nevertheless, low energy consumption is also one of the important design issues for embedded systems, especially for those used in battery driven mobile/handheld devices, where large and heavy batteries are not feasible. A low power/energy design is, therefore, mostly desired for a long battery life.

In this paper, we aim to reduce energy overhead in instruction prefetching by using a simple prefetching hardware/software design and an efficient prefetching operation scheme. We investigate two approaches: the decoded loop instruction cache-based prefetching (DLICP) and the enhanced DLICP.

The decoded loop instruction cache (DLIC) originates from the decoded instruction buffer (DIB) proposed in Bajwa et al. (1997). It is a small tag-less cache residing between the instruction decoder and the execution unit in the microprocessor to store decoded loop instructions so that fetching and decoding the same set of instructions for the following loop iterations can be avoided, hence reducing energy dissipation in the processor.

We extend this energy-saving technique to instruction prefetching by overlapping the execution of decoded loops with fetching instructions to the cache from memory so that most instructions are available in the cache when they are executed. This approach is effective for loop intensive applications. For applications with a small amount of loops, we enhance the design with the existing Next Line Prefetching (NLP) scheme, which has

been proved efficient in cache miss reduction for applications with a dominant sequential instruction execution flow (Smith & Hsu, 1992).

INSTRUCTION PREFETCHING TECHNIQUES

The proliferation of handheld, mobile, and ubiquitous devices has led to the research boom on embedded systems. Various design issues and approaches have been proposed. Bisdikian et al. (1998) present an experimental platform to research technologies and applications that enable ubiquitous, environment-aware, and low-cost computing of handheld devices in the wireless personal access networks. Medvidovic et al. (2003) developed a software-architecture-based scheme to support distributed computation on handheld devices. Several power management schemes at the hardware level have also been proposed to reduce power consumption in different mobile system components such as the displayer (Min & Cha, 2007), the graphic processing unit (Nam, Lee, Kim, Lee, & Yoo, 2008).

Our proposal in this paper is an energy-efficient instruction prefetching scheme for embedded processors that can be used in the mobile/handheld devices for high system performance and low energy consumption. This section reviews some existing instruction prefetching methods for cache performance optimization.

Existing instruction prefetching techniques can be classified as software based prefetching and hardware based prefetching. Software prefetching schemes (Gornish, Granston, & Veidenbaum, 1990; Luk & Mowry, 1998; Cristal et al., 2005) rely on the compiler to insert prefetch instructions into the program code before the application is executed, which requires a known memory access behavior and a dedicated compiler.

The hardware-based prefetching is transparent to software and exploits the status of the program execution to dynamically prefetch instructions for future use. It is more flexible than the software based approach but incurs hardware overhead and increases the complexity of the processor architecture. The hardware approaches mainly include sequential prefetching and non-sequential prefetching.

Next Line Prefetching (Smith & Hsu, 1992) that utilizes the spatial locality of the program execution is one of the sequential prefetch approaches. On an instruction cache miss, it fetches the current cache miss line and sequentially prefetches the next lines to reduce possible cache misses. Based on the similar idea but with some flexibility, the adaptive sequential prefetching method (Dahlgren, Dubois, & Stenstroem, 1995) allows varying number of cache lines to be fetched.

The stream buffer prefetching (Jouppi, 1990) is another type of sequential prefetching designed specifically for the direct-mapped cache (where conflict cache misses may become a key problem). This approach places the prefetched cache line into a stream buffer and only writes it to the cache when it is actually referenced by the processor, to reduce possible conflict cache misses. A downside of this approach is that, in case a referenced data item is missing in both the cache and buffer, the buffer will be flushed by the next cache line. This may waste many prefetched cache lines and make the prefetching scheme ineffective.

The sequential prefetching is efficient for programs with sequential execution. To handle applications with a large number of branches, Pierce et al. (1996) proposed a non-sequential Wrong-Path scheme that prefetches instructions for all branch directions. Stride-directed prefetching (Fu, Patel, & Janssens, 1992; Kim & Veidenbaum, 1997) is another non-sequential approach, which is based on the observation that if a memory address is accessed; the memory location some stride away from the address is likely to be accessed soon. This method examines the memory access behavior for such a potential stride. If the stride is found, cache lines to be prefetched are offset by such a distance. A third example of non-sequential prefetching

is the shadow directory prefetching (Charney & Puzak, 1997); it associates each cache line with a shadow address that points to the next possible cache line. When a cache line is accessed and hit, its shadow-addressed cache line will be prefetched.

Some other non-sequential schemes utilize cache miss prediction for instruction prefetch. Joseph and Grunwald (1999) proposed a prediction based a prefetching technique, where a Markov model is used to correlate a stream of instruction misses. The predicted miss addresses are stored in the prediction table indexed by the related miss address. The instruction prefetching is triggered when a cache miss occurs. The tag-correlating prefetching (Hu, Martonosi, & Kaxiras, 2003) is a similar technique. However, it only stores the tag bits of the missing instruction addresses to reduce its size. The fetch directed instruction prefetching scheme (Reinman, Calder, & Austin, 1999) uses a branch predictor to predict the program execution stream. The branch predictor generates a queue of prefetching targets and the prefetched cache lines are initially stored in an additional prefetch buffer. The prefetched instructions are only written into the cache when they are referenced. A cache miss due to the misprediction has to flush the prefetching target queue and the prefetched instruction buffer.

In the branch history-guided prefetching Srinivasan et al. (2001) propose to correlate the instruction cache misses with branch instructions based on their execution history. They store the correlations in the prefetching table indexed by the address of branch instruction. The prefetching is triggered by the branch instructions when the same correlations are found later during the program execution. Zhang et al. (2002) propose the execution history guided prefetching where they correlate the cache misses with every instruction according to the execution history. This scheme has finer granularity than Srinivasan et al. (2001); any instruction (not only the branch instructions) can potentially be the prefetching trigger to allow more prefetching opportunities and effectiveness.

The above hardware-based prefetching schemes are popular techniques for cache performance improvement of the general purpose computer architecture. Most schemes are impractical in mobile or ubiquitous embedded systems, where low energy consumption is of ultimate importance.

In this paper, we aim to reduce energy consumption in instruction prefetching and further improve the prefetching efficiency by effectively paralleling instruction prefetching with the processor execution. We compared our approach with the Next Line Prefetching, which is the most power effective approach among existing prefetching schemes. Experimental results show that our design is more efficient in both power reduction and performance improvement. The structure and working principle of our DLICP scheme are introduced in the next section, where the hardware/software codesign and the prefetching operation scheme are discussed in detail.

DLIC-BASED INSTRUCTION PREFETCHING

The system architecture with our proposed DLIC-based prefetching scheme is illustrated in Figure 1. It contains (1) a five-stage pipeline processor, with the decoded loop instruction cache (DLIC) sitting between the instruction decode stage (ID) and execution stage (EXE); and (2) a two-level memory hierarchy, with the separate on-chip instruction cache (I-Cache) and data cache (D-Cache), an off-chip main memory, and a memory controller. The memory controller controls memory access in two fashions each operated by Fetcher and Prefetcher, respectively. For a normal processor execution, Fetcher retrieves instructions from cache, or from memory if there is a cache miss; During the execution of a decoded instruction loop, Prefetcher fetches instructions that are to be executed after the loop but are not yet in the cache.

It is worth noting that the DLIC structure implemented in this paper has a more ability than the normal decoded instruction buffer. It allows to cache loops of indeterminate loop counts, rather than only cache loops of a known number

Figure 1. Architecture for DLIC-based instruction prefetching

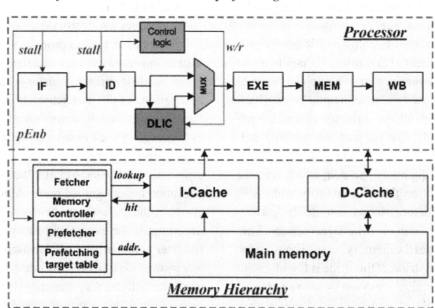

of iterations in the traditional decoded instruction buffer design.

The hardware/software design for the decoded loop instruction cache and an efficient strategy for prefetching operations are elaborated in the following subsections.

Hardware/Software Design of Decoded Loop Instruction Cache

For a given application, its loop execution behavior can be easily extracted, which makes it possible to use software approach to storing decoded loop instructions so that associated hardware component can be simplified, hence reducing hardware area cost and energy consumption. We aim at basic loops that are frequent and small in size, which is often the case in most embedded and ubiquitous systems (Villarreal, Lysecky, Cotterell, & Vahid, 2002). For such loops, we define two special instructions:

$$\begin{cases} slp \quad f, \quad \#iterations \\ elp \quad f, \quad rs, \quad rt \end{cases} \quad (1)$$

Instruction *slp* will be inserted at the top of loops, and instruction *elp* the end of loops, to control to store and execute the decoded loop instructions.

The formats of these two instructions (based on SimpleScalar PISA (Burger & Austin, 1997) architecture) are illustrated in Figure 2a (more explanation will follow). The hardware components related to the cache operation are given in Figure 2b.

Apart from the multiple entries in the DLIC cache for storing decoded instructions, there are two special registers: R_{LC} (for loop count in terms of the number of loop iterations) and R_{LS} (for loop size in terms of the number of instructions in the loop), and two special counters: C_{instr} (for counting number of executed instructions in a loop) and C_{iter} (for counting number of executed loop iterations). The counters are initialized for each loop execution.

Depending on whether the loop count is known at the compile time, the instruction pair will have different treatment for two different cases, which is controlled by flag f in the instruction.

CASE 1: LOOPS WITH DETERMINATE LOOP COUNTS

For the loop whose iteration count is known before execution, the flag f of the *slp* instruction is set to *0xFF*, and the flag in instruction *elp* set to *0x00*. Examples of such loops are given in Figure 3a and b, where both *while-loop* and *for-loop* have a determinate loop count at the compile time. Figure 3c demonstrates the corresponding *slp/elp* instructions with the loop count equal to 10.

When executing *slp* with the *0xFF* flag value, the processor saves the *#iterations* value given in the instruction in register R_{LC} and enables the

Figure 2. Formats of special instructions (a); components in DLIC cache (b)

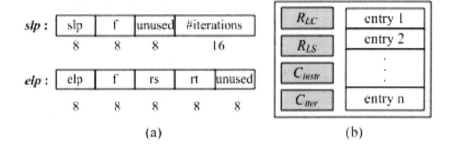

Figure 3. Determinate loop count: (a) while-loop; (b) for-loop; (c) special instruction pair

i = 10;
while (i--) {
  sum += i;
}
	for (i = 0; i < 10; i++)
	{
	  sum += i;
	}
		slp 0xff, 10
		.
		.
		.
		elp 0x00

(a)	(b)	(c)

decoded instruction caching function. For each instruction executed in the first loop iteration, its decoded instruction value is sequentially stored in the DLIC cache and counter C_{instr} is incremented by 1. Therefore, at the end of the first iteration when instruction *elp* is encountered, C_{instr} records the number of instructions in the loop. This number is then saved in register R_{LS} (the register for loop size) and C_{instr} is reset to 0 for the next loop iteration. For each loop iteration, counter C_{iter} is incremented by 1 until it reaches the loop count value stored in register R_{LC}, which means the decoded instruction loop is finished and the processor is back to the normal execution state.

CASE 2: LOOPS WITH INDETERMINATE LOOP COUNTS

For loops with unknown loop counts at the compile time, as the examples shown in Figure 4a and b, the *f* flag for instruction *slp* is set to *0x00* and for instruction *elp* instruction, it is set to *0xFF*.

During executing *slp* with the *0x00* flag value, register R_{LC} and counter C_{iter} are not used. But register R_{LS} and counter C_{instr} work in the same way as in Case 1. R_{LS} stores the total number of decoded instructions that should be executed for each of the loop iterations, and C_{instr} counts the number of executed instructions for the current iteration. Unlike in Case 1, where *elp* is executed only once for a loop, *elp* in Case 2 will be executed at the end of each iteration to determine whether the loop is finished. When the condition that registers *rs* and *rt* have the same value is satisfied, the loop execution is terminated.

It is worth noting that by using the special instructions and related hardware design, the loop control instructions in the original program code can be removed, reducing the total instruction count of the application, hence improving performance.

Figure 4. Indeterminate loop count: (a) while-loop; (b) for-loop; (c) special instruction pair

i = 1 << (s - 1);
while (*retval < i)
{ i = (-1 << s) - 1;
  *retval = *retval + i;
}
	if (offset > 0)
	for (i = 0; i < (offset + 7) / 8; i++)
	{
	  *outbuf++ = buf[i];
	}
		slp 0x00
		.
		.
		.
		elp 0xff, rs, rt

(a)	(b)	(c)

Instruction Prefetching Control Scheme

Due to different program control flows, some prefetched instructions may not be actually used, which not only wastes time but also incurs unnecessary energy lost.

To improve the prefetching efficiency, we try to prefetch instructions on the execution path of high operating frequency. Take the execution control flow shown in Figure 5a as an example. It contains 7 basic blocks; each block consists of a sequence of instructions. Blocks *B1* and *B6* are loops (*L1*, *L2*), whose decoded instructions will be cached; between the two loops are four basic blocks connected by two branches, which leads to various execution paths. The frequencies of the branches to different targets are shown in the flow. Blocks *B2*, *B4*, *B6* form an execution path with a higher execution frequency. We, therefore, want to prefetch the instructions in those blocks during the *L1* execution.

The frequent execution path for each decoded loop can be found by profiling and is stored in a table, called prefetching target table (PTT). Each entry in the table associates with a decoded loop and holds the information of the instruction blocks on the frequent execution path. For each instruction block, its start address and size in terms of the number of instructions are provided so that all instructions of the block in the memory can be located by *Prefetcher* (Figure 1). Figure 5b illustrates the PTT table for the execution flow in Figure 5a. The first block in each table entry is always the immediate block of the related loop.

During execution of a decoded loop, we use the PTT table to find the instructions on the frequent execution path and to prefetch them from memory if they are not available in the cache. To explain, we use the execution of *L1* as an example (see Figure 5c for the execution timing diagram).

Assume *L1* has 8 iterations. After the first iteration, all instructions in the loop have been decoded and saved in the DLIC cache. The processor is

Figure 5. (a) An example of execution control flow; (b) the corresponding prefetching target table; (c) execution diagram of the DLICP scheme

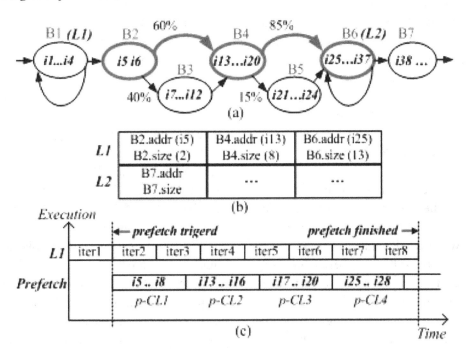

now in the state for the decoded loop execution, where the Program Counter (PC) is temporarily disabled with its value statically pointing to the instruction immediately after the loop (i.e., instruction i5 in the example) during the whole decode loop execution.

When the L1 execution enters the second iteration, the instruction prefetching is triggered. The *Prefetcher* searches the PTT table (based on the current PC value) for the first instruction block on the frequent execution path. The block is then checked to see whether instructions of the block are available in the cache, if not, the related cache line(s) will be prefetched; otherwise, continue to the next basic block. This process is repeated for the rest of the instruction blocks in the PTT entry until the execution of the decoded loop is finished, as illustrated in Figure 5c, where a 4-word cache line is assumed and each instruction is one-word long. Here we also assume all instructions on the frequent execution path are not available in the cache and are prefetched in four cache lines (denoted by $p\text{-}CL_1$ to $p\text{-}CL_4$ in Figure 5c).

As can be seen, the cache line size and duration of the decoded loop execution affect the number of instructions that can be prefetched. The larger the cache line and the longer the decoded loop execution, the more instructions can be fetched. But the large cache line may allow *Prefetcher* to fetch instructions that are not needed; for example, instructions i7, i8 in block B3 were brought along by prefetching the first cache line $p\text{-}CL_1$.

Enhanced DLIC Prefetching

With the DLICP design, the prefetching operation is restricted by the availability of basic loops and distribution of these loops in the program. For an application with a small number of such loops or loops are mainly run at the end of execution, only limited prefetching operations can be performed. This limitation can be circumvented by incorporating the existing NLP prefetching scheme that always prefetches instructions on a cache miss.

It must be emphasized that the cache miss saving from DLICP does not incur performance overhead because of the parallel prefetching operations (Figure 5c), while the saving from NLP is accompanied with the cache miss performance penalty. With DLIC, such penalties can be reduced. Therefore, combining both DLICP and NLP, we can achieve a higher cache miss reduction with a smaller performance overhead. This NLP scheme is implemented in the Fetcher memory control component in our system (Figure 1). The prefetching control of the enhanced DLICP is summarized as the algorithm in Figure 6. Next section presents the experimental setup, simulation results, and related discussions.

EXPERIMENTL RESULTS

To examine the effectiveness of our DLIC-based prefetching scheme, we applied it to a set of applications from Motorola's Powerstone (Scott, Lee, Arends, & Moyer, 1998) and MiBench (Guthaus et al., 2001) benchmark suites, which are widely used in the embedded application domain for automotive control, image processing, audio/video coding. The reference input data of each program are used in our experiments. Table 1 gives a short description of each benchmark.

Experimental Setup

The system setup for our experiments is shown in Figure 7. We selected the Simplescalar PISA (Burger & Austin, 1997) as the target processor instruction set architecture. An in-house VHDL model of the PISA processor, which was generated by the commercial tool ASIPMeister© (Itoh et al., 2000), was used as the platform for the application simulation.

The experiment started with a given application written in C, compiled by the simplescalar-gcc cross compiler and then simulated on the VHDL model. The loop behavior of each application was

Figure 6. Algorithm of prefetching control

```
/* if the IF pipeline stage has an instruction miss */
if IF-Stage-has-ICache-Miss then
    /* pipeline is stalled */
    fetching related cache line from memory;
    prefetching following cache lines;
    /* else if the whole loop is cached from the ID pipeline stage */
else if ID-Stage-encounters-elf-instruction then
    /* pipeline is active, and instruction execution is in parallel with instruction prefetching */
    prefetching following cache lines while executing decoded loop instructions;
    /* otherwise, normal pipeline execution */
else
    execution-on-instruction-cache;
end if
```

extracted from the execution instruction trace, based on which the frequent basic loops were modified with the special instruction pairs for decoded loop instruction caching. Hardware designs of the related prefetching schemes were then integrated to the processor model for evaluating their logic cost and energy overhead with Synopsys Design Compiler©. The area and energy consumption for the I-cache and main memory were obtained from CACTI 5 (Thoziyoor, Muralimanohar, Ahn, & Jouppi, 2008), which is a

Table 1. Description of the benchmarks

Benchmark	Description
blit	Graphics application
crc	Cyclic redundancy check
dijkstra	Graph search algorithm for networking
g3fax	Group three fax decode
jpeg	JPEG 24-bit image decompression
qsort	Quick sort
rc4	Stream cipher used in network protocols
rijndael	Standard encryption algorithm for AES
salsa	Data encryption
seal	Software-optimized encryption algorithm
sha	Cryptographic hash functions

widely used for evaluation of access time, area, and energy consumption of different cache and memory models.

In our experiment, we assumed the on-chip I-cache was 2-way set associative of 2K bytes, with the line size of 32 bytes. The small 2K I-cache is suitable for ubiquitous embedded systems where the costs are very restrained.

Performance Improvement

Performance can be evaluated in terms of total execution time, which is the product of the total number of clock cycles used and the clock cycle time when running an application. Cache performance affects the execution clock cycles. Therefore, we first investigate the cache performance.

With prefetching, apart from the two normal cache states (cache hit and cache miss), there is an extra case – the data requested is not yet in the cache, but is on the bus, being transferred from the memory to the cache by prefetching. We refer to this special case as false miss since it does not incur a new memory access.

Unlike a real cache miss, which has a fixed miss penalty, the false miss has a reduced and varying miss penalty due to the parallel operations of

Figure 7. Experimental setup

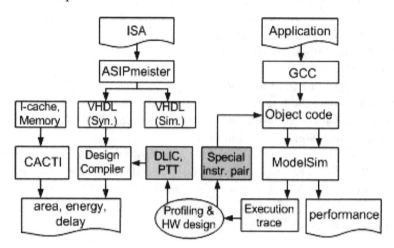

processor and prefetching, as illustrated in Figure 8, where on a real miss, the miss penalty is fixed (T); for the false misses, the miss penalty varies (t1 and t2 in the example), depending on the relative time needed for the processor to finish available instructions during the next line prefetching.

Table 2 lists the measurements of real cache misses (namely, false misses being excluded) when running different applications under the three prefetching schemes (columns 3-5). For a comparison, the baseline design without prefetching is also given in the table (column 2). The baseline is a 2-way associative instruction cache

of 2K bytes and has a cache line of 32 bytes. The cache access time is assumed as one clock cycle. The last row shows the average value. The normalized cache miss ratios as compared to the baseline design, are plotted in Figure 9.

As can be seen from the measurements, NLP provides a better cache-miss reduction (an average of 34.6%) than DLICP (22.1%). This is due to the insufficient basic loops available. Therefore, fewer prefetching operations in DLICP were performed, hence less cache misses were reduced. Two exceptions are the salsa and seal benchmarks, where there are a large amount of basic loops

Figure 8. Cache miss penalty

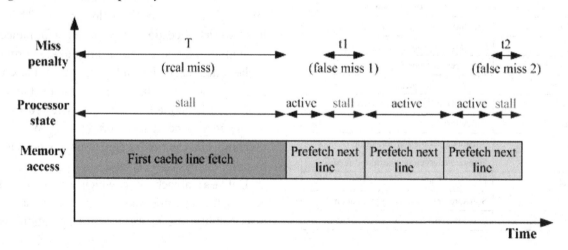

Table 2. Cache misses

	Baseline	**NLP**	**DLICP**	**Ehd. DLICP**
blit	59	34	49	31
crc	57	33	45	29
dijkstra	19774	12440	17845	17828
g3fax	1072	914	1028	992
jpeg	66186	52192	64849	64796
qsort	83	55	62	43
rc4	98	53	69	62
rijndael	1020	600	684	656
salsa	446	280	265	242
seal	1908	1278	1123	1114
sha	3197	2163	2581	2563
AVG	8536.4	6367.5	8054.5	8032.4

The DLICP and enhanced DLICP, however, need extra special instructions inserted in the program at compile time and thus the numbers of executed instructions are different from the baseline design. The system performance of the three designs can be, therefore, compared in terms of the total execution clock cycles.

Figure 10 shows the programs' total execution cycles normalized to the baseline design without instruction prefetching, and the cache miss penalty is assumed as 32 clock cycles. As can be seen, both DLICPs present better results than NLP for all applications. For some applications, such as blit, crc, and rijndael, the enhanced DLICP achieves remarkably higher performance improvement than DLICP. Up to 21% performance can be improved by the enhanced DLICP, as compared to the maximal 11% improvement from NLP. On average, the performance improvements of NLP, DLICP and enhanced DLICP are 4.2%, 8.2% and 8.9%, respectively. The enhanced DPLCP is two times better than NLP in terms of performance improvement. The performance improvement is largely due to the savings of false cache misses which are significant in NLP. False cache misses suffer a longer processor stall than cache hits.

evenly distributed in the program such that DLICP can reduce more cache misses than NLP by prefetching. With combined DLICP and NLP, we can, however, improve the cache miss reduction by an average of 32.4%.

Since prefetching does not affect the processor instruction set architecture and organization, all designs can have the same clock cycle time. NLP executes an equal number of instructions for a given application as the baseline design.

Figure 9. Normalized I-cache miss ratio

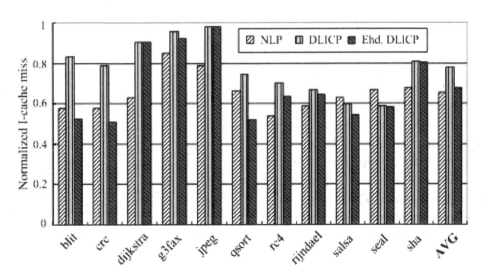

Figure 10. Performance improvement

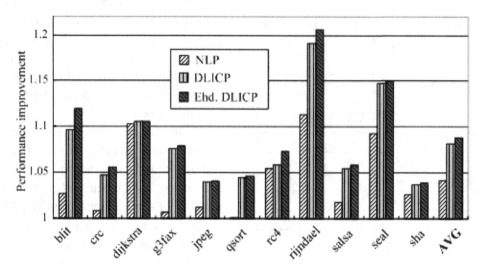

Implementation Costs

To evaluate the area costs and power consumption of prefetching, we have modeled the NLP, DLICP and enhanced DLICP designs in VHDL and each design is estimated using Synopsys Design Compiler©. For the cache and memory, we first obtain their area costs and the energy consumption per access from CACTI 5 (Thoziyoor et al., 2008), based on which we then estimate the total energy overhead of the prefetching. Both Design Compiler© and CACTI simulations are based on the 65nm technology.

Table 3 lists the simulation results. Rows 2 to 4 show the area cost of each prefetching logic and their energy consumption per memory access. The costs of the decoded instruction cache (DLIC) used by the two DLIC prefetching schemes are given in row 5, followed by the costs of the prefetching target table (PTT) that is used by the enhanced DLICP. It is worth to mention that because most applications have loops with less than 32 instructions and the number of decoded control signals for each instruction is less than 192 in our instruction architecture, we therefore set the decoded cache size as 32×192 bits. The last two rows (Rows

7 and 8) in Table 3 present the area and energy consumption of cache and memory per memory access measured from CACTI.

As can be seen from the table the NLP scheme shows small overheads as compared to our two DLICP approaches, in terms of area and energy consumption per memory access of the hardware logic (see data in rows 2-4). However, energy per access of the off-chip instruction memory (last two rows) is much higher than that of the on-chip I-cache, about 45x times higher for a 2M memory over the 2KB cache, which results in the savings on the overall energy overhead.

Table 3. Area cost and energy consumption

	Area [μm²]	Energy/Access [pJ]
NLP	1070	0.54
DLICP	3561	1.31
Ehd. DLICP	3932	1.61
DLIC (32 x 192 bits)	118958	20.41
PTT (32 x 40 bits)	18915	3.62
Instruction cache (2KB)	184255	42.66
Instruction memory (2MB)	3652402	1871.44

Energy Overhead Reduction

Some instructions prefetched may never be used, namely never be accessed by the processor before being flushed from the cache. Such useless prefetches do not aid performance improvement rather than waste valuable energy.

Table 4 shows our measurements of the total prefetches and useless prefetches when executing each application. As can be seen from the table, both DLICP schemes demonstrate low useless prefetches (336 and 350, respectively) as compared to the 4199 found in NLP.

Figure 11 gives the percentages of useless prefetches over the total prefetch number, which shows the DLICP is most effective – of all prefetches, 39.5% are useless and the rest contribute to the cache hits (data accessed from cache instead of memory), hence performance improvement and energy reduction.

To calculate the energy overhead of the three prefetching schemes, we use the runtime profile of the I-cache, main memory, and the prefetching logic activities (number of accesses, number of hits/misses, number of useless prefetches) collected during simulation, together with the energy per access values as given in Table 3.

Figure 12 shows the results when using the main memories of different sizes ranging from 64K to 4M bytes, where the energy overheads (displayed in lines) are normalized to the energy consumption of main memory access (shown in columns) of the baseline design without the prefetching function.

It can be seen from the plots, NLP consumes much higher energy than the other two schemes, consistently about 49% energy consumption over all different memory configurations. This is because such energy overhead is decided largely by the useless prefetching, where the number of useless prefetchings of NLP is about 49% of the I-cache miss in the baseline design. On the other hand, the energy overheads of the DLICP and enhanced DLICP are under 3.5%. When the main memory size is reduced to below 256KB, even energy savings from the two DLICPs can be observed. For example, a saving of 4.5% can be achieved when the main memory is as small as

Table 4. Useless prefetches

	NLP		DLICP		Ehd. DLICP	
	prefetches	useless	prefetches	useless	prefetches	useless
blit	34	9	12	2	34	6
crc	33	9	14	2	37	9
dijkstra	12440	5106	3108	1179	3130	1184
g3fax	914	756	162	118	215	135
jpeg	52192	38198	2339	1002	2414	1024
qsort	55	27	33	12	57	17
rc4	53	8	47	18	66	30
rijndael	600	180	564	228	627	263
salsa	280	114	415	234	454	250
seal	1278	648	1288	503	1301	507
sha	2163	1129	1012	396	1060	426
AVG	6368	4199	818	336	854	350

Figure 11. Normalized useless prefetches

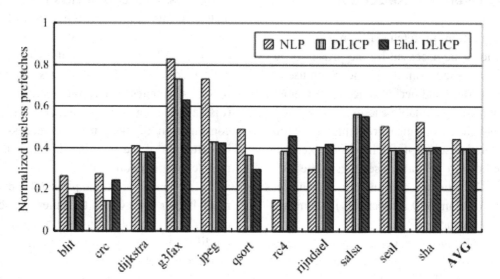

Figure 12. Energy penalties of the three schemes normalized to the energy consumption of the instruction memory of the baseline design

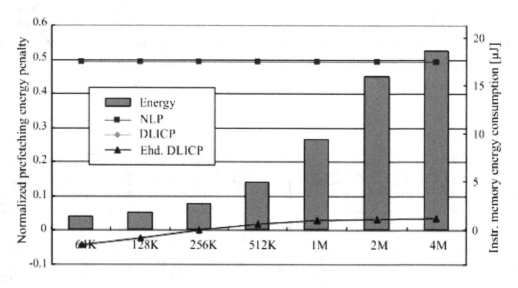

64KB. This is because the energy overhead of DLICP prefetching can be canceled out by the energy savings due to fetching decoded loop instructions from the energy-efficient DLIC instead from the energy-expensive I-cache during execution.

Design Efficiency

We use the ratio of the performance improvement over the energy overhead, that is, performance improvement per energy overhead, to evaluate the overall design efficiency of the three prefetching schemes and the results are plotted in Figure 13. The data used to calculate the ratio values can

Figure 13. Design efficiency of the three schemes

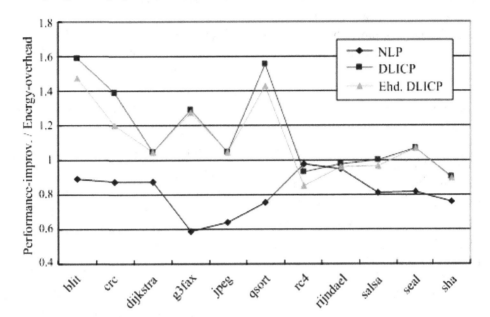

be seen in Table 5, which lists the performance improvement (P-IP), the relative energy overhead (E-OH), and the design efficiency (Ratio) for each of the applications under the three different prefetching designs.

As can be seen, both our two approaches show a higher efficiency than the NLP scheme for almost all the benchmarks, with only one exception, for application rc4, where rare loops are available for the decoded loop instruction pretching. The design efficiency of DLICP is slightly better than the

Table 5. The ratio of performance improvement over energy overhead

	NLP			DLICP			Ehd. DLICP		
	P-IP	E-OH	Ratio	P-IP	E-OH	Ratio	P-IP	E-OH	Ratio
blit	1.027	1.153	0.89	1.096	0.690	1.59	1.119	0.758	1.48
crc	1.009	1.158	0.87	1.048	0.756	1.39	1.057	0.879	1.20
dijkstra	1.104	1.259	0.88	1.106	1.057	1.05	1.106	1.059	1.04
g3fax	1.006	1.706	0.59	1.076	0.832	1.29	1.079	0.848	1.27
jpeg	1.013	1.578	0.64	1.040	0.993	1.05	1.041	0.994	1.05
qsort	1.001	1.326	0.75	1.044	0.671	1.56	1.046	0.731	1.43
rc4	1.054	1.082	0.97	1.058	1.139	0.93	1.073	1.262	0.85
rijndael	1.113	1.177	0.95	1.191	1.221	0.97	1.206	1.256	0.96
salsa	1.018	1.256	0.81	1.055	1.057	0.99	1.059	1.093	0.97
seal	1.093	1.340	0.82	1.147	1.074	1.07	1.149	1.076	1.07
sha	1.027	1.354	0.76	1.037	1.146	0.90	1.038	1.155	0.90
AVG			0.81			1.16			1.11

enhanced DLICP in terms of performance improvement per energy overhead. This is largely due to the smaller useless prefetches from the DLICP, as has been demonstrated in Table 4.

CONCLUSION

Our experiment results show that even the Next Line Prefetching (NLP), an existing low cost prefetching scheme, incurs a high energy overhead (around 49% of memory energy consumption), which is impractical for energy-aware embedded and ubiquitous systems.

In this paper, we presented an energy efficient instruction prefetching design for ubiquitous embedded systems with two-level memory hierarchy (on-chip cache and off-chip memory). We exploit the decoded loop cache and maximally parallelize the instruction prefetching with the decoded loop execution to reduce instruction cache misses while at a low energy overhead. The decoded loop cached based prefetching (DLICP) can be enhanced with the NLP approach for applications, where limited loops available for the decoded loop cache.

Our experiments show that both DLICP schemes outperform NLP with improved performance and much less energy overhead, an average of 3.5% extra energy consumption as compared to 49% extra energy consumed by NLP. For some applications, the enhanced DLICP scheme offers a markedly better performance than DLICP. Up to 21% performance can be improved by the enhanced DLCIP, as compared to the 11% performance improvement by NLP.

As for the future work, we will extend the decoded loop cached design so that it can handle more complex loop cases such as the loops with procedure calls and conditional branches. Such a design extension will enable more instructions to be prefetched, hence even higher performance and lower energy consumption can be achieved.

REFERENCES

Bajwa, R. S., Hiraki, M., Kojima, H., Gorny, D. J., Nitta, K., & Shridhar, A. (1997). Instruction buffering to reduce power in processors for signal processing. *IEEE Transactions on Very Large Scale Integration Systems*, 5(4), 417–424. doi:10.1109/92.645068

Bisdikian, C., Bhagwat, P., Gaucher, B. P., Janniello, F. J., Korpeoglu, I., & Naghshineh, M. (1998). WISAP: A wireless personal access network for handheld computing devices. *IEEE Personal Communications*, 5(6), 18–25. doi:10.1109/98.736474

Burger, D. C., & Austin, T. M. (1997). *The simplescalar tool set, version 2.0* (Tech. Rep. No. CS-TR-1997-1342). Madison, WI: University of Wisconsin.

Charney, M. J., & Puzak, T. R. (1997). Prefetching and memory system behavior of the SPEC95 benchmark suite. *IBM Journal of Research and Development*, 41(3), 265–286. doi:10.1147/rd.413.0265

Cristal, A., Santana, O., Cazorla, F., Galluzzi, M., Ramirez, T., & Pericas, M. (2005). Kilo-instruction processors: Overcoming the memory wall. *IEEE Micro*, 25(3), 48–57. doi:10.1109/MM.2005.53

Dahlgren, F., Dubois, M., & Stenstroem, P. (1995). Sequential hardware prefetching in shared-memory multiprocessors. *IEEE Transactions on Parallel and Distributed Systems*, 6(7), 733–746. doi:10.1109/71.395402

Fu, J. W. C., Patel, J. H., & Janssens, B. L. (1992). Stride directed prefetching in scalar processors. *ACM SIGMICRO Newsletter*, 23(1-2), 102–110. doi:10.1145/144965.145006

Gornish, E. H., Granston, E. D., & Veidenbaum, A. V. (1990). Compiler-directed data prefetching in multiprocessors with memory hierarchies. In *Proceedings of the 4th International Conference on Supercomputing* (pp. 354-368).

Guthaus, M. R., Ringenberg, J. S., Ernst, D., Austin, T. M., Mudge, T., & Brown, R. B. (2001). Mibench: A free, commercially representative embedded benchmark suite. In *Proceedings of the IEEE 4th Annual Workshop on Workload Characterization* (pp. 83-94).

Hennessy, J. L., & Patterson, D. A. (2003). *Computer architecture: A quantitative approach* (3rd ed.). Amsterdam, The Netherlands: Elsevier Science.

Hu, Z., Martonosi, M., & Kaxiras, S. (2003). TCP: Tag correlating prefetchers. In *Proceedings of the 9th International Symposium on High-Performance Computer Architecture* (pp. 317-326).

Itoh, M., Higaki, S., Takeuchi, Y., Kitajima, A., Imai, M., Sato, J., et al. (2000). Peas-iii: An asip design environment. In *Proceedings of the IEEE International Conference on Computer Design* (pp. 430-436).

Joseph, D., & Grunwald, D. (1999). Prefetching using Markov predictors. *IEEE Transactions on Computers, 48*(2), 121–133. doi:10.1109/12.752653

Jouppi, N. P. (1990). Improving direct-mapped cache performance by the addition of a small fully-associative cache and prefetch buffers. In *Proceedings of the 17th Annual International Symposium on Computer Architecture* (pp. 364-373).

Kim, S., & Veidenbaum, A. V. (1997). Stride-directed prefetching for secondary caches. In *Proceedings of the International Conference on Parallel Processing* (pp. 314-321).

Luk, C.-K., & Mowry, T. C. (1998). Cooperative prefetching: Compiler and hardware support for effective instruction prefetching in modern processors. In *Proceedings of the 31st Annual ACM/IEEE International Symposium on Microarchitecture* (pp. 182-194).

Medvidovic, N., Mikic-Rakic, M., Mehta, N. R., & Malek, S. (2003). Software architectural support for handheld computing. *IEEE Computer, 36*(9), 66–73.

Min, J., & Cha, H. (2007). Reducing display power in dvs-enabled handheld systems. In *Proceedings of the International Symposium on Low Power Electronics and Design* (pp. 395-398).

Nam, B.-G., Lee, J., Kim, K., Lee, S., & Yoo, H.-J. (2008, April). Cost-effective low-power graphics processing unit for handheld devices. *IEEE Communications Magazine, 46*(4), 152–159. doi:10.1109/MCOM.2008.4481355

Pierce, J., & Mudge, T. (1996). Wrong-path instruction prefetching. In *Proceedings of the 29th Annual ACM/IEEE International Symposium on Microarchitecture* (pp. 165-175).

Reinman, G., Calder, B., & Austin, T. (1999). Fetch directed instruction prefetching. In *Proceedings of the 32nd Annual ACM/IEEE International Symposium on Microarchitecture* (pp. 16-27).

Scott, J., Lee, L. H., Arends, J., & Moyer, B. (1998). Designing the low-power M-CORE architecture. In *Proceedings of the International Symposium on Computer Architecture Power Driven Microarchitecture Workshop* (pp. 145-150).

Smith, A. J. (1978). Sequential program prefetching in memory hierarchies. *Computer, 11*(12), 7–21. doi:10.1109/C-M.1978.218016

Smith, J. E., & Hsu, W.-C. (1992). Prefetching in supercomputer instruction caches. In *Proceedings of the ACM/IEEE Conference on Supercomputing* (pp. 588-597).

Srinivasan, V., Davidson, E. S., Tyson, G. S., Charney, M. J., & Puzak, T. R. (2001). Branch history guided instruction prefetching. In *Proceedings of the 7th International Conference on High Performance Computer Architecture* (pp. 291-300).

Thoziyoor, S., Muralimanohar, N., Ahn, J. H., & Jouppi, N. P. (2008). *Cacti: An integrated cache and memory access time, cycle time, area, leakage, and dynamic power model* (Tech. Rep. No. HPL-2008-20). Palo Alto, CA: HP Laboratories.

Villarreal, J., Lysecky, R., Cotterell, S., & Vahid, F. (2002). *A study on the loop behavior of embedded programs* (Tech. Rep. No. UCR-CSE-01-03). Riverside, CA: University of California.

Zhang, Y., Haga, S., & Barua, R. (2002). Execution history guided instruction prefetching. In *Proceedings of the 16th International Conference on Supercomputing* (pp. 199-208).

This work was previously published in the International Journal of Handheld Computing Research (IJHCR), Volume 2, Issue 4, edited by Wen-Chen Hu, pp. 42-58, copyright 2011 by IGI Publishing (an imprint of IGI Global).

Chapter 19
Interactive Rendering of Indoor and Urban Environments on Handheld Devices by Combining Visibility Algorithms with Spatial Data Structures

Wendel B. Silva
University of Utah, USA

Maria Andréia F. Rodrigues
Universidade de Fortaleza – UNIFOR, Brazil

ABSTRACT

This work presents a comparative study of various combinations of visibility algorithms (view-frustum culling, backface culling, and a simple yet fast algorithm called conservative backface culling) and different settings of standard spatial data structures (non-uniform Grids, BSP-Trees, Octrees, and Portal-Octrees) for enabling efficient graphics rendering of both indoor and urban 3D environments, especially suited for low-end handheld devices. Performance tests and analyses were conducted using two different mobile platforms and environments in the order of thousands of triangles. The authors demonstrate that navigation at interactive frame rates can be obtained using geometry rather than image-based rendering or point-based rendering on the cell phone Nokia n82.

INTRODUCTION

Interactive rendering of indoor and urban environments is a crucial problem and still a challenge for the field of computer graphics. Basically, the visualization pipeline describes the essential steps to generate an image from data stored in memory. The pipeline may be broken into four critical stages: retrieval from storage, processing in main memory, rendering, and display. It has to be computed for every frame to visualize an interactive motion (Akenine-Möller, Haines, &

DOI: 10.4018/978-1-4666-2785-7.ch019

Hoffman, 2008). The possible frame rate depends mainly on the 3D data. Several are the factors that are influential in this problem, such as: number of objects in an environment and its geometric complexity, object distribution, object dynamics, level of realism of the objects, number of pixels to be painted, etc. For example, the resulting rendering time usually grows with the number of triangles in the scene.

The display of 3D interactive graphics requires that each polygon is clipped against the viewing frustum (Cohen-Or, Chrysanthou, Silva, & Durand, 2002). In other words, rendering the objects that lie outside the field of view of the camera would be a waste of time since they are not visible on the screen. The fact that the users have control over their movements around the 3D environment means that rendering has to happen instantaneously in response to user interaction. As we know, real-time rendering relies on the graphics processors on the local execution platform to make all the rendering calculations. So, it can make some significant demands on the users' software and hardware capabilities.

Visibility algorithms are used to speed up image generation by rejecting invisible geometry before actual hidden-surface removal is performed. Visibility is not a trivial problem since a small change in the viewpoint might cause large changes in the visibility (Cohen-Or, Chrysanthou, Silva, & Durand, 2002).

The challenge of optimizing visibility algorithms for interactive rendering of polygons becomes even more complex when the execution platform is a handheld device (non-interactive graphics can be simply rendered as simple bitmaps). Actually, implementing fully 3D graphics on handheld devices is particularly difficult because the majority of them, that form the largest part of the market share, still present important limitations when compared to traditional personal computers: low processing power; little storage memory; lack of dedicated software support; etc. Also, there is an order of magnitude difference

between high and low-end mobile devices in graphics processing and computational capacity (Pulli, Aarnio, Miettinen, Roimela, & Vaarala, 2008; Capin, Pulli, & Akenine-Möller, 2008). Moreover, although some hardware characteristics are similar between our handheld devices of today and our personal computers from the nineties, there are some important and distinct features too, for example, the velocity of memory information allocation.

While some research has been done on the use of handheld computer technology, the full potential within the context of interactive rendering is still unknown. A visibility algorithm, for example, may run efficiently in one handheld device but be inefficient on another. Hence, there is an evident demand for proposals of optimization that propitiate the efficient use of the different technological resources available in each type of mobile device, in such a way as to ensure the generation of compact and simultaneously realistic implementations. This means that new studies and algorithms are required. The key question is to know what the best choice of visibility algorithm is in this context.

In our previous work (Silva & Rodrigues, 2009) we presented a detailed view of a 3D visualization and navigation system we have implemented on handheld devices, using the Open Graphics Library for Embedded Systems (OpenGL ES API) (KhronosGroup, 2008). Our prototype system initially presented only the traditional view frustum culling algorithm and two implementations of the backface culling algorithm: one based on the traditional algorithm (Clark, 1976) and another one available in the OpenGL ES API (Vincent, 2009), associated with Octrees for geometric rendering of an indoor environment containing 6199 polygons.

In this work, we have considerably extended our system for enabling efficient 3D graphics rendering of both indoor and urban environments on handheld devices. Firstly, we propose to examine several standard visibility algorithms we have implemented (view-frustum culling and backface culling), and a simple, yet fast algorithm, we designed and called conservative backface culling, to analyze how these approaches apply to handheld devices. Secondly, we associate these visibility algorithms with different settings of standard spatial data structures (Samet & Webber, 1988), such as non-uniform Grids, BSP-Trees, Octrees, and Portal-Octrees, and compare their performance through extensively testings of camera paths through four different 3D environments in the order of thousands of triangles on two different mobile platforms. We demonstrate in our results that navigation at interactive frame rates on low-end handheld devices can be obtained using geometry rather than image-based rendering or point-based rendering.

The remainder of this paper is organized as follows. The next section presents related work. We then conduct an overview of basic concepts with regard to some visibility algorithms and spatial subdivision schemes, as well as describe details of our implementation. The tests and results are detailed and discussed along with special notes on performance-related aspects. Finally, a conclusion and future work are summarized.

RELATED WORK

Some representative studies related to the visualization of 3D graphics on mobile devices are discussed in this section. When pertinent, they are compared or contrasted with ours.

The amount of current published work on local 3D visualization on handheld devices is limited because few mobile devices target graphical applications (Huang, Bue, Pattath, Ebert, & Thomas, 2007). Acceleration techniques have been developed to increase visualization speed in complex graphical environments, composed of a large quantity of polygons (Akenine-Möller, Haines, & Hoffman, 2008). Visibility algorithms are among these techniques. In this context, Cohen-Or et al. (2002) conducted a quite detailed compara-

tive study between different existing methods of visualization (Cohen-Or, Chrysanthou, Silva, & Durand, 2002).

Visibility algorithms seek the efficient removal of non-visible parts that compose a scene so that they are not processed by the rendering pipeline. Among the best-known algorithms are the methods that are executed in a pre-processing (offline) phase, and those that are executed in the application execution time (online). The visibility algorithms can also be classified as to their working space. There are algorithms that work in the space of the object, using 3D information from the environment; and those that operate in the space of the image, using a 2D representation (Cohen-Or, Rich, Lerner, & Shenkar, 1996) of the 3D space (Coorg & Teller, 1996), (Bittner, Havran, & Slavík, 1998). In this work we use 3D information from the environment, and the visibility algorithms implemented for walkthrough and related applications are executed at software level.

In terms of positioning, we can define two types of virtual environments: indoor and outdoor (urban environments, landscape and forest scenes). Visibility algorithms describe specific solutions for these types of environments (indoor, outdoor, or both). For example, Luebke and Georges partitioned an indoor environment into cells and portals, identifying the visible area for each cell (Luebke & Georges, 1995). Wonka and Schmalstieg developed a solution in real time to identify the visible parts of a scene in an urban environment (Wonka & Schmalstieg, 1999).

In a recent work, Mulloni et al. (2007) created and OpenGL ES based building walkthrough system that resembles ours in some aspects. However, they used the hierarchical view-frustum culling and the portal culling for their culling algorithms to investigate an indoor interactive walkthrough method of a 3D building, while we use the view-frustum culling, backface culling, and a simple yet fast algorithm we named conservative backface culling, for enabling efficient graphics rendering of both indoor and urban 3D environments. In an indoor environment, the cells can be divided typically into rooms and corridors. However, urban environments present another case of densely occluded scenes. In this case, the environment is open, where occlusion is formed from individual objects, such as buildings (Nurminen, 2009). While Mulloni et al. (2007) split an X3D indoor model (a modern successor of VRML format) into parts (rooms and corridors connected by portals) and viewed them with manually constructed visibility lists, we use a group of spatial data structures (non-uniform Grids, BSP-Trees, Octrees, and Portal-Octrees). Frequently, only part of large models can fit in memory at any time on handheld devices. We use these spatial data structures to maintain the dynamic scene and only the 3D scenes in the viewing area are loaded into the mobile device memory. Due to slow I/O operations, these authors discarded all textures from their 3D model. Further, that work does not present any detailed study on the performance of the developed system and uses a 3D building model consisting of 28608 triangles, whereas ours presents performance tests and analyses using environments in the order of thousands of triangles.

Lluch et al. (2005) proposed a client/server application where the hidden part of the environment is removed in the server, and the scene is shown on the handheld device using an open-source 3D library, targeted for PDAs and mobile phones, very similiar to that of OpenGL and OpenGL ES. In our work, however, removal of the hidden geometry occurs on the client itself, not requiring obligatory access to the server. Chang and Ger developed an application for handheld devices (Chang & Ger, 2002), for viewing 3D graphical environments, however, contrary to our work, the application is image-based rendering and does not use the OpenGL ES API. Duguet and Drettakis (2004) proposed an efficient way to display complex geometries. However, the tested geometric model is point-based rendering, and the spatial data structure implemented is the P-Grids. While various authors have used image-based rendering

(Chang & Ger, 2002), point-based rendering (Duguet & Drettakis, 2004), and volume rendering algorithms (Moser & Weiskopf, 2008), we have implemented and combined geometry rendering algorithms.

Hudson et al. (1997) consider that the objects positioned in the shadow generated by the obstacles are not visible to the observer. Starting from this idea, they described a method of visibility based on the dynamic choice of a set of obstacles, and on the calculation of their respective shadows. In contrast, we use several different structures of spatial partitioning, in the case, non-uniform Grids, BSP-Trees, Octrees, and Portal-Octrees (these last two structures with different depth levels) to make use of the spatial coherence.

Nurminen developed a 3D application of virtual map navigation for exterior environments on mobile devices, for which only the occlusion culling algorithm is implemented (Nurminen, 2006). In this work, apart from implementing various geometry rendering algorithms (view-frustum culling, backface culling, and conservative backface culling) and analyzing system performance for both indoor and urban environments, we combined these visibility algorithms and hierarchical data structures with the objective of obtaining interactive rendering rates. Moser and Weiskopf (2008) developed a technique for rendering and visualization of volumes on mobile devices. Differently from our work, they have not used visibility algorithms. Besides, their proposed solution is specific for volume rendering.

Some authors have carried out comparative studies of 3D applications in mobile devices. For example, Pulli (2006) elaborated quite a detailed study concerning some APIs more largely used in handheld devices. Very recently, Capin et al. presented a survey that concentrates on the state of the art in mobile interactive graphics (Pulli, Aarnio, Miettinen, Roimela, & Vaarala, 2008). Hwang et al. (2006) explored camera control parameters to generate different possibilities of scene views. Hachet et al. (2006) developed methods of interaction for navigating in virtual environments in mobile devices. Barbosa and Rodrigues (2006) developed two applications based on a framework they have implemented: a virtual rescue training for fire fighters, and a 3D virtual tour where multiple clients navigate and interact with each other in a shared 3D model (Rodrigues & Barbosa, 2006). However, in none of these works were visibility algorithms combined with any spatial data structure to optimize and demonstrate the performance gains of the application.

Finally, graphical systems in general have specific implementation details, dependent on the computing platforms being used. One example is the free 3D graphics application Blender which has been recently ported to the Windows Mobile platform (Russo, 2010), requiring modification to be able to run on PocketPCs. In contrast, the current version of our system provides the functionalities necessary to create graphical applications for running on several computing platforms (smartphones, cell phones, Pocket PCs, and desktop computers), without the need of porting. In this paper, however, we focus on performance analysis of our system using two computing platforms: cell phone and PocketPC.

OVERVIEW

Before describing the experiments conducted and the results obtained, we would like to give an overview of the algorithms in which the performance analysis of our system is based on. This will form a basis for the more detailed discussion in later sections. First, some rendering algorithms, followed by spatial subdivision schemes, are presented in the next subsections. Basically, the aim of the visibility algorithms is to reduce the number of triangles sent to the rendering pipeline, discarding the polygons that are not visible from a point in space where the observer is positioned, and spatial data schemes correspond to structures that manipulate spatial data. In general, both the

visibility algorithms and the spatial data are used to speed up rendering.

Visibility Algorithms

With regard to the visibility algorithms, we have 1) implemented and explored two algorithms (the traditional view-frustum culling, and another one, namely, in this work, conservative backface culling); 2) used and tested the backface culling algorithm available in the OpenGL ES API (Vincent, 2009); and 3) combined these four visibility algorithms with different settings of standard spatial subdivision schemes.

View-frustum culling algorithms can discard many polygons in the scene from the rendering process, removing those that are outside the viewing frustum. They are generally light, relatively simple to implement, and can be associated with other visibility algorithms and spatial data structures to enable faster rendering. They can also provide efficient intersection tests for frustum against bounding volumes, as well as mechanisms for speed-ups based on temporal coherence, resulting in important savings in computational resources. Therefore, making this visibility algorithm a possible candidate to be examined for mobile devices.

Basically, back-face culling algorithms avoid the rendering of polygons that face away from the viewer. Motivated by the fact that two different API OpenGL ES backface culling implementations (Vincent, 2009; Nokia, 2009) tested in our work have not contributed towards a rendering performance gain, we propose and we have implemented a very simple yet fast conservative backface culling method. In the traditional backface culling algorithm, for each triangle, we need to calculate the dot product between the normal vector of each of its face and the look-at vector of the camera. In our work, we opted for using a conservative approach to trade accuracy for speed by avoiding, in a short time interval, the rendering of a maximum number of polygons that face away from the viewer. To speed up backface culling, we execute all the calculations in pre-processing time, i.e., we divide the *x,z* coordinate system into regions (the number of regions changes according to the field of view) and identify the region in which the normal vector of the face of the triangle falls. In execution time, we calculate the region in which the camera look-at is located. If the region of the normal to the face of the triangle is identical to the camera look-at region (this verification uses only one byte comparison for each visible face), the triangle will be discarded; otherwise, it will be rendered.

In processing time, due to the possibility of navigation through the environments, we use the camera coordinate system instead of the fixed *z*-coordinate in the viewing coordinate system for backface culling. In pre-processing time, however, we use the *z*-coordinate in the viewing coordinate system for partitioning the space into regions.

Some other approaches may produce better culling than our algorithm does, as they consider more regions for culling (generally trading speed for accuracy). For example, Kumar *et al.* proposed a culling method that achieves sublinear speedup by discarding groups of polygons as a whole instead of visiting each polygon (Kumar, Manocha, Garrett, & Lin, 1996), whereas Zhang and Hoff III (1997) described culling operations per polygon on the pixel level, without imposing structural requirements on the database. Their approach has high complexity data structure and the memory requirement for their two-phase image-space tests is also extensive. The method implemented by Zhang and Hoff III is closely related to ours. It reduces the backface test to one logical operation per polygon (it uses a bitmask to represent groups of normals in the normal space partitioning) while requiring two bytes extra storage per polygon. However, our method needs one extra byte per polygon and creates the normal space partitions in the plane *x,z* to generate the regions that makes it suitable for low-end mobile and handheld devices.

Spatial Subdivision Schemes

As for the spatial data structures, we have implemented non-uniform Grids, Octrees, Portal-Octrees and BSP-Trees (Samet & Webber, 1988). A brief description of each of the data structures, as well as the specific details of each implementation will be presented next.

The non-uniform Grid is a structure that does not use much computing resources. Also, in choosing the number and size of the partitions one can consider the previous knowledge of the geometry and distribution of the static objects within the environment. In our implementation, for each environment, the developer defines the number and the size of partitions in each of the coordinate axes *x,y,z*. This offers more flexibility on choosing the grid structure that best meets the performance goals of the application.

An Octree is an easy-to-compute and easy-to-maintain structure for 3D polygonal objects. In our implementation, during the construction of the Octree, the environment is recursively subdivided, starting with a cube in which the entire data set fits until the depth of the tree reaches a limit predefined by the developer. This limit value can be set according to the environment characteristics, during the creation of the structure. We store only the indexes of the polygons in the Octree leaf nodes, avoiding redundant data storage. Additionally, we do not divide the polygons simultaneously present in more than one leaf node, in which case the division would increase the quantity of stored polygons, impairing the performance of the application.

Portal-Octree is a hybrid structure which combines the division of the environment into cells and the partition of the cell contents usually using Octrees (Dalmau, 2003). An indoor environment, for instance, can be divided into rooms with each one being a cell. The geometry contained within each cell is partitioned using an Octree. Considering that each room is connected to another one through a portal, one can easily use this data structure in conjunction with a visibility algorithm of cells-and-portals. An Octree does not need to be very deep and can be used for calculating the visible objects. In dynamic environments, it is only necessary to recalculate the current room's Octree. Our implementation of the Portal-Octree uses the non-uniform Grid to partition the environment. Each voxel belonging to the Grid corresponds to a cell. Each cell is then subdivided using an Octree of different depth levels. By knowing the geometry of the environment a priori, specific partitions can be chosen such that the Grid better adapts to the structure of the scene. For each cell of the Grid a spatial partition is undertaken using Octrees. Similarly to what occurs in the Octree, in the Portal-Octree the developer can also previously define the maximum depth level of the tree.

BSP-Trees divide a region of space into two subregions at each node. In our BSP-Tree implementation (a general oriented tree), to minimize the number of polygons stored in the structure (as in the Octree), the polygons located in more than one partition are not divided, and their references are stored in the nodes of both sides of the partition. We use the polygons of the environment as partition planes and choose the most balanced node. It is important to observe that, as the depth of a hierarchical structure is increased, a growing number of nodes will need to be tested. The number of triangles in each node will be less than, or equal to, the quantity of triangles present in its father node. This means that, the deeper the structure, the longer the processing time to conduct the tests in all the nodes. Furthermore, the processing time necessary to render a node may be less than the time spent to run through the hierarchical structure to this specific node.

It should be emphasized that the combination of visibility algorithms with different spatial partitioning structures may result in varying performance. The verification of intersection between an object and the viewing frustum also requires different intersection tests, depending on the spatial data structure being used. For example, in the

Octree, the intersection tests are realized between the bounding boxes of each node and the frustum; on the other hand, in the BSP-Tree, the intersection tests are done between the viewing frustum and the partition plane. In our view-frustum culling algorithm, we have implemented the intersection tests between the viewing frustum and (1) a point, (2) a sphere, and (3) a cube.

Choosing a good combination of visibility algorithms and hierarchical representations for 3D mobile systems is a hard, yet crucial decision. Actually, it has a very large impact on the final performance characteristics of your graphical application.

TESTS AND RESULTS

For the performance analysis tests, we have used two different mobile platforms: the iPaq hx2490b PocketPC which has an Intel PXA270 520MHz processor, with 192 MB total available memory (128 MB ROM and 64MB SDRAM), and without GPU; and the cell phone Nokia n82 which has an ARM 11 332MHz processor, with 100 MB internal memory, 128 MB SDRAM Memory, up to 90 MB Free Executable RAM Memory, and with GPU. All the tests were executed with the depth buffer (one of OpenGL ES features, also called z-buffer, which is a buffer that associates a depth value with each pixel in the frame buffer) of the mobile device enabled (Van Verth & Bishop, 2004). The depth of a pixel is determined based on the size of the view and projection matrix selected for rendering. If the depth buffer is enabled the objects in the scene will be drawn according to their z coordinate. Otherwise, according to the order of rendering. Thus, as each object in the scene is rendered the pixels that are closest to the camera are kept, as those objects block the view of the objects behind them.

Using these two mobile device models, our performance goal was to achieve the highest possible frame rates when rendering either indoor or urban environments. More specifically, the four 3D environments (E_1, E_2, E_3, and E_4) shown in Figure 1 and Figure 2 were used. These environments vary both in size and geometry complexity, with each one having different numbers of triangles and different types of objects (tables, penguins, cows, and dragons) spatially distributed in different ways. For each environment, three directional light sources (that simulate very distant point light sources) and Gouraud shading (which is used to achieve smooth lighting on low-polygon surfaces without the heavy computational requirements of calculating lighting for each pixel) were used, without texture maps applied to the surfaces of the polygons. In pre-processing time, the four environments were partitioned using the non-uniform Grid, BSP-Tree, Octree, and Portal-Octree structures (the last two data structures with 3, 4, 5, 6

Figure 1. The indoor environments E_1 (on the left) and E_2 (on the right). E_1 has two stories with 40 rooms aligned with the coordinate axes plus 6 objects (tables), totalizing 6199 triangles, and 214 KB of size. E_2 has the same stories as E_1 does, with 12 objects (penguins and tables) added, totalizing 30199 triangles, and 1048 KB of size.

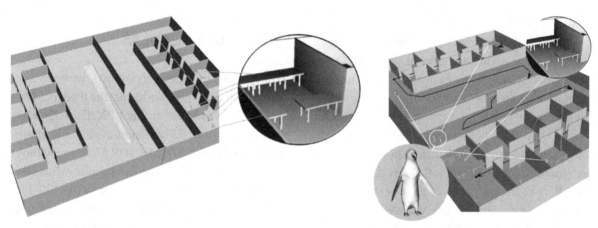

Figure 2. The urban environments E_3 (on the left) and E_4 (on the right) created using the procedural 3D model generator application CityGen (Leyvand, 2009). E_3 has 9 city blocks containing 16 non-aligned buildings each, with different orientations and heights uniformly distributed (2548 triangles, and 108 KB of size). E_4 has the same number of city blocks as E3 does, with 5 objects (cows and dragons) added, totalizing 10040 triangles and 511 KB of size.

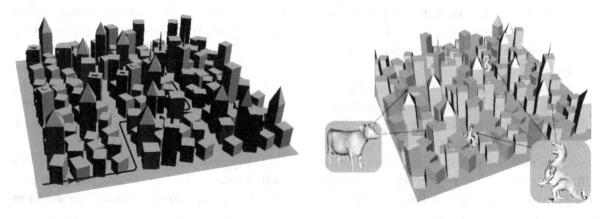

and 1, 2, 3, 4 depth levels, respectively). Octrees and Portal-Octrees with more depth are used to preserve the detailed features of some 3D scenes. However, memory consumption can become large when these structures have a maximum depth greater than a certain thresholding. In this work, the depths chosen for Octree and Portal-Octree are, therefore, intuitive memory versus performance tuning parameters. In practice, for example, we observed that maintaining a maximum depth level 4 for the Octree was a good trade-off.

We specified and prerecorded two different camera paths through the modeled environments to guarantee a better control in the experiments being conducted, as well as to make it easier to reproduce the tests: one for the indoor environments (E_1 and E_2), and another for the urban ones (E_3 and E_4). The former camera path contains 1476 frames, and the latter, 1655.

In order to possibly diminish the influence of the chosen trajectories on the results obtained, we have defined and complied with the following requirements: (i) the trajectories should travel over a vast region of the 3D environments; (ii) similar distances should be covered, containing an equivalent number of frames, to make comparison of obtained results easier; (iii) along the trajectory, the camera look-at should often change, allowing that, at each frame, a considerably varying number of polygons be contained in the viewing frustum.

Initially, for each of the four environments and using an Octree depth 4 for the representation of the 3D scenes, five types of tests were conducted with different combinations of the visibility algorithms: (1)view-frustum culling (VFC); (2) view-frustum culling and backface culling (VFC+BC); (3) view-frustum culling and conservative backface culling (VFC+CBC); (4) view-frustum culling, conservative backface culling, and backface culling (VFC+CBC+BC); and (5) without using any visibility algorithm.

Although not explored here, in all of our performance tests, the occlusion culling algorithm in its traditional form has achieved poor results (the processing time necessary to execute it was higher than the time spent to render the discarded triangles), and because of this reason we do not address it here. Also, due to space restrictions, only the most significant results are shown graphically.

Due to processing time variations and to guarantee an evaluation of better quality, each of the tests was run five times and, for each frame, the average processing times (in milliseconds) were calculated. In the majority of the tests, the curves generated by the combination VFC+CBC+BC obtained the best performance results. For each environment, we also calculated the standard deviation (SD) for each combination of visibility algorithms, and used these values as measures of their stability. Summary statistics presented in Table 1 confirms that, for E_1, E_2, E_3, and E_4 environments, the combined VFC+CBC+BC algorithm has a much smaller standard deviation than the other three algorithms. Overall, there is good evidence to suggest that the VFC+CBC+BC algorithm is faster and more stable than the others tested.

Finally, we conducted additional performance tests, at this time to identify the best spatial data structure for each one of the four environments. For this, we used the VFC+CBC+BC algorithms, and the following spatial data structures (some of them with different depths): (1) Non-Uniform Grid; (2) BSP-Tree; (3) Octree; and (4) Portal-Octree.

In all of our set of experiments, the best performance results, i.e. the highest frame rates, were achieved by the mobile phone Nokia n82 (the results of the visibility algorithms and spatial data structures can be seen in Table 1 and Table 2). In the next two subsections we will focus on detailing and discussing these results for the indoor and urban environments.

Table 1. Average processing time and standard deviation (SD) for rendering combinations of visibility algorithms associated with the Octree depth 4 data structure along the trajectories in the indoor (E_1 and E_2) and urban (E_3 and E_4) environments, using the cell phone Nokia n82

Visibility Algorithms	E_1 Indoor 6199 triangles 214 KB of size		E_2 Indoor 30199 triangles 1048 KB of size		E_3 Urban 2548 triangles 108 KB of size		E_4 Urban 10040 triangles 511 KB of size	
	Mean (ms)	SD	Mean (ms)	SD	Mean (ms)	SD	Mean (ms)	SD
VFC	38	10.6	101	43.8	28	0.7	60	0.0
VFC+BC	37	21.9	101	45.3	28	11.3	60	11.3
VFC+CBC	37	10.6	96	66.5	28	0.0	56	11.3
VFC+CBC+BC	36	0.7	93	33.2	26	0.0	56	0.0

Table 2. Average processing time and frame/s (FPS) obtained along the trajectories in the indoor (E_1 and E_2) and urban (E_3 and E_4) environments to render the scenes using different spatial data structures and the VFC+CBC+BC algorithms, on the cell phone Nokia n82

Spatial Data Structures	E_1 Indoor 6199 triangles 214 KB of size		E_2 Indoor 30199 triangles 1048 KB of size		E_3 Urban 2548 triangles 108 KB of size		E_4 Urban 10040 triangles 511 KB of size	
	Mean (ms)	FPS	Mean (ms)	FPS	Mean (ms)	FPS	Mean (ms)	FPS
Non-Uniform Grid	93.4	11	416.0	2	25.3	44	102.0	10
Octree 4	36.0	31	96.2	18	26.0	40	56.0	21
BSP-Tree	73.3	15	231.0	5	46.7	22	103.0	11
Portal-Octree 3	88.8	12	242.0	5	41.2	25	97.0	11

Indoor Environments

For the indoor environment E_1, we see similar results for the four combinations of visibility algorithms (first column, Table 1). More specifically, the processing times for rendering this environment vary from 36ms to 38ms. The combination of VFC+CBC+BC (solid black curve on the top side of Figure 3) with the Octree depth 4 obtained the best results for both iPaq PocketPC and cell phone Nokia n82, requiring, respectively, 129.1ms and 36ms, on average, to render each frame along the trajectory.

As the number of triangles is increased (from 6199 in E_1 to 30199 in E_2), performance of the

visibility algorithms rapidly degrades, while the VFC+CBC+BC maintains consistent results, still delivering the best performance. More specifically, for the indoor environment E_2, the VFC+CBC+BC with the Octree depth 4 obtained the best results, requiring on average 93ms (third column, Table 1, and solid black curve on the bottom side of Figure 3) to render each frame along the trajectory on the cell phone Nokia n82. Tests were also executed with different depth levels for Octree and Portal-Octree and the best depths found were 4 and 3, respectively. As for the tests using different spatial data structures, processing times to render the indoor environments appear very sensitive to them. The non-uniform

Figure 3. Processing time to render the Octree and BSP-tree data structures combined with the VFC+CBC+BC algorithms, along the trajectory in the indoor environments E_1 (on the top side) and E_2 (on the bottom side), using the cell phone Nokia n82

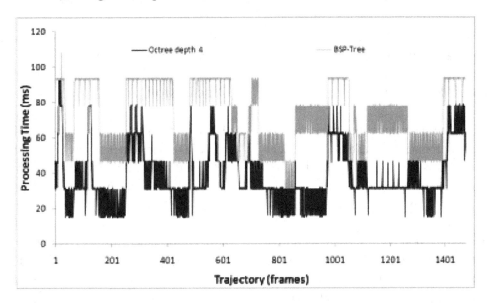

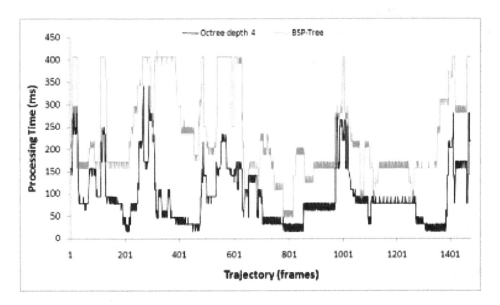

grid, BSP-Tree, and Portal-Octree structures achieved very poor performances with a frame rate of less than 15 frames/s (second and fourth columns, Table 2). However, for E_1 and E_2 the Octree structure depth 4 performs fairly well, with stable and interactive average rates close to 31 frames/s and 18 frames/s, respectively.

Actually, one can observe that for both E_1 and E_2 indoor environments, the processing time necessary to render each frame has a relationship with the number of triangles sent to the rendering pipeline. In other words, the processing time increases proportionately to the increase in the number of triangles sent to the rendering pipeline (the correlation values calculated for the Octree depth 4 in E_1 and E_2 are +0.95). Also, the number of triangles sent to the rendering pipeline has a greater influence on the application's performance than the time necessary to execute the searches in the spatial partitioning structures.

Urban Environments

The results for the algorithms and data structures in the urban environments E_3 and E_4 can be seen in the fifth and seventh columns of Table 1, respectively.

For E_3, similarly to the tests conducted in the indoor environments, the VFC+CBC+BC algorithms obtained the best performance. As for the tests with the spatial partitioning structures, on the cell phone Nokia n82, the non-uniform Grid (fifth column in Table 2 and solid black curve on the top side of Figure 4) obtained the best performance, requiring 25.3ms, on average, to render each frame along the trajectory (whereas on iPaq PocketPC was spent 107ms, on average). As for the structures, the depth levels influenced the performance of Octree and Portal-Octree; however, they did not produce results competitive enough to outdo the performance of the non-uniform Grid. For this environment, both the number of triangles sent to the rendering pipeline, and the search time in the spatial partitioning structure influenced in the performance differences of the spatial partitioning structures.

For the urban environment E_4 that contains 10,040 triangles, with the Octree depth 4, the VF+CBC has comparable performance to the VFC+CBC+BC (seventh column in Table 1). For the cell phone Nokia n82, these algorithms require 56ms (solid black curve on the bottom side of Figure 4), on average, to render each frame along the trajectory (just for comparison, on iPaq PocketPC they required 435.3ms, on average). As for the tests with the spatial partitioning structures, the Octree depth 4 obtains the best results, offer-

ing clear superior performance (seventh column in Table 2).

For this environment, the number of triangles sent to the rendering pipeline demonstrated direct influence in the performance differences of the spatial partitioning structures.

Discussion

The four best performances obtained in each one of the 3D environments using the cell phone Nokia n82 are discussed in more detail in this section.

For the indoor environments E_1 and E_2, the VFC+CBC+BC with the Octree depth 4 (see black and grey curves in Figure 5) obtained the best performance results. Average rates of 31 frames/s (second column, Table 2) and 18 frames/s (fourth column, Table 2) are achieved in E1 and E2, respectively. For the urban environment E_3, the VFC+CBC+BC with the non-uniform Grid (grey curve in Figure 6) obtained average rates of 44 frames/s (sixth column, Table 2). Finally, for the urban environment E_4, the VFC+CBC with the Octree 4 (black curve in Figure 6) achieved average rates of 21 frames/s (eighth column, Table 2).

For all algorithms and data structures, performance consistently decreases from approximately 44 frames/s (E_3) to 31 frames/s (E_1) to 21 frames/s (E_4) to 18 frames/s (E2). One can observe that, although E_1 and E_3 correspond to different types of environments (indoor and urban, respectively) and have different numbers of triangles (E_1 has 6199 and E_3 has 2548), their performances are similar in terms of their respective frame rates (31 and 44 frames/s). A similar behavior can be observed, when we compare the environments E_2 and E_4. The number of triangles of E_2 is approximately 3 times bigger than the number of triangles of E_4, however, both environments obtained similar frame rates (18 and 21 frames/s, respectively).

The visibility algorithms as well as the spatial data structures seem to be more effective in the

Figure 4. Processing time using the cell phone Nokia n82 to render the Non-uniform Grid and Octree data structures combined with the VFC+CBC+BC algorithms, and to render the Octree and Portal-Octree structures with the VFC+CBC algorithms, along the trajectory in the urban environments E_3 (on the top side) and E_4 (on the bottom side), respectively

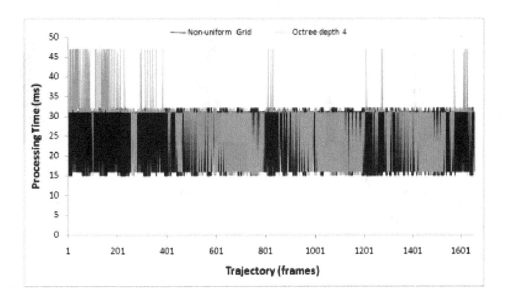

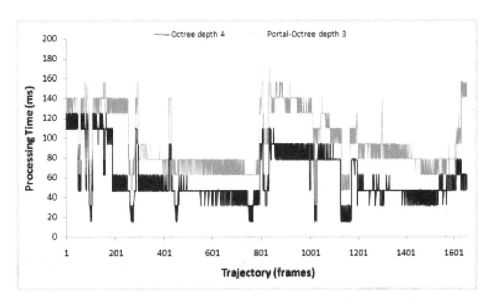

indoor environments. We suspect that this may be due to the use of depth buffers. We also hypothesize that in situations in which the 3D environments consist of hundreds (e.g., E_1) or even a few million of triangles (e.g., E_3), the backface culling algorithm could be disabled. Since the number of triangles to be discarded in each of these environments is small, it seems that it is not worth spending extra processing time to find out which triangles discard. Finally, we can also conclude that, if the built-in backface culling algorithm does not contribute significantly to improve the

Figure 5. Frame rate/s obtained, using the cell phone Nokia n82, along the trajectory in the indoor environments E1 and E2

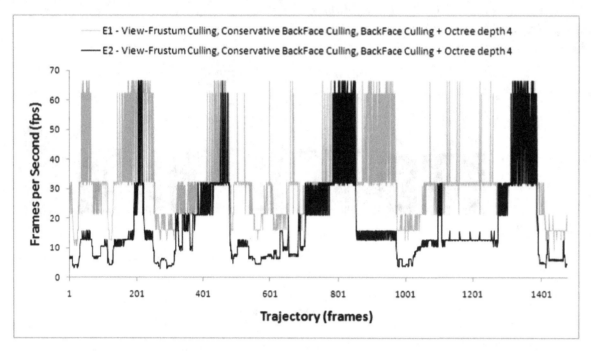

Figure 6. Frame rate/s obtained, using the cell phone Nokia n82, along the trajectory in the urban environments E3 and E4

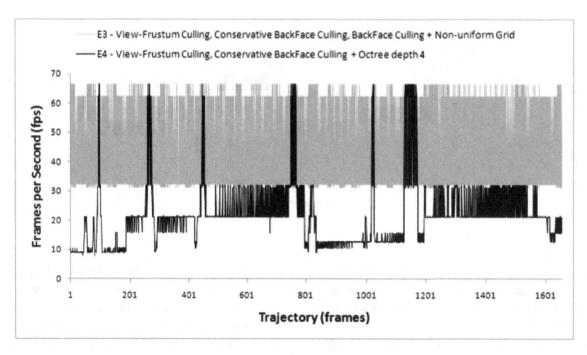

performance, but a software backface culling does, it almost certainly means they are triangle limited and not pixel limited in their graphics processing. The software version of the backface culling sends fewer triangles to the OpenGL ES API, while the hardware version appears to process those triangles but avoids rasterizing and shading them.

We run tests with different 3D environments and concluded that our system scales well as the size and geometric complexity of the 3D scene increase. We address important constraints of mobile handheld devices in an efficient way, by exploiting a combination of visibility algorithms and spatial data structures to minimize the non-visible geometry that is processed during walkthroughs (consequently, the processing time to render the scenes and the memory consumption). The advantage of our approach is also its generality: depending on the application, it can be applied to indoor as well as urban environments.

CONCLUSION AND FUTURE WORK

We examined and compared various standard visibility algorithms we have implemented (view-frustum culling and backface culling), as well as a simple, yet fast algorithm, we designed and called conservative backface culling, to analyze how these approaches apply to handheld devices. Also, we associated these visibility algorithms with different settings of standard spatial data structures (non-uniform Grids, BSP-Trees, Octrees, and Portal-Octrees) for enabling efficient graphics rendering of both indoor and urban 3D environments. We conducted several system performance tests and analyses using two mobile platforms and trajectories of camera locomotion on environments in the order of thousands of triangles. We demonstrated in our results that our system is able to render four different example environments at interactive frame rates from most view points, using geometry rather than image-based rendering or point-based rendering on the

cell phone Nokia n82. The visibility algorithms as well as the spatial data structures examined appear to be more effective in the indoor than in the urban environments, because similar performances in terms of frame rates were obtained by different types of environments, with different numbers of triangles (E_1 and E_3 with 6K and 2K triangles, and E_2 and E_4 with 30K and 10K triangles, respectively).

Currently, there is still a lack of empirical knowledge to prescribe how existing rendering algorithms and spatial data structures, that use geometry rather than image-based or point-based rendering, could be applied to different models of handheld devices. In this context, we believe that the lessons we learned with our analysis of various combinations of culling and spatial subdivision algorithms for enabling efficient graphics rendering of both indoor and urban environments may be beneficial to others developers in future graphical application modeling efforts, especially suited for low-end handheld devices that still comprise the overwhelming majority of the world mobile market (reaching out the largest group of mobile phone users rather than the fewer smartphones currently out in the market). If we want to reach more people by mobile, we need to offer a good low-end mobile experience, too. And that includes strategies for enabling efficient graphics rendering at interactive frame rates of environments with 3D graphics-rich contents. It is therefore becoming increasingly evident that the handheld domain can offer many opportunities for the application of interactive computer graphics. It includes innumerable variety of applications ranging from interactive design of buildings, to virtual-reality systems, to computer games.

Finally, the next stage of this work is to evolve the current implementation to reduce the number of accesses to the memory of the mobile devices (an alternative to be implemented is to exploit sequential access schemes of triangle meshes). Among the research to be performed we plan to extend the system by means of implementing other algorithms, such as occluder fusion for occlusion culling, potentially visible sets (PVS) BSP-Tree, portal culling for Portal-Octree, a well as other spatial partitioning structures, such as Kd-Tree. Another possibility may be to implement some object-centric data structure. We also plan to develop an adaptive algorithm that can be used to verify different Octree depths automatically. This can be done, for example, using some cost function or the number of objects per leaf as a metric for terminating the Octree. Lastly, more complex and realistic 3D environments can still be explored to verify system performance.

ACKNOWLEDGMENT

Maria Andréia Formico Rodrigues is supported by the Brazilian Agency CNPq under grant No. 473933/2008-0 and would like to thank for its financial support.

REFERENCES

Akenine-Möller, T., Haines, E., & Hoffman, N. (2008). *Real-time rendering*. Boston: A. K. Peters, Ltd.

Barbosa, R. G., & Rodrigues, M. A. F. (2006). Supporting guided navigation in mobile virtual environments. In *Proceedings of the 13th ACM Symposium on Virtual Reality Software and Technology (VRST)*, Limassol, Cyprus (pp. 220-226).

Bittner, J., Havran, V., & Slavík, P. (1998). Hierarchical visibility culling with occlusion trees. In *Proceedings of the 1998 Computer Graphics International (CGI)* (pp. 207-219).

Capin, T. K., Pulli, K., & Akenine-Möller, T. (2008). The state of the art in mobile graphics research. *IEEE Computer Graphics and Applications, 28*(4), 74–84. doi:10.1109/MCG.2008.83

Chang, C.-F., & Ger, S.-H. (2002). Enhancing 3D graphics on mobile devices by image-based rendering. In *Proceedings of the 3rd IEEE Pacific Rim Conference on Multimedia (PCM)*, London (pp. 1105-1111).

Clark, J. H. (1976). Hierarchical geometric models for visible surface algorithms. *Communications of the ACM, 19*(10), 547–554. doi:10.1145/360349.360354

Cohen-Or, D., Chrysanthou, Y., Silva, C. T., & Durand, F. (2002). A survey of visibility for walkthrough applications. *IEEE Transactions on Visualization and Computer Graphics, 9*(3), 412–431. doi:10.1109/TVCG.2003.1207447

Cohen-Or, D., Rich, E., Lerner, U., & Shenkar, V. (1996). A real-time photo-realistic visual flythrough. *IEEE Transactions on Visualization and Computer Graphics, 2*(3), 255–265. doi:10.1109/2945.537308

Coorg, S. R., & Teller, S. J. (1996). Temporally coherent conservative visibility. In *Proceedings of the 1996 Symposium on Computational Geometry* (pp. 78-87).

Dalmau, D. S.-C. (2003). *Core techniques and algorithms in game programming*. New Riders Games.

Duguet, F., & Drettakis, G. (2004). Flexible point-based rendering on mobile devices. *IEEE Computer Graphics and Applications, 24*(4), 57–63. doi:10.1109/MCG.2004.5

Hachet, M., Decle, F., & Guitton, P. (2006). Z-Goto for efficient navigation in 3D Environments from Discrete Inputs. In *Proceedings of the 13th Symposium on Virtual Reality Software and Technology (VRST)*, Limassol, Cyprus (pp. 236-239).

Huang, J., Bue, B., Pattath, A., Ebert, D. S., & Thomas, K. M. (2007). Interactive illustrative rendering on mobile devices. *IEEE Computer Graphics and Applications*, *27*(3), 48–56. doi:10.1109/MCG.2007.63

Hudson, T., Manocha, D., Cohen, J., Lin, M. C., Hoff, K. E., III, & Zhang, H. (1997). Accelerated occlusion culling using shadow frusta. In *Proceedings of the 1997 ACM Symposium on Computational Geometry* (pp. 1-10).

Hwang, J., Jung, J., & Kim, G. J. (2006). Hand-held virtual reality: a feasibility study. In *Proceedings of the 13th Symposium on Virtual Reality Software and Technology (VRST)*, Limassol, Cyprus (pp. 356-363).

KhronosGroup. (2008). *OpenGL ES - Embedded System*. Retrieved October 16, 2008, from http://www.khronos.org/opengles/

Kumar, S., Manocha, D., Garrett, B., & Lin, M. (1996). Hierarchical back-face culling. In *Proceedings of the 7th Eurographics Workshop on Rendering* (pp. 231-240).

Leyvand, T. (2009). *CityGen - A procedural model generator*. Retrieved April 10, 2009, from http://leyvand.com/research/citygen

Lluch, J., Gaitán, R., Camahort, E., & Vivó, R. (2005). Interactive three-dimensional rendering on mobile computer devices. In *Proceedings of the 18th ACM SIGCHI International Conference on Advances in Computer Entertainment Technology (ACE)*, Valencia, Spain (pp. 254-257).

Luebke, D., & Georges, C. (1995). Portals and mirrors: simple, fast evaluation of potentially visible sets. In *Proceedings of the 1995 Symposium on Interactive 3D Graphics (SI3D)*, Monterey, CA (pp. 105-106).

Moser, M., & Weiskopf, D. (2008). Interactive volume rendering on mobile devices. In *Proceedings of the Vision, Modeling, and Visualization Conference (VMV)*, Konstanz, Germany (pp. 217-226).

Mulloni, A., Nadalutti, D., & Chittaro, L. (2007). Interactive walkthrough of large 3D models of buildings on mobile devices. In *Proceedings of the 12th International Conference on 3D Web Technology (Web3D)*, Dijon, France (pp. 17-25).

Nokia. (2009). *OpenGL ES 1.1 Plug-in. Built on S60* (3rd ed.). Retrieved April 2, 2009, from http://www.forum.nokia.com/

Nurminen, A. (2006). m-LOMA - a mobile 3D city map. In *Proceedings of the 11th International Conference on 3D Web Technology (Web3D)*, Columbia, MD (pp. 7-18).

Nurminen, A. (2009). *Mobile three-dimensional city maps*. Unpublished doctoral dissertation, Helsinki University of Technology, Helsinki, Finland.

Pulli, K. (2006). New APIs for mobile graphics. In *Proceedings of the 2006 SPIE Electronic Imaging Multimedia on Mobile Devices II*, San Jose, CA (pp. 1-13).

Pulli, K., Aarnio, T., Miettinen, V., Roimela, K., & Vaarala, J. (2008). *Mobile 3D graphics with OpenGL ES and M3G*. San Francisco: Morgan Kauffman.

Rodrigues, M. A. F., Barbosa, R. G., & Mendonça, N. C. (2006). Interactive mobile 3D graphics for on-the-go visualization and walkthroughs. In *Proceedings of the 2006 ACM Symposium on Applied Computing (SAC)*, Dijon, France (pp. 1002-1007).

Russo, S. (2010). *Blender pocket - digital creation in your pocket*. Retrieved March 9, 2010, from http://russose.free.fr/BlenderPocket/

Samet, H., & Webber, R. E. (1988). Hierarchical data structures and algorithms for computer graphics. Part I. *IEEE Computer Graphics and Applications, 8*(3), 48–68. doi:10.1109/38.513

Silva, W. B., & Rodrigues, M. A. F. (2009). A lightweight 3D visualization and navigation system on handheld devices. In *Proceedings of the 2009 ACM Symposium on Applied Computing (SAC),* Honolulu, HI (pp. 162-166).

Van Verth, J. M., & Bishop, L. M. (2004). *Essential mathematics for games and interactive applications: a programmer's Guide.* San Francisco: Morgan Kaufmann.

Vincent, H.-M. W. (2009). *Vincent 3D rendering library - open source graphics libraries for mobile and embedded devices. Vincent ES 1.x library based on the OpenGL (R) ES 1.1 API specification (common lite).* Retrieved April 2, 2009, from http://www.vincent3d.com/Vincent3D/index.html

Wonka, P., & Schmalstieg, D. (1999). Occluder shadows for fast walkthroughs of urban environments. *Computer Graphics Forum, 18*(3), 51–60. doi:10.1111/1467-8659.00327

Zhang, H., & Hoff, K. E., III. (1997). Fast backface culling using normal masks. In *Proceedings of the 1997 ACM Symposium on Interactive 3D Graphics,* Providence, RI (pp. 103-106).

This work was previously published in the International Journal of Handheld Computing Research (IJHCR), Volume 2, Issue 1, edited by Wen-Chen Hu, pp. 55-71, copyright 2011 by IGI Publishing (an imprint of IGI Global).

Chapter 20
Design and Implementation of Binary Tree Based Proactive Routing Protocols for Large MANETS

Pavan Kumar Pandey
Aricent Technologies, India

G. P. Biswas
Indian School of Mines, India

ABSTRACT

The Mobile Ad hoc Network (MANET) is a collection of connected mobile nodes without any central-ized administration. Proactive routing approach is one of those categories of proposed routing protocol which is not suitable for larger network due to their high overhead to maintain routing table for each and every node. The novelty of this approach is to form a binary tree structure of several independent sub-networks by decomposing a large network to sub-networks. Each sub-network is monitored by an agent node which is selected by several broadcasted regulations. Agent node maintains two routing in-formation; one for local routing within the sub-network and another for routing through all other agent node. In routing mechanism first source node checks for destination within sub-network then source sends destination address to respective parent agent node if destination is not available in local routing, this process follows up to the destination node using agent mode. This approach allowed any proactive routing protocol with scalability for every routing mechanism. The proposed approach is thoroughly analyzed and its justification for the connectivity through sub-networks, routing between each source to destination pair, scalability, etc., are given, which show expected performance.

DOI: 10.4018/978-1-4666-2785-7.ch020

INTRODUCTION

Mobile ad hoc networks (MANETs) (Spojme-novic, 2002) are autonomous system, multi-hop wireless network without requiring any existing infrastructure. This type of networks is designed by number of mobile nodes connected through wireless links. In MANETs, two nodes can directly communicate with each other when they are within radio range. Otherwise, intermediate nodes are required to relay messages to neighboring nodes along the path from source to destination. Ad hoc networks (Spojmenovic, 2002) are characterized by frequent changes and mobile nodes may join the network, disconnect or move at any time. These problems make the routing problem in ad hoc networks more difficult than traditional wired networks.

This kind of network is most usable for environments where nodes are frequently changing and low bandwidth available, such as moving vehicles in battlefield. Many restrictions should be well considered, such as limited power and bandwidth. Mobile ad hoc networks are suitable for proper communication in military and also used to play an important role in many fields without the use of a fixed infrastructure such as disaster search-and-rescue operations, data acquisition in remote areas, conference and convention centers etc.

One of the major issues related to MANET implementation is the routing. The major routing approaches are obtained into this area are the well-known DSDV (Perkins, 1994) in proactive and the AODV (Chiang & Gerla, 1997) in reactive routing protocols. In MANET, every node operated as a router as well as node also, maintaining individually routes to other nodes. If the number of nodes increases, the routing overhead will increase rapidly. In order to support multi-hop routing (Chiang & Gerla, 1997), there are several different protocols have been proposed. There are different standards to divide these routing protocols in different category: proactive rout-

ing versus reactive routing or flat routing versus cluster based routing and so on.

In proactive protocols (Spojmenovic, 2002; Royer & Toh, 1999) routes between every pair of nodes are established in advance even if no data transmission is required. Routes to all destinations are updated periodically. These routing protocols maintain consistent, up-to-date routing information for every pair of source and destination in the network. These protocols require to maintaining one or more tables for routing information, and they respond to frequently changes in network topology by propagating updates throughout the network to maintaining a consistent updates of network. In contrast, on demand (reactive) protocols (Spojmenovic, 2002; Royer & Toh, 1999) establish the route to a destination only when data transmission is required. Thus, a node broadcast route request packet to the network and waits for the route reply message to form a route to the destination node. This routing protocol reduces the routing load as compared to the proactive protocols.

In flat routing protocol, each pair of source and destination communicate through peer-to-peer relationship and participating in routing equally. Routing algorithm is easily implemented. However, it is having poor performance in terms of scalability (Iwata, Chiang, Pei, Gerla, & Chen, 1999) in a certain extent. For example, DSDV (Perkins, 1994; Rahman & Zukarnain, 2004), DSR (Broch & Malts, 1998) and AODV (Rahman & Zukarnain, 2004) are typical flat routing protocols. The hierarchical (Spojmenovic, 2002; Pei, Gerla, & Hong, 1999) routing protocol communicates through so many sub-networks forming by dividing the whole network into many logical areas. There are different routing strategies are used inside and outside the logical area. Compared with flat routing protocol, the hierarchical routing protocol possesses better performance and is propitious to support scalable networks. At present, the growing interest in wireless ad hoc network techniques has resulted in many hierarchical routing protocols such as ZRP (Haas & Peariman, 2001), each routing protocol having its own advantages and drawbacks. Research is continued on commendably satisfying the demands of multi-hop wireless ad hoc networks on the aspects of scalability, less routing overhead, and complexity.

The novelty of this protocol to provide better scalability, low delay, and low normalized routing overhead in multi-hop wireless ad hoc networks. The main feature of this protocol is to divide large network in different independent clusters and use the proactive routing approach to construct routing table for each and every node. The network is divided in several sub-networks having feasible number of nodes and selects the agent node for each sub-network to interconnect all these sub-networks.

In this protocol proactive routing approaches is used in both within sub network and inter sub-network routing. This approach provide the low route acquisition delay and provide the better scalability in any proactive protocol by decomposing of large network and choosing an agent node, agent node is used for connecting the independent sub-networks. This approach can be used in organization having several different co-organizations which are required to communicate among them. So for better study of this paper we must aware about proactive routing protocols and some related previous works which is discussed in detail in next section.

The rest of this paper is organized as follows: Next, Proactive routing approach, a detailed description of this protocol, illustrating the main features of the protocol's operation, is given, the correctness of protocol and we analyze its performance complexity, finally last section presents our conclusion.

Proactive Routing Protocols and Related Works

Proactive routing protocol periodically updating its routing information for network topology and every node eventually has consistent and up-to-date global routing information for the entire network. This approach has the advantages of timely exchanging network information such as available bandwidth, delay, topology etc. and supporting real-time services. Although it is not suitable for large scale networks since many unnecessary routes still need to be established and the periodic updating may increase routing overhead and communication overhead. In proactive routing family basically Destination Sequenced Distance Vector (DSDV) Routing (Perkins, 1994) and Optimized Link State Routing (OLSR) Protocol (Clausen, Jacquet, Laouiti, Minet, Muhlethaler, Qayyum, & Viennot, 2001) are included. The on-demand protocol is more efficient because each node tries to reduce routing overhead by only sending routing packets when needed for data transmission and a route is released when the data transmission is completed. As to flat routing and hierarchical routing, this is a classification according to network structure underlying routing protocols.

Destination-Sequenced Distance Vector routing protocol (Perkins, 1994; Rahman & Zukarnain, 2004) (DSDV) is a typical routing protocol for MANETs, which is based on the Distributed Bellman-Ford algorithm. In DSDV, every mobile node in the network maintains a routing table that records all possible destinations within the network, next hop to be visited and the total number of hops to each destination. Each route is tagged with a sequence number which is originated by the destination, indicating how old the route is. Each node manages its own sequence number by increasing by two (always even number) every time. In route selection process, when a route update with a higher sequence number is obtained, the old route is replaced with new one. In case of two or more routes with the same sequence number, the route with better metric is used. The routing table is maintained by each and every node so it is known as table driven routing protocol.

Updates are transmitted periodically or immediately when any significant topology change is detected. There are two ways of performing routing update: "full dump," in which a node transmits the full information regarding to updating of routing table, and "incremental update," in which a node sends only some updated entries that have changed from last change. In DSDV, broken link may be detected by the data link layer protocol. If a node detects that a link has broken, it puts the metric to infinity, and issues a route update to the other nodes regarding the link status with new sequence number. All other nodes repeat this process until they receive an update with a higher sequence number to provide it with a fresh route again. To avoid fluctuations in route updates, DSDV employs a "settling time" data, which is used to predict the time when route becomes stable.

The major advantage of DSDV is that it provides loop-free routes at all instants. However, the performance of DSDV becomes significantly worse when it uses link breakage detection from the Media Access Control (MAC) protocol, Each node (except the destination) that hears this update will record an infinite metric for the destination, and it has to wait until it receives the next route update originated by the destination before it can rebuild its route to the destination. In this period, a lot of routing packets are generated, and many data packets may be dropped. It tries to create a new loop-free route by resolving invalid route information in a restricted area when link breaks, without affecting the entire network.

An Optimized Link-State Protocol (OLSR) (Clausen, Jacquet, Laouiti, Minet, Muhlethaler, Qayyum, & Viennot, 2001) another proactive routing protocol is already proposed. It is also suitable for smaller network, there are also scalability problems is available on larger networks. To improve the performance of OLSR routing algorithms there is need to combine features

of distance-vector and link-state schemes. This protocol is known as wireless routing protocol (WRP), which removes the counting-to-infinity problem and eliminates temporary loop without increasing the amount of control traffic.

There are numerous modifications in DSDV already have been proposed so far to improve the performance of DSDV routing protocol in wireless ad hoc network. The primary attribute of DSDV are simplicity in terms of implementation, loop free to avoid the routing overhead and the storage overhead due to routing tables should be low.

An efficient DSDV (Eff-DSDV) (Pei, Gerla, & Chen, 2008) Protocol proposed to improve the performance of DSDV which overcomes the problem of stale routes. This protocol establish the temporary route through some neighbor nodes which nodes having valid route to particular destination when the link breakage is detected. Each entry in the routing table has an additional entry for route update time. This update time is associated with the ROUTE-ACK packet and it is used in selecting a temporary route. In case of receiving multiple ROUTE-ACK with the same number of minimum hops, ad hoc host chooses that route which has the latest update time.

DSDV-MC (Lee, 2003) protocol extends the DSDV routing protocol into a multiple-channel version. To maintain the consistency of routing tables in a dynamically varying topology, each node periodically transmits updates. The routing table contains all available destinations, number of hops to each destination, and the channel indices of neighboring nodes. Nodes also transmit updates immediately when new updated information is available, such as a topology change or a channel switch.

A new proactive routing protocol called Multipath Destination Sequenced Distance Vector (MDSDV) protocol (Peter, King, & Etorban, 2003) is totally based on the well defined single path protocol known as DSDV. This protocol extends the DSDV protocol up to multiple separate paths to each destination in the network.

Two new fields called second hop and link-id are added in routing table. Both second hop and link-id which is generated by the destination are used to get separate paths from any pair of source destination. MDSDV provide a single protocol to forming routing tables keeping the optimal paths to every destination. Since MDSDV, like DSDV, keeps consistent updating information of whole network, every node is always having route to every destination node in the network. MDSDV maintains two tables; routing table and neighbors table. By using neighbors table, any node can determine if it is isolated or not.

One another approach to routing in the multi-hop wireless ad hoc networks is based on the Sub-area Tree (Liu, Shan, Wei, & Wang, 1997; Pandey & Biswas, 2010) which is originated by some root nodes. When a subarea tree is established, a logical subarea has already been established. Namely a subarea tree contains several logical subarea connected through interconnect node. Therefore, the whole network is composed into one subarea tree in the end. In last each node either joins in a subarea tree or become an interconnect node. STR is hierarchical routing approach and does not attempt to consistently maintain routing information in every node. The strategy in intra subarea as well as among root nodes and inter-connect nodes is proactive routing; whereas the strategy of inter subarea is on-demand routing means reactive routing protocol.

PROPOSED BINARY TREE BASED PROACTIVE ROUTING PROTOCOL

Proposed Binary tree-based (BTB) proactive routing approach is to overcome the problem of scalability of any proactive routing protocol using binary tree decomposition for multi-hop wireless ad hoc network. In this approach the whole network is divided in several sub-networks, those are connected in binary tree architecture and each sub-network is monitored by agent node. Every

agent node maintains two routing information one is for local routing within sub-network and another for routing to other parent-child agent nodes in binary tree architecture. All agent nodes monitor its own two children agent node also after developing the binary tree decomposition of large network.

A route for every source to destination pair can be defined using any proactive routing protocol. According to this protocol first the network is divided in equal two parts and select agent nodes for those sub-networks then those sub-networks are again divided in two equal sub-networks and again select agent nodes for these sub-networks. In forming binary tree architecture agent nodes of current sub-networks are connected to the agent node of previously divided sub-network, in this way the whole network is converted in binary tree of several sub-networks. In routing mechanism and updating information regarding to dynamically changed topology of ad-hoc network any table driven routing approach can be used.

Deployment of Network in Several Sub-Networks

This is the first step of proposed approach for proactive routing protocol. During this step the network is divided in several sub-networks which are connected in binary tree architecture. This step is purely oriented to scalability of proactive routing protocol because when the number of nodes in network is very large then routing process is very difficult in this type of network. Therefore to provide the scalability of routing approach the large network must be divided in sub-networks of feasible number of nodes. Therefore this step plays very important role in whole approach. Each sub-network should be independent and contains nearly equal number of nodes in same level of tree architecture.

In this step first, the large network is divided in two nearly equal independent sub-networks. No nodes should be common in two or more sub-networks. If number of nodes in sub-networks is not feasible to perform routing efficiently using any proactive routing approach then the sub-network is again divided in two nearly equal sub-networks. This process follows up to each sub-network contains feasible number of nodes to perform routing mechanism perfectly. Feasible value of nodes decides how many nodes in network can be operated using any table driven routing protocol, i.e., how many nodes can be maintained in single routing table. This value is not any standard value and it is not fixed, therefore it can be varied from network to network. This step gives the output of many independent sub-networks containing feasible number of nodes.

Selection of Agent Node for Each Sub-Network

This is the next step of proposed decomposition approach. This step is responsible to select the node to monitor the sub-network. Agent node is selected by auto discovery process, auto discovery process is little complex because it contains following steps first broadcasting a regulation which is basically a condition after satisfying this condition particular node is selected as a agent node this regulation is dependent on network characteristics it may be ID Number, number of neighbor nodes, transaction capability, residual energy, stability or other measure values. In this way the agent node of each sub-network is selected for monitoring the whole sub-network and other agent nodes.

This node is also part of the sub network. Every sub-network should have an agent node which connects its own sub-network to other sub-network. This node plays key role in this approach because it is responsible for routing among several sub-networks. In this decomposition approach every node keeps the information regarding to all nodes in sub-network. Agent node maintains two types of information first information is same as other nodes in sub-networks and another table is regarding to other selected agent node of several

sub-networks. This approach is restricted for proactive routing approach therefore each node maintains information in the form of table so agent node maintain two tables. First table is restricted to number of destinations available in sub-networks and second table restricted to agent nodes.

So after dividing network in several sub-networks every node maintain a routing table according any proactive routing table format. Table 1 describe all field included in table maintained by each nodes in MANETs.

After maintaining routing table every node having destinations of all nodes in particular sub-network with next hop to be visited and total cost up to those destination. Every route has a sequence number which is originated by destination. Always route with higher sequence number is preferred for routing if two nodes having same sequence number then better metric route is preferred. Upon receiving the table a node may either update its table or may wait for another table from its next neighbor. Based on the sequence number of the table update, it may forward or reject the table. In this way the agent node of each sub-network is selected for monitoring the whole sub-network and other agent nodes. This node is also part of the sub network. Every sub-network should have agent node which.

Data is also kept about the length of time between arrival of the first and the arrival of the

Table 1. Field description of routing table of proposed routing approach

Field	Description
Destination Address	All nearest neighbors satisfying Key factor
Next Hop	Next node to be visited for particular destination.
Metric	Total no. of nodes to be visited up to destination.
Sequence No.	No assigned by destination represent how the route is stale
Settling Time	Time when route to particular destination is stabilized.

best route for each particular destination. Based on this data a decision may be made to delay advertising routes which are about to change soon thus damping fluctuations of the route tables The advertisement of routes which may not have stabilized yet is delayed in order to reduce the number of rebroadcasts of possible route entries that normally arrive with the same sequence number. To overcome this problem we use settling time for route to each destination.

Establishment of Sub-Networks Binary Tree

In the process of establishing binary tree architecture, each agent node of network is connected to the agent nodes of its sub-network we can see that every agent node has acquired the routing information of other nodes included in its sub-network and others agent node. So through agent node every node in network has its own proper routing information. In binary tree architecture one network is divided in two equal independent part so agent node of network is connected to agent nodes of both sub-networks, then again this process is followed up to the feasible limit of nodes in this way it will form binary tree architecture of several sub-networks. In Figure 1 binary tree architecture of several sub-networks which are connected through several agent nodes are presented. Binary tree concept is always followed because any network or sub-network is always divided in two parts and all agent node of divided sub-network is connected to agent node of parent

Figure 1. Binary tree architecture with connected agent node

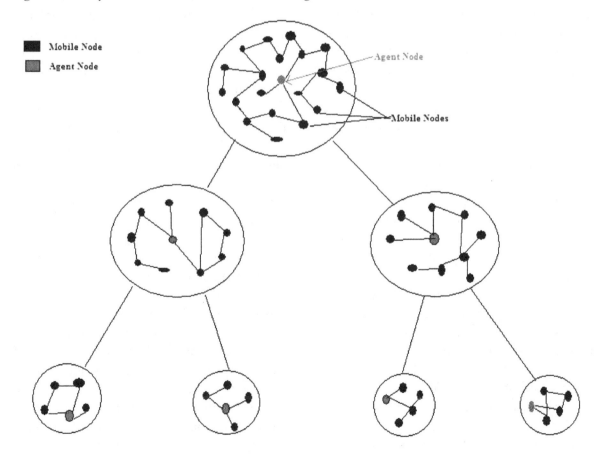

sub-network therefore all agent nodes in network is connected with parent-child relationship in tree architecture.

Meanwhile, after having finished the process of establishing tree architecture, independent hierarchical architecture is formed. Every node in network is connected to all other nodes through several agent nodes of sub-networks. In network every node directly connected with other nodes within sub-networks and connected with all nodes of sibling sub-network through parent agent node, therefore every pair of nodes connected in network.

Maintenance of Whole Network Topology and Routing Mechanism

Our proposed approach is basically related to providing scalability of a network; therefore it is based on partition of network and then applies any proactive routing approach for routing and updating mechanism. So for a routing mechanism there are two steps followed for finding the proper route form source to destination. First, a source node will try to find destination within sub-network if it is available, then source is directly connected to destination; otherwise, source sends this destination address to agent node of this sub-network and agent node communicates with parent agent node regarding route to destination. The same process is followed up to destination. Therefore send responsibility of finding destination to agent node is considered in second step of routing mechanism. All type of routing in proposed protocol is performed according to table driven routing protocol (proactive routing protocol) whether routing within sub-network or among several sub-networks.

Based on the above-mentioned description, the routing mechanism proceeds as follows:

If the destination node of data packet is in sub-network of source node, source node will forward the packet to the destination directly. If the destination node of data packet is not in sub-network of source node, the node will forward

the packet to its agent node. Then agent node sends this data packet to agent node of sibling sub-network through parent agent node. Then current agent node will work as previous agent node, this process repeat many times up to the process when destination node match in the sub-network of any agent node then find route from source to destination. In this route one source node then all agent nodes up to destination is considered. The source node forwards the data packet to the any node or agent node which knows the optimal routing. Source node or agent node will forward data packet to the destination node according to the known routing information after it receives the data packet. The whole route from supposed source-destination pair is described in Figure 2.

In updating information every node identifies whether its neighbor node still exists or not through detecting the routing information sent periodically by its neighbor node, and vice versa. One of the most important parameters to be chosen is the time between broadcasting the routing information packets However when any new or substantially modified route information is received by a Mobile Hosts the new information will be retransmitted soon subject to constraints imposed for damping route fluctuations effecting the most rapid possible dissemination of routing information among all the cooperating Mobile Hosts This quick rebroadcast introduces a new requirement for our protocols to converge as soon as possible.

The broken link may be detected by the layer protocol or it may instead be inferred if no broadcasts have been received for a while from a former neighbor a broken link is described by a metric infinite (i.e., any value greater than the maximum allowed metric). When a link to a next hop has broken any route through that next hop is immediately assigned an infinite metric and assigned an updated sequence number. Since this qualifies as a substantial route change, such modified routes are immediately disclosed in a broadcast routing information packet. Building information to describe broken links is the only

Figure 2. Routing from source node to destination

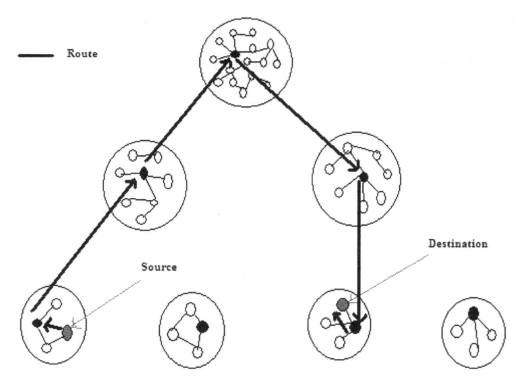

situation when the sequence number is generated by any mobile host other than the destination mobile host.

Sequence numbers defined by the originating Mobile Hosts are defined to be even numbers, and sequence numbers generated to indicate infinite metrics are odd numbers. In this way any "real" sequence numbers will supersede an infinite metric. When a node receives an infinite metric, and it has a later sequence number with a infinite metric, it triggers a route update broadcast to disseminate the important news about that destination. In a very large population of Mobile Hosts adjustments will likely be made in the time between broadcasts of the routing information packets. In order to reduce the amount of information carried in these packets two types will be defined. One will carry all the available routing information called a "full dump." The other type will carry only information changed since the last full dump called an "incremental."

Box 1 displays the pseudo-code of the binary tree based proactive routing protocol.

Example

A simplified example is given in Figure 3 and there are 16 mobile nodes. At first all the nodes are same and each node is with a unique ID. The line between two nodes denotes a wireless link and these two nodes can communicate directly if both are present in same sub-network. Assume the condition of feasible number of node of MANET for proactive routing protocol is 4 so the network having 16 nodes is divided up to 4 nodes and binary tree formed among several sub-networks.

Let *node 6* want to send data packet to *node 13* then firstly *node 6* checks for *node 13* in its own sub-network node 6 having node 1, 2, 4, 6, 7 in its sub-network so node 6 send this packet to corresponding agent node 4, then node 4 checks in other sub-network of agent node 9 also because

Box 1. Procedure BTB-proactive (G, S, D, R) Ref. 3

```
//visualize network as graph G (V,E), S is source node, D is destination node and R represent route from S to D node.
{
Procedure divide-network (V, v1, v2, k)                    Ref. 3.1
// v1 & v2 is number of nodes in divided sub-network and k indicates feasible number of nodes
{
If (V < =k)
v1 = v2 = V;
Else
{
If (V%2 = = 0)
v1 = v2 = V/2
Else
v1 = (V+1) / 2;
v2 = (V / 2);
} }
Procedure select-agent (v, A, R)                          Ref. 3.2
// v number of nodes in sub-network, A is agent node and R is broadcasted regulation
{
        For (i=1; i=< v; i++)
{
     If (v[j].feature = Regulation (R))
A= v[i];
End if
} }
Procedure Update (R)                                      Ref. 3.4
// procedure for updating the route R due to change in topology.
{
        Update.Metric(R);
// New[ ]:Best route for current seq,
// Old[ ]: Best route for previous seq
    if (R.seq == New[R.dest].Seq && R.Metric < New[R.dest].Metric)
{
        New[R.dest] = R;
        New[R.dest].BestTime = Now;
}
    else if (R.Seq > New[R.dest].Seq)
    else if (R.Seq > New[R.dest].Seq)
{
        Old[R.dest] = New[R.dest];
// save best route of last seq no
        New[R.dest] = r
        curr[r.dest].first_time = now;
        curr[r.dest].best_time = now;
}
end if
   end if
Procedure Routing (G, S, D, R)                            Ref. 3.4
//visualize network as graph G (V,E), S is source node, D is destination node and R represent route from S to D node.
{
    For (j=1; j=< K1; j++)
// K1 is number of nodes in sub-network of source
        If (v[j] = D)
R = {there exists (S, D) belongs to G: for every S, D belong to G}
        Else
        v[j] = A;
        Call Routing (G, A, D, R1)
// new route is R1when N become source node.
            R= R U R1
End if
}
```

Figure 3. Network with proper route for example of proposed approach

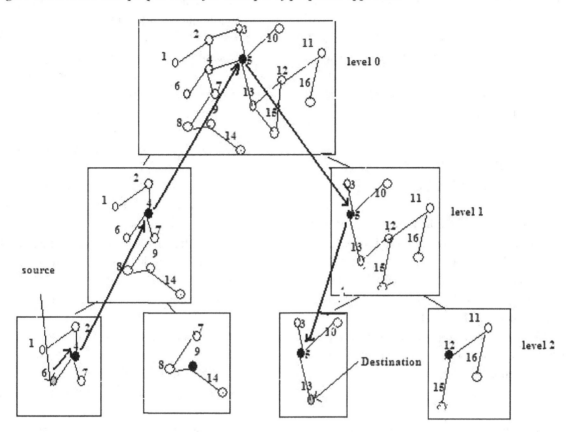

it is also agent node of parent sub-network. it is not

available then node 4 send this packet to node 5, then node 5 work as a source node and find route to node 13 according to proposed approach and combined with previous path. Then finally this process stops and route from node 6 to node 13 is $6 \leftrightarrow 4 \leftrightarrow 5 \leftrightarrow 13$.

PERFORMANCE ANALYSIS OF BTB-PROACTIVE APPROACH

The binary tree based proactive routing protocol is basically used for large MANET to provide scalability by dividing the whole network in several independent sub-networks. This approach can be used in organization where several different ad hoc networks are connected for communication

therefore the performance our proposed routing protocol play important role in routing approach. The performance is measured in terms of time complexity.

There are some assumptions to determine the actual performance of algorithm. Suppose the number of nodes in network is n, and the feasible number of nodes in each MANET for proper routing using proactive routing protocol is k. this value also play important role in size of network because the network is divided up to the k value. There are mainly two operations are performed in procedure BTB-Proactive one is divide-network and other one is select-agent. Therefore the performance proposed protocol is dependent on number of sub-networks. Suppose we are not considering the value of k so the complexity of the proposed routing protocol is $O\ (log2\ n)$ because this protocol is based on binary tree architecture so the complete network is divided in two parts then again in two parts so the complexity will be same as complexity of binary tree. Selection of agent node is also performed in same way of binary tree operation so the total complexity of algorithm in terms of time complexity is *O (log2 n)* for n number of nodes in network.

JUSTIFICATION OF PROPOSED ROUTING PROTOCOL

According to the protocol, initially every node in network having its own unique ID Then Can this approach act on the network and develop a numerous independent sub-network including all nodes in network and can it extended up to any number of nodes? This question is concerned with the effectiveness of proposed approach and the correctness wants to be proved. The following theorem assures it.

Theorem: We visualize the network as a graph *G (V, E)*, where *V* is the set of vertices and *E* is the set of edges. Assume *G* is connected then there are following situation is true for the network.

◦ Through BTB-Proactive every $v \in V$ of the divide-network *(V, v1, v2, k)* so $v \in v1 \cup v2$ *(i.e., $\notin v1 \cap v2$).*

◦ After adding some nodes *N* total nodes become *V+N* then case i must be true for nodes *V+N* also.

◦ There exists a route *R= {S Ai..., D}* where *S*: source node, *D*: destination node, *Ai*: agent node. *S, D, Ai* belongs to *V* and *i =1, 2, 3..., n* where *n* is number of sub-network.

Proof:

1. In the case we want to prove for the process of divide-network every node of network must be included in either of two sub-networks and any node cannot be included in both sub-networks because divided sub-networks are independent. This case is proved by mathematical induction.

For *V=1* so no need to divide further then first situation is correct for *V=1.*

Suppose above case is true for *V= n*, i.e., *v[i] V* for *i=1, ,.., n* and first case of given theorem is true for n nodes.

So for *V=n+1* above situation must be true because if n is odd so n+1 is even number and if n is even so *n+1* is odd number for both case is already considered in procedure BTB-proactive. *G* is connected so this node will be considered in any sub-network so case 1 is also true for *n+1* node also, i.e., in this approach no nodes is skipped and no nodes are common in two sub-networks.

2. After adding some nodes *N*, total nodes become *(V+N)*, if all *V* nodes are participated in some sub-network so all *(V+N)* nodes is also participated in some sub-network. In case of 1, it is already proved that all *V* Nodes are participated in some sub-network; no nodes in *V* are skipped.

After adding *N* nodes, using forming independent sub-network of whole network having *(V+N)* nodes, we can say that there exists $y \in V$ participate divide-network *(V+N, v1, v2, k)* where $z \in N$, therefore there exists $z \in v1 \cup v2$.

Through the agent node in the sub-network of one node is connected to the sub-network of another node without skipping any node in total *(V+N)* nodes.

3. According to this approach Route *(S, D)= {S, A1, A2, A3..., D}* exists, there into *Aj (j= 1, 2, 3,..., n)* are agent nodes. This situation proves that every pair of nodes having a route using mathematical induction.

For *V=1* so there are no route are required so this case is true for *V= 1.*

Suppose this case is true for *V=n*, i.e., source and destination form a route form source node *S* to destination node *D* through several agent nodes *Aj (j=1, 2,..., n).*

So for $V=n+1$ so this extra node is neighbor of some node and it is also connected some nodes which are included in n number of nodes so after dividing the network this node must be included in any sub-network with previous n nodes therefore path must be established in $n+1$ nodes also through some other agent nodes.

CONCLUSION

A scalable approach of any proactive routing protocol for multi-hop wireless ad hoc network is proposed. This approach divides the large network in several independent sub-networks which are connected in binary tree fashion through several agent nodes and thus supports any proactive routing schemes in dense and large scale ad hoc networks. It is also observed that the performance is better especially when the number of nodes in the network is higher by reducing routing overhead caused by high nodal density. The selection of agent node shows a great gains when the network is dense due to provide modularity in network. The correctness of the protocol indeed show that this approach provides a flexible routing framework for scalable routing over mobile ad hoc networks while keeping all the advantages introduced by the associated proactive routing scheme.

REFERENCES

Broch, D. J. J., & Malts, D. (1998). *The dynamic source routing protocol for mobile ad hoc networks.* Retrieved from http://www.ietf.org/rfc/rfc4728.txt

Chiang, C. C., & Gerla, M. (1997). Routing and multicast in multihop, mobile wireless networks. In *Proceedings of IEEE International Conference on Universal Personal Communication Record*, San Diego, CA (pp. 546-551).

Clausen, T., Jacquet, P., Laouiti, A., Minet, P., Muhlethaler, P., Qayyum, A., & Viennot, L. (2001). *Optimized link state routing protocol.* Retrieved from http://www.ietf.org/rfc/rfc3626.txt

Haas, Z. J., & Peariman, M. R. (2001). The performance of query control schemes for the zone routing protocol. *IEEE/ACM Transactions on Networking, 9*(4), 427–438. doi:10.1109/90.944341

Iwata, C.-C., Chiang, G., Pei, G., Gerla, M., & Chen, T.-W. (1999). Scalable routing strategies for ad hoc wireless networks. *IEEE Journal on Selected Areas in Communications*, 1369–1379. doi:10.1109/49.779920

Lee, U. (2003). *A proactive routing protocol for multi-channel wireless adhoc networks (DSDV-MC).* Unpublished doctoral dissertation, Virginia Polytechnic University, Blacksburg, VA.

Liu, G., Shan, C., Wei, G., & Wang, H. (1997). *Subarea tree routing (STR) in multi-hop wireless ad hoc networks.* Retrieved from http://202.194.20.8/proc/iccs2008/papers/339.pdf

Pandey, P. K., & Biswas, G. P. (2010). Design of scalable routing protocol for wireless ad hoc network based on DSDV. In *Proceedings of the International Conference on Mobile Information Systems* (pp. 164-174).

Pei, G., Gerla, M., & Chen, T. W. (2008). An efficient destination sequenced distance vector routing protocol for mobile ad hoc networks. In *Proceedings of the International Conference on Computer Science and Information Technology* (pp. 467-471).

Pei, G., Gerla, M., & Hong, X. Y. (1999). A wireless hierarchical routing protocol with group mobility. In *Proceedings of the IEEE Conference on Wireless Communications and Networking*, New Orleans, LA (pp. 1538-1542).

Perkins, C. E. (1994). Highly dynamic destination-sequenced distance-vector routing (DSDV) for mobile computers. In *Proceedings of the ACM SIGCOMM International Conference on Communications Architectures, Protocols and Applications* (pp. 234-244).

Peter, J. B., King, A., & Etorban, I. S. I. (2003). A DSDV-based multipath routing protocol for mobile ad-hoc networks. In *Proceedings of the International Conference on Wireless Networks*.

Rahman, A. H. A., & Zukarnain, Z. A. (2004). Performance comparison of AODV, DSDV and I-DSDV routing protocols in mobile ad hoc networks. *European Journal of Scientific Research, 31*(4), 566–576.

Royer, E. M., & Toh, C.-K. (1999). A review of current routing protocols for ad-hoc wireless mobile networks. *IEEE Personal Communications*, 46-55.

Spojmenovic, I. (2002). *Handbook of wireless network and mobile computing*. New York, NY: John Wiley & Sons. doi:10.1002/0471224561

This work was previously published in the International Journal of Handheld Computing Research (IJHCR), Volume 2, Issue 4, edited by Wen-Chen Hu, pp. 82-94, copyright 2011 by IGI Publishing (an imprint of IGI Global).

Compilation of References

Abdul Karim, N. S., Darus, S. H., & Hussin, R. (2006). Mobile phone applications in academic library services: a students' feedback survey. *Campus-Wide Information Systems, 23*(1), 35–51. doi:10.1108/10650740610639723

Abowd, G., Atkeson, C., & Essa, I. (1998). *Ubiquitous smart spaces.* Paper presented at the DARPA Workshop.

Adhikari, S., Paul, A., & Ramachandran, U. (2002). *D-stampede: Distributed programming system for ubiquitous computing* (Tech. Rep. No. GIT-CC-01-04). Atlanta, GA: Georgia Institute of Technology.

Agarwal, R., & Karahanna. (2000). Time flies when you are having fun: Cognitive absorption and beliefs about information technology usage. *Management Information Systems Quarterly, 24*(4), 665–694. doi:10.2307/3250951

Agarwal, R., & Prasad, J. (1998). A conceptual and operational definition of personal innovativeness in the domain of information technology. *Information Systems Research, 9*(2), 204–216. doi:10.1287/isre.9.2.204

Ailisto, H., Alahuhta, P., Hataja, V., Kylloenen, V., & Lindholm, M. (2002, September). *Structuring context aware applications: Five-layer model and example case.* Paper presented at the Workshop on Concepts and Models for Ubiquitous Computing, Goteborg, Sweden.

Aitenbichler, E., Kangasharju, J., & Mühlhäuser, M. (2007). MundoCore: A Light-weight Infrastructure for Pervasive Computing. *Pervasive and Mobile Computing, 3*(4), 332–361. doi:10.1016/j.pmcj.2007.04.002

Akenine-Möller, T., Haines, E., & Hoffman, N. (2008). *Real-time rendering.* Boston: A. K. Peters, Ltd.

Alberts, C. J., & Dorofee, A. J. (2001). *OCTAVE criteria, version 2.0* (Tech. Rep. No. CMU/SEI-2001-TR-016). Pittsburgh, PA: Carnegie Mellon University.

Al-Khaldi, M. A., & Al-Jabri, I. M. (1998). The relationship of attitudes to computer utilization: new evidence from a developing nation. *Computers in Human Behavior, 14*(1), 23–42. doi:10.1016/S0747-5632(97)00030-7

Al-Khatib, J., & Malshe, A., & AbdulKader, M. (2008). Perception of unethical negotiation tactics: A comparative study of US and Saudi managers. *International Business Review, 17*(1), 78–102. doi:10.1016/j.ibusrev.2007.12.004

Amoretti, M., & Zanichelli, F. (2009). The multi-knowledge service-oriented architecture: Enabling collaborative research for e-health. In *Proceedings of the 42nd Hawaii International Conference on System Sciences* (pp. 1-8).

Anderson, P. (2007). *What is Web 2.0? Ideas, technologies and implications for education.* Retrieved from http://www.jisc.ac.uk/media/documents/techwatch/tsw0701b.pdf

Anderson, B. D. O., & Moore, J. (1979). *Optimal filtering.* Upper Saddle River, NJ: Prentice Hall Publishing.

Anderson, R. (2001). *Security engineering: A guide to building dependable distributed systems.* New York, NY: John Wiley & Sons.

Andreasen, M. S., Nielsen, H. V., Schroder, S. O., & Stage, J. (2007). What happened to remote usability testing?: An empirical study of three methods. In *Proceedings of the SIGCHI Conference on Human Factors in Computing Systems* (pp. 1405-1414). New York, NY: ACM Press.

Asthana, A., Cravatts, M., & Krzyzanowski, P. (1994). An Indoor Wireless System for Personalized Shopping Assistance. In *Proceedings of the IEEE Workshop on Mobile Computing Systems and Applications*, Santa Cruz, CA (pp. 69-74).

Aumuller, D., Do, H., Massmann, S., & Rahm, E. (2005). Schema and ontology matching with COMA++. In *Proceedings of the ACM SIGMOD International Conference on Management of Data* (pp. 906-908).

Avoine, G., & Tchamkerten, A. (2009). An efficient distance bounding RFID authentication protocol: Balancing false-acceptance rate and memory requirement. In P. Samarati, M. Yung, F. Martinelli, & C. A. Ardagna (Eds.), *Proceedings of the 12th International Conference on Information Security* (LNCS 5735, pp. 250-261).

Bajwa, R. S., Hiraki, M., Kojima, H., Gorny, D. J., Nitta, K., & Shridhar, A. (1997). Instruction buffering to reduce power in processors for signal processing. *IEEE Transactions on Very Large Scale Integration Systems, 5*(4), 417–424. doi:10.1109/92.645068

Balasubramaniyan, J., & Fernandez, G. (1998). An architecture for intrusion detection using autonomous agents. In *Proceedings of the annual computer security applications conference (ACSAC)*.

Baldauf, M., Dustdar, S., & Rosenberg, F. (2007). A survey on context-aware systems. *International Journal of Ad Hoc and Ubiquitous Computing, 2*(4), 263–277. doi:10.1504/IJAHUC.2007.014070

Ballagas, R., Memon, F., Reiners, R., & Borchers, J. (2007). iStuff mobile: Rapidly prototyping new mobile phone interfaces for ubiquitous computing. In *Proceedings of the SIGCHI Conference on Human Factors in Computing Systems* (pp. 1107-1116). New York, NY: ACM Press.

Barbosa, R. G., & Rodrigues, M. A. F. (2006). Supporting guided navigation in mobile virtual environments. In *Proceedings of the 13th ACM Symposium on Virtual Reality Software and Technology (VRST)*, Limassol, Cyprus (pp. 220-226).

Bardram, J. E. (2004). Applications of context-aware computing in hospital work: Examples and design principles. In *Proceedings of the ACM Symposium on Applied Computing*. New York, NY: ACM Press.

Barrett, L. F., & Barrett, D. J. (2001). An introduction to computerized experience sampling in psychology. *Social Science Computer Review, 19*(2), 175–185. doi:10.1177/089443930101900204doi:10.1177/089443930101900204

Barrett, L. F., & Barrett, D. J. (2005). *ESP: The experience sampling program.* Retrieved from http://www.experience-sampling.org

Bartelt, C., Fischer, T., Niebuhr, D., Rausch, A., Seidl, F., & Trapp, M. (2005). Dynamic integration of heterogeneous mobile devices. In *Proceedings of the Workshop on Design and Evolution of Autonomic Application Software* (pp. 1-7).

Bass, T. (2000). Intrusion detection systems and multisensor data fusion. *Communications of the ACM, 43*(4), 99–105. doi:10.1145/332051.332079

Baumeister, R. F., & Leary, M. R. (1995). The Need to belong: Desire for interpersonal attachment as a fundamental human motivation. *Psychological Bulletin, 117*(3), 497–529. doi:10.1037/0033-2909.117.3.497

Beckett, D. (Ed.). (2004). *RDF/XML syntax specification (Revised).* Retrieved from http://www.w3.org/TR/REC-rdf-syntax/

Bellare, M., Canetti, R., & Krawczyk, H. (1996). Keying hash functions for message authentication. In N. Koblitz (Ed.), *Proceedings of the 16th Annual International Cryptology Conference on Advances in Cryptology* (LNCS 1109, pp. 1-15).

Bendavid, Y., Wamba, S. F., & Lefebvre, L. A. (2006). Proof of concept of an RFID-enabled supply chain in a B2B e-commerce environment. In *Proceedings of the 8th international Conference on Electronic Commerce: the New E-Commerce: innovations For Conquering Current Barriers, Obstacles and Limitations to Conducting Successful Business on the internet (ICEC '06)* (Vol. 156, pp. 564-568). New York: ACM.

Berenson, H., Bernstein, P., Gray, J., Melton, J., O'Neil, E., & O'Neil, P. (1995). A critique of ANSI SQL isolation levels. In *Proceedings of the ACM SIGMOD International Conference on Management of Data*, San Jose, CA (pp. 1-10).

Bernard, J. J., & Tracy, M. (2008). Sponsored search: An overview of the concept, history, and technology. *International Journal of Electronic Business, 6*(2), 114–131. doi:10.1504/IJEB.2008.018068

Bernstein, P. A. (1996). Middleware: A model for distributed system services. *Communications of the ACM*, *39*(2), 86–98. doi:10.1145/230798.230809

Biegel, G., & Cahill, V. (2004). A framework for developing mobile, context-aware applications. In *Proceedings of the 2nd IEEE Conference on Pervasive Computing and Communication* (pp. 361-365).

Bisdikian, C., Bhagwat, P., Gaucher, B. P., Janniello, F. J., Korpeoglu, I., & Naghshineh, M. (1998). WISAP: A wireless personal access network for handheld computing devices. *IEEE Personal Communications*, *5*(6), 18–25. doi:10.1109/98.736474

Bittner, J., Havran, V., & Slavík, P. (1998). Hierarchical visibility culling with occlusion trees. In *Proceedings of the 1998 Computer Graphics International (CGI)* (pp. 207-219).

Black, A. D., Car, J., Pagliari, C., Anandan, C., Cresswell, K., & Bokun, T. (2011). The impact of ehealth on the quality and safety of health care: A systematic overview. *PLoS Medicine*, *8*(1). doi:10.1371/journal.pmed.1000387

Blobel, B. (2002). Architecture of secure portable and interoperable electronic health records. In. *Proceedings of the International Conference on Computational Science-Part, II*, 982–994.

Blobel, B. (2004). Authorisation and access control for electronic health record systems. *International Journal of Medical Informatics*, 251–257. doi:10.1016/j.ijmedinf.2003.11.018

Bojars, U., Breslin, J., Finn, A., & Decker, S. (2008). Using the Semantic Web for linking and reusing data across Web 2.0 communities. *Journal of Web Semantics*, *6*(1), 21–28. doi:10.1016/j.websem.2007.11.010

Bojars, U., Breslin, J., Peristeras, V., Tummarello, G., & Decker, S. (2008). Interlinking the social web with semantics. *IEEE Intelligent Systems*, *23*(3), 29–40. doi:10.1109/MIS.2008.50

Bottaro, A., & Gérodolle, A. (2008). Home Soa: Facing protocol heterogeneity in pervasive applications. In *Proceedings of the 5th International Conference on Pervasive Services* (pp. 73-80). New York, NY: ACM Press.

Bouchard, T., Hemon, M., Gagnon, F., Gravel, V., & Munger, O. (2008, March). Mobile telephones used as boarding passes: Enabling technologies and experimental results. In *Proceedings of the Forth International Conference on Autonomic and Autonomous Systems* (pp. 255-259).

Bouwman, H., De Vos, H., & Haaker, T. (Eds.). (2008). *Mobile service innovation and business models*. New York, NY: Springer. doi:10.1007/978-3-540-79238-3

Bowlby, J. (1969). *Attachment and loss. Vol. 1: Attachment*. NewYork: Basic Books.

Brands, S., & Chaum, D. (1994). Distance-bounding protocols. In T. Helleseth (Ed.), *Proceedings of the Workshop on the Theory and Application of Cryptographic Techniques* (LNCS 765, pp. 344-359).

Brauner, S. *(2009)*. Untersuchung des internationalen Marktpotentials eines interaktiven, kontextsensitiven und multimodalen Kommissionierungstools für die SAP AG mit abschliessender Empfehlung für das weitere strategische Vorgehen (Investigation of the international market potential of an interactive, context-sensitive and multi-modal picking tool for SAP AG with final recommendation for further strategic proceeding). *Unpublished Master's Thesis*.

Bray, T., Paoli, J., Sperberg-McQueen, C. M., Maler, E., Yergeau, F., & Cowan, J. (Eds.). (2006). *Extensible markup language (xml) 1.1 (Second edition)*. Retrieved from http://www.w3.org/TR/xml11/

Breese, J. S., Heckerman, D., & Kadie, C. (2008). Empirical analysis of predictive algorithms for collaborative filtering. In *Proceedings of the Fourteenth Conference on Uncertainty in Artificial Intelligence* (pp. 43-52).

Breitbart, Y., Garcia-Molina, H., & Silberschatz, A. (1992). Overview of multidatabase transaction management. In. *Proceedings of the Conference of the Centre for Advanced Studies on Collaborative Research*, *2*, 23–56.

Brewster, S. (2002). Overcoming the lack of screen space on mobile computers. *Personal and Ubiquitous Computing*, *6*(3), 188–205. doi:10.1007/s007790200019

Broch, D. J. J., & Malts, D. (1998). *The dynamic source routing protocol for mobile ad hoc networks*. Retrieved from http://www.ietf.org/rfc/rfc4728.txt

Broder, A., Fontoura, M., Josifovski, V., & Riedel, L. (2007). A semantic approach to contextual advertising. In *Proceedings of the 30th Annual International ACM SIGIR Conference on Research and Development in Information Retrieval* (pp. 559-566). New York, NY: ACM Press.

Broström, R., Engström, J., Agnvall, A., & Markkula, G. (2006). Towards the next generation intelligent driver information system (IDIS): The VOLVO car interaction manager concept. In *Proceedings of the ITS World Congress* (p. 47).

Brumitt, B., Meyers, B., Krumm, J., Kern, A., & Shafer, S. (2000). Easyliving: Technologies for intelligent environments. In P. Thomas & H.-W. Gellerson (Eds.), *Proceedings of the Second International Symposium on Handheld and Ubiquitous Computing*, Bristol, UK (LNCS 1927, pp. 12-29).

Bulut, E., Khadraoui, D., & Marquet, B. (2007). Multi-agent based security assurance monitoring system for telecommunication infrastructures. In *Proceedings of the Fourth IASTED International Conference on Communication, Network and Information Security* (pp. 90-95).

Bureau of Labor Statistics. (2009). *Packers and packagers, hand.* National Employment Matrix.

Burger, D. C., & Austin, T. M. (1997). *The simplescalar tool set, version 2.0* (Tech. Rep. No. CS-TR-1997-1342). Madison, WI: University of Wisconsin.

Burigat, S., Chittaro, L., & Gabrielli, S. (2008). Navigation techniques for small-screen devices: An evaluation on maps and web pages. *International Journal of Human-Computer Studies, 66*(2), 78–97. doi:10.1016/j.ijhcs.2007.08.006

Büyüközkan, G. (2009). Determining the mobile commerce user requirements using an analytic approach. *Computer Standards & Interfaces, 31*(1), 144–152. doi:10.1016/j.csi.2007.11.006

Buzeto, F. N., Filho, C. B. P., Castanho, C. D., & Jacobi, R. P. (2010). DSOA: A service oriented architecture for ubiquitous applications. In P. Bellavista, R.-S. Chang, H.-C. Chao, S.-F. Lin, & P. M. A. Sloot (Eds.), *Proceedings of the 5th International Conference of Advances in Grid and Pervasive Computing* (LNCS 6104, pp. 183-192).

Cai, M., & Frank, M. (2004). RDFPeers: A scalable distributed RDF repository based on a structured peer-to-peer network. In *Proceedings of the 13th International Conference on World Wide Web* (pp. 650-657). New York, NY: ACM Press.

Capin, T. K., Pulli, K., & Akenine-Möller, T. (2008). The state of the art in mobile graphics research. *IEEE Computer Graphics and Applications, 28*(4), 74–84. doi:10.1109/MCG.2008.83

Carter, S., & Mankoff, J. (2005). When participants do the capturing: The role of media in diary studies. In *Proceedings of the SIGCHI Conference on Human Factors in Computing Systems* (pp. 899-908).

Carter, S., Mankoff, J., & Heer, J. (2007). Momento: Support for situated ubicomp experimentation. In *Proceedings of the SIGCHI Conference on Human Factors in Computing Systems* (pp. 125-134).

Castelluccia, C., & Avoine, G. (2006). Noisy Tags: A Pretty Good Key Exchange Protocol for RFID Tags. In *Proceedings of the 7th IFIP WG 8.8/11.2 International Conference*, Tarragona, Spain.

Cerqueira, R. (2001). *Gaia: A development infrastructure for active spaces.* Paper presented at the Joint Workshop of UBICOMP and Application Models and Programming Tools for Ubiquitous Computing, Atlanta GA.

Certicom Research. (2000). *Standards for efficient cryptography – SEC1: Elliptic curve cryptography.* Retrieved from http://www.secg.org/collateral/sec1_final.pdf

Chang, C. (2009). *Design and implementation of an ontology-based distributed RDF store based on Chord network.* Unpublished doctoral dissertation, Tatung University, Taipei, Taiwan.

Chang, C.-F., & Ger, S.-H. (2002). Enhancing 3D graphics on mobile devices by image-based rendering. In *Proceedings of the 3rd IEEE Pacific Rim Conference on Multimedia (PCM)*, London (pp. 1105-1111).

Charney, M. J., & Puzak, T. R. (1997). Prefetching and memory system behavior of the SPEC95 benchmark suite. *IBM Journal of Research and Development, 41*(3), 265–286. doi:10.1147/rd.413.0265

Chatterjee, P., Hoffman, D. L., & Novak, T. P. (2006). Modeling the clickstream: Implications for web-based advertising efforts. *Marketing Science*, *22*(4), 520–541. doi:10.1287/mksc.22.4.520.24906

Chen, A. (2005). Context-aware collaborative filtering system: Predicting the user's preference in the ubiquitous computing environment. In T. Strang & C. Linnhoff-Popien (Eds.), *Proceedings of the International Workshop Location- and Context-Awareness* (LNCS 3479, pp. 244-253).

Chen, G., & Kotz, D. (2000). *A survey of context-aware mobile computing research* (Tech. Rep. No. TR2000-381). Hanover, NH: Dartmouth College.

Chen, H., Perich, F., Finin, T., & Joshi, A. (2004). SOUPA: Standard ontology for ubiquitous and pervasive applications. In *Proceedings of the First Annual International Conference on Mobile and Ubiquitous Systems: Networking and Services* (pp. 258-267).

Chen, Y., Tsai, F. S., & Chan, K. L. (2007). Blog search and mining in the business domain. In *Proceedings of the International Workshop on Domain Driven Data Mining* (pp. 55-60).

Chen, L., & Nath, R. (2008). A socio-technical perspective of mobile work. *Information Knowledge Systems Management*, *7*(1-2), 41–60.

Cheong, J. H., & Park, M. (2005). Mobile internet acceptance in Korea. *Internet Research*, *15*(2), 125–140. doi:10.1108/10662240510590324

Chesta, C., Paternò, F., & Santoro, C. (2004). Methods and tools for designing and developing usable multiplatform interactive applications. *PsychNology Journal*, *2*(1), 123–139.

Cheung, S., Lindqvist, U., & Fong, M. W. (2003). Modeling multistep cyber attacks for scenario recognition. In *Proceedings of the third DARPA information survivability conference and exposition.*

Chia, Y., Tsai, F. S., Ang, W. T., & Kanagasabai, R. (2011). Context-aware mobile learning with a semantic service-oriented infrastructure. In *Proceedings of the 7th International Symposium on Web and Mobile Information Services.*

Chiang, C. C., & Gerla, M. (1997). Routing and multicast in multihop, mobile wireless networks. In *Proceedings of IEEE International Conference on Universal Personal Communication Record*, San Diego, CA (pp. 546-551).

Chien, B. C., Tsai, H. C., & Hsueh, Y. K. (2009). CADBA: A context-aware architecture based on context database for mobile computing. In *Proceedings of the International Workshop on Pervasive Media and the Sixth International Conference on Ubiquitous Intelligence and Computing* (pp. 367-372).

Chien, B. C., He, S. Y., Tsai, H. C., & Hsueh, Y. K. (2010). An extendible context-aware service system for mobile computing. *Journal of Mobile Multimedia*, *6*(1), 49–62.

Clarke, I. (2001). Emerging value propositions for m-commerce. *The Journal of Business Strategy*, *18*(2), 133–148.

Clark, J. H. (1976). Hierarchical geometric models for visible surface algorithms. *Communications of the ACM*, *19*(10), 547–554. doi:10.1145/360349.360354

Clausen, T., Jacquet, P., Laouiti, A., Minet, P., Muhlethaler, P., Qayyum, A., & Viennot, L. (2001). *Optimized link state routing protocol.* Retrieved from http://www.ietf.org/rfc/rfc3626.txt

Clulow, J., Hancke, G. P., Kuhn, M. G., & Moore, T. (2006). So near and yet so far: Distance bounding attacks in wireless networks. In L. Buttyán, V. D. Gligor, & D. Westhoff (Eds.), *Proceedings of the 3rd European Workshop on Security and Privacy in Ad Hoc and Sensor Networks* (LNCS 4357, pp. 83-97).

Cohen-Or, D., Chrysanthou, Y., Silva, C. T., & Durand, F. (2002). A survey of visibility for walkthrough applications. *IEEE Transactions on Visualization and Computer Graphics*, *9*(3), 412–431. doi:10.1109/TVCG.2003.1207447

Cohen-Or, D., Rich, E., Lerner, U., & Shenkar, V. (1996). A real-time photo-realistic visual flythrough. *IEEE Transactions on Visualization and Computer Graphics*, *2*(3), 255–265. doi:10.1109/2945.537308

Common Criteria. (2006). *Common criteria for information technology, part 1: Introduction and general model version 3.1.* Retrieved from http://www.commoncriteriaportal.org/files/ccfiles/CCPART1V3.1R1.pdf

Consolvo, S., & Walker, M. (2003). Using the experience sampling method to evaluate ubicomp applications. *IEEE Pervasive Computing/IEEE Computer Society and IEEE Communications Society, 2*(2), 24–31. doi:10.1109/MPRV.2003.1203750doi:10.1109/MPRV.2003.1203750

Constantiou, I. D., Papazafeiropoulou, A., & Vendelø, M. T. (2009). Does culture affect the adoption of advanced mobile services? A comparative study of young adults' perceptions in Denmark and the UK. *SIGMIS Database, 40*(4), 132–147. doi:10.1145/1644953.1644962

Coorg, S. R., & Teller, S. J. (1996). Temporally coherent conservative visibility. In *Proceedings of the 1996 Symposium on Computational Geometry* (pp. 78-87).

Cote, M., Suryn, W., Laporte, C., & Martin, R. (2005). The evolution path for industrial software quality evaluation methods applying ISO/IEC 9126:2001 quality model: Example of MITRE's SQAE method. *Software Quality Journal, 13*(1), 17–30. doi:10.1007/s11219-004-5259-6

Coursaris, C., & Hassanein, K. (2002). Understanding m-commerce. *Quarterly Journal of Electronic Commerce, 3*(3), 247–271.

Cristal, A., Santana, O., Cazorla, F., Galluzzi, M., Ramirez, T., & Pericas, M. (2005). Kilo-instruction processors: Overcoming the memory wall. *IEEE Micro, 25*(3), 48–57. doi:10.1109/MM.2005.53

Csikszentmihalyi, M. (1990). *Flow: The Psychology of Optimal Experience*. New York: Harper and Row.

Cuppens, F. (2001). Managing alerts in a multi-intrusion detection environment. In *Proceedings of the 17th annual computer security applications conference.*

Cziekszentimihalyi, M., & Larson, R. (1987). Validity and reliability of the experience-sampling method. *The Journal of Nervous and Mental Disease, 56*, 5–18.

D'Anjou, J., Fairbrother, S., Kehn, D., Kellerman, J., & McCarthy, P. (2005). *The Java developer's guide to Eclipse* (2nd ed.). Reading, MA: Addison-Wesley.

Dahlberg, T., Mallat, N., Ondrus, J., & Zmijewska, A. (2008). Past, present and future of mobile payments research: A literature review. *Electronic Commerce Research and Applications, 7*(2), 165–181. doi:10.1016/j.elerap.2007.02.001

Dahlgren, F., Dubois, M., & Stenstroem, P. (1995). Sequential hardware prefetching in shared-memory multiprocessors. *IEEE Transactions on Parallel and Distributed Systems, 6*(7), 733–746. doi:10.1109/71.395402

Dai, H., & Palvi, P. C. (2009). Mobile commerce adoption in China and the United States: A cross-cultural study. *SIGMIS Database, 40*(4), 43–61. doi:10.1145/1644953.1644958

Dalmau, D. S.-C. (2003). *Core techniques and algorithms in game programming*. New Riders Games.

d'Aquin, M., Motta, E., Dzbor, M., Gridinoc, L., Heath, T., & Sabou, M. (2008). Collaborative semantic authoring. *IEEE Intelligent Systems, 23*(3), 80–83. doi:10.1109/MIS.2008.43

Davis, A. M. (1992). Operational prototyping: A new development approach. *IEEE Software, 9*, 70–78. doi:10.1109/52.156899

Davis, F. D. (1989). Perceived usefulness, perceived ease of use, and user acceptance of information technology. *Management Information Systems Quarterly, 13*(3), 319–340. doi:10.2307/249008

Davis, F. D., Bagozzi, R. P., & Warshaw, P. R. (1989). User acceptance of computer technology: A comparison of two theoretical models. *Management Science, 35*, 982–1003. doi:10.1287/mnsc.35.8.982

de Sá, M., & Carriço, L. (2009). Mobile support for personalized therapies - OmniSCOPE: Richer artefacts and data collection. In *Proceedings of the 3rd International Conference on Pervasive Computing Technologies for Healthcare* (pp. 1-8). New York, NY: ACM Press.

de Sá, M., Carriço, L., Duarte, L., & Reis, T. (2008). A framework for mobile evaluation. In *Proceedings of the SIGCHI Conference on Human Factors in Computing Systems* (pp. 2673-2678).

Dean, J., & Ghemawat, S. (2004). MapReduce: Simplified data processing on large clusters. In *Proceedings of the Sixth Symposium on Operating System Design and Implementation* (pp. 137-150).

Desmedt, Y. (1988). Major security problems with the ``unforgeable'' (Feige)-Fiat-Shamir proofs of identity and how to overcome them. In *Proceedings of SecuriCom* (pp. 15-17).

Desmedt, Y., & Frankel, Y. (1991). Shared generation of authenticators and signatures. In J. Feigenbaum (Ed.), *Proceedings of Advances in Cryptology* (LNCS 576, pp. 457-469).

Desmedt, Y., & Jajodia, S. (1997). *Redistributing secret shares to new access structures and its applications* (Tech. Rep. No. ISSE-TR-97-01). Washington, DC: George Mason University.

Deutsch, M., & Willis, R. (1988). *Software Quality Engineering: A Total Technical and Management Approach.* Upper Saddle River, NJ: Prentice-Hall.

Dey, A. K., & Abowd, G. D. (1999). The context toolkit: Aiding the development of context-aware applications. In *Proceedings of the SIGCHI Conference on Human Factors in Computing Systems* (pp. 434-441).

Dey, A. K., Futakawa, M., Salber, D., & Abowd, G. D. (1999). The Conference Assistant: Combining Context-Awareness with Wearable Computing. In *Proceedings of the 3rd International Symposium on Wearable Computers (ISWC)*, San Francisco (pp. 21-28).

Dey, A. (2001). Understanding and using context. *Personal and Ubiquitous Computing, 5*(1), 4–7. doi:10.1007/s007790170019

Dey, A. K., Abowd, G. D., & Salber, D. (2001). A conceptual framework and a toolkit for supporting the rapid prototyping of context-aware applications. *Human-Computer Interaction, 16*, 97–166. doi:10.1207/S15327051HCI16234_02

DigitalHealth. (2008). *Project description in Danish of shared medication record.* Retrieved from http://www.sdsd.dk/det_goer_vi/Faelles_Medicinkort.aspx

Dimitriou, T. (2008). Proxy Framework for Enhanced RFID Security and Privacy. In *Proceedings of the Consumer Communications and Networking Conference*, Las Vegas, NV.

Do, H., & Rahm, E. (2002). COMA - a system for flexible combination of schema matching approaches. In *Proceedings of the 28th Conference on Very Large Data Bases* (pp. 610-621).

Dobrev, A., Jones, T., Stroetmann, K., Vatter, Y., & Peng, K. (2009). *The socio-economic impact of interoperable electronic health record (EHR) and ePrescribing systems in Europe and beyond.* Retrieved from http://www.ehealthnews.eu/publications/latest/1839-the-socio-economic-impact-of-interoperable-ehr-and-eprescribing-systems-in-europe-and-beyond

Douligeris, C., & Serpanos, D. N. (2007). *Network security: Current status and future directions.* New York: Wiley.

Drimer, S., & Murdoch, S. (2007). Keep your enemies close: Distance bounding against smartcard relay attacks. In *Proceedings of the 16th USENIX Security Symposium* (pp. 87-102).

Duguet, F., & Drettakis, G. (2004). Flexible point-based rendering on mobile devices. *IEEE Computer Graphics and Applications, 24*(4), 57–63. doi:10.1109/MCG.2004.5

Eckel, G. (2001). LISTEN - augmenting everyday environments with interactive soundscapes. In *Proceedings of the 13 Spring Days Workshop Moving between the physical and the digital: exploring and developing new forms of mixed reality user experience*, Porto, Portugal.

Eisenhauer, M., Lorenz, A., Zimmermann, A., Duong, T., & James, F. (2005). Interaction by movement - one giant leap for natural interaction in mobile guides. In *Proceedings of the International Workshop on Artificial Intelligence in Mobile Systems.*

European Committee for Standardization. (1999). *Health informatics: Electronic healthcare record communication - Part 1: Extended architecture (Tech. Rep. No. CEN TC251/WG1).* Brussels, Belgium: European Committee for Standardization.

Evans, D. L., Bond, P. J., & Bement, A. L. (2004). *Standards for security categorization of federal information and information systems.* Gaithersburg, MD: NIST.

Falas, T., & Kashani, H. (2007). Two-dimensional barcode decoding with camera-equipped mobile phones. In *Proceedings of the Fifth Annual IEEE International Conference on Pervasive Computing and Communication Workshops* (pp. 297-600).

Fang, J., Shao, P., & Lan, G. (2009). Effects of innovativeness and trust on web survey participation. *Computers in Human Behavior*, *25*, 144–152. doi:10.1016/j.chb.2008.08.002

Feldhofer, M., Dominikus, S., & Wolkerstorfer, J. (2004). Strong Authentication for RFID Systems Using the AES Algorithm. In M. Joye & J. Quisquater (Eds.), *Cryptographic Hardware and Embedded Systems* (LNCS 3156, pp. 357-370). New York: Springer.

Ferguson, N., Whiting, D., & Schneier, B. (2003). *Helix: Fast encryption and authentication in a single cryptographic primitive* (LNCS 2887, pp. 330-346). New York: Springer.

Ferguson, N., Schenier, B., & Kohno, T. (2010). *Cryptography engineering: Design principles and practical applications*. Indianapolis, IN: Wiley.

Fishbein, M., & Ajzen, I. (1975). *Belief, Attitude, Intention and Behavior: An Introduction to Theory and Research*. Reading, MA: Addison-Wesley.

Frank, L. (2003, July 1-2). Data cleaning for datawarehouses built on top of distributed databases with approximated acid properties. In *Proceedings of the International Conference on Computer Science and its Applications*, San Diego, CA (pp. 160-164).

Frank, L., & Andersen, S. K. (2010). Evaluation of different database designs for integration of heterogeneous distributed electronic health records. In *Proceedings of the IEEE/ICME International Conference on Complex Medical Engineering* (pp. 204-209).

Frank, L., & Mukherjee, A. *(2002, June). Distributed electronic patient encounter with high performance and availability. In* Proceedings of the 15th IEEE Symposium on Computer-Based Medical Systems *(pp. 373-376).*

Frank, L., & Munck, S. (2008). An overview of architectures for integrating distributed electronic health records. In *Proceeding of the 7th International Conference on Applications and Principles of Information Science* (pp. 297-300).

Frankel, Y., Gemmell, P., MacKenzie, P. D., & Yung, M. (1997). Optimal resilience proactive public-key cryptosystems. In *Proceedings of the IEEE Symposium on Foundations of Computer Science* (pp. 384-393). Washington, DC: IEEE Computer Society.

Frank, L. (2005). Replication methods and their properties. In Rivero, L. C., Doorn, J. H., & Ferraggine, V. E. (Eds.), *Encyclopedia of database technologies and applications*. Hershey, PA: IGI Global. doi:10.4018/978-1-59140-560-3.ch092

Frank, L. (2010). *Design of distributed integrated heterogeneous or mobile databases*. Saarbrücken, Germany: Lambert Academic Publishing.

Frank, L., & Zahle, T. U. (1998). Semantic acid properties in multidatabases using remote procedure calls and update propagations. *Software, Practice & Experience*, *28*(1), 77–98. doi:10.1002/(SICI)1097-024X(199801)28:1<77::AID-SPE148>3.0.CO;2-R

Franklin, T., & van Harmelen, M. (2007). *Web 2.0 for content for learning and teaching in higher education*. Retrieved from http://www.jisc.ac.uk/publications/reports/2007/web2andpolicyreport.aspx

Froehlich, J., Chen, M. Y., Consolvo, S., Harrison, B., & Landay, J. A. (2007). MyExperience: A system for in situ tracing and capturing of user feedback on mobile phones. In *Proceedings of the 5th International Conference on Mobile Systems, Applications and Services* (pp. 57-70).

Froehlich, J., Chen, M., Smith, I., & Potter, F. (2006). Voting with your feet: An investigative study of the relationship between place visit behavior and preference. In *Proceedings of the Conference on Ubiquitous Computing* (pp. 333-350).

Fu, J. W. C., Patel, J. H., & Janssens, B. L. (1992). Stride directed prefetching in scalar processors. *ACM SIGMICRO Newsletter*, *23*(1-2), 102–110. doi:10.1145/144965.145006

Gans, G. K., Hoepman, J. H., & Garcia, F. D. (2008). A practical attack on the MIFARE classic. In G. Grimaud & F.-X. Standaert (Eds.), *Proceedings of the 8th IFIP WG 8.8/11.2 International Conference on Smart Card Research and Advanced Applications* (LNCS 5189, pp. 267-282).

Gao, J. Z., Kulkarni, V., Ranavat, H., Chang, L., & Hsing, M. (2009). A 2D barcode-based mobile payment system. In *Proceedings of the 3rd International Conference on Multimedia and Ubiquitous Engineering* (pp.320-329).

Gao, J. Z., Prakash, L., & Jagatesan, R. (2007). Understanding 2D-barcodes technology and applications in m-commerce – design and implementation of a 2D barcode processing solution. In *Proceedings of 31st Annual International Computer Software and Applications Conference,* Beijing, China (Vol. *2,* pp. 49-56).

Garcia-Molina, H., & Salem, K. (1987). Sagas. In *Proceedings of the SIGMOD Conference on Management of Data* (pp. 249-259).

Garlan, D. (2002). Project aura: Toward distraction-free pervasive computing. *IEEE Pervasive Computing/IEEE Computer Society and IEEE Communications Society,* *1*(2), 22–31. doi:10.1109/MPRV.2002.1012334

Garofalakis, J., Stefani, A., Stefanis, V., & Xenos, M. (2007). Quality attributes of consumer-based m-commerce systems. In *Proceedings of the ICETE-Business Conference* (pp. 130-136).

Ghinea, G., & Angelides, M. C. (2004). A user perspective of quality of service in m-commerce. *Multimedia Tools and Applications,* *22*(2), 187–206. doi:10.1023/B:MTAP.0000011934.59111.b5

Gillet, D., Helou, S., Yu, C., & Salzmann, C. (2008). Turning Web 2.0 social software into versatile collaborative learning solutions. In *Proceedings of the First International Conference on Advances in Computer-Human Interaction* (pp. 170-176). Washington, DC: IEEE Computer Society.

Gillet, D., Ngoc, A., & Rekik, Y. (2005). Collaborative web-based experimentation in flexible engineering education. *IEEE Transactions on Education,* *48*(4), 696–704. doi:10.1109/TE.2005.852592

Gillmann, L. *(2008).* Wirtschaftliche Evaluierung eines innovativen multimodal interaktiven Systems zur Unterstützung von Kommissionierprozessen (Economic evaluation of an innovative multi-modal interactive system in support of picking processes). *Unpublished Master's Thesis.*

Gopalakrishna, R., & Spafford, E. (2004). A framework for distributed intrusion detection using interest-driven cooperating agents. In *Proceedings of international symposium on recent advances in intrusion detection.*

Gornish, E. H., Granston, E. D., & Veidenbaum, A. V. (1990). Compiler-directed data prefetching in multiprocessors with memory hierarchies. In *Proceedings of the 4th International Conference on Supercomputing* (pp. 354-368).

Graff, C., & de Weck, O. (2006). A modular state-vector based modeling architecture for diesel exhaust system design, analysis and optimization. In *Proceedings of 11th AIAA/ISSMO Multidisciplinary Analysis and Optimization Conference* (pp. 6-8).

Gray, J., & Reuter, A. (1993). *Transaction processing: Concepts and techniques.* San Francisco, CA: Morgan Kaufmann.

Gu, T., Wang, X. H., Pung, H. K., & Zhang, D. Q. (2004). An ontology-based context model in intelligent environments. In *Proceedings of Communication Networks and Distributed Systems Modeling and Simulation* (pp. 270-275).

Gudivada, V. N., & Raghavan, V. V. (1995). Design and evaluation of algorithms for image retrieval by spatial similarity. *ACM Transactions on Information Systems,* *13*(2), 115–144. doi:10.1145/201040.201041

Gunasekaran, A., & McGaughey, R. E. (2009). Mobile commerce: Issues and obstacles. *International Journal of Business Information Systems,* *4*(2), 245–261. doi:10.1504/IJBIS.2009.022826

Gunter, T. D., & Terry, N. P. (2005). The emergence of national electronic health record architectures in the United States and Australia: Models, costs, and questions. *Journal of Medical Internet Research,* *7*(1), 3. doi:10.2196/jmir.7.1.e3

Gu, T., Pung, H. K., & Zhang, D. Q. (2004). Toward an OSGi-based infrastructure for context-aware applications. *IEEE Pervasive Computing/IEEE Computer Society and IEEE Communications Society, 3*(4), 66–74. doi:10.1109/MPRV.2004.19

Guthaus, M. R., Ringenberg, J. S., Ernst, D., Austin, T. M., Mudge, T., & Brown, R. B. (2001). Mibench: A free, commercially representative embedded benchmark suite. In *Proceedings of the IEEE 4th Annual Workshop on Workload Characterization* (pp. 83-94).

Haas, Z. J., & Peariman, M. R. (2001). The performance of query control schemes for the zone routing protocol. *IEEE/ACM Transactions on Networking, 9*(4), 427–438. doi:10.1109/90.944341

Hachet, M., Decle, F., & Guitton, P. (2006). Z-Goto for efficient navigation in 3D Environments from Discrete Inputs. In *Proceedings of the 13ᵗʰ Symposium on Virtual Reality Software and Technology (VRST)*, Limassol, Cyprus (pp. 236-239).

Ha, I., Yoon, Y., & Choi, M. (2007). Determinants of adoption of mobile games under mobile broadband wireless access environment. *Information & Management, 44*(3), 276–286. doi:10.1016/j.im.2007.01.001

Hair, J. F., Anderson, R. E., Tatham, R. L., & Black, W. C. (1998). *Multivariate Data Analysis* (5th ed.). Upper Saddle River, NJ: Prentice-Hall.

Häkkilä, J., & Mäntyjärvi, J. (2006). Developing design guidelines for context-aware mobile applications. In *Proceedings of the 3rd International Conference on Mobile Technology, Applications and Systems* (p. 24). New York, NY: ACM Press.

Hall, D. L. (2001). *Handbook of Multisensor Data Fusion* (1st ed.). Boca Raton, FL: CRC Publishing.

Hancke, G., & Kuhn, M. (2005). An RFID distance bounding protocol. In *Proceedings of the 1st International Conference on Security and Privacy for Emerging Areas in Communications Networks* (pp. 67-73). Washington, DC: IEEE Computer Society.

Handschuh, S., Staab, S., & Ciravegna, F. (2002). S-CREAM - semi-automatic CREAtion of metadata. In A. Gomez-Perez & V. R. Benjamins (Eds.), *Proceedings of 13th International Conference Knowledge Engineering and Knowledge Management Ontologies and the Semantic Web* (LNCS 2473, pp. 358-372).

Handschuh, S., & Staab, S. (2003). CREAM: CREAting metadata for the Semantic Web. *Computer Networks, 42*(5), 579–598. doi:10.1016/S1389-1286(03)00226-3

Hansen, F. A., & Grønbæk, K. (2008). Social web applications in the city: A lightweight infrastructure for urban computing. In *Proceedings of the Nineteenth ACM Conference of Hypertext and Hypermedia* (pp. 175-180).

Hartmann, B., Klemmer, S. R., Bernstein, M., Abdulla, L., Burr, B., Robinson-Mosher, A., et al. (2006). Reflective physical prototyping through integrated design, test, and analysis. In *Proceedings of the 19th Annual ACM Symposium on User Interface Software and Technology* (pp. 299-308). New York, NY: ACM Press.

Hart, P. E., Nilsson, N. J., & Raphael, B. (1968). A formal basis for the heuristic determination of minimum cost paths. *IEEE Transactions on Systems Science and Cybernetics, 4*(2), 100–107. doi:10.1109/TSSC.1968.300136

Head, M. R., Govindaraju, M., Slominski, A., Liu, P., Abu-Ghazaleh, N., van Engelen, R., et al. (2005). A benchmark suite for soap-based communication in grid web services. In *Proceedings of the ACM/IEEE Conference on Supercomputing* (p. 19). Washington, DC: IEEE Computer Society.

Hegering, H. G., Küpper, A., Linnhoff-Poien, C., & Reiser, H. (2003). Management challenges of context-aware services in ubiquitous environments. In M. Brunner & A. Keller (Eds.), *Proceedings of the 14th IFIP/IEEE International Workshop on Distributed Systems: Operations and Management* (LNCS 2867, pp. 321-339).

Heijden, D. H. V. (2004). User acceptance of hedonic information systems. *Management Information Systems Quarterly, 28*(4), 695–704.

Helal, S., Mann, W., El-Zabadani, H., King, J., Kaddoura, Y., & Jansen, E. (2005). The gator tech smart house: A programmable pervasive space. *IEEE Computer Magazine, 38*(3), 50.

Hennessy, J. L., & Patterson, D. A. (2003). *Computer architecture: A quantitative approach* (3rd ed.). Amsterdam, The Netherlands: Elsevier Science.

Henrici, D., & Muller, P. (2004). Hash-based Enhancement of Location Privacy for Radio-frequency Identification Devices using Varying Identifiers. In *Proceedings of the Second IEEE Annual Conference on Pervasive Computing and Communications Workshops*, Orlando, FL.

Henricksen, K., Indulska, J., & Rakotonirainy, A. (2002). Modeling context information in pervasive computing system. In *Proceedings of the First International Conference on Pervasive Computing* (pp. 167-180).

Hernando, M. E., Gómez, E. J., García-Olaya, A., Torralba, V., & Pozo, F. (2006). A mobile telemedicine workspace for diabetes management. In Micheli-Tzanakou, E. (Ed.), *Topics in biomedical engineering* (pp. 587–599). New York, NY: Springer.

Hiltunen, M., Schlichting, R., Ugarte, C., & Wong, G. (2000). Survivability through Customization and Adaptability: The Cactus Approach. In *Proceedings of the DARPA Information Survivability Conference and Exposition* (pp. 294-307).

HL7. (2005). *Health level seven (HL7).* Retrieved from http://www.hl7.org/

Hofer, T., Schwinger, W., Pichler, M., Leonhartsberger, G., Altmann, J., & Retschitzegger, W. (2002). Context-awareness on mobile devices - the hydrogen approach. In *Proceedings of the 36th Annual Hawaii International Conference on System Sciences* (pp. 292-302).

Hofstede, G. (2009). *Geert Hofstede cultural dimensions.* Retrieved June 20, 2009, from http://www.clearlycultural.com/geert-hofstede-cultural-dimensions

Holleis, P., & Schmidt, A. (2008). MakeIt: Integrate user interaction times in the design process. In *Proceedings of the 6th International Conference on Pervasive Computing* (pp. 56-74).

Holleis, P., Otto, F., Hussmann, H., & Schmidt, A. (2007). Keystroke-level model for advanced mobile phone interaction. In *Proceedings of the Conference on Human Factors in Computing Systems* (pp. 1505-1514). New York, NY: ACM Press.

Holmquist, L. E. (2005). Prototyping: Generating ideas or cargo cult designs? *Interaction, 12*(2), 48–54. doi:10.1145/1052438.1052465

Holstein, D. K. (2009). A systems dynamics view of security assurance issues: The curse of complexity and avoiding chaos. In *Proceedings of the 42nd Hawaii International Conference on System Sciences* (pp. 1-9).

Holzinger, A. (2005). Usability engineering methods for software developers. *Communications of the ACM, 48*(1), 71–74. doi:10.1145/1039539.1039541

Honeywell Imaging and Mobility. (2008). *Mobile ticketing technology.* Retrieved from http://www.airport-int.com/categories/mobile-ticketing/mobile-ticketing-choosing-the-right-technology-platform-is-critical-to-your-programs-success.asp

Hong, S. J., & Lerch, F. J. (2002). A laboratory study of customers' preferences and purchasing behavior with regards to software components. *The Data Base for Advances in Information Systems, 33*(3), 23–37.

Hong, S. J., Thong, J. Y., Moon, J., & Tam, K. (2008). Understanding the behavior of mobile data services consumers. *Information Systems Frontiers, 10*(4), 431–445. doi:10.1007/s10796-008-9096-1

Hopper, N., & Blum, M. (2001). Secure Human Identification Protocols. In C. Boyd (Ed.), *Advances in Cryptology – ASIA CRYPT 2001* (LNCS 2248, pp. 52-66). Berlin: Springer Verlag.

Houde, S., & Hill, C. (1997). What do prototypes prototype? In Helander, M., Landauer, T., & Prabhu, P. (Eds.), *Handbook of human-computer interaction.* Cambridge, MA: Elsevier Science.

Howard, A., Hafeez-Baig, A., Howard, S., & Gururajan, R. (2006). A framework for the adoption of wireless technology in healthcare: An indian study. In *Proceedings of the 17th Australasian Conference on Information Systems*, Adelaide.

Hsieh, G., Li, I., Dey, A., Forlizzi, J., & Hudson, S. E. (2008). Using visualizations to increase compliance in experience sampling. In *Proceedings of the Conference on Ubiquitous Computing* (pp. 164-167).

Hu, Z., Martonosi, M., & Kaxiras, S. (2003). TCP: Tag correlating prefetchers. In *Proceedings of the 9th International Symposium on High-Performance Computer Architecture* (pp. 317-326).

Huang, J., Bue, B., Pattath, A., Ebert, D. S., & Thomas, K. M. (2007). Interactive illustrative rendering on mobile devices. *IEEE Computer Graphics and Applications, 27*(3), 48–56. doi:10.1109/MCG.2007.63

Huang, W. W., Wang, Y., & Day, J. (2007). *Global mobile commerce: Strategies, implementation and case studies.* Hershey, PA: IGI Global. doi:10.4018/978-1-59904-558-0

Hudson, J. M., Christensen, J., Kellogg, W. A., & Erickson, T. (2002). I'd be overwhelmed, but it's just one more thing to do: Availability and interruption in research management. In *Proceedings of the SIGCHI Conference on Human Factors in Computing Systems* (pp. 97-104).

Hudson, T., Manocha, D., Cohen, J., Lin, M. C., Hoff, K. E., III, & Zhang, H. (1997). Accelerated occlusion culling using shadow frusta. In *Proceedings of the 1997 ACM Symposium on Computational Geometry* (pp. 1-10).

Hull, R., Clayton, B., & Melamed, T. (2004). Rapid authoring of mediascapes. In *Proceedings of the 6th International Conference of Ubiquitous Computing* (pp. 125-142).

Hung, S. Y., & Chang, C. M. (2005). User acceptance of wap services: Test of competing theories. *Computer Standards & Interfaces, 28,* 359–370. doi:10.1016/j.csi.2004.10.004

Hung, S., Ku, C., & Chang, C. (2003). Critical factors of WAP services adoption: An empirical study. *Electronic Commerce Research and Applications, 2,* 42–62. doi:10.1016/S1567-4223(03)00008-5

Hu, W.-C., Zuo, Y., Chen, L., & Yang, H.-J. (2010). Contemporary issues in handheld computing research. *International Journal of Handheld Computing Research, 1*(1), 1–23. doi:10.4018/jhcr.2010090901

Hwang, J., Jung, J., & Kim, G. J. (2006). Hand-held virtual reality: a feasibility study. In *Proceedings of the 13th Symposium on Virtual Reality Software and Technology (VRST)*, Limassol, Cyprus (pp. 356-363).

IATA. (2007). *Standard paves way for global mobile phone check-in.* Retrieved from http://www.iata.org/pressroom/pr/2007-11-10-01.htm

IEEE Computer Society. (2005). The 2D data matrix barcode. *IEEE Computing & Control Engineering Journal, 16*(6), 39.

IEEE 802.11. (2007). IEEE Standard for Information technology-Telecommunications and information exchange between systems-Local and metropolitan area networks-Specific requirements: Part 11: Wireless LAN Medium Access Control (MAC) and Physical Layer (PHY) Specifications.

International Organization for Standardization. (1994). *ISO/IEC 9797: Information technology - security techniques - data integrity mechanisms using a cryptographic check function employing a block cipher algorithm.* Retrieved from http://www.iso.org/iso/iso_catalogue/catalogue_tc/catalogue_detail.htm?csnumber=22053

International Organization for Standardization. (2000). *ISO/IEC 16022: Information technology – international symbology specification – data matrix.* Retrieved from http://www.iso.org/iso/catalogue_detail.htm?csnumber=29833

International Organization for Standardization. (2004). *ISO/IEC 9126: Software product evaluation –quality characteristics and guidelines for the user.* Geneva, Switzerland: International Organization for Standardization.

International Organization for Standardization. (2005). *ISO/IEC 9798-6: Information technology - security techniques - entity authentication - part 6: Mechanisms using manual data transfer.* Retrieved from http://www.iso.org/iso/iso_catalogue/catalogue_tc/catalogue_detail.htm?csnumber=39721

International Organization for Standardization. (2009). *ISO/IEC 27004: Information technology - Security techniques - Information security management measurements.* Geneva, Switzerland: International Organization for Standardization.

Intille, S. S., Rondoni, J., Kukla, C., Ancona, I., & Bao, L. (2003). A context-aware experience sampling tool. In *Proceedings of the SIGCHI Conference on Human Factors in Computing Systems* (pp. 972-973).

ISO 13407. (1999). *Human-centred design processes for interactive systems.*

ISO 9241. (1998). ISO 9241-10: Ergonomic requirements for the design of dialogs - part 10 guidance on usability.

ISO 9241. (1998). ISO 9241-11: Ergonomic requirements for office work with visual display terminals - part 11 guidance on usability.

ISO/IEC 18000-6. (2004). Information technology - Radio frequency identification for item management - Part 6: Parameters for air interface communications at 860 MHz to 960 MHz.

Isomursu, M., Tähti, M., Väinämö, S., & Kuutti, K. (2007). Experimental evaluation of five methods for collecting emotions in field settings with mobile applications. *International Journal of Human-Computer Studies*, *65*(4), 404–418. doi:10.1016/j.ijhcs.2006.11.007doi:10.1016/j.ijhcs.2006.11.007

Issarny, V., Sacchetti, D., Chibout, R., Dalouche, S., & Musolesi, M. (2005). *WSAMI: A middleware infrastructure for ambient intelligence based on web services*. Retrieved from https://www-roc.inria.fr/arles/index.php/software/64-wsami-a-middleware-infrastructure-for-ambient-intelligence-based-on-web-services.html

Itoh, M., Higaki, S., Takeuchi, Y., Kitajima, A., Imai, M., Sato, J., et al. (2000). Peas-iii: An asip design environment. In *Proceedings of the IEEE International Conference on Computer Design* (pp. 430-436).

Iwata, C.-C., Chiang, G., Pei, G., Gerla, M., & Chen, T.-W. (1999). Scalable routing strategies for ad hoc wireless networks. *IEEE Journal on Selected Areas in Communications*, 1369–1379. doi:10.1109/49.779920

Jahns, V. (2009). Mobile computing and urban systems: A literature review. In *Proceedings of the Conference on Techniques and Applications for Mobile Commerce* (pp. 17-26).

Jansen, W. (2009). *Directions in security metrics research* (Tech. Rep. No. NISTIR 7564). Gaithersburg, MD: National Institute of Standards and Technology.

Jennings, N. R. (1999). An agent-based software engineering. In *Proceedings of the 9th European Workshop on Modelling Autonomous Agents in a Multi-Agent World*.

Jen, W.-Y., Chao, C.-C., Hung, M. C., Li, Y.-C., & Chi, Y. P. (2007). Mobile information and communication in the hospital outpatient service. *International Journal of Medical Informatics*, *76*(8), 565–574. doi:10.1016/j.ijmedinf.2006.04.008

Joseph, D., & Grunwald, D. (1999). Prefetching using Markov predictors. *IEEE Transactions on Computers*, *48*(2), 121–133. doi:10.1109/12.752653

Jouppi, N. P. (1990). Improving direct-mapped cache performance by the addition of a small fully-associative cache and prefetch buffers. In *Proceedings of the 17th Annual International Symposium on Computer Architecture* (pp. 364-373).

Juels, A. (2004). Minimalist Cryptography for Low-cost RFID Tags. In *Proceedings of the Fourth Conference on Security in Communication Networks*, Amalfi, Italy (pp. 149-153).

Juels, A., & Weis, S. (2005). Authenticating Pervasive Devices with Human Protocols. In V. Shoup (Ed.), *Advances in Cryptography – Crypto 05* (LNCS 3126, pp. 198-293). Berlin: Springer Verlag.

Juels, A., Rivest, R., & Szydlo, M. (2003). The Blocker Tag: Selective Blocking of RFID Tags for Consumer Privacy. In *Proceedings of the ACM Conference on Computer and Communication Security* (pp. 103-111).

Julisch, K. (2004). Clustering intrusion detection alarms to support root cause analysis. *ACM Transactions on Information and System Security*, *6*(4), 443–471. doi:10.1145/950191.950192

Junglas, I. (2007). On the usefulness and ease of use of location-based services: Insights into the information system innovator's dilemma. *International Journal of Mobile Communications*, *5*(4), 389–408. doi:10.1504/IJMC.2007.012787

Kabasakal, H., & Bodur, M. (2002). Arabic cluster: A bridge between east and west. *Journal of World Business*, *37*(1), 40–54. doi:10.1016/S1090-9516(01)00073-6

Kahan, J., & Koivunen, M. Prud'Hommeaux, E., & Swick, R. (2001). Annotea: An open RDF infrastructure for shared web annotations. In *Proceedings of the International Conference on the World Wide Web*, Hong Kong.

Kahneman, D., Krueger, A. B., Schkade, D. A., Schwarz, N., & Stone, A. A. (2004). A survey method for characterizing daily life experience: The day reconstruction method. *Science*, *306*, 1776. doi:10.1126/science.1103572doi:10.1126/science.1103572

Kalra, D., Lewalle, P., Rector, A., Rodrigues, J. M., Stroetmann, K., Surjan, G., et al. (2009). *Semantic interoperability for better health and safer healthcare.* Retrieved from http://ec.europa.eu/information.../health/.../2009semantic-health-report.pdf

Kapoor, A., & Horvitz, E. (2008). Experience sampling for building predictive user models: A comparative study. In *Proceedings of the SIGCHI Conference on Human Factors in Computing Systems* (pp. 657-666).

Kappel, G., Retschitzegger, W., Kimmerstorfer, E., Pröll, B., Schwinger, W., & Hofer, T. (2002). Towards a generic customisation model for ubiquitous Web applications. In *Proceedings of the 2nd International Workshop on Web Oriented Software Technology*, Málaga, Spain.

Karim, R. (2006). An Efficient Collaborative Intrusion Detection System for MANET Using Bayesian Approach. In *Proceedings of the MSWiM.*

Karygiannis, T., Eydt, B., Bunn, L., & Phillips, T. (2007). *Guidelines for Securing Radio Frequency Identification (RFID) Systems.* National Institute of Standard and Technology.

Kato, H., & Tan, K. T. (2007). First read rate analysis of 2D-barcodes for camera phone applications as a ubiquitous computing tool. In *Proceedings of the IEEE Region 10 Conference* (pp. 1-4).

Kato, H., & Tan, K. T. (2007). Pervasive 2D barcodes for camera phone applications. *IEEE Pervasive Computing/ IEEE Computer Society and IEEE Communications Society, 6*(4), 76–85. doi:10.1109/MPRV.2007.80

Katz, J., & Shin, J. (2006). Parallel and Concurrent Security of the HB and HB++ Protocols. In *Proceedings of the Advances in Cryptology (EUROCRYPT 2006)* (LNCS 4004, pp. 73-87). New York: Springer.

Kaufmann, O., Lorenz, A., Oppermann, R., Schneider, A., Eisenhauer, M., & Zimmermann, A. (2007). Implicit interaction for pro-active assistance in a context-adaptive warehouse application. In *Proceedings of the 4th international conference on mobile technology, applications, and systems and the 1st international symposium on Computer human interaction in mobile technology*, Singapore.

Kemmerer, R. A., & Giovanni, V. (2005). Hi-DRA: intrusion detection for internet security. *Proceedings of the IEEE, 93*(10), 1848–1857. doi:10.1109/JPROC.2005.853547

Khan, V. J., Markopoulos, P., & Eggen, B. (2007). On the role of awareness systems for supporting parent involvement in young children's schooling. *International Federation for Information Processing, 241*, 91–101.

Khan, V. J., Markopoulos, P., de Ruyter, B., & IJsselsteijn, W. (2007). Expected information needs of parents for pervasive awareness systems. In B. Schiele, A. K. Dey, H. Gellersen, B. de Ruyter, M. Tscheligi, R. Wichert et al. (Eds.), *Proceedings of the European Conference on Ambient Intelligence* (LNCS 4794, pp. 332-339).

Khan, V. J., Markopoulos, P., Mota, S., IJsselsteijn, W., & de Ruyter, B. (2006). Intra-family communication needs: How can awareness systems provide support? In *Proceedings of the 2nd International Conference on Intelligent Environments* (Vol. 2, pp. 89-94).

Khedr, M., & Karmouch, A. (2005). ACAI: Agent-based context-aware infrastructure for spontaneous applications. *Journal of Network and Computer Applications*, 19–44. doi:10.1016/j.jnca.2004.04.002

KhronosGroup. (2008). *OpenGL ES - Embedded System.* Retrieved October 16, 2008, from http://www.khronos.org/opengles/

Kim, C., Avoine, G., Koeune, F., Standaert, F., & Pereira, O. (2009). The Swiss-knife RFID distance bounding protocol. In *Proceedings of Information Security and Cryptology* (pp. 98-115).

Kim, S., & Veidenbaum, A. V. (1997). Stride-directed prefetching for secondary caches. In *Proceedings of the International Conference on Parallel Processing* (pp. 314-321).

Kim, G., & Ong, S. M. (2005). An exploratory study of factors influencing m-learning success. *Journal of Computer Information Systems*, 92–97.

Kim, S. Y., Choi, H. C., Won, W. J., & Oh, S. Y. (2009). Driving environment assessment using fusion of in- and our-of-vehicle vision systems. *International Journal of Automotive Technology, 10*(1), 103–113. doi:10.1007/s12239-009-0013-5

Kim, S., & Garrison, G. (2009). Investigating mobile wireless technology adoption: An extension of the technology acceptance model. *Information Systems Frontiers, 11*(3), 323–333. doi:10.1007/s10796-008-9073-8

King, W. R., & He, J. (2006). A meta-analysis of the technology acceptance model. *Information & Management, 43*, 740–755. doi:10.1016/j.im.2006.05.003

Kipp, M. (2001). Anvil - a generic annotation tool for multimodal dialogue. In *Proceedings of the 7th European Conference on Speech Communication and Technology* (pp. 1367-1370). New York, NY: ACM Press.

Kirakowski, J., & Corbett, M. (1993). SUMI: the software usability measurement inventory. *British Journal of Educational Technology, 24*(3), 210–212. doi:10.1111/j.1467-8535.1993.tb00076.x

Klemmer, S. R., Sinha, A. K., Chen, J., Landay, J. A., Aboobaker, N., & Wang, A. (2000). Suede: A Wizard of Oz prototyping tool for speech user interfaces. In *Proceedings of the 13th Annual ACM Symposium on User Interface Software and Technology* (pp. 1-10). New York, NY: ACM Press.

Klevinsky, T. J., Laliberte, S., & Gupta, A. (2002). *Hack I.T.—Security through penetration testing*. Reading, MA: Addison-Wesley.

Knight, J., Strunk, E., & Sullivan, K. (2003). Towards a Rigorous Definition of Information System Survivability. In *Proceedings of the DARPA Information Survivability Conference and Exposition*, Washington, DC.

Koch, S. (2006). Home Telehealth –Current state and future trends. *International Journal of Medical Informatics, 8*, 565–576. doi:10.1016/j.ijmedinf.2005.09.002

Kourouthanassis, P., & Roussos, G. (2003). Developing Consumer-Friendly Pervasive Retail Systems. *IEEE Pervasive Computing / IEEE Computer Society and IEEE Communications Society, 2*(2), 32–39. doi:10.1109/MPRV.2003.1203751

Krasner, J. (2004). *Using elliptic curve cryptography (ECC) for enhanced embedded security financial advantages of ECC over RSA or Diffie-Hellman (DH).* Retrieved from http://embeddedforecast.com/EMF-ECC-FINAL1204.pdf

Krötzsch, M., & Vrandecic, D. (2009). Semantic Wikipedia. *Social Semantic Web*, 393-421.

Krötzsch, M., Vrandecic, D., & Völkel, M. (2006). Semantic MediaWiki. In I. Cruz, S. Decker, D. Allemang, C. Preist, D. Schwabe, P. Mika et al. (Eds.), *Proceedings of the 5th International Semantic Web Conference* (LNCS 4273, pp. 935-942).

Küçükay, F., & Bergholz, J. (2004). Driver assistant systems. In *Proceedings of the International Conference on Automotive Technologies*.

Kumar, S., Manocha, D., Garrett, B., & Lin, M. (1996). Hierarchical back-face culling. In *Proceedings of the 7th Eurographics Workshop on Rendering* (pp. 231-240).

Kwee, A. T., & Tsai, F. S. (2009). Mobile novelty mining. *International Journal of Advanced Pervasive and Ubiquitous Computing, 1*(4), 43–68. doi:10.4018/japuc.2009100104

Kwon, O. B., & Sadeh, N. (2004). Applying case-based reasoning and multi-agent intelligent system to context-aware comparative shopping. *Decision Support Systems, 37*(2), 199–213.

Laboratories, R. S. A. (2002). *PKCS #1 v2.1: RSA cryptography standard*. Retrieved from ftp://ftp.rsa.com/pub/pkcs/pkcs-1/pkcs-1v2-1d2.pdf

Larusson, J. A., & Alterman, R. (2009). Wikis to support the "collaborative" part of collaborative learning. *International Journal of Computer-Supported Collaborative Learning, 4*(4), 371–402. doi:10.1007/s11412-009-9076-6

Lassar, W. M., Manolis, C., & Lassar, S. S. (2005). The relationship between consumer innovativeness, personal characteristics, and online banking adoption. *International Journal of Bank Marketing, 23*(2), 176–199. doi:10.1108/02652320510584403

Lawrence, R. D., Almasi, G. S., Kotlyar, V., Viveros, M. S., & Duri, S. S. (2001). Personalization of supermarket product recommendations. *Data Mining and Knowledge Discovery, 5*, 11–32. doi:10.1023/A:1009835726774

Le Grand, C. H. (2005). *Software security assurance: A framework for software vulnerability management and audit*. Longwood, FL: CHL Global Associates and Ounce Labs, Inc.

Lee, U. (2003). *A proactive routing protocol for multi-channel wireless adhoc networks (DSDV-MC)*. Unpublished doctoral dissertation, Virginia Polytechnic University, Blacksburg, VA.

Lee, M., Cheung, K. O., Christy, M. K., & Chen, Z. (2003). Acceptance of internet-based learning medium: The role of extrinsic and intrinsic motivation. *Information & Management, 42*, 1094–1104.

Lefebvre, L. A., Lefebvre, E., Bendavid, Y., Wamba, S. F., & Boeck, H. (2006). RFID as an Enabler of B-to-B e-Commerce and Its Impact on Business Processes: A Pilot Study of a Supply Chain in the Retail Industry. In *Proceedings of the 39th Annual Hawaii International Conference on System Sciences*.

Legris, P., Ingham, J., & Collerette, P. (2003). Why do people use information technology? A critical review of the technology acceptance model. *Information & Management, 40*, 191–204. doi:10.1016/S0378-7206(01)00143-4

Leichtenstern, K., & André, E. (2009). Studying multi-user settings for pervasive games. In *Proceedings of the 11th International Conference on Human-Computer Interaction with Mobile Devices and Services* (pp. 190-199). New York, NY: ACM Press.

Leichtenstern, K., & André, E. (2009). The assisted user-centred generation and evaluation of pervasive. In *Proceedings of the Third European Conference on Ambient Intelligence* (pp. 245-255).

Leichtenstern, K., & André, E. (2010). MoPeDT - features and evaluation of a user-centred prototyping tool. In *Proceedings of the 2nd ACM SIGCHI Symposium on Engineering Interactive Computing Systems* (pp. 93-102). New York, NY: ACM Press.

Leichtenstern, K., André, E., & Rehm, M. (2010). Using the hybrid simulation for early user evaluations. In *Proceedings of the 6th Nordic Conference on Human-Computer Interaction* (pp. 315-324). New York, NY: ACM Press.

Lewis, W., Agarwal, R., & Sambamurthy, V. (2003). Sources of influence on beliefs about information technology use: An empirical study of knowledge workers. *Management Information Systems Quarterly, 27*(4), 657–678.

Leyvand, T. (2009). *CityGen - A procedural model generator*. Retrieved April 10, 2009, from http://leyvand.com/research/citygen

Li, Y., Hong, J. I., & Landay, J. A. (2004). Topiary: A tool for prototyping location-enhanced applications. In *Proceedings of the 17th Annual ACM Symposium on User Interface Software and Technology* (pp. 217-226). New York, NY: ACM Press.

Lian, J., & Lin, T. (2008). Effects of consumer characteristics on their acceptance of online shopping: Comparisons among different product types. *Computers in Human Behavior, 24*(1), 48–65. doi:10.1016/j.chb.2007.01.002

Liao, C., Palvia, P., & Chen, J. (2009). Information technology adoption behavior life cycle: Toward a Technology Continuance Theory (TCT). *International Journal of Information Management, 29*, 309–320. doi:10.1016/j.ijinfomgt.2009.03.004

Li, D., Chau, P. Y. K., & Lou, H. (2005). Understanding individual adoption of instant messaging: An empirical investigation. *Journal of the Association for Information Systems, 6*(4), 102–126.

Lifton, J., & Paradiso, J. A. (2009). Dual reality: Merging the real and virtual. In *Proceedings of the First International ICST Conference on Facets of Virtual Environments* (pp. 12-18).

Li, H., Xu, M., & Li, Y. (2008). 802.11-based Wireless Mesh Networks. In *Proceedings of Complex, Intelligent and Software Intensive Systems conference*. The Research of Frame and Key Technologies for Intrusion Detection System in IEEE. doi:10.1109/CISIS.2008.42

Lim, C., & Kwon, T. (2006). Strong and Robust RFID Authentication Enabling Perfect Ownership Transfer. In *Proceedings of the 8th Conference on Information and Communications Security*, Raleigh, NC.

Lim, W., Nguyen, T. V., Yu, M., & Choi, D. (2007). Converting sensor data into ontology using DOM model in home network. In *Proceedings of the International Symposium on Information Technology Convergence* (pp. 101-105).

Lin, T., Ho, T., Chan, Y., & Chung, Y. (2008). M-ring: A distributed, self-organized, load-balanced communication method on super peer network. In *Proceedings of the 9th International Symposium on Parallel Architectures, Algorithms, and Networks*, Sydney, Australia (pp. 59-64). Washington, DC: IEEE Computer Society.

Lin, H., & Wang, Y. (2003). An examination of the determinants of customer loyalty in mobile commerce contexts. *Information & Management, 43*, 271–282. doi:10.1016/j.im.2005.08.001

Lin, Y.-M., & Shih, D.-H. (2008). Deconstructing mobile commerce service with continuance intention. *International Journal of Mobile Communications, 6*(1), 67–87. doi:10.1504/IJMC.2008.016000

Liu, G., Shan, C., Wei, G., & Wang, H. (1997). *Subarea tree routing (STR) in multi-hop wireless ad hoc networks.* Retrieved from http://202.194.20.8/proc/iccs2008/papers/339.pdf

Liu, Y., Comaniciu, C., & Man, H. (2006). A Bayesian Game Approach for Intrusion Detection in Wireless Ad Hoc Networks. In *Proceedings of the GameNets conference.*

Li, W., & McQueen, R. J. (2008). Barriers to mobile commerce adoption: An analysis framework for a country-level perspective. *International Journal of Mobile Communications, 6*(2), 231–257. doi:10.1504/IJMC.2008.016579

Lluch, J., Gaitán, R., Camahort, E., & Vivó, R. (2005). Interactive three-dimensional rendering on mobile computer devices. In *Proceedings of the 18th ACM SIGCHI International Conference on Advances in Computer Entertainment Technology (ACE)*, Valencia, Spain (pp. 254-257).

Loch, K. D., Straub, D. W., & Kamel, S. (2003). Diffusing the internet in the Arab world: The role of social norms and technological culturation. *IEEE Transactions on Engineering Management, 50*(1), 45–63. doi:10.1109/TEM.2002.808257

Lolling, A. (2003). *Analyse der menschlichen Zuverlässigkeit bei Kommissioniertätigkeiten* (Analysis of human reliability in picking processes). Shaker.

Lorenz, A., & Zimmermann, A. (2006). User modelling in a distributed multi-modal application. In *Proceedings of the Workshop on Ubiquitous User Modeling*, Riva del Garda, Italy.

Lorenz, A., Zimmermann, A., & Eisenhauer, M. (2005). Enabling natural interaction by approaching objects. In *Proceedings of the Workshop on Adaptivity and User Modeling in Interactive Systems.* DFKI.

Losavio, F., Chirinos, L., Matteo, A., Levy, N., & Ramdane, A. (2004). ISO quality standards for measuring architectures. *Journal of Systems and Software, 72*, 209–223. doi:10.1016/S0164-1212(03)00114-6

Lucas, H. C. Jr, Weinberg, C. B., & Clowes, K. W. (1975). Sales response as a function of territorial potential and sales representative workload. *JMR, Journal of Marketing Research, 12*, 298–305. doi:10.2307/3151228

Luebke, D., & Georges, C. (1995). Portals and mirrors: simple, fast evaluation of potentially visible sets. In *Proceedings of the 1995 Symposium on Interactive 3D Graphics (SI3D)*, Monterey, CA (pp. 105-106).

Luimula, M., Sääskilahti, K., Partala, T., Pieskä, S., & Alaspää, J. (2010). Remote navigation of a mobile robot in an RFID-augmented environment. *Personal and Ubiquitous Computing, 14*(2), 125–136. doi:10.1007/s00779-009-0238-3

Lu, J., Yao, J. E., & Yu, C.-S. (2005). Personal innovativeness, social influences and adoption of wireless internet services via mobile technology. *The Journal of Strategic Information Systems, 14*, 245–268. doi:10.1016/j.jsis.2005.07.003

Luk, C.-K., & Mowry, T. C. (1998). Cooperative prefetching: Compiler and hardware support for effective instruction prefetching in modern processors. In *Proceedings of the 31st Annual ACM/IEEE International Symposium on Microarchitecture* (pp. 182-194).

Lu, Y., Zhang, L., & Wang, B. (2009). A multidimensional and hierarchical model of mobile service quality. *Electronic Commerce Research and Applications, 8*(5), 228–240. doi:10.1016/j.elerap.2009.04.002

Lu, Y., Zhou, T., & Wang, B. (2009). Exploring Chinese users' acceptance of instant messaging using the theory of planned behavior, the technology acceptance model, and the flow theory. *Computers in Human Behavior, 25,* 29–39. doi:10.1016/j.chb.2008.06.002

MacKenzie, C. M., Laskey, K., McCabe, F., Brown, P. F., Metz, R., & Hamilton, B. A. (2006). *Reference model for service oriented architecture 1.0.* Retrieved from http://www.oasis-open.org/committees/download.php/19679/soa-rm-cs.pdf

Madhavan, J., Bernstein, P. A., & Rahm, E. (2001). Generic schema matching with Cupid. In *Proceedings of the 27th International Conference of Very Large Data Bases* (pp. 49-58).

Malhotra, N. K., Kim, S. S., & Agarwal, J. (2004). Internet users' information privacy concerns (IUIPC): The construct, the scale, and a causal model. *Information Systems Research, 15*(4), 336–355. doi:10.1287/isre.1040.0032

Mallat, N., Rossi, M., Tuunainen, K., & Öörni, A. (2009). The impact of use context on mobile services acceptance: The case of mobile ticketing. *Information & Management, 46,* 190–195. doi:10.1016/j.im.2008.11.008

Markopoulos, P. (2005). Designing ubiquitous computer human interaction: The case of the connected family. In H. Isomaki, A. Pirhonen, C. Roast, & P. Saariluoma (Eds.), *Future interaction design* (pp. 125–150). New York, NY: Springer. doi:10.1007/1-84628-089-3_8doi:10.1007/1-84628-089-3_8

McCoy, S., Galletta, D. F., & King, W. R. (2007). Applying TAM across cultures: the need for caution. *European Journal of Information Systems, 16,* 81–90. doi:10.1057/palgrave.ejis.3000659

McGee-Lennon, M. R., Ramsay, A., McGookin, D., & Gray, P. (2009). User evaluation of OIDE: A rapid prototyping platform for multimodal interaction. In *Proceedings of the 1st ACM SIGCHI Symposium on Engineering Interactive Computing Systems* (pp. 237-242). New York, NY: ACM Press.

Medvidovic, N., Mikic-Rakic, M., Mehta, N. R., & Malek, S. (2003). Software architectural support for handheld computing. *IEEE Computer, 36*(9), 66–73.

Mehl, M. R., Pennebaker, J. W., Crow, M. D., Dabbs, J., & Price, J. H. (2001). The electronically activated recorder (EAR): A device for sampling naturalistic daily activities and conversations. *Behavior Research Methods, Instruments, & Computers, 33,* 517–523. doi:10.3758/BF03195410doi:10.3758/BF03195410

Mehrotra, S., Rastogi, R., Silberschatz, A., & Korth, H. F. (2002). A transaction model for multidatabase systems. In *Proceedings of the 12th International Conference on Distributed Computing Systems* (pp. 56-63).

Melnik, S., Garcia-Molina, H., & Rahm, E. (2002). Similarity flooding: A versatile graph matching algorithm and its application to schema matching. In *Proceedings of the 18th International Conference on Data Engineering* (pp. 117-128).

Menezes, A., Oorschot, P., & Vanstone, S. (1996). *Handbook of applied cryptography.* Boca Raton, FL: CRC Press.

Menger, K. (1932). Botenproblem (Messenger Problem). *Ergebnisse eines Mathematischen Kolloquium, 2,* 11-12.

Meso, P., Musa, P., & Mbarika, V. (2005). Towards a model of consumer use of mobile information and communication technology in LDCs: The case of Sub-Saharan Africa. *Information Systems Journal, 15*(2), 119–146. doi:10.1111/j.1365-2575.2005.00190.x

Miller, M. (2004). *Technology: Cost per error and return on investment.* Retrieved from http://www.vocollect.com/np/documents/CostPerErrorWhitePaper.pdf

Min, J., & Cha, H. (2007). Reducing display power in dvs-enabled handheld systems. In *Proceedings of the International Symposium on Low Power Electronics and Design* (pp. 395-398).

Mitrokotsa, A., Rieback, M., & Tanenbaum, A. (2009). Classification of RFID Attacks. *Information System Frontiers: A Journal for Innovation and Research.*

Mobiqa. (2007). *Case study: Scotland rugby league world cup qualifier: Mobiqa team up with PayPal mobile to offer the world's first end-to-end mobile ticket purchase and delivery service.* Retrieved from http://www.mobiqa.com/live/files/RugbyLeaguecasestudy.pdf

Mobiqa. (2007). *Case study: Village cinemas Czech Republic: Village cinemas became the first cinema chain in Europe to offer the mobile ticketing service to film fans.* Retrieved from http://www.mobiqa.com/cinema/files/VillageCinemacasestudy.pdf

Mobiqa. (2008). *Case study: Northwest Airlines revolutionise air travel.* Retrieved from http://www.mobiqa.com/airlines/files/NWAcasestudy.pdf

Modahl, M., Bagrak, I., Wolenetz, M., Hutto, P., & Ramachandran, U. (2004). Mediabroker: An architecture for pervasive computing. In *Proceedings of the 2nd Annual Conference on Pervasive Computing and Communication* (pp. 253-262). Washington, DC: IEEE Computer Society.

Moite, S. (1992). How smart can a car be. In *Proceedings of the Intelligent Vehicles Symposium* (pp. 277-279).

Morgan, R. M., & Hunt, S. D. (1994). The commitment-trust theory of relationship marketing. *Journal of Marketing, 58*(3), 20–38. doi:10.2307/1252308

Morla, R., & Davies, N. (2004). Evaluating a location-based application: A hybrid test and simulation environment. *IEEE Pervasive Computing/IEEE Computer Society and IEEE Communications Society, 3*(3), 48–56. doi:10.1109/MPRV.2004.1321028

Moser, M., & Weiskopf, D. (2008). Interactive volume rendering on mobile devices. In *Proceedings of the Vision, Modeling, and Visualization Conference (VMV)*, Konstanz, Germany (pp. 217-226).

Mouratidis, H., & Giorgini, P. (2007). Secure Tropos: A security-oriented extension of the tropos methodology. *International Journal of Software Engineering and Knowledge Engineering, 17*(2), 285–309. doi:10.1142/S0218194007003240

Mowday, R. T., Porter, L. W., & Steers, R. M. (1982). *Employees-Organization linkage: The psychology of commitment, absenteeism, and turnover.* New York: Academic Press.

Mulloni, A., Nadalutti, D., & Chittaro, L. (2007). Interactive walkthrough of large 3D models of buildings on mobile devices. In *Proceedings of the 12th International Conference on 3D Web Technology (Web3D)*, Dijon, France (pp. 17-25).

Murata, T. (1989). Petri nets: Properties, analysis and applications. *Proceedings of the IEEE, 77*(4), 541–580. doi:10.1109/5.24143

Myers, B. A. (1995). User interface software tools. *ACM Transactions on Computer-Human Interaction, 2*(1), 64–103. doi:10.1145/200968.200971

Nakazawa, J., Tokuda, H., Edwards, W. K., & Ramachandran, U. (2006). A bridging framework for universal interoperability in pervasive systems. In *Proceedings of the 26th IEEE International Conference on Distributed Computing Systems* (pp. 3-3).

Nam, B.-G., Lee, J., Kim, K., Lee, S., & Yoo, H.-J. (2008, April). Cost-effective low-power graphics processing unit for handheld devices. *IEEE Communications Magazine, 46*(4), 152–159. doi:10.1109/MCOM.2008.4481355

Ngai, E. W. T., & Gunasekaran, A. (2007). A review for mobile commerce research and applications. *Decision Support Systems, 43*, 3–15. doi:10.1016/j.dss.2005.05.003

Nielsen, J., & Molich, R. (1990). Heuristic evaluation of users interfaces. In *Proceedings of the SIGCHI Conference on Human Factors in Computing Systems* (pp. 249-256).

Nielsen, J., Clemmensen, T., & Yssing, C. (2002). Getting access to what goes on in people's heads?: reflections on the think-aloud technique. In *Proceedings of the second Nordic conference on Human-computer interaction (NordiCHI '02)* (pp. 101-110). New York: ACM.

Nielsen, J. (1994). *Usability engineering.* San Francisco, CA: Morgan Kaufmann.

Ning, P., Cui, Y., Reeves, D. S., & Xu, D. (2004). Techniques and tools for analyzing intrusion alerts. *ACM Transactions on Information and System Security, 7*(2), 274–318. doi:10.1145/996943.996947

Ning, P., Jajodia, S., & Wang, S. (2001). Abstraction-based intrusion detection in distributed environments. *ACM Transactions on Information and System Security (TISSEC), 4*(9), 407–452. doi:10.1145/503339.503342

Nokia. (2009). *OpenGL ES 1.1 Plug-in. Built on S60* (3rd ed.). Retrieved April 2, 2009, from http://www.forum.nokia.com/

Nurminen, A. (2006). m-LOMA - a mobile 3D city map. In *Proceedings of the 11ᵗʰ International Conference on 3D Web Technology (Web3D)*, Columbia, MD (pp. 7-18).

Nurminen, A. (2009). *Mobile three-dimensional city maps.* Unpublished doctoral dissertation, Helsinki University of Technology, Helsinki, Finland.

Nysveen, H., Pedersen, P. E., & Thorbornsen, H. (2005). Explaining intention to use mobile chat services: Moderating effects of gender. *Journal of Consumer Marketing, 22*(5), 247–256. doi:10.1108/07363760510611671

Nysveen, H., Pedersen, P. E., & Thorbornsen, H. (2005). Intention to use mobile services: antecedents and cross-service comparison. *Journal of the Academy of Marketing Science, 33*(3), 330–346. doi:10.1177/0092070305276149

O'Hara, K., Kindberg, T., Glancy, M., Baptista, L., Sukumaran, B., Kahana, G., et al. (2007). Social practice in location-based collecting. In *Proceedings of the SIGCHI Conference on Human Factors in Computing Systems* (pp. 1225-1234).

O'Hara, K., Kindberg, T., Glancy, M., Baptista, L., Sukumaran, B., & Kahana, G. (2007). Collecting and sharing location-based context on mobile phones in a zoo visitor experience. *Computer Supported Cooperative Work, 16*(1-2), 11–44. doi:10.1007/s10606-007-9039-2

O'Reily, T. (2005). *What is Web 2.0: Design patterns and business models for the next generation of software.* Retrieved from http://oreilly.com/web2/archive/what-is-web-20.html

Oaks, S., Traversat, B., & Gong, L. (2002). *JXTA in a nutshell.* Sebastopol, CA: O'Reilly Media.

Okazaki, S. (2006). What do we know about mobile Internet adopters? A cluster analysis. *Information & Management, 43*, 127–141. doi:10.1016/j.im.2005.05.001

OLF. (2009). *OLF Guideline No 123: Classification of process control, safety and support ICT systems based on criticality.* Retrieved from http://www.olf.no/Documents/Retningslinjer/100-127/123%20-%20Classification%20of%20process%20control,%20safety%20and%20support.pdf?epslanguage=no

Oppermann, R., & Specht, M. (2000). A Context-sensitive Nomadic Information System as an Exhibition Guide. In *Proceedings of the Second Symposium on Handheld and Ubiquitous Computing* (LNCS, pp. 127-142). New York: Springer.

Ouedraogo, M., Mouratidis, H., Khadraoui, D., & Dubois, E. (2009). A probe capability metric taxonomy for assurance evaluation. In *Proceedings of the UEL's AC&T Conference.*

Ouedraogo, M., Savola, R., Mouratidis, H., Preston, D., Khadraoui, D., & Dubois, E. (2010). Taxonomy of quality metrics for security verification process. *Journal of Software Quality.*

Owen, M., Grant, L., Sayers, S., & Facer, K. (2006). *Social software and learning.* Retrieved from http://www.futurelab.org.uk/resources/documents/opening_education/Social_Software_report.pdf

Pandey, P. K., & Biswas, G. P. (2010). Design of scalable routing protocol for wireless ad hoc network based on DSDV. In *Proceedings of the International Conference on Mobile Information Systems* (pp. 164-174).

Peeters, R., Kohlweiss, M., & Preneel, B. (2009). Threshold things that think: Authorisation for resharing. In J. Camenisch & D. Kesdogan (Eds.), *Proceedings of the IFIP WG 11.4 International Workshop on Open Research Problems in Network Security* (LNCS 309, pp. 111-124).

Peeters, R., Sulmon, N., Kohlweiss, M., & Preneel, B. (2009). Threshold things that think: Usable authorisation for resharing. In *Proceedings of the 5ᵗʰ Symposium on Usable Privacy and Security* (p. 18). New York, NY: ACM Press.

Pei, G., Gerla, M., & Chen, T. W. (2008). An efficient destination sequenced distance vector routing protocol for mobile ad hoc networks. In *Proceedings of the International Conference on Computer Science and Information Technology* (pp. 467-471).

Pei, G., Gerla, M., & Hong, X. Y. (1999). A wireless hierarchical routing protocol with group mobility. In *Proceedings of the IEEE Conference on Wireless Communications and Networking*, New Orleans, LA (pp. 1538-1542).

Peris-Lopez, P., Hernandex-Castro, C., Estevez-Tapiador, J., & Ribagorda, A. (2006). RFID Systems: A Survey on Security Threats and Proposed Solutions. In *Proceedings of the International Conference on Personal Wireless Communications* (LNCS 4217).

Perkins, C. (1998). *SLP white paper topic.* Retrieved from http://playground.sun.com/srvloc/slp_white_paper.html

Perkins, C. E. (1994). Highly dynamic destination-sequenced distance-vector routing (DSDV) for mobile computers. In *Proceedings of the ACM SIGCOMM International Conference on Communications Architectures, Protocols and Applications* (pp. 234-244).

Peter, J. B., King, A., & Etorban, I. S. I. (2003). A DSDV-based multipath routing protocol for mobile ad-hoc networks. In *Proceedings of the International Conference on Wireless Networks.*

Pierce, J., & Mudge, T. (1996). Wrong-path instruction prefetching. In *Proceedings of the 29th Annual ACM/IEEE International Symposium on Microarchitecture* (pp. 165-175).

Porras, P., & Neumann, P. (1997). EMERALD: Event monitoring enabling responses to anomalous live disturbances. *In Proceedings of the 20th NIS security conference.*

Pulli, K. (2006). New APIs for mobile graphics. In *Proceedings of the 2006 SPIE Electronic Imaging Multimedia on Mobile Devices II*, San Jose, CA (pp. 1-13).

Pulli, K., Aarnio, T., Miettinen, V., Roimela, K., & Vaarala, J. (2008). *Mobile 3D graphics with OpenGL ES and M3G.* San Francisco: Morgan Kauffman.

Qatar, I. C. T. (2009). *Launch of New Telecommunications Licenses for Fixed and Mobile Services.* Retrieved June 6, 2009, from www.ict.gov.qa/files/marketoverview.pdf

Rabin, T. (1998). A simplified approach to threshold and proactive RSA. In H. Krawczyk (Ed.), *Proceedings of the 18th International Cryptology Conference on Advances in Cryptology* (LNCS 1462, pp. 89-104).

Raento, M., Oulasvirta, A., Petit, R., & Toivonen, H. (2005). ContextPhone: A prototyping platform for context-aware mobile applications. *IEEE Pervasive Computing / IEEE Computer Society and IEEE Communications Society, 4*(2), 51–59. doi:10.1109/MPRV.2005.29

Rahman, A. H. A., & Zukarnain, Z. A. (2004). Performance comparison of AODV, DSDV and I-DSDV routing protocols in mobile ad hoc networks. *European Journal of Scientific Research, 31*(4), 566–576.

Rahm, E., & Bernstein, P. A. (2001). A survey of approaches to automatic schema matching. *International Journal on Very Large Data Bases, 10*(4), 334–350. doi:10.1007/s007780100057

Rasmussen, K., & Capkun, S. (2009). Location privacy of distance bounding protocols. In *Proceedings of the ACM Conference on Computer and Communications Security* (pp. 149-160). New York, NY: ACM Press.

Rasmussen, K., & Capkun, S. (2010). Realization of RF distance bounding. In *Proceedings of the 19th Usenix Conference on Security* (p. 25).

Rau, P. S. (1998). A heavy vehicle drowsy driver detection and warning system: scientific issues and technical challenges. In *Proceeding of the 16th International Technical Conference on the Enhanced Safety of Vehicles.*

Regelson, M., & Fain, D. (2006). Predicting click-through rate using keyword clusters. In *Proceedings of the Second Workshop on Sponsored Search Auctions.*

Reinman, G., Calder, B., & Austin, T. (1999). Fetch directed instruction prefetching. In *Proceedings of the 32nd Annual ACM/IEEE International Symposium on Microarchitecture* (pp. 16-27).

Rhodes, B. J. (1997). The Wearable Remembrance Agent: A System for Augmented Memory. *Personal Technologies Special Issue on Wearable Computing, 1*(1), 218–224.

Ribeiro, B., Gondim, J., Jacobi, R., & Castanho, C. (2009). *Autenticação mútua entre dispositivos no middleware uos.* SBSEG – Simpósio Brasileiro em Segurança da Informação e de Sistemas Computacionais.

Ribeiro-Neto, B., Cristo, M., Golgher, P. B., & de Moura, E. S. (2005). Impedance coupling in content-targeted advertising. In *Proceedings of the 28th Annual International ACM SIGIR Conference on Research and Development in Information Retrieval* (pp. 496-503). New York, NY: ACM Press.

Rieback, M., Crispo, B., & Tanenbaum, A. (2005). RFID Guardian: A Battery-powered Mobile Device for RFID Privacy Management. In. *Proceedings of the Australian Conference on Information Security and Privacy, 3574,* 184–194. doi:10.1007/11506157_16

Rieman, J. (1993). The diary study: A workplace-oriented research tool to guide laboratory efforts. In *Proceedings of the SIGCHI Conference on Human Factors in Computing Systems* (pp. 321-326).

Rivest, R., Shamir, A., & Adleman, L. (1978). A method for obtaining digital signatures and public-key cryptosystems. *Communications of the ACM, 210*(2), 120–126. doi:10.1145/359340.359342

Robertson, S., & Robertson, J. (1999). *Mastering the requirements process*. Reading, MA: Addison-Wesley.

Rodrigues, M. A. F., Barbosa, R. G., & Mendonça, N. C. (2006). Interactive mobile 3D graphics for on-the-go visualization and walkthroughs. In *Proceedings of the 2006 ACM Symposium on Applied Computing (SAC)*, Dijon, France (pp. 1002-1007).

Rodriguez, M. D., Favela, J., Martinez, E. A., & Munoz, M. A. (2004). Location-aware access to hospital information and services. *IEEE Transactions on Information Technology in Biomedicine, 8*(4), 448–455. doi:10.1109/TITB.2004.837887

Rogers, E. M. (1983). *Diffusion of Innovations*. New York: Free Press.

Rogers, Y., Sharp, H., & Preece, J. (2002). *Interaction design: Beyond human-computer interaction*. New York, NY: John Wiley & Sons.

Roman, M., & Campbell, R. H. (2001). *A model for ubiquitous applications*. Retrieved from http://gaia.cs.uiuc.edu/papers/ubicomp01-c.pdf

Roman, M., Hess, C., Cerqueira, R., Ranganathan, A., Campbell, R. H., & Nahrstedt, K. (2002). Gaia: A middleware infrastructure for active spaces. *IEEE Pervasive Computing/IEEE Computer Society and IEEE Communications Society, 1*(4), 74–83. doi:10.1109/MPRV.2002.1158281

Rosenkrantz, D. J., Stearns, R. E., & Lewis, P. M. (1977). An analysis of several heuristics for the traveling salesman problem. *Fundamental Problems in Computing*, 45-69.

Rouibah, K., & Abbas, H. (2006). *Modified Technology Acceptance Model for Camera Mobile Phone Adoption: Development and validation*. Paper Presented for the 17th Australasian Conference on Information System Web. Retrieved from http://www.acis2006.unisa.edu.au/

Rouibah, K. (2008). Social Usage of Instant Messaging by individuals outside the workplace in Kuwait: A structural Equation Model. *IT & People, 21*(1), 34–68. doi:10.1108/09593840810860324

Rouibah, K., & Hamdy, H. (2009). Factors Affecting Information Communication Technologies Usage and Satisfaction: Perspective From Instant Messaging in Kuwait. *Journal of Global Information Management, 17*(2), 1–29.

Rouibah, K., Ramayah, T., & May, O. S. (2009). User acceptance of internet banking in Malaysia: Test of three acceptance models. *International Journal of E-Adoption, 1*(1), 1–19.

Royer, E. M., & Toh, C.-K. (1999). A review of current routing protocols for ad-hoc wireless mobile networks. *IEEE Personal Communications*, 46-55.

Rukzio, E., Leichtenstern, K., Callaghan, V., Schmidt, A., Holleis, P., & Chin, J. (2006). An experimental comparison of physical mobile interaction techniques: Touching, pointing and scanning. In *Proceedings of the Eighth International Conference on Ubiquitous Computing* (pp. 87-104).

Russo, S. (2010). *Blender pocket - digital creation in your pocket*. Retrieved March 9, 2010, from http://russose.free.fr/BlenderPocket/

Sacramento, V., Endler, M., Rubinsztejn, H. K., Lima, L. S., Gonçalves, K., Nascimento, F. N. et al. (2004). MOCA: A middleware for developing collaborative applications for mobile users. *IEEE Distributed Systems Online, 5*(10).

Saint-Andre, P. (2004). *RFC 3920: Extensible messaging and presence protocol (xmpp): Core. Request for Comments, IETF*. Retrieved from http://tools.ietf.org/html/rfc3920

Samet, H., & Webber, R. E. (1988). Hierarchical data structures and algorithms for computer graphics. Part I. *IEEE Computer Graphics and Applications, 8*(3), 48–68. doi:10.1109/38.513

Sanchez-Franco, M. J., Ramos, A. F. V., & Velicia, F. A. M. (2009). The moderating effect of gender on relationship quality and loyalty toward Internet service providers. *Information & Management, 46,* 196–202. doi:10.1016/j.im.2009.02.001

Sarker, S., & Wells, J. D. (2003). Understanding mobile handheld device use and adoption. *Communications of the ACM, 46*(12), 35–40. doi:10.1145/953460.953484

Saunders, S., Ross, M., Staples, G., & Wellington, S. (2006). The software quality challenges of service oriented architectures in e-commerce. *Software Quality Journal, 14,* 65–75. doi:10.1007/s11219-006-6002-2

Savola, R. M. (2007). Towards a taxonomy for information security metrics. In *Proceedings of the International Conference on Software Engineering Advances,* Cap Esterel, France.

Schafer, J. B., Konstan, J., & Riedl, J. (1999). Recommender systems in e-commerce. In *Proceedings of the 1st ACM Conference on Electronic Commerce* (pp. 158-166). New York, NY: ACM Press.

Schepers, J., & Wetzels, M. (2007). A meta-analysis of the technology acceptance model: Investigating subjective norm and moderation effects. *Information & Management, 44,* 90–103. doi:10.1016/j.im.2006.10.007

Schilit, B., Adams, N., & Want, R. (1994). Context aware computing applications. In *Proceedings of the IEEE Workshop on Mobile Computing Systems and Applications,* Santa Cruz, CA (pp. 85-90). Washington, DC: IEEE Computer Society.

Schilit, A., & Theimer, M. (1994). Disseminating active map information to mobile hosts. *IEEE Network, 8*(5), 22–32. doi:10.1109/65.313011

Schlosser, A. E., Shavitt, S., & Kanfer, A. (1999). Survey of Internet users' attitudes toward Internet advertising. *Journal of Interactive Marketing, 13*(3), 34–54. doi:10.1002/(SICI)1520-6653(199922)13:3<34::AID-DIR3>3.0.CO;2-R

Schmidt, A., Aidoo, K. A., Takaluoma, A., Tuomela, U., Laerhoven, K. V., & de Velde, W. V. (1999). Advanced interaction in context. In *Proceedings of the First International Symposium on Handheld and Ubiquitous Computing,* Karlsruhe, Germany (pp. 89-101).

Schmitt, J., Kropff, M., Reinhardt, A., Hollick, M., Schafer, C., Remetter, F., et al. (2008). *An extensible framework for context-aware communication management using heterogeneous sensor networks* (Tech. Rep. No. TR-KOM-2008-08). Darmstadt, Germany: Technische Universität Darmstadt.

Schmohl, R., & Baumgarten, U. (2008). A generalized context-aware architecture in heterogeneous mobile computing environments. In *Proceedings of the Fourth International Conference on Wireless and Mobile Communications* (pp. 118-124).

Schneider, A., Lorenz, A., Zimmermann, A., & Eisenhauer, M. (2006). Multimodal Interaction in Context-Adaptive Systems. In *Proceedings of the Second Workshop on Context Awareness for Proactive Systems* (p. 101).

Schneier, B. (1999). Attack Trees. *Dr. Dobb's Journal of Software Tools, 24,* 12–29.

Schnell, R., & Kreuter, F. (2003). *Separating interviewer and sampling-point effects. In UC Los Angeles: Department of Statistics, UCLA.* Retrieved from http://www.escholarship.org/uc/item/7d48q754

Schnorr, C. P. (1989). Efficient identification and signatures for smart cards. In G. Brassard (Ed.), *Proceedings of Advances in Cryptology* (LNCS 435, pp. 239-252).

Scott, J., Lee, L. H., Arends, J., & Moyer, B. (1998). Designing the low-power M-CORE architecture. In *Proceedings of the International Symposium on Computer Architecture Power Driven Microarchitecture Workshop* (pp. 145-150).

Seddigh, N., Pieda, P., Matrawy, A., Nandy, B., Lambadaris, L., & Hatfield, A. (2004). Current trends and advances in information assurance metrics. In *Proceedings of the Conference on Privacy, Trust Management and Security* (pp. 197-205).

Selim, H. M. (2003). An empirical investigation of student acceptance of course websites. *Computers & Education, 40*(4), 343–360. doi:10.1016/S0360-1315(02)00142-2

Shamir, A. (1979). How to share a secret. *Communications of the ACM, 220*(11), 612–613. doi:10.1145/359168.359176

Shoup, V. (2000). Practical threshold signatures. In B. Preneel (Ed.), *Proceedings of the International Conference on the Theory and Application of Cryptographic Techniques* (LNCS 1807, pp. 207-220).

Silva, W. B., & Rodrigues, M. A. F. (2009). A lightweight 3D visualization and navigation system on handheld devices. In *Proceedings of the 2009 ACM Symposium on Applied Computing (SAC),* Honolulu, HI (pp. 162-166).

Singelée, D., & Preneel, B. (2007). Distance bounding in noisy environments. In F. Stajano, C. Meadows, S. Capkun, & T. Moore (Eds.), *Proceedings of the 4th European Workshop on Security and Privacy in Ad Hoc and Sensor Networks* (LNCS 4572, pp. 101-115).

Sklar, B. (2002). *Reed-Solomon codes*. Retrieved from http://www.informit.com/content/images/art_sklar7_reed-solomon/elementLinks/art_sklar7_reed-solomon.pdf

Smith, J. E., & Hsu, W.-C. (1992). Prefetching in supercomputer instruction caches. In *Proceedings of the ACM/IEEE Conference on Supercomputing* (pp. 588-597).

Smith, A. J. (1978). Sequential program prefetching in memory hierarchies. *Computer, 11*(12), 7–21. doi:10.1109/C-M.1978.218016

Song, B. (2008). RFID Tag Ownership Transfer. In *Proceedings of the 4th Workshop on RFID Security*, Budapest, Hungary.

Souzis, A. (2005). Building a semantic wiki. *IEEE Intelligent Systems, 20*(5), 87–91. doi:10.1109/MIS.2005.83

Spojmenovic, I. (2002). *Handbook of wireless network and mobile computing*. New York, NY: John Wiley & Sons. doi:10.1002/0471224561

Srinivasan, V., Davidson, E. S., Tyson, G. S., Charney, M. J., & Puzak, T. R. (2001). Branch history guided instruction prefetching. In *Proceedings of the 7th International Conference on High Performance Computer Architecture* (pp. 291-300).

Stahl, G., & Hesse, F. (2009). Paradigms of shared knowledge. *International Journal of Computer-Supported Collaborative Learning, 4*(4), 365–369. doi:10.1007/s11412-009-9075-7

Stajano, F., & Anderson, R. (1999). The Resurrecting Duckling: Security Issues for Ad-hoc Wireless Networks. In *Proceedings of the 7th International Workshop on Security Protocols* (LNCS 1796, pp. 172-194). Berlin: Springer Verlag.

Staniford, S., Hoagland, J., & McAlerney, J. (2002). Practical automated detection of stealthy portscans. *Journal of Computer Security, 1*(10), 105–136.

Stefani, A., & Xenos, M. (2008). E-commerce system quality assessment using a model based on ISO 9126 and belief networks. *Software Quality Control, 16*(1), 107–129.

Stevens, W., Myers, G., & Constantine, L. (1979). *Structured design: Fundamentals of a discipline of computer program and systems design* (pp. 205–232). Upper Saddle River, NJ: Prentice Hall.

Stoica, I., Morris, R., Liben-Nowell, D., Karger, D., Kaashoek, M., & Dabek, F. (2003). Chord: A scalable peer-to-peer lookup protocol for internet applications. *IEEE Transactions on Networking, 11*(1), 17–32. doi:10.1109/TNET.2002.808407

Stoneburner, G. (2001). *Underlying technical models for information technology security*. Gaithersburg, MD: National Institute of Standards and Technology.

Strang, T., & Popien, C. L. (2004). A context modeling survey. In *Proceedings of the Workshop on Advanced Context Modelling, Reasoning and Management as part of UbiComp* (pp. 33-40).

Strang, T., Popien, C. L., & Frank, K. (2003). CoOL: A context ontology language to enable contextual interoperability. In *Proceedings of the 4th IFIP International Conference on Distributed Applications and Interoperable Systems* (pp. 236-247).

Strunk, E. A., & Knight, J. C. (2006, May 23). The essential synthesis of problem frames and assurance cases. In *Proceedings of the Second International Workshop on Applications and Advances in Problem Frames*.

Stui, M. (2005). *The use of bar code SMS in mobile marketing, advertising, CRM*. Retrieved from http://www.adazonusa.com/theuseofbarcodesmsinmobilemarketingadvertisingcrm-a-3.html

Sun, H., & Zhang, P. (2006). Causal relationships between perceived enjoyment and perceived ease of use: An alternative approach. *Journal of the Association for Information Systems, 7*(9), 618–645.

Swanson, M., Nadya, B., Sabato, J., Hash, J., & Graffo, L. (2003). *Security metrics guide for information technology systems* (Tech. Rep. No. NIST-800-55). Gaithersburg, MD: National Institute of Standards and Technology.

Tan, C., Sheng, B., & Li, Q. (2006). Secure and Serverless RFID Authentication and Search Protocol. *IEEE Transactions on Wireless Communications, 7*(3).

Tang, Y., & Chen, S. (2005). Defending against internet worms: A signature-based approach. In *Proceedings of the IEEE Infocom conference.*

Tang, S. M., Wang, F. Y., & Miao, Q. H. (2006). ITSC 05: Current issues and research trends. *IEEE Intelligent Systems, 21*(2), 96–102. doi:10.1109/MIS.2006.31

Tarasewich, P. (2003). Designing mobile commerce applications. *Communications of the ACM, 46*(12), 57–60. doi:10.1145/953460.953489

Tarvainen, P. (2004). Survey of the Survivability of IT Systems. In *Proceedings of the 9th Nordic Workshop on Secure IT-systems*, Helsinki, Finland.

Teo, T. S. H., Lim, V. K. G., & Lai, R. Y. C. (1999). Intrinsic and extrinsic motivation in Internet usage. *Omega, 27*(1), 25–37. doi:10.1016/S0305-0483(98)00028-0

The Mobile World. (2009). Retrieved from www.themobileworld.com

The OSGi Alliance. (2009). *OSGi service platform core specification.* Retrieved from http://www.osgi.org/download/r4v41/r4.core.pdf

Thing, V., Subramaniam, P., Tsai, F. S., & Chua, T.-W. (2011). Mobile phone anomalous behaviour detection for real-time information theft tracking. In *Proceedings of the Second International Conference on Technical and Legal Aspects of the e-Society.*

Thompson, R., Compeau, D., & Higgins, C. (2006). Intentions to use information technologies: An integrative model. *Journal of Organizational and End User Computing, 18*(3), 25–46.

Thoziyoor, S., Muralimanohar, N., Ahn, J. H., & Jouppi, N. P. (2008). *Cacti: An integrated cache and memory access time, cycle time, area, leakage, and dynamic power model* (Tech. Rep. No. HPL-2008-20). Palo Alto, CA: HP Laboratories.

Tompkins, J. A., & Smith, J. D. (1998). *Warehouse Management Handbook* (2nd ed.). New York: Tompkins Associates.

Toth, T., & Kruegel, C. (2002). Connection-history based anomaly detection. In *Proceedings of the IEEE workshop on information assurance and security.*

Trappey, C. V., Trappey, A. J., & Liu, C. S. (2009). Develop patient monitoring and support system using mobile communication and intelligent reasoning. In *Proceedings of the IEEE International Conference on Systems, Man and Cybernetics* (pp. 1195-1200).

Tsai, F. S. (2006). *Object-oriented software engineering.* Singapore: McGraw-Hill.

Tsai, F. S. (2010). Security issues in e-healthcare. *Journal of Medical and Biological Engineering, 30*(4), 209–214. doi:10.5405/jmbe.30.4.04

Tsai, F. S. (2011). Web-based geographic search engine for location-aware search in Singapore. *Expert Systems with Applications, 38*(1), 1011–1016. doi:10.1016/j.eswa.2010.07.129

Tsai, F. S., Etoh, M., Xie, X., Lee, W.-C., & Yang, Q. (2010). Introduction to mobile information retrieval. *IEEE Intelligent Systems, 25*(1), 11–15. doi:10.1109/MIS.2010.22

Tsai, F. S., Han, W., Xu, J., & Chua, H. C. (2009). Design and development of a mobile peer-to-peer social networking application. *Expert Systems with Applications, 36*(8), 11077–11087. doi:10.1016/j.eswa.2009.02.093

Tsudik, G. (2006). YA-TRAP: Yet Another Trivial RFID Authentication Protocol. In *Proceedings of the 4th Annual IEEE International Conference on Pervasive Computing and Communications*, Pisa, Italy.

Tucker, S. (1963). *The Break-Even System: A Tool for Profit Planning.* Upper Saddle River, NJ: Prentice Hall.

Tuyls, P., & Batina, L. (2006). RFID-Tags for Anti-Counterfeiting. In *Proceedings of the Cryptographer's Track at the RSA Conference*, San Jose, CA.

Vajda, I., & Buttyan, L. (2003). Lightweight Authentication Protocols for Low-cost RFID Tags. In *Proceedings of the Second Workshop on Security in Ubiquitous Computing*, Seattle, WA.

Valdes, A., & Skinner, K. (2001). Probabilistic alert correlation. *In Proceedings of the 4th international symposium on recent advances in intrusion detection.*

Van Verth, J. M., & Bishop, L. M. (2004). *Essential mathematics for games and interactive applications: a programmer's Guide.* San Francisco: Morgan Kaufmann.

Vandenhouten, R., & Seiz, M. (2007, September). Identification and tracking goods with the mobile phone. In *Proceedings of the International Symposium on Logistics and Industrial Informatics* (pp. 25-29).

Vaughn, R. B., Henning, R., & Siraj, A. (2002). Information assurance measures and metrics – state of practice and proposed taxonomy. In *Proceedings of the IEEE International Hawaii Conference on System Sciences* (p. 331.3).

Venkatesh, V., & Ramesh. (2003). Understanding usability in mobile commerce. *Communications of the ACM*, *46*(12), 53–56. doi:10.1145/953460.953488

Venkatesh, V., & Davis, F. D. (2000). A theoretical extension of the technology acceptance model: Four longitudinal field. *Management Science*, *46*, 186–204. doi:10.1287/mnsc.46.2.186.11926

Venkatesh, V., Morris, M. G., & Ackerman, P. L. (2000). A longitudinal field study of gender differences in individual technology adoption decision making processes. *Organizational Behavior and Human Decision Processes*, *83*, 33–60. doi:10.1006/obhd.2000.2896

Venkatesh, V., Morris, M. G., Davis, G. B., & Davis, F. D. (2003). User acceptance of information technology: Toward a unified view. *Management Information Systems Quarterly*, *27*(3), 425–478.

Villarreal, J., Lysecky, R., Cotterell, S., & Vahid, F. (2002). *A study on the loop behavior of embedded programs* (Tech. Rep. No. UCR-CSE-01-03). Riverside, CA: University of California.

Vincent, H.-M. W. (2009). *Vincent 3D rendering library - open source graphics libraries for mobile and embedded devices. Vincent ES 1.x library based on the OpenGL (R) ES 1.1 API specification (common lite).* Retrieved April 2, 2009, from http://www.vincent3d.com/Vincent3D/index.html

W3C. (2008). *Mobile Web best practices 1.0.* Retrieved from http://www.w3.org/TR/mobile-bp/

W3C. (2010). *Mobile Web application best practices.* Retrieved from http://www.w3.org/TR/2010/CR-mwabp-20100211/

Wang, C., Zhang, P., Choi, R., & Eredita, M. (2002). Understanding consumers attitude toward advertising. In *Proceedings of the Eighth Americas Conference on Information System* (pp. 1143-1148).

Wang, X. H., Zhang, D. Q., Gu, T., & Pung, H. K. (2004). Ontology based context modeling and reasoning using OWL. In *Proceedings of the Second IEEE Annual Conference on Pervasive Computing and Communications Workshops* (pp. 18-22).

Wang, F. Y. (2006). Driving into the future with ITS. *IEEE Intelligent Systems*, *21*(3), 94–95. doi:10.1109/MIS.2006.45

Wang, F. Y., Zeng, D., & Yang, L. Q. (2006). Smart cars on smart roads, an IEEE intelligent transportation systems society update. *IEEE Pervasive Computing/IEEE Computer Society and IEEE Communications Society*, *5*(4), 68–69. doi:10.1109/MPRV.2006.84

Wang, W. L., & Lin, C. H. (2008). A study of two-dimensional barcode prescription system for pharmacists' activities of NHI contracted pharmacy. *Yakugaku Zasshi*, *128*(1), 123–127. doi:10.1248/yakushi.128.123

Want, R. (2004). Enabling ubiquitous sensing with RFID. *IEEE Computer*, *37*(4), 84–86.

Weikum, G., & Schek, H. J. (1992). Concepts and applications of multilevel transactions and open nested transactions. In *Proceedings of the Database Transaction Models for Advanced Applications* (pp. 515-553).

Weinstein, R. (2005). RFID: A Technical Overview and Its Application to the Enterprise. *IT Professional*, *7*(3), 27–33. doi:10.1109/MITP.2005.69

Weis, S., Sarma, S., Rivest, R., & Engels, D. (2003). Security and Privacy Aspects of Low-cost Radio Frequency Identification Systems. In *Proceedings of the 1ˢᵗ International Conference on Security in Pervasive Computing*, Boppard, Germany.

Weiser, M. (1991). The computer for the 21st century. *Scientific American.*

Weiser, M., & Brow, J. S. (1995). *Designing calm technology.* Retrieved from http://nano.xerox.com/weiser/calmtech/calmtech.htm

Weiser, M. (1991). The computer for the 21st century. *Scientific American*, 94–104. doi:10.1038/scientificamerican0991-94

Weiser, M. (1993). The world is not a desktop. *Interactions (New York, N.Y.)*, *1*(1), 7–8. doi:10.1145/174800.174801

Weiser, M., & Brown, J. (1997). The coming age of calm technology. *Beyond Calculation. The Next Fifty Years of Computing*, *8*, 75–85.

Weiss, H. M., Beal, D. J., Lucy, S. L., & MacDermid, S. M. (2004). *Constructing EMA studies with PMAT: The Purdue momentary assessment tool user's manual.* Retrieved from.http://www.ruf.rice.edu/~dbeal/pmatusermanual.pdf

Weiss, G. (2002). Welcome to the (almost) digital hospital. *IEEE Spectrum*, 44–49. doi:10.1109/6.988704

Weister, M. (1993). Hot topics: Ubiquitous computing. *IEEE Computer*, *26*(10), 71–72.

White, G. B., Fisch, E. A., & Pooch, U. W. (1996). Cooperating security managers: A peer-based intrusion detection system. *IEEE Network*, *5*(3), 20–23. doi:10.1109/65.484228

Wong, T. M., Wang, C., & Wing, J. M. (2002). *Verifiable secret redistribution for threshold sharing schemes* (Tech. Rep. No. CMU-CS-02-114). Pittsburgh, PA: Carnegie Mellon University.

Wonka, P., & Schmalstieg, D. (1999). Occluder shadows for fast walkthroughs of urban environments. *Computer Graphics Forum*, *18*(3), 51–60. doi:10.1111/1467-8659.00327

Wool, A. (2004). A quantitative study of firewall configuration errors. *Computer*, *37*(6), 62–67. doi:10.1109/MC.2004.2

Wooldridge, M. (2002). *An introduction to multi-agent systems.* New York, NY: John Wiley & Sons.

World development Indicator. (2009). *Internet based database.*

Wright, A. (2009). Get smart. *Communications of the ACM*, *52*(1), 15–16. doi:10.1145/1435417.1435423

Yan, H., & Selker, T. (2000). Context-Aware Office Assistant. In *Proceedings of the 5ᵗʰ Infernational Conference on Intelligent User Interfaces*, New Orleans, LA (pp. 276-279). New York: ACM.

Yang, K. (2007). Exploring factors affecting consumer intention to use mobile advertising in Taiwan. *Journal of International Consumer Marketing*, *20*(1), 33–49. doi:10.1300/J046v20n01_04

Yang, S. (2006). Context aware ubiquitous learning environments for peer-to-peer collaborative learning. *Journal of Educational Technology & Society*, *9*(1), 188–201.

Yang, S., Chen, I., & Shao, N. (2004). Ontological enabled annotations and knowledge management for collaborative learning in virtual learning community. *Journal of Educational Technology & Society*, *7*(4), 70–81.

Yan, X., Gong, M., & Thong, Y. L. (2006). Tow tales of one service: User acceptance of short message service (SMS) in Hong Kong and China. *Info*, *8*(1), 16–28. doi:10.1108/14636690610643258

Yao, M. Z., & Flanagin, A. J. (2006). A self-awareness approach to computer-mediated communication. *Computers in Human Behavior*, *22*, 518–544. doi:10.1016/j.chb.2004.10.008

Yee, K. Y., Tiong, A. W., Tsai, F. S., & Kanagasabai, R. (2009). OntoMobiLe: A generic ontology-centric service-oriented architecture for mobile learning. In *Proceedings of the IEEE Tenth International Conference on Mobile Data Management: Systems, Services and Middleware, Workshop on Mobile Media Retrieval* (pp. 631-636).

Yergeau, F. (2003). *Utf-8, a transformation format of ISO 10646.* Retrieved from http://www.ietf.org/rfc/rfc2279.txt

Yih, W., Goodman, J., & Carvalho, V. R. (2006). Finding advertising keywords on web pages. In *Proceedings of the 15th International Conference on World Wide Web* (pp. 213-222). New York, NY: ACM Press.

Yi, M. Y., Jackson, J. D., Park, J. S., & Probst, J. C. (2006). Understanding information technology acceptance by individual professionals: Toward an integrative view. *Information & Management*, *43*, 350–363. doi:10.1016/j.im.2005.08.006

Zhai, S., Morimoto, C., & Ihde, S. (1999). Manual and gaze input cascaded (MAGIC) pointing. In *Proceedings of the Conference on Human Factors in Computing Systems (CHI'99)* (pp. 246-253). New York: ACM.

Zhang, A., Nodine, M., Bhargava, B., & Bukhres, O. *(1994). Ensuring relaxed atomicity for flexible transactions in multidatabase systems. In* Proceedings of the ACM SIGMOD International Conference on Management of Data *(p. 78).*

Zhang, H., & Hoff, K. E., III. (1997). Fast backface culling using normal masks. In *Proceedings of the 1997 ACM Symposium on Interactive 3D Graphics*, Providence, RI (pp. 103-106).

Zhang, J., & Varadharajan, V. A. (2008). New Security Scheme for Wireless Sensor Networks. In *Proceedings of the IEEE Global Telecommunications Conference.*

Zhang, Y., & Lee, W. (2000). Intrusion Detection in Wireless Ad-Hoc Networks. In *Proceedings of the Sixth Annual International Conference on Mobile Computing and Networking.*

Zhang, Y., Haga, S., & Barua, R. (2002). Execution history guided instruction prefetching. In *Proceedings of the 16th International Conference on Supercomputing* (pp. 199-208).

Zhou, X., Cai, Y., Godavari, G. K., & Chow, C. E. (2004). An adaptive process allocation strategy for proportional responsiveness differentiation on Web servers. In *Proceedings IEEE international conference on web services.*

Zuo, Y., Pimple, M., & Lande, S. (2009). A Framework for RFID Survivability Requirement Analysis and Specification. In *Proceedings of International Joint Conference on Computing, Information and Systems Sciences and Engineering*, Bridgeport, CT.

About the Contributors

Wen-Chen Hu received a BE, an ME, an MS, and a PhD, all in Computer Science, from Tamkang University, Taiwan, the National Central University, Taiwan, the University of Iowa, Iowa City, and the University of Florida, Gainesville, in 1984, 1986, 1993, and 1998, respectively. He is currently an associate professor in the Department of Computer Science of the University of North Dakota, Grand Forks. He was an Assistant Professor in the Department of Computer Science and Software Engineering at the Auburn University, Alabama, for years. He is the Editor-in-Chief of the *International Journal of Handheld Computing Research* (IJHCR) and an associate editor of the *Journal of Information Technology Research* (JITR), and has acted as editor and editorial advisory/review board member for over 30 international journals/books and served more than 30 tracks/sessions and program committees for international conferences. He has also won a couple of awards of best papers, best reviewers, and community services. Dr. Hu has been teaching more than 10 years at the US universities and over 10 different computer/IT-related courses, and advising more than 50 graduate students. He has published over 100 articles in refereed journals, conference proceedings, books, and encyclopedias, edited eight books and conference proceedings, and solely authored a book entitled *"Internet-enabled handheld devices, computing, and programming: mobile commerce and personal data applications."* His current research interests include handheld/mobile/smartphone/tablet computing, location-based services, Web-enabled information system such as search engines and Web mining, electronic and mobile commerce systems, and Web technologies. He is a member of the IEEE Computer Society and ACM (Association for Computing Machinery).

S. Hossein Mousavinezhad received his Ph.D. in Electrical Engineering from Michigan State University, East Lansing, Michigan. He is currently a Professor and the Chair of the Department of Electrical Engineering Computer Science (EECS), Idaho State University, Pocatello, Idaho. His research interests include digital signal processing, bioelectromagnetics, and communication systems. Dr. Mousavinezhad is a recipient of the Institute of Electrical and Electronics Engineers (IEEE) Third Millennium Medal. He received American Society for Engineering Education (ASEE) Electrical and Computer Engineering Division's Meritorious Service Award in June 2007. Professor Mousavinezhad is a program evaluator for the Accreditation Board for Engineering and Technology (ABET).

* * *

T. Abbas H. is an assistant professor of Management Information Systems at Kuwait University. He received his B.S. in Economics from Kuwait University, and received M.S. and Ph.D. from Illinois Institute of Technology. His research interests include information ethics, information security, artificial intelligence, and expert systems.

Elisabeth André is a full professor of Computer Science at Augsburg University, Germany, and Chair of the Laboratory for Human-Centered Multimedia. Her current work focuses on multimodal interfaces, social signal processing and embodied conversational agents. Elisabeth André is on the Editorial Board of Cognitive Processing (International Quarterly of Cognitive Science), Universal Access to the Information Society: An Interdisciplinary Journal, AI Communications (AICOM), Journal of Autonomous Agents and Multi-Agent Systems (JAAMAS), Journal on Multimodal Interfaces, Editorial Board of IEEE Transactions on Affective Computing (TAC), ACM Transactions on Intelligent Interactive Systems (TIIS) and International Journal of Synthetic Emotions (IJSE). In 2007, Elisabeth André was appointed an Alcatel Research Fellow at Internationales Zentrum für Kultur- und Technikforschung of Stuttgart University (IZKT). In 2010, she was elected to be a member of the German Academy of Sciences Leopoldina and Academia Europaea.

G. P. Biswas is the Associate Professor in the Department of Computer Science & Engineering, Indian school of Mines (ISM), Dhanbad since 1999. Before joining ISM, he was engaged in a research project on the "Designing of CA/PCA based BIST structure for VLSI (soft/hard) cores," sponsored by M/s Fujitsu Microelectronics, USA with the Department of Computer Science & Technology, BE College, Sibpur, Howrah, WB.

Fabricio Nogueira Buzeto received both B.S. and M.S. degrees from Brasília University, Brazil, in 2007 and 2010, respectively. He is currently a PhD candidate in the Department of Computer Science at Brasília University, Brazil, and the CIO at Intacto Engenharia de Sistemas Co. His research interests are Ubicomp and smart space systems.

Yu Cai is an assistant professor at School of Technology in Michigan Technological University. His research interests include green computing, distributed systems and network security. He received his Ph.D. in Computer Science from University of Colorado in 2005. He is a member of IEEE and ACM.

Carla Denise Castanho is an associate professor in the Department of Computer Science at Universidade de Brasília, Brazil, since 2005. She received her M.E. and Ph.D. degrees from the Department of Electrical and Computer Engineering at Nagoya Institute of Technology, Japan, in 1998 and 2001, respectively. Her current research interests include Game Development, Ubiquitous and Pervasive Computing.

Chun-Fu Chang received MSc degree in 2009 from the Department of Computer Science and Engineering, Tatung University. His research focus on developing distributed RDF store based structured P2P network.

Lee Chang received his Ph.D. in Electrical Engineering from the University of Texas at Austin in 1972. From 1972 to 1984 he worked for IBM in New York and San José, leading several worldwide information system projects in France, Germany, Italy, Japan, and the US. In 1986 he joined the engineering faculty at San Jose State University (SJSU). He is also an information technology consultant/educator for companies and universities in China, Hong Kong, Malaysia, and Taiwan. At SJSU, he has served as the Associate Chair of Computer Engineering Department in charge of the curricula development of computer and software engineering programs. His teaching and research interests include Software Systems Engineering and Service-Oriented Architecture. Currently, he is leading a team of SJSU faculty members and Silicon Valley industry experts to develop and offer a MS in Software Engineering program with Emphasis in Cloud Computing and Virtualization.

Been-Chian Chien received the B.S. in Computer Engineering from National Chiao Tung University, Hsin-Chu, Taiwan, in 1987, the M.S. and the Ph.D. in Computer Science and Information Engineering in 1989 and 1992 from National Chiao Tung University, Hsin-Chu, Taiwan, respectively. He became an associate professor of the Department of International Trade at Nan-Tai Institute of Technology from August 1994 to July 1996 and an associate professor of the Department of Information Engineering, I-Shou University, Kaohsiung from August 1996 to July 2004. He was invited to be the visiting scholar of the BISC (Berkeley Initiative on Soft Computing) at U. C. Berkeley, C.A. in 1999 and the School of Electrical and Computer Engineering, Georgia Institute of Technology, Atlanta, GA in 2008. Now, he is a professor of the Department of Computer Science and Information Engineering, National University of Tainan, Tainan, Taiwan. Dr. Chien's major research interest includes computational intelligence, artificial intelligence, database systems and the design and analysis of algorithms. He has published over 120 articles in various journals, national and international conference proceedings. His current research activities involve machine learning, knowledge discovery and data mining, context-aware systems and context data management.

Eric Dubois received the M.S. degree in computer science from the University of Namur, Belgium, in 1981 and the degree of *"Docteur-Ingenieur en Informatique"* from the Institut National Polytechnique de Lorraine, Nancy, France, in 1984. Since 2000, in Luxembourg he works at the Public Research Centre Henri Tudor where he is the managing director of the "Service Science and Innovation" department (www.ssi.tudor.lu/). Besides management activities, Dr. E. Dubois is active in the software engineering and information system fields for about 25 years. His specific focus is on the requirements engineering (RE) topic where he published over 100 papers with specific interests in business services and security requirements engineering. He is member of the IFIP 2.9 RE working group and of the editorial board of the REJ Spinger journal. He was program co-chair of the international IEEE RE'02 conference, general chair of the conference on advanced information systems engineering (CaiSE'06) and co general chair of the 13th IEEE Conference on Commerce and Enterprise Computing (CEC 2011). He is member of the steering committee of the REFSQ series of conferences. Recent interests are in the new discipline of "Service Science" where he led the EU Erasmus DELLIISS project (www.delliiss.eu). Eric Dubois is member of the ERCIM (European Research Consortium in Informatics and Mathematics) board of directors. He is visiting professor at the Universities of Namur and Luxembourg.

Markus Eisenhauer has a PhD in cognitive science and is head of the business unit Mobile Knowledge at the Fraunhofer Institute of Applied Information Technology. Markus is expert in cognitive and context modelling, ubiquitous and calm computing as well as in participatory system development, ergonomic evaluation methodology as well as cognitive experiments. He has extensive experience in managing international and European projects including: Etracking (Strep), SAiMotion (funded by the BMBF), GiGaMobile (Cooperation with Telematica Institute), MICA (funded by SAP), PROLEARN (NoE), InterMedia (NoE), Hydra (IP) and MACE (eContent+), Intrepid (IP).

Jianbo Fan is a full professor of the School of Electronic and Information Engineering at Ningbo University of Technology, Ningbo, China. He obtained his Master degree in Computer Science from Zhejiang University, Hangzhou, China, in 2002. From 1989 to 1990, he worked as a Research Scientist in Aachen Junior Technology College, Germany. From 2005 to 2006, he worked as a Research Fellow with Department of Computer Science at Zhejiang University, China. His past and present research interests include: Information processing, Database and Pattern recognition. He has published more than 30 papers in the relevant fields.

Carlos Botelho de Paula Filho received the B.S. degree from Brasília University, Brazil, in 2007. He is currently is a researcher and technical leader at Intacto Engenharia de Sistema Co. and is enrolled in the Master Program of the Department of Computer Science at Brasília University, Brazil. His research interests include ubiquitous computing and digital television convergence.

Lars Frank has for 20 years been a database consultant for both private companies and organizations in the public sector. Since 1994 he has been Associate professor at the Department of Informatics in Copenhagen Business School. His research areas are system integration, data warehousing, ERP architectures, distributed health records, e-commerce, mobile databases, transaction models, work flow management, multidatabases, and data modeling. In 2008 he received a Dr. Merc. Degree from Copenhagen Business School for a dissertation on integration of heterogeneous IT-systems.

Jerry Gao is a tenured professor at the Computer Engineering Department in San Jose State University. In 1995, Dr. Gao graduated with Ph.D. degree in Computer Science in The University of Texas at Arlington. Before he joined San Jose State University in 1998, he had over 10 years of industry development and management experience on software engineering and Internet applications. His current research interests include SOA-based/model-based software engineering, cloud computing and mobile computing. Dr. Gao has published extensive research results in component-based software engineering, software validation, wireless computing and mobile commerce. Now, he has co-authored three technical books, published over 100 publications in IEEE/ACM journals, International conferences. He has published numerous book chapters, co-edited a number of journal special issues, and served as editorial board members for several journals. Moreover, he has organized numerous international conferences and workshops as co-chairs and co-program chairs.

John D. Garofalakis (http://athos.cti.gr/garofalakis/) is Associate Professor at the Department of Computer Engineering and Informatics, University of Patras, Greece, and Director of the applied research department "Telematics Center" of the Research Academic Computer Technology Institute (RACTI). He is responsible and scientific coordinator of several European and national IT and Telematics Projects (ICT, INTERREG, etc.). His publications include more than 120 articles in refereed International Journals and Conferences. His research interests include Web and Mobile Technologies, Performance Analysis of Computer Systems, Computer Networks and Telematics, Distributed Computer Systems, Queuing Theory.

Ji Gu received the BSc degree in computer science from the East China University of Science and Technology, Shanghai, China, in 2002, and the MSc degree in computer science from the Leiden University, Leiden, The Netherlands, in 2005, respectively. He has worked in the telecommunication industry as a design engineer and is currently pursuing the PhD degree in the School of Computer Science and Engineering at the University of New South Wales, Australia. His work has focused on computer architecture, digital circuit, and VLSI design, with particular emphasis on low-power and high-performance design, at system or architecture level, of the application-specific memory hierarchy, processor architecture, and embedded systems.

Hui Guo received her BE and ME degrees in Electrical and Electronic Engineering from Anhui University (China) and PhD in Electrical and Computer Engineering from the University of Queensland (Australia). Prior to start her academic career, she had worked in a number of companies/organizations for information systems design and enhancement. She is now a senior lecturer in the School of Computer Science and Engineering at the University of New South Wales, Australia. Her research interests include Application Specific Processor Design, Low Power System Design, and Embedded System Optimization.

Shiang-Yi He received the B.S. in Computer Science and Information Engineering from Tamkang University, Taipei, Taiwan, in 2007, and the M.S in Computer Science and Information Engineering from National University of Tainan in 2009, respectively. His major research interest includes computational intelligence, Context aware computing, and databases.

Ricardo Pezzuol Jacobi is an associate professor in the Computer Science Department, Universidade de Brasilia, Brazil. Jacobi has a M.S. in electrical engineering from Universidade Federal do Rio Grande do Sul, in 1986, Brazil, and a PhD in Applied Sciences from Université Catholique de Louvain, Belgium, in 1993. He is currently vice-director at the UnB Gama Faculty, a new technology campus created in 2008 at Gama, Brazil. He has published over 60 papers in refereed journals, conference proceedings and book chapters and has coordinated several projects in Microelectronics and Embedded Systems, including international cooperations. His research interests include Embedded System Design, Bioinformatics, Digital TV, Reconfigurable Architectures, and Ubiquitous Computing.

Marc Jentsch completed an apprenticeship as IT specialist at the Bertelsmann AG in Gütersloh in 2003. After that, he studied Geoinformatics at the University of Münster and received his diploma in 2008. Simultaneously, he worked in the field of output management at arvato systems Technologies in Gütersloh and as web developer at max&partner in Düsseldorf. Since 2008, he is working as researcher at

Fraunhofer FIT in Sankt Augustin in the industrial project MICA as well as the research projects Hydra, AILB, ZfS and B-IT. Marc is currently responsible for the international research project Intermedia. His research interests include mobile and pervasive applications, location-based services, energy efficiency and augmented reality.

Nan Jing is currently a Senior Staff of the R&D department at the Bloomberg L. P. in New York, where he is working on the mobile software platforms and applications for financial services. Prior to Bloomberg, Nan worked as mobile specialist at Antenna Software and Industrial Scientific, where he helped in designing the customer relationship management and behavior-based safety management applications respectively, on mobile computing platforms including Blackberry, Android and Windows Mobile. Before that he was the first engineer in the mobile player team at DivX and led the development of the first few versions of DivX Mobile Player. Nan has published a number of papers and one book in software design, web service, decision science and business process management. Nan is a frequent reviewer for conferences and journals in information system and mobile technologies, and he is presently on the editorial review boards of *International Journal of Handheld Computing Research* and *International Journal of Software Engineering and Knowledge Engineering*. Nan has earned a PhD degree and a Master degree from University of Southern California and a Bachelor degree from Peking University, China, all in Computer Science.

Djamel Khadraoui is a research program manager at CRP Henri Tudor. He is also a scientific coordinator of many EU and national projects, and supervising PhD thesis. He is active in the areas of enterprise IT security, intelligent systems and software engineering. Other interests are related to Multi-Agent Systems and mobile computing. He is expert and DC COST member in the domain of ICT and Transport. He was a general chairman of AISTA 04 (IEEE International Conference in Advances on Intelligent Systems) and UBIROADS 09 (IEEE Workshop), and involved in many conferences and workshops as a member of technical committee. He coordinated the publishing of a book entitled: Advances in Enterprise IT Security in coordination with IDEA Group Publishing. He mainly participated and coordinated several previous and current EU projects: LINKALL, as a technical project manager, BUGYO, CARLINK, RED, €- CONFIDENTIAL, as a national coordinator. His recent important projects are: TITAN, WISAFECAR, BUGYO-beyond and MICIE. Dr. Khadraoui is also providing lectures in IT Security (Distributed Systems Security, DB security and Dependability) at the master level at the University of Luxembourg.

Vassilis-Javed Khan is a senior lecturer at NHTV Breda University of Applied Sciences, in the Netherlands. He completed his PhD and P.D. Eng. in Eindhoven University of Technology in the Netherlands and his MSc from the University of Patras in Greece. He has worked as a post-doctoral researcher at Eindhoven University of Technology, visiting researcher at Philips Research in Eindhoven, research intern at Vodafone R&D in Maastricht and as a software engineer at the Computer Technology Institute in Patras, Greece. His research interests fall under the umbrella of HCI and include pervasive computing, location based services, awareness systems and game evaluation. An entrepreneurial side of his research is KidzFrame which he has co-founded.

David Kuo is a professional software engineer at a software company in Silicon Valley. He has 10 years of working experience in the software industry on software development, relational database systems, computer-aided software engineering (CASE) tools, product trainings, software project coordination, technical support, and system and network administration. Besides his full-time job, most recently he has been involved in developing mobile commerce and security prototype services and systems for a startup company. Just for fun, he has developed and released a personal finance tracking application for the Android platform. His research and development interests include mobile applications, cloud computing technologies, web services, enterprise applications, client/server systems, software development processes, design patterns, and code construction. He received his master's degree in Software Engineering from San Jose State University in 2009, and has been invited to give lectures at Software Engineering graduate courses at San Jose State University.

Karin Leichtenstern is a PhD student and research assistant at Augsburg University. Before that, she graduated in Media Informatics at the University of Munich in 2006 and stayed a half of a year at the Intelligent Inhabited Environments Group at the University of Essex, UK. Her main research interests include HCI-related aspects of Pervasive and Mobile Computing as well as Usability Engineering. Karin's PhD aims at tool-support for interface designers of mobile applications when running through the human-centred design process.

Po-Shen Lin received MSc degree in 2009 from the Department of Computer Science and Engineering, Tatung University. His research is focusing on integrating Semantic Web technology with social web sites to make Web 2.0 services more intelligent.

Panos Markopoulos studied electrical engineering and computer science in the National Technical University of Athens and human-computer interaction in Queen Mary University of London, where he did a PhD on formal methods in human computer interaction. He has held research positions in Queen Mary, University of London and Philips Research in Eindhoven and is currently a Professor in the Department of Industrial Design of the Eindhoven University of Technology on the topic of Awareness Systems and Ambient Experiences. His interests extend in several areas of the fields of human-computer interaction and interaction design such as model based design, privacy and trust, social games for children, evaluation methodology for children, connectedness between family and friends, persuasive technologies, and healthcare. Panos Markopoulos has published extensively in all these topics, and is the first co-author of a book on evaluating children's interactive products, published by Morgan Kaufmann in 2008.

Haralambos Mouratidis holds a B.Eng. (Electronics with Computing Science) from the University of Wales, Swansea, U.K. and M.Sc. (Data Communications) and PhD (Computer Science) degrees from the University of Sheffield, U.K. Dr. Mouratidis is Principal Lecturer in Secure Systems and Software Development at the School of Computing, Information Technology and Engineering (CITE) at the University of East London and Co-director of the Distributed Software Engineering. His research interests are related to security requirements engineering, secure information systems development and methodologies and methods for secure software systems. He has published more than 80 papers in refereed journal and papers and he has been involved in various Programme Committees as General

Chair, Programme Chair and members. He is Co-Chair of the 23ʳᵈ International Conference on Advanced Information Systems Engineering and the editor of a forthcoming book on Software Engineering for Secure Systems: Industrial and Research Perspectives from IGI Publishing.

Moussa Ouedraogo holds an MSc in Computer systems engineering from the University of East London. He has been involved in a number of national project within the Grand Duchy of Luxembourg (TITAN project) and European project dealing with security and assurance such as Celtic €-CONFI-DENTIAL, BUGYO-BEYOND and FP 7 MICIE. His main research interests include security metrics and evaluation, assurance, trust, risks management and Requirement Engineering. He is also a reviewer for the Springer journal of network and systems management and acted as a program committee member for the MESSA 2010 workshop. He is currently pursuing his research activity at the Public research Centre Henri Tudor in Luxembourg.

Pavan Kumar Pandey is working at Aricent Technologies (Telecom) as a software engineer since June 30 2010. He has completed M. Tech from ISM Dhanbad Jharkhand after he completed B. Tech degree in 2007 from computer Science & Engineering.

Louise Pape-Haugaard is currently doing research in architectural challenges in healthcare. Since 2007 she has been a PhD-stipend in Medical Informatics at Aalborg University, with a main focus on creating interoperability across a plethora of complex clinical it-systems targeting different requirements. Creating realistic interoperability between complex clinical it-systems requires an architectural solution which embraces and supports security and privacy, reusing existing infrastructure, reduces efforts of vendors and end-users, and maximizes performance. Research areas are Electronic Health Records and complex clinical it systems, privacy/security, data modeling, and semantic interoperability.

Roel Peeters obtained an Electrical Engineering degree at the Katholieke Universiteit Leuven (Belgium) in 2007. After his studies he started a PhD at the COSIC research group at that same university. His research interests lie in the fields of threshold cryptography and mobile security.

Christian R. Prause received his diploma in computer science and psychology from the University of Bonn (Germany) in 2006. He has since worked at Fraunhofer FIT as a researcher in the transnational EU projects MACE and HYDRA, and is a certified UML expert. He was the project manager of the MICA Pilot at FIT and headed the software development there. Christian's research interests include software engineering and its processes, knowledge management, the open source phenomenon, and the collaboration of distributed teams. With his work on software quality he wants to reduce total development costs in research projects, and optimize the transfer of research results into industry. Marc Jentsch completed an apprenticeship as IT specialist at the Bertelsmann AG in Gütersloh in 2003. After that, he studied Geoinformatics at the University of Münster and received his diploma in 2008. Simultaneously, he worked in the field of output management at arvato systems Technologies in Gütersloh and as web developer at max&partner in Düsseldorf. Since 2008, he is working as researcher at Fraunhofer FIT in Sankt Augustin in the industrial project MICA as well as the research projects Hydra, AILB, ZfS and B-IT. Marc is currently responsible for the international research project Intermedia. His research interests include mobile and pervasive applications, location-based services, energy efficiency and augmented reality.

Bart Preneel received the Electrical Engineering degree and the Doctorate in Applied Sciences from the Katholieke Universiteit Leuven (Belgium), where he is currently full professor (*gewoon hoogleraar*) and head of the research group COSIC. He is president of the IACR (International Association for Cryptologic Research) and a member of the Editorial Board of the Journal of Cryptology, the IEEE Transactions on Information Forensics and Security, and the International Journal of Information and Computer Security. He is also a Member of the Accreditation Board of the Computer and Communications Security Reviews (ANBAR, UK). His main research interests are cryptology and information security.

Matthias Rehm is an associate professor at CREATE, the Department of Architecture, Design and Media Technology at Aalborg University. He received his doctoral degree from Bielefeld University for a thesis on multimodal learning in virtual agents and has since worked in the area of social interactions focusing on embodied agents, multimodal and mobile interaction as well as cultural aspects of computing. He has been involved in a number of international research projects on a European level and beyond in the area of multimodal interactive systems and has over 50 publications in peer-reviewed journals and conferences.

Maria Andréia F. Rodrigues is a Full Professor at the University of Fortaleza (UNIFOR), Fortaleza-CE, Brazil. She holds a Ph.D. degree in Computer Science from Imperial College, University of London, U.K. (1999). Prof. Andréia has co-organized the 15th Brazilian Symposium on Computer Graphics and Image Processing (SIBGRAPI) in 2002. She has co-chaired the SIBGRAPI'05, and served as a PC member in several international and national conferences such as the 22nd ACM Symposium on Applied Computing (SAC'07), Special Tracks on Handheld Computing and Computer Applications in Health Care, as well as Tutorial Chair of SAC'08. Currently, she is serving as a PC member of the 31st Annual Conference of the European Association for Computer Graphics (EUROGRAPHICS'10), Special Track on Mobile Visual Computing, and SIBGRAPI'10, among others. Her research interests are in interactive computer graphics, virtual reality, and handheld computing.

Kamel Rouibah is an Associate Professor of information systems, College of Business Administration (CBA), Kuwait University. He holds a PhD in Information Systems from Ecole Polytechnique of Grenoble, France. Before joining CBA, he worked at the Faculty of Technology Management at Eindhoven (Netherlands) and Institut National de la Recherche Scientifique (France). His research interests include Design of Information Systems, Management Information Systems, Engineering Data Management, Workflow Management, Information System and Information Technology Acceptance. He has authored/coauthored over 50 research publications in peer-reviewed reputed journals and conference proceedings. He was involved in several European projects. His publications appeared in several leading journals: such as Journal of Strategic Information System, IT & People, Journal of Global Information Management, Computers in Industry; He has received the excellence younger researcher award from Kuwait

University for the academic year 2001/2002. Dr Rouibah has directed many funded research projects, and has served as the program committee member of various international conferences and reviewer for various international journals. He is the co-editor of "Emerging markets and e-commerce in developing economies" book. He has taught many information systems courses in France, Netherlands, and Kuwait.

Yanbo Ru received his BS degree from Huazhong University of Science and Technology, and his MS and PhD degrees from University of Southern California, all in computer science. Dr. Ru's areas of expertise include web technology, information retrieval, online advertising, parallel processing, and cloud computing. He applies his research to problems of building business search engine and directory and pay-per-click advertising network, focusing on information extraction and classification and list generation. Dr. Ru is also a core developer of CloudBase project - an open source data warehouse system built on top of Map-Reduce architecture. Dr. Ru currently works at Business.com Inc.

Wendel Bezerra Silva is currently a Ph.D. student at the University of Utah. He received his MS degree from the University of Fortaleza (UNIFOR), Fortaleza-CE, Brazil. His research interest centers on interactive computer graphics, mobile technology and scientific visualization. Wendel is a member of the Scientific Computing and Imaging Institute (SCI), University of Utah.

Dave Singelée received the Master's degree of Electrical Engineering and a PhD in Applied Sciences in 2002 and 2008 respectively, both from the Katholieke Universiteit Leuven (Belgium). He worked as an ICT security consultant at PricewaterhouseCoopers Belgium, and is currently a postdoctoral researcher at the research group COSIC (Computer Security and Industrial Cryptography, Katholieke Universiteit Leuven). His main research interests are cryptography, security and privacy of wireless communication networks, cryptographic authentication protocols for RFID, and secure localization schemes. He has authored and co-authored more than 20 scientific publications, and participated in various research projects.

Jie Sun is a lecturer of the School of Electronic and Information Engineering at Ningbo University of Technology, Ningbo, China. She obtained her PhD. degree in Computer Science from Zhejiang University, Hangzhou, China, in 2009. She received the BS and MS degrees both in Computer Science from Lanzhou University in 1998 and 2003, respectively. From 1998 to 2002, she worked as an assistant lecture of Department of Computer Science and technology at Shenyang Aviation Industrial Institute. In 2005, she worked as a Research Assistant at The University of Hongkong. Her past and present research interests include: Pervasive computing and context-aware computing. She has published more than 20 papers in the relevant fields.

Antonia Stefani graduated from the University of Patras, Department of Mathematics in 1999. She received her MSc degree in Computer Science, "Mathematics on Computers and Decision Making", from the department of Mathematics and the department of Computer Engineering and Informatics of the University of Patras, in 2001. Her scientific area was "Foundations of Computer Science and its Applications in Automated Decision Making". She received the PhD degree in 2008 at Hellenic Open University, School of Science and Technology. Her research interests include software quality, software metrics, e-commerce and m-commerce systems.

Vassilios Stefanis (http://www.stefanis.net) received his diploma from the University of Patras, Computer Engineering and Informatics Department in 2005. He also received an MSc degree in "Computer Science and Engineering" from the same department in 2008 and now he is a PhD candidate. From 2005 he works as a researcher and software engineer to the research academic computer technology institute (RA-CTI) in Patras. Also, from 2008 he teaches as a Lecturer at TEI of Messolonghi, Department Of Applied Informatics in Management & Finance. His research interests include internet technologies, mobile web, mobile applications and p2p technologies.

Flora S. Tsai is currently with the School of Electrical & Electronic Engineering, Nanyang Technological University (NTU), Singapore. She is a graduate of MIT, Columbia University, and NTU with degrees in Electrical Engineering and Computer Science. Dr. Tsai's current research focuses on developing intelligent techniques for data mining in text and social media, which aims to balance the technical significance and business concerns to create techniques that are useful in real-world scenarios. In particular, Dr. Tsai has pioneered research in blog data mining, novelty mining, and mobile information retrieval. Other research interests include software engineering, mobile application development, cyber security, electronic healthcare, and machine learning techniques for bioinformatics. Dr. Tsai was a recipient of the 2005 IBM Faculty Award, 2007 IBM Real-time Innovation Award, 2010 IBM Faculty Innovation Award, and has published over 50 international journal and conference papers. She is a senior member of IEEE and member of ACM.

Daniel Wong received a BS in Computer Science in 2002 and a MS in Software Engineering in 2008 from San Jose State University. He took interested in mobile technology while studying in San Jose State University. He worked at PayPal, Inc (an eBay company) during his graduate studies and developed a complete Automation Testing Management system for PayPal's QA organization. Daniel Wong currently works at Netflix, Inc in Los Gatos as a Senior Software Engineer working for the Movie Meta Data team, developing Web based application and Restful API using various Java technologies and Amazon Cloud technologies. He continues to work different mobile projects (iPhone, Android, Blackberry, etc) on his spare time.

Yong Yao received his BS degree from Peking University, and his M.S. and PhD degrees from Cornell University, all in computer science. Dr. Yao's areas of expertise include wireless sensor networks, distributive query processing and optimization, and XML databases. Dr. Yao built one of the first two wireless sensor network database systems, Cougar, with a full-fledge support to declarative query processing in wireless sensor networks. He also made breakthrough contributions on adaptive query processing and power-efficient data retrieval techniques in ad-hoc networks. Dr. Yao has many publications in the form of book chapters, major international conference papers and elite journal papers. His papers have received world-wide recognition, and have been referenced more than 1,500 times by scientists around the world. Dr. Yao currently works at IBM Silicon Valley Lab.

Ching-Long Yeh was born in Taipei, Taiwan. He is an associate professor of the Computer Science and Engineering Department, Tatung University, Taipei, Taiwan. He received PhD degree in Artificial Intelligence from the University of Edinburgh in 1995. His research interests include web technology, knowledge engineering and knowledge management, electronic business, e-learning and natural language processing.

Yongping Zhang is a full professor of the School of Electronic and Information Engineering at Ningbo University of Technology, Ningbo, China. He obtained his PhD. degree in Control Science and Engineering from Xi'an Jiaotong University, Xi'an, China, in 1998. From 2006 to 2007, he worked as a Research Fellow with Department of Computer Science at National University of Singapore. From 2003 to 2005, he worked as a Research Scientist with Bioengineering Institute and UniServices Limited at The University of Auckland, Auckland, New Zealand. From 2001 to 2003, he worked as a Postdoctoral Fellow at Massey University, New Zealand. His past and present research interests include: BioMedical Image processing and analysis, Bioengineering, Computer Vision, Pattern Recognition, Wavelet Analysis and Neural Networks. He has published more than 50 papers in the relevant fields.

Yanjun Zuo received a Ph.D. in Computer Science from the University of Arkansas, Fayetteville, USA in 2005. He also holds a master degree in Computer Science from the University of Arkansas and a master degree in Business Administration from the University of North Dakota, Grand Forks, USA. Currently he is an assistant professor at the University of North Dakota. His research interests include survivable and self-healing systems, security in pervasive computing, database security, and information privacy. He has published numerous articles in major conference proceedings and referred journals such as Decision Support Systems, International Journal of Information and Computer Security, Information Systems Frontiers—a Journal of Research and Innovation, and IEEE Transactions on Systems, Man, and Cybernetics.

Index